Lecture Notes in Computer Science 6390

Commenced Publication in 1973
Founding and Former Series Editors:
Gerhard Goos, Juris Hartmanis, and Jan van Leeuwen

Oded Goldreich (Ed.)

Property Testing

Current Research and Surveys

 Springer

Volume Editor

Oded Goldreich
Weizmann Institute of Science
Faculty of Mathematics and Computer Science
76100 Rehovot, Israel
E-mail: oded.goldreich@weizmann.ac.il

Library of Congress Control Number: 2010936638

CR Subject Classification (1998): F.2, I.2, F.1, I.3.5, H.3, G.2

LNCS Sublibrary: SL 1 – Theoretical Computer Science and General Issues

ISSN   0302-9743
ISBN-10   3-642-16366-1 Springer Berlin Heidelberg New York
ISBN-13   978-3-642-16366-1 Springer Berlin Heidelberg New York

springer.com

© Springer-Verlag Berlin Heidelberg 2010
Printed in Germany

Typesetting: Camera-ready by author, data conversion by Scientific Publishing Services, Chennai, India
Printed on acid-free paper   06/3180

Preface

Property testing is the study of super-fast (randomized) algorithms for approximate decision making. These algorithms are given direct access to items of a huge data set, and determine whether this data set has some predetermined (global) property or is far from having this property. Remarkably, this approximate decision is made by accessing a small portion of the data set.

Property testing has been a subject of intensive research in the last couple of decades, with hundreds of studies conducted in it and in closely related areas. Indeed, property testing is closely related to probabilistically checkable proofs (PCPs), and is related to coding theory, combinatorics, statistics, computational learning theory, computational geometry, and more.

The current volume provides a taste of the area of property testing. It grew out of a mini-workshop on property testing that took place in January 2010 in the Institute for Computer Science (ITCS) at Tsinghua University (Beijing). The mini-workshop brought together a couple of dozen leading researchers in property testing and related areas. At the end of this mini-workshop it was decided to compile a collection of extended abstracts and surveys that reflects the program of the mini-workshop.

Property Testing at a Glance

Property testing is a relaxation of decision problems and it focuses on algorithms that can only read parts of the input. Thus, the input is represented as a function (to which the tester has oracle access) and the tester is required to accept functions that have some predetermined property (i.e., reside in some predetermined set) and reject any function that is "far" from the set of functions having the property. Distances between functions are defined as the fraction of the domain on which the functions disagree, and the threshold determining what is considered far is presented as a proximity parameter, which is explicitly given to the tester.

An asymptotic analysis is enabled by considering an infinite sequence of domains, functions, and properties. That is, for any n, we consider functions from D_n to R_n, where $|D_n| = n$. (Often, one just assumes that $D_n = [n] \stackrel{\text{def}}{=} \{1, 2, ..., n\}$.) Thus, in addition to the input oracle, representing a function $f : D_n \to R_n$, the tester is explicitly given two parameters: a size parameter, denoted n, and a proximity parameter, denoted ϵ.

Definition: *Let $\Pi = \bigcup_{n \in \mathbb{N}} \Pi_n$, where Π_n contains functions defined over the domain D_n. A* tester *for a property Π is a probabilistic oracle machine T that satisfies the following two conditions:*

1. *The tester accepts each $f \in \Pi$ with probability at least 2/3; that is, for every $n \in \mathbb{N}$ and $f \in \Pi_n$ (and every $\epsilon > 0$), it holds that $\Pr[T^f(n, \epsilon) = 1] \geq 2/3$.*

2. *Given $\epsilon > 0$ and oracle access to any f that is ϵ-far from Π, the tester rejects with probability at least 2/3; that is, for every $\epsilon > 0$ and $n \in \mathbb{N}$, if $f : D_n \to R_n$ is ϵ-far from Π_n, then $\Pr[T^f(n, \epsilon) = 0] \geq 2/3$, where f is ϵ-far from Π_n if, for every $g \in \Pi_n$, it holds that $|\{e \in D_n : f(e) \neq g(e)\}| > \epsilon \cdot n$.*

If the tester accepts every function in Π with probability 1, then we say that it has one-sided error; that is, T has one-sided error if for every $f \in \Pi$ and every $\epsilon > 0$, it holds that $\Pr[T^f(n, \epsilon) = 1] = 1$. A tester is called non-adaptive if it determines all its queries based solely on its internal coin tosses (and the parameters n and ϵ); otherwise it is called adaptive.

This definition does not specify the query complexity of the tester, and indeed an oracle machine that queries the entire domain of the function qualifies as a tester (with zero error probability...). Needless to say, we are interested in testers that have significantly lower query complexity.

Research in property testing is often categorized according to the type of functions and properties being considered. In particular, algebraic property testing focuses on the case in which the domain and range are associated with some algebraic structures (e.g., groups, fields, and vector spaces) and studies algebraic properties such as being a polynomial of low degree. In the context of testing graph properties, the functions represent graphs or rather allow certain queries to such graphs; for example, in the adjacency matrix model, graphs are represented by their adjacency relation and queries correspond to pairs of vertices where the answers indicate whether or not the two vertices are adjacent in the graph. (In an alternative model, known as the incidence-list model, graphs are represented by functions that assign to the pair (v, i) the ith neighbor of vertex v.)

Current research in property testing focuses mainly on query (and/or sample) complexity, while either ignoring time complexity or considering it a secondary issue. The current focus on these information-theoretic measures is justified by the fact that even the latter are far from being understood. (Indeed, this stands in contrast to the situation in, say, PAC learning.)

The representation of problems' instances is crucial to any study of computation, since the representation determines the type of information that is explicit in the input. This issue becomes much more acute when one is only allowed partial access to the input (i.e., making a number of queries that result in answers that do not fully determine the input). An additional issue, which is unique to property testing, is that the representation may effect the distance measure (i.e., the definition of distances between inputs). This is crucial because property testing problems are defined in terms of this distance measure.

The Contents of This Volume

This volume contains extended abstracts of almost all works presented at the workshop as well as a large number of surveys. The surveys refer to various sub-areas of property testing and/or to research directions in property testing. Some of these surveys correspond to presentations that took place in the workshop, and others were written for this volume by some of the workshop's participants. The list of surveys includes:

- Eli Ben-Sasson: Limiting the Rate of Locally Testable Codes
- Eric Blais: Testing Juntas
- Artur Czumaj and Christian Sohler: Sublinear-Time Algorithms
- Oded Goldreich: A Brief Introduction to Property Testing
- Oded Goldreich: Locally Testing Codes and Proofs
- Oded Goldreich: Testing Graph Properties
- Ilan Newman: Property Testing in the "Massively Parameterized" Model
- Krzysztof Onak: Sublinear Graph Approximation Algorithms
- Sofya Raskhodnikova: Transitive-Closure Spanners
- Rocco Servedio: Testing by Implicit Learning
- Madhu Sudan: Invariance in Property Testing

The list of extended abstracts includes:

- Noga Alon: On Fast Approximation of Graph Parameters
- Victor Chen: Testing Linear-Invariant Non-Linear Properties
- Victor Chen: A Hypergraph Dictatorship Test with Perfect Completeness
- Artur Czumaj: Testing Monotone Continuous Distributions on Real Cubes
- Oded Goldreich: Algorithmic Aspects of Property Testing in the Dense Graphs Model
- Prahladh Harsha: Composition of Low-Error 2-Query PCPs
- Tali Kaufman: Symmetric LDPC Codes and Local Testing
- Swastik Kopparty: Optimal Testing of Reed-Muller Codes
- Michael Krivelevich: Comparing the Strength of Query Types
- Michael Krivelevich: Hierarchy Theorems for Property Testing
- Kevin Matulef: Testing (Subclasses of) Linear Threshold Functions
- Krzysztof Onak: External Sampling
- Krzysztof Onak: The Query Complexity of Edit Distance
- Ronitt Rubinfeld: Maintaining a Large Matching or a Small Vertex Cover
- Michael Saks: Local Monotonicity Reconstruction
- Shubhangi Saraf: Some Recent Results on Testing of Sparse Linear Codes
- Asaf Shapira: Testing Linear Invariant Properties
- Christian Sohler: Testing Euclidean Spanners

The surveys and extended abstracts appearing in this volume were not refereed. The extended abstracts refer to papers that have either appeared or are likely to appear in peer-reviewed conferences and journals.

Acknowledgments

I wish to thank all the authors who have contributed to the current volume as well as all researchers who have contributed to the research being surveyed in it.

July 2010   Oded Goldreich

Table of Contents

A Brief Introduction to Property Testing

Oded Goldreich

Department of Computer Science, Weizmann Institute of Science, Rehovot, Israel
oded.goldreich@weizmann.ac.il

Abstract. This short article provides a brief description of the main
issues that underlie the study of property testing. It is meant to serve as
a general introduction to a collection of surveys and extended abstracts
that cover various specific subareas and research directions in property
testing.

1 Introduction

Property Testing is the study of super-fast (randomized) algorithms for approximate decision making. These algorithms are given direct access to items of a
huge data set, and determine whether this data set has some predetermined
(global) property or is far from having this property. Remarkably, this approximate decision is made by accessing a small portion of the data set.

Property Testing has been a subject of intensive research in the last couple of
decades, with hundreds of studies conducted in it and in closely related areas.

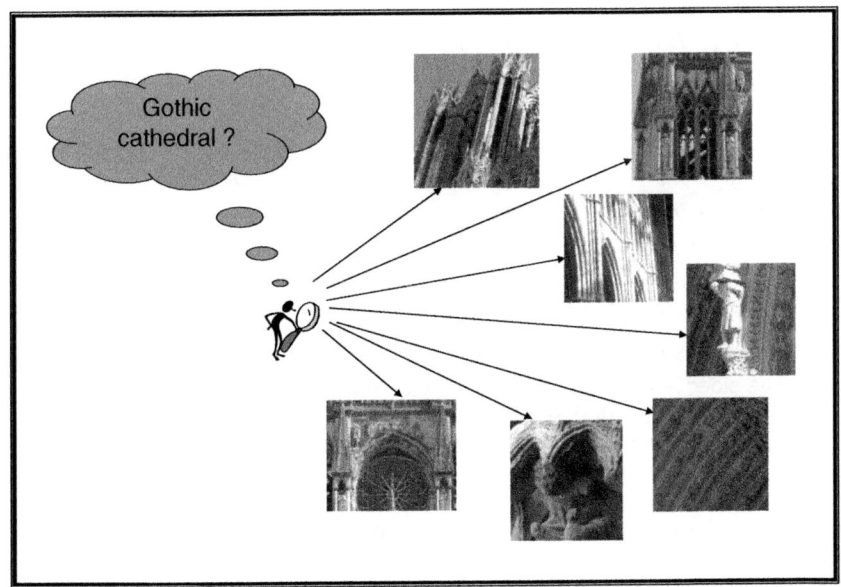

Fig. 1. Property Testing – An illustration

O. Goldreich (Ed.): Property Testing, LNCS 6390, pp. 1–5, 2010.

Indeed, Property Testing is closely related to Probabilistically Checkable Proofs (PCPs), and is related to Coding Theory, Combinatorics, Statistics, Computational Learning Theory, Computational Geometry, and more.

This brief introduction to the area of Property Testing is confined to conceptual issues; that is, it focuses on the main notions and models being studied, while hardly mentioning the numerous results obtained in the various models. This deficiency of the current article is corrected by the various surveys and extended abstracts presented in the current volume. In addition, we refer the interested reader to two recent surveys of Ron [9,10].

2   The Issues

Property testing is a relaxation of decision problems and it focuses on algorithms that can only read parts of the input. Thus, the input is represented as a function (to which the tester has oracle access) and the tester is required to accept functions that have some predetermined property (i.e., reside in some predetermined set) and reject any function that is "far" from the set of functions having the property. Distances between functions are defined as the fraction of the domain on which the functions disagree, and the threshold determining what is considered far is presented as a proximity parameter, which is explicitly given to the tester.

An asymptotic analysis is enabled by considering an infinite sequence of domains, functions, and properties. That is, for any n, we consider functions from D_n to R_n, where $|D_n| = n$. (Often, one just assumes that $D_n = [n] \overset{\text{def}}{=} \{1, 2, ..., n\}$.) Thus, in addition to the input oracle, representing a function $f : D_n \to R_n$, the tester is explicitly given two parameters: a size parameter, denoted n, and a proximity parameter, denoted ϵ.

Definition 1. *Let $\Pi = \bigcup_{n \in \mathbb{N}} \Pi_n$, where Π_n contains functions defined over the domain D_n. A* tester for a property *Π is a probabilistic oracle machine T that satisfies the following two conditions:*

1. *The tester accepts each $f \in \Pi$ with probability at least $2/3$; that is, for every $n \in \mathbb{N}$ and $f \in \Pi_n$ (and every $\epsilon > 0$), it holds that $\Pr[T^f(n, \epsilon) = 1] \geq 2/3$.*
2. *Given $\epsilon > 0$ and oracle access to any f that is ϵ-far from Π, the tester rejects with probability at least $2/3$; that is, for every $\epsilon > 0$ and $n \in \mathbb{N}$, if $f : D_n \to R_n$ is ϵ-far from Π_n, then $\Pr[T^f(n, \epsilon) = 0] \geq 2/3$, where f is ϵ-far from Π_n if, for every $g \in \Pi_n$, it holds that $|\{e \in D_n : f(e) \neq g(e)\}| > \epsilon \cdot n$.*

If the tester accepts every function in Π with probability 1, then we say that it has one-sided *error; that is, T has one-sided error if for every $f \in \Pi$ and every $\epsilon > 0$, it holds that $\Pr[T^f(n, \epsilon) = 1] = 1$. A tester is called* non-adaptive *if it determines all its queries based solely on its internal coin tosses (and the parameters n and ϵ); otherwise it is called* adaptive.

Definition 1 does not specify the query complexity of the tester, and indeed an oracle machine that queries the entire domain of the function qualifies as a tester (with zero error probability...). Needless to say, we are interested in testers that have significantly lower query complexity.

Research in property testing is often categorized according to the type of functions and properties being considered. In particular, algebraic property testing focuses on the case that the domain and range are associated with some algebraic structures (e.g., groups, fields, and vector spaces) and studies algebraic properties such as being a polynomial of low degree (see, e.g., [3,11]). In the context of testing graph properties (see, e.g., [4]), the functions represent graphs or rather allow certain queries to such graphs (e.g., in the adjacency matrix model, graphs are represented by their adjacency relation and queries correspond to pairs of vertices where the answers indicate whether or not the two vertices are adjacent in the graph).[1]

Ramifications. While most research in property testing refers to distances with respect to the uniform distribution on the function's domain, other distributions and even distribution-free models were also considered. That is, for a (known or unknown) distribution μ on the domain, we say that f is ϵ-far from g (w.r.t μ) if $\mathrm{Pr}_{e \sim \mu}[f(e) \neq g(e)] > \epsilon$. Indeed, Definition 1 refers to the case that μ is uniform over the domain (i.e., D_n).

A somewhat related model is one in which the tester obtains random pairs $(e, f(e))$, where each sample e is drawn (independently) from the aforementioned distribution. Such random (f-labeled) example can be either obtained on top of the queries to f or instead of them. This is also the context of testing distributions, where the examples are actually unlabeled and the aim is testing properties of the underlying distribution (rather than properties of the labeling which is null here).

A third ramification refers to the related notions of *tolerant testing* and *distance approximation* (cf. [8]). In the latter, the algorithm is required to estimate the distance of the input (i.e., f) from the predetermined set of instances having the property (i.e., Π). Tolerant testing usually means only a crude distance approximation that guarantees that inputs close to Π (rather than only inputs in Π) are accepted while inputs that are far from Π are rejected (as usual).

On the current focus on query complexity. Current research in property testing focuses mainly on query (and/or sample) complexity, while either ignoring time complexity or considering it a secondary issue. The current focus on these information theoretic measures is justified by the fact that even the latter are far from being understood. (Indeed, this stands in contrast to the situation in, say, PAC learning.)

On the importance of representation. The representation of problems' instances is crucial to any study of computation, since the representation determines the type of information that is explicit in the input. This issue becomes much more acute when one is only allowed partial access to the input (i.e., making a number of queries that result in answers that do not fully determine the input). An additional issue, which is unique to property testing, is that the representation may effect the distance measure (i.e., the definition of distances between inputs).

---

[1] In an alternative model, known as the incidence-list model, graphs are represented by functions that assign to the pair (v, i) the ith neighbor of vertex v.

This is crucial because property testing problems are defined in terms of this distance measure.

The importance of representation is forcefully demonstrated in the gap between the complexity of testing numerous natural graph properties in two natural representations: the adjacency matrix representation (cf. [4]) and the incidence lists representation (cf. [5]).

Things get to the extreme in the study of locally testable codes, which may be viewed as evolving around testing whether the input is "well formed" with respect to some fixed error correcting code. Interestingly, the general study of locally testable codes seeks an arbitrary succinct representation (i.e., a code of good rate) such that well-formed inputs (i.e., codewords) are far apart and testing well-formness is easy (i.e., there exists a low complexity codeword test).

3   A Brief Historical Perspective

Property testing first appeared as a tool towards program checking (see the linearity tester of [3]) and the construction of PCPs (see the low-degree tests and their relation to locally testable codes, as discussed in [11]). In these settings it was natural to view the tested object as a function, and this convention continued also in [4], which defined property testing in relation to PAC learning. More importantly, in [4] property testing is promoted as a new type of computational problems, which transcends all its natural applications.

While [3,11] focused on algebraic properties, the focus of [4] was on graph properties. From this perspective the choice of representation became less obvious, and oracle access was viewed as allowing local inspection of the graph rather than being the graph itself.[2] The distinction between objects and their representations became more clear when an alternative representation of graphs was studied in [5,6]. At this point, query complexity that is polynomially related to the size of the object (e.g., its square root) was no longer considered inhibiting. This shift in scale is discussed next.

Recall that initially property testing was viewed as referring to functions that are implicitly defined by some succinct programs (as in the context of program checking) or by "transcendental" entities (as in the context of PAC learning). From this perspective the yardstick for efficiency is being polynomial in the length of the query, which means being polylogarithmic in the size of the object. However, when viewing property testing as being applied to (huge) objects that may exist in explicit form in reality, it is evident that any sub-linear complexity may be beneficial.

The realization that property testing may mean any algorithm that does not inspect its entire input seems crucial to the study of testing distributions, which emerged with [2]. In general, property testing became identified as a study of a special type of sublinear-time algorithms.

---

[2] That is, in this case the starting point is the (unlabeled) graph itself, and its representation as a (labeled) graph by either its adjacency matrix or incidence list is an auxiliary conceptual step.

Another consequence of the aforementioned shift in scale is the decoupling of the representation from the query types. In the context of graph properties, this culminated in the model of [7].

Nevertheless, the study of testing properties within query complexity that only depends on the proximity parameter (and is thus totally independent of the size of the object) remains an appealing and natural direction. A remarkable result in this direction is the characterization of graph properties that are testable within such complexity in the adjacency matrix model [1].

References

1. Alon, N., Fischer, E., Newman, I., Shapira, A.: A Combinatorial Characterization of the Testable Graph Properties: It's All About Regularity. In: 38th STOC, pp. 251–260 (2006)
2. Batu, T., Fortnow, L., Rubinfeld, R., Smith, W.D., White, P.: Testing that Distributions are Close. In: 41st FOCS, pp. 259–269 (2000)
3. Blum, M., Luby, M., Rubinfeld, R.: Self-Testing/Correcting with Applications to Numerical Problems. JCSS 47(3), 549–595 (1993); Extended abstract in 22nd STOC (1990)
4. Goldreich, O., Goldwasser, S., Ron, D.: Property testing and its connection to learning and approximation. Journal of the ACM, 653–750 (July 1998); Extended abstract in 37th FOCS (1996)
5. Goldreich, O., Ron, D.: Property Testing in Bounded Degree Graphs. Algorithmica 32(2), 302–343 (2002); Extended abstract in 29th STOC (1997)
6. Goldreich, O., Ron, D.: A Sublinear Bipartitness Tester for Bounded Degree Graphs. Combinatorica 19(3), 335–373 (1999); Extended abstract in 30th STOC (1998)
7. Kaufman, T., Krivelevich, M., Ron, D.: Tight Bounds for Testing Bipartiteness in General Graphs. In: Arora, S., Jansen, K., Rolim, J.D.P., Sahai, A. (eds.) RANDOM 2003 and APPROX 2003. LNCS, vol. 2764, pp. 341–353. Springer, Heidelberg (2003)
8. Parnas, M., Ron, D., Rubinfeld, R.: Tolerant Property Testing and Distance Approximation. JCSS 72(6), 1012–1042 (2006); Preliminary version in ECCC (2004)
9. Ron, D.: Property Testing: A Learning Theory Perspective. Foundations and Trends in Machine Learning 1(3), 307–402 (2008)
10. Ron, D.: Algorithmic and Analysis Techniques in Property Testing. Foundations and Trends in TCS 5(2), 73–205 (2010)
11. Rubinfeld, R., Sudan, M.: Robust Characterization of Polynomials with Applications to Program Testing. SIAM Journal on Computing 25(2), 252–271 (1996)

The Program of the Mini-Workshop

Oded Goldreich

Department of Computer Science, Weizmann Institute of Science, Rehovot, Israel
oded.goldreich@weizmann.ac.il

Abstract. This article provides an annotated version of the program of the mini-workshop on propoerty testing that took place in January 2010 in the Institute for Computer Science (ITCS) at Tsinghua University (Beijing). The mini-workshop brought together a couple of dozen of leading researchers in Property Testing and related areas.

Editor's note. The original program is annotated by brief comments that represent the author's subjective perspective on the various presentation. In order to emphasize the subjective nature of these comments, their original informal style was maintained below.

Session 1 (Friday, January 8, 9:00-10:15) [chair: Oded Goldreich]

- Amy Wang: Welcoming comments.
- Rocco Servedio: Testing by Implicit Learning.
- Kevin Matulef: Testing (subclasses of) Linear Threshold Functions.

Kevin was ill, and so Rocco gave both talks. I am fascinated by the **implicit learning** paradigm, which is pivoted on emulating learning algorithms for k-variable functions by using n-bit long samples with k influential variables. The emulation proceeds by randomly partitioning the variables to $O(k^2)$ sets, identifying k sets that are likely to each contain a single influential variable, and converting the n-bit long samples to corresponding k-bit samples. Specifically, each n-bit sample is converted by determining for each influential set whether the subset of variables labeled 1 or the subset labeled 0 is influential, and setting the corresponding bit accordingly. Once the learning algorithm outputs a hypothesis for the k-bit function, it is tested using adequate queries (which are again emulated correspondingly). The second work shows that, while the class of all halfspaces is testable in query complexity that is independent of n, testing some natural subclasses of halfspaces (i.e., "unate reorientations of majority" where the weights are ± 1) requires query complexity that depends on n.

Session 2 (Friday, January 8, 10:45-12:00) [chair: Avrim Blum]

- Eric Blais: Testing juntas and function isomorphism.
- Michael Krivelevich: Hierarchy Theorems for Property Testing

O. Goldreich (Ed.): Property Testing, LNCS 6390, pp. 6–12, 2010.

Eric motivated the problem of optimizing the complexity of testing juntas by referring to the wide applicability of juntas (i.e., functions that depend on few of their inputs). The algorithm presented in Eric's work is very appealing, and has query complexity $\widetilde{O}(k/\epsilon)$, where ϵ is the proximity parameter. In addition, Eric advocated a study of the following class of parameterized problems. For a fixed function g, which is known to the tester, the task is testing whether the given oracle/function is isomorphic to g (i.e., f equals g under a relabeling of the indices of the variables). In a sharp shift from the concrete to the general, Michael presented an overview of results that assert the existence of properties with arbitrary (reasonable) query complexity bound.

Session 3 (Friday, January 8, 14:00-15:30) [chair: Bernard Chazelle]

- Ilan Newman: A Survey on Property Testing in the "Underlying Graph" Model or the "Massively Parameterized" Model.
- Sofya Raskhodnikova: Transitive-Closure Spanners with Applications to Monotonicity Testing.
- Christian Sohler: Testing Euclidean Spanners.

Ilan surveyed works that refer to "massively parameterized" testing problems. The parameters are huge structures (e.g., graphs) and the problem refers to a substructure of it (e.g., an assignment to the graph's edges). In the case of graphs, the assignment may be viewed as an orientation of the fixed graph or a subgraph of it. Sofya discussed the notion of transitive-closure spanners (i.e., spanners of the transitive closure graph), which are implicit in some monotonicity testers. Christian discussed the notion of geometric graphs, which are graphs embedded in a Euclidean space $\mathbb{R}^d$, and the corresponding notion of spanners (which are geometric graphs in which the graphical distance between pairs of vertices is approximated by their Euclidean distance).

Panel 1 (Friday, January 8, 16:00-17:00) On the Connection of Property Testing to Computational Learning Theory and Computational Geometry. Panelists: Avrim Blum, Bernard Chazelle, and Rocco Servedio.

Bernard advocated an attempt to provide a sublinear time analysis of dynamic systems, which may consist of selecting few objects and tracing their activity during time. Avrim emphasized the high cost of queries (rather than random examples) in typical learning applications, and I suggested to consider a two-parameter complexity measure that separates the number of random examples from the number of actual queries. Rocco suggested to try to relate property testing to agnostic learning (rather than to standard learning), and highlighted the open problem of tolerant testing of half planes.

Session 4 (Saturday, January 9, 9:00-10:15) [chair: Shafi Goldwasser]

- Madhu Sudan: Invariance in Property Testing.
- Victor Chen: Testing Linear-Invariant Non-Linear Properties.

Madhu mused on what makes problems easily testable and suggested that closure under a rich permutation group plays a major role. This seems supported by the numerous results regarding algebraic functions (which are all invariant under suitable affine transformations) and graph properties (which, by definition, are closed under isomorphism). He surveyed a unified approach that refers to the former, where the affine transformations refer to the identification of the domain with a vector space. Victor demonstrated that the approach extends also to non-linear properties, and Noga commented that an alternative demonstration may be provided by a property that is a union of two disjoint linear spaces.

Session 5 (Saturday, January 9, 10:45-12:00) [chair: Michael Saks]

- Asaf Shapira: Testing Linear Invariant Properties.
- Noga Alon: On Constant-Time Approximation of Invariants of Bounded Degree Graphs.
- Ronitt Rubinfeld: Maintaining a Large Matching or a Small Vertex Cover.
- Krzysztof Onak: External Sampling.

Asaf's study refers to the paramertized problem of whether a given subset of $[n]$ is free from containing any solution to a fixed set of linear equations. This was studied before with respect to a single equation (i.e., $x + y = z$), and Asaf's work treats any linear system. Noga's study refers to approximating quantities such as the independence number of bounded-degree graphs. He shows that constant-time algorithms can almost match the best approximation bounds that are previously known for PTAS, and that better bounds cannot be achieved in constant-time. (This is related to Krzysztof's presentation in Session 6.) Ronitt's work demonstrates that techniques employed in the context of distributed algorithms can be applied to the context of dynamic graph algorithms. This augments prior connections between constant-round distributed algorithms and constant-time approximation algorithms (discovered by Parnas and Ron [4]). Ronitt asked whether a direct relation can be shown between dynamic graph algorithms and constant-time approximation algorithms. Krzysztof advocated applying "external memory cost" measures to property testing algorithms. He showed such a result for the problem of element distinctness, presenting an algorithm that samples $\sqrt{nB/\epsilon}$ random B-bit blocks.

Session 6 (Saturday, January 9, 14:00-15:30) [chair: Noga Alon]

- Krzysztof Onak: Sublinear Graph Approximation Algorithms.
- Michael Saks: Local Monotonicity Reconstruction.
- Ronitt Rubinfeld: Testing Properties of Distributions (a survey).

A natural setting in which sublinear time algorithms may be useful is in approximating the value of various graph theoretic quantities, especially when these quantities are hard to determine (e.g., minVC). Parnas, Ron, and Marko [4,3] focused on maximal matching and minVC, while observing a correspondence between constant-time approximation algorithms for the size of the maximal independent set and distributed algorithms that find such sets. Krzysztof surveyed further developments along these lines, focusing on the extension of the study to many other problems (e.g., maximum matching, minimum dominating sets, etc) and the introduction of new techniques and improved results. The new techniques are also applicable to various testing problems; see the improvement obtained over [2] in the context of testing minor-free graphs in the bounded-degree model. Michael discussed the local reconstruction problem applied to objects for which there are exponentially many adequate solutions. The problem in this case is to determine some fixed solution as a function of the corrupted oracle and the random bits (used in the reconstruction), where in some applications it is desirable to use few random bits. Ronitt surveyed the study of testing properties of distributions, starting with the problem of testing identity to some known distribution (i.e., a parameterized problem), which is solvable by $\sqrt{n}$ samples when n is the domain's size (and the estimate is via collision probabilities). Testing that two given distributions (i.e., both distributions are "given" by samples) requires 3-way collisions and so $n^{2/3}$ samples. In contrast, approximating the distance to the uniform distribution requires an "almost linear" number of samples.

Panel 2 (Saturday, January 9, 16:00-17:00) On the Connection of Property Testing to Coding Theory, Combinatorics, and Statistics. Panelists: Madhu Sudan, Noga Alon, and Ronitt Rubinfeld.

Noga focused on the "global vs local" nature of property testing that is closely related to a main theme in combinatorics initiated by Erdos in the 1950's. He noted that a tester of k-colorability is implicit in work of the 1980's, but the query bound obtained there is a tower of exponents in the proximity parameter. Answering a question, Noga kind of suggested the challenge of testing triangle-freeness in $2^{O(1/\epsilon)^2}$ queries. (Recall that the currently best known bound is a tower of exponents in $\mathrm{poly}(1/\epsilon)$.) Ronitt reported on the awakening of interest of statisticians in the effect of the size of the domain (of values). (Traditionally, Statistics is concerned with the effect of the size of the universe (sample space), whereas we are concerned with the effect of the number of values actually obtained by samples in this space.) Still there are significant cultural differences, since in Statistics one often makes (implicit) assumptions about the distribution and/or assumes that the distribution itself is selected at random among some possibilities (hence "likelihood" is relevant). Ronitt also pointed out possible applications to areas that use Statistics such as natural language processing and data bases. Madhu pointed out that sublinear time algorithms arise naturally in the context of coding theory. Shafi advocated the study of property testing of algebraic (and number theoretic) problems beyond linearity and low-degree properties. A concrete example may be testing that a polynomial (given by its evaluations) is an irreducible polynomial. Can a tester be significantly more efficient than polynomial interpolation?

Session 7 (Sunday, January 10, 9:00-10:15) [chair: Ronitt Rubinfeld]

- Prahladh Harsha: Composition of Low-Error 2-Query PCPs using Decodable PCPs.

- Victor Chen: A Hypergraph Dictatorship Test with Perfect Completeness.
- Michael Krivelevich: Comparing the Strength of Query Types in Property Testing.

Prahladh started by relating two-query PCPs to (highly) robust PCPs, and proceeded in terms of the latter. to two-query PCPs. The new composition of robust PCPs starts by observing that there is no need to keep the outer proof for consistency, since consistency can be tested by comparing two related inner proofs (which refer to the same position in the outer proof). This, however, changes nothing because still consistency can be achieved by modifying less than half of the total length of both scanned parts. The new idea is to check consistency among d such parts, which may allow to reduce robustness error to approximately $1/d$. Indeed, this is doable, and is what the new composition theorem does, while relying on the fact that the inner PCP is decodable. My view is that the resulting two-query low-error PCPs are interesting mainly because decreasing the error to an arbitrarily small constant (and even below) does not result in increasing the proof length. Moving from PCPs to codeword tests, Victor surveyed the state of the art regarding the amortized query complexity of dictatorship testing (with perfect completeness), which currently stands on soundness error of $O(q \cdot 2^{-q})$ per q non-adaptive queries (see [6]). Michael discussed a general framework for testing graph properties, where distances are normalized by the actual number of edges and various types of queries are considered. In the work, "group queries" (akin of "group testing") are shown to be strictly stronger than the combination of the standard vertex-pair and neighbor queries.

Session 8 (Sunday, January 10, 10:45-12:00) [chair: Michael Krivelevich]

- Shubhangi Saraf: Some Recent Results on Testing of Sparse Linear Codes.
- Swastik Kopparty: Optimal Testing of Reed-Muller Codes.

Shubhangi advocated viewing locally testable linear codes as a special case of *tolerant* testing the Hadamard codewords under some adequate distributions (specifically, the distribution that is uniform on the corresponding positions of the Hadamard code). The tolerance condition prevents rejecting a codeword based on positions that are assigned zero probability. She presented two results, relating to two notion of tolerant testing (or rather distance approximation (cf. [5]): the more general result yields a weak notion of approximation and is obtained under the so-called *uniformly correlated condition*, whereas a strong notion of approximation requires a stronger correlated condition. A distribution μ is called k-uniformly correlated if there is a joint distribution of k-tuples such that each element (in the k-tuple) is distributed as μ but their sum is uniformly distributed. The stronger condition requires that this holds when these k elements are independently drawn from μ, which is equivalent to requiring that the code be sparse and have small Fourier coefficients. Swastik presented an improved analysis of the AKKLR linearity test (for degree d multi-linear polynomials over GF(2)) [1], showing that the basic test rejects far away function with constant probability (rather than with probability $\Omega(2^{-d})$).

Session 9 (Sunday, January 10, 14:00-15:30) [chair: Madhu Sudan]

- Eli Ben-Sasson: Limiting the Rate of Locally Testable Codes.
- Tali Kaufman: Symmetric LDPC Codes and Local Testing.
- Artur Czumaj: Testing Monotone Continuous Distributions on High-Dimensional Real Cubes.
- Krzysztof Onak: The Query Complexity of Edit Distance.
- Sofya Raskhodnikova: Testing and Reconstruction of Lipschitz Functions with Applications to Privacy.
- Oded Goldreich: Algorithmic Aspects of Property Testing in the Dense Graphs Model.

Eli surveyed several results that assert that various parameters of LTCs imply an exponentially vanishing rate. One such result refers to LTCs that are tested by a small set of constraints (which is only somewhat larger than the dimension of the dual code). Tali advocated the study of codes that are each characterized by a short constraint and its orbit under a suitable group. Arthur advocated the study of testing continuous distributions, focusing on the case that they are actually discrete (i.e., assume a finite number of possible values). Krzysztof contrasted the query complexity of testing the edit distance of a given string (from a fixed string) to the corresponding problem for Ulam distance. Sofya presented an application of reconstruction procedures (as presented by Michael Saks in Session 2) to data privacy; specifically, if data privacy is maintained for functions of a certain type, then we may want to modify the given function to that class. I tried to call attention to the work "Algorithmic Aspects of Property Testing in the Dense Graphs Model" (co-authored by Dana), while highlighting several perspectives and questions that arise.

Panel 3 (Sunday, January 10, 16:00-17:00)

- Prahladh Harsha on decoding in the low-error regime
- Ron Rivest on possible applications of property testing to the security evaluation of hashing functions.

Prahladh mused on whether decoding in the low-error regime may find additional applications in property testing. Ron asked whether property testing techniques can be employed to the evaluation of the quality of various cryptographic compression functions.

References

1. Alon, N., Krivelevich, M., Kaufman, T., Litsyn, S., Ron, D.: Testing Reed-Muller codes. IEEE Transactions on Information Theory 51(11), 4032–4038 (2005); An extended abstract appeared in the proceedings of RANDOM 2003 (under the title Testing Low-Degree Polynomials over GF(2)
2. Benjamini, I., Schramm, O., Shapira, A.: Every Minor-Closed Property of Sparse Graphs is Testable. In: 40th STOC, pp. 393–402 (2008)

3. Marko, S., Ron, D.: Approximating the Distance to Properties in Bounded-Degree and General Sparse Graphs. ACM Transactions on Algorithms 5(2) (2009); Extended abstarct in the proceedings of Random (2006)
4. Parnas, M., Ron, D.: Approximating the Minimum Vertex Cover in Sublinear Time and a Connection to Distributed Algorithms. TCS 381(1-3), 183–196 (2007); Preliminary version in ECCC (2005)
5. Parnas, M., Ron, D., Rubinfeld, R.: Tolerant Property Testing and Distance Approximation. JCSS 72(6), 1012–1042 (2006); Preliminary version in ECCC (2004)
6. Tamaki, S., Yoshida, Y.: A Query Efficient Non-Adaptive Long Code Test with Perfect Completeness. In: ECCC, TR09-074 (2009)

Limitation on the Rate of Families of Locally Testable Codes

Eli Ben-Sasson

Computer Science Department, Technion — Israel Institute of Technology, Haifa,
32000, Israel
eli@cs.technion.ac.il

Abstract. This paper describes recent results which revolve around the
question of the rate attainable by families of error correcting codes that
are locally testable. Emphasis is placed on motivating the problem of
proving upper bounds on the rate of these codes and a number of interesting open questions for future research are suggested.

Keywords: Locally testable codes, error correcting code, probabilistically checkable proofs.

1 Introduction

A locally testable code (LTC) is an error correcting code for which membership in
the code can be ascertained, to a high degree of confidence, by a random process
that queries a negligible fraction of a purported codeword. Locally testable codes
lie at the core of all known constructions of probabilistically checkable proofs
(PCPs), from [1,2] to [3], their discovery has inspired the study of property
testing [4], and the construction of such codes has been of great interest to
theoretical computer science in the recent past. Several surveys describe the
concepts around which these codes revolve [5,6], and a number of distinct ways
to obtain such codes are known by now (see Section 1.2). The purpose of this
brief survey, which assumes familiarity with the basic notion of an LTC, is to
explain what is known about the *limitations* of constructions of such codes, or,
in plain words, what kinds of LTCs are mathematically impossible to obtain.

When studying locally testable codes we are interested in both the classically
studied parameters of error correcting codes, such as *rate* and *relative distance*,
as well as in the local-testability parameters of the code, the *query complexity* or
number of entries read by the testing process, and the *completeness* and *soundness* which measure the probability of correctness of this process (these concepts
are defined in the next subsection). We intend to study the interplay between
these two kinds of code-related parameters so let us informally explain what kind
of trade-offs we expect to see. Better local-testability parameters, like smaller
query complexity and larger completeness and soundness parameters should be
expected to negatively affect the classical coding parameters, decreasing the rate
and/or relative distance of the code. We can show that this intuition does indeed

O. Goldreich (Ed.): Property Testing, LNCS 6390, pp. 13–31, 2010.

hold for certain families of codes, as surveyed later on. But for all the effort that has gone into the study of LTCs, the fundamental question that motivates our study (Question 1), regarding the existence of an asymptotically good family of LTCs, remains wide open. Before we continue we pause to recall the definition of a locally testable code and the reader familiar with this definition and the associated notation is encouraged to skip the following subsection.

1.1   Defining Locally Testable Codes

We assume familiarity with the basic definitions of error correcting codes, which can be found, e.g., in [7]. A *code* $\mathcal{C}$ over alphabet Σ of *blocklength* n, *message-length* k and *minimal distance* d will be called an $(n, k, d)_\Sigma$-code. It is a subset of Σ^n of size at least $|\Sigma|^k$ which satisfies the condition that for any pair of distinct codewords $w, w' \in \mathcal{C}$ their *Hamming distance*, defined as the number of entries on which w and w' disagree, is at least d. We shall reserve the letter w to denote codewords and r to denote "received" words, words which are not known to belong to $\mathcal{C}$. The ith entry of r will be denoted by r_i.

Two fundamental parameters of a code are its *rate* $\rho(\mathcal{C}) = k/n$ which measures the ratio of message to codeword length and the *relative distance* $\delta(\mathcal{C}) = d/n$ which dictates the noise-resilience of the code. We shall be interested in *families of codes* $\{\mathcal{C}_n \subset \Sigma^n \mid n \in \mathbb{Z}\}$. A family of codes is said to be *asymptotically good* if all members of it have positive rate and relative distance, i.e., there exist constants $\rho, \delta > 0$ such that each $\mathcal{C}_n$ satisfies $\rho(\mathcal{C}_n) \geq \rho$ and $\delta(\mathcal{C}_n) \geq \delta$. Given $\mathcal{C}$ and $r \in \Sigma^n$ let $\delta_\mathcal{C}(r)$ denote the *relative (Hamming) distance* between r and $\mathcal{C}$, defined as the minimal fraction of entries of r that need to be changed in order to obtain a word in $\mathcal{C}$. When $\delta_\mathcal{C}(r) \geq \epsilon$ we say r is δ-*far* from $\mathcal{C}$ and otherwise say r we say ϵ-*close* to it.

When Σ is the q-element finite field $\mathbb{F}_q$ (when the size of $\mathbb{F}$ is known or insignificant we use $\mathbb{F}$ to denote it) and $\mathcal{C}$ is a *linear code*, i.e., a k-dimensional subspace of $\mathbb{F}^n$, we shall say $\mathcal{C}$ is an $[n, k, d]_\mathbb{F}$-code. In this case the distance of the code is equal to the minimal weight of a nonzero codeword, where the *weight* of a word $r \in \mathbb{F}^n$ is the number of nonzero entries in r.

A *locally testable code* is an error correcting code — we expect it to have large relative distance — which comes with a randomized algorithm, called a *tester*, that samples a small number of entries of a received word $r \in \Sigma^n$ and is capable of distinguishing with nontrivial probability between the "good" case that r is an uncorrupted codeword, i.e., that r belongs to $\mathcal{C}$ (so $\delta_\mathcal{C}(r) = 0$) and the "bad" case that r is ϵ-far from $\mathcal{C}$. Since the definition of an LTC is tied to that of a tester we give both of them together.

Definition 1 (Tester and locally testable code). *Let $\mathcal{C}$ be an $(n, k, d)_\Sigma$-code. A (q, ϵ, s, c)-tester for $\mathcal{C}$ is a randomized algorithm T with oracle access to a purported codeword $r \in \Sigma^n$ which operates as follows. The tester T uses randomness to sample at most q entries of r and outputs a verdict which is either* accept *or* reject. *Denote by $T^r[R]$ the output of T on oracle r and random coins R. We say that T is a q-query tester, or, simply, a q-tester.*

The code $\mathcal{C}$ is said to be (q, ϵ, s, c)-locally testable *if it has a q-tester that satisfies the following* completeness *and* soundness *requirements. It the tester satisfies the (stronger) requirement of* strong soundness *we say $\mathcal{C}$ is a (q, s, c)-strong locally testable* code.

Completeness. *If $r \in \mathcal{C}$ then*

$$\Pr_{R} [T^r[R] = \mathsf{accept}] \geq c.$$

Soundness. *For every $r \notin \mathcal{C}$ that is ϵ-far from $\mathcal{C}$*

$$\Pr_{R} [T^r[R] = \mathsf{reject}] \geq s.$$

Strong Soundness. *For every $r \notin \mathcal{C}$*

$$\Pr_{R} [T^r[R] = \mathsf{reject}] \geq s \cdot \delta_{\mathcal{C}}(r).$$

The parameters q, ϵ, s, c are known respectively as the query complexity, distance threshold, soundness *and* completeness.

When $c = 1$ we say the code and tester have perfect completeness *and in such cases will often, for simplicity, omit reference to c.*

Remark 1 (Distance threshold and high-error, or list-decoding, LTCs). To get nontrivial LTCs the distance threshold ϵ should be less than half the relative distance of the code. Otherwise, it could be the case that there simply are no words ϵ-far from it, in which case the trivial tester that accepts all words shows that the code is $(0, \epsilon, 1, 1)$-LTC. We shall set the distance threshold to be one third the minimal distance of the code[1] and refer to such a $(q, \delta(\mathcal{C})/3, s, c)$-LTCs as a LTC for the *low-error regime*, or, simply, a *low-error* LTC. The choice of this name is because if $r \in \Sigma^n$ is accepted by the tester with probability greater than $1 - s$, we know that r is $\delta(\mathcal{C})/3$-close to $\mathcal{C}$, i.e., it has a *low fraction* of errors. Another common name for such a LTC is a *unique decoding LTC* because in the case just described there is a unique codeword that is closest to r.

For values of ϵ greater than half the minimal distance of $\mathcal{C}$, we say that $\mathcal{C}$ is a LTC in the *high-error*, or *list-decoding* regime. This is because a word accepted with probability greater than $1 - s$, which is known to be ϵ-close to $\mathcal{C}$, can in fact be ϵ-close to *list* of codewords. In the setting of high-error LTCs the kind of questions that are of interest revolve around understanding the connection between the acceptance probability of a received word and its proximity to the code. We shall not discuss these questions in this survey, due to scarcity of relevant results on rate limitations of such codes.

---

[1] Some of the LTC rate limitations surveyed here, like [8,9,10], require the distance threshold to be less than one third the minimal distance. This is due to technical reasons arising in the proofs. In any case, all known LTC constructions work for any sufficiently small distance parameter and the standard assumption in property testing settings is that the distance threshold is an arbitrarily small nonzero constant.

Remark 2 (Non-adaptivity and perfect completeness). A tester is said to be *non-adaptive* if the codeword-entries queried by it depend only on the value of the random coins (in particular, they do not depend on answers given to earlier queries). All known LTC constructions are nonadaptive, i.e., the tester associated with them is nonadaptive. For a family of LTCs with perfect completeness and constant query complexity adaptivity can be assumed without loss of generality, by incurring at most a constant factor reduction in the soundness parameter. Furthermore, almost all known LTCs are linear and consequently can be assumed to be nonadaptive and with perfect completeness (cf. Theorem 2), the notable exception to both linearity and perfect completeness is the "long code" of [11].

Remark 3 (Soundness and completeness). To get a meaningful definition we must require s to be greater than $1 - c$. Otherwise every code can be seen to be a $(0, 0, s, c)$-LTC, the tester associated with it rejects all words with probability s, hence accepts all words, and, in particular, all codewords, with probability $\geq c$.

Remark 4 (Running time). Our definition of a tester does not put any limitation on the running time of the tester. For families of codes with constant query complexity this is not a severe restriction because the tester can always be assumed to run in (nonuniform) time that is at most polynomial in the blocklength, and under reasonable assumptions the running time is quasi-linear, i.e., bounded by $n\mathrm{poly}\log n$ (cf. [12]). Families of linear codes — almost all known LTCs fall in this category — can similarly be assumed to require (nonuniform) quasi-linear running time because they can be tested by "linear testers" (as explained in Section 2.1). The main advantage to not putting a running-time constraint on the tester is that it allows us to focus on the code structure and avoid questions about computational complexity.

1.2   A Brief Survey of Known LTC Constructions

The purpose of this section is to display the abundance and variance of LTC constructions which should motivate both the search for a common denominator to all the different ways LTCs are constructed, as well as the study of limitations of these codes.

LTCs based on low-degree polynomials. The first family of LTCs, due to [13], is the family of homomorphisms from a finite group G to a subgroup H of G. Formally, $\mathcal{C}(G, H) \subset H^G$ has one codeword corresponding to each group-homomorphism $\phi : G \to H$ and this codeword is the evaluation of ϕ on all elements of G. This family was shown to be a low-error LTC in [13]. The special case of G being the additive group $\mathbb{F}^n$ and $H = \mathbb{F}$ for a prime field $\mathbb{F}$ was shown in [14] to be a locally testable in the high-error, or list-decoding, regime. The codes thus obtained are called Hadamard codes and correspond to the code of evaluations of n-variate, degree 1, homogenous polynomials. The generalization to arbitrary degree d polynomials was carried out promptly for the case of $d < |\mathbb{F}|$. This family of codes, known as Reed-Muller codes, was shown to be locally testable in the low-error regime in

[15,2], and in the high-error regime by [16,17]. Later on the case of $d \geq |\mathbb{F}|$ was analyzed for the low-error regime by [18,19] and for the high-error regime by [20] for the special case of $d = 2$. High-error LTCs based on polynomials of degree $d \geq 3$ and $d \geq |\mathbb{F}|$ remains as an interesting open problem.

Group invariant LTCs. An "invariance-based" approach to the construction of LTCs was implicitly suggested by [18] and explicitly undertaken, for the special case of affine-invariant codes, by [21] (see also [22,23,24]). More on this approach can be found in Section 3 and in the survey [25]. Roughly speaking, this approach is based on finding codes that are invariant under a "sufficiently rich" group of permutations, and additionally contain some local constraints that all codewords satisfy. The group-invariance of the code then implies a multitude of local constraints that all codewords satisfy, and this leads the way to prove local-testability.

Composed LTCs. Another way to construct LTCs, which among other things leads to the LTCs achieving the best known rate, relies on the use of probabilistically checkable proofs of proximity (PCPPs) [26,27] (see also [28]). Another approach that is also described as "combinatorial", because it relies neither on properties of low-degree polynomials, nor on group theory, is based on taking a repeated tensor-product of codes [29]. It should be pointed out that the codes arising from these methods are low-error LTCs and it remains to see what kind of LTCs in the high-error regime can emerge from high-soundness PCP composition techniques like those of [30,31].

Sparse unbiased LTCs. The final family of LTCs we are aware of consists of *sparse, unbiased* binary linear codes, i.e., linear codes over $\mathbb{F}_p$ for prime p that have a number of codewords that is only polynomial in the blocklength and for which all nonzero codewords have relative weight that is very close to $1 - \frac{1}{p}$ [32,33] (see also [34]).

1.3   Why Study Limitations of LTCs?

Before explaining why we think LTC limitations are worth pursuing we post the fundamental problem underlying our quest.

Question 1 (Do asymptotically good LTCs exist?). Prove or refute the following statement: There exists an asymptotically good family of binary error correcting codes $\{C_n \subseteq \{0,1\}^n \mid n \in \mathbb{Z}\}$ with relative distance δ that is a family of $(q, \delta/3, s, c)$-LTC, for some integer q and soundness and completeness parameters satisfying $c + s > 1$ (see Remark 3).

The main reason to study limits of LTCs is because this seems to be the most meaningful way to understand the limits of basic PCP-related parameters, most notably the *rate* of PCP proofs which we define as the ratio between the length of an **NP**-witness for an **NP**-instance ϕ, and the length of a probabilistically checkable proof for ϕ. The problem with the direct approach to bounding the rate of PCPs is that any nontrivial lower bound on the rate — even one that proves that PCP proof length is greater than zero — implies $\mathbf{P} \neq \mathbf{NP}$. Since all proofs of the PCP theorem make use of LTCs, and moreover the rate of the LTC is an upper bound on the rate of the PCP constructed from it, giving a negative answer to Question 1 would imply that PCP proofs constructed by current techniques will not attain constant rate. Anticipating future practical applications use of PCPs in cryptography and security-related protocols [35,36,37], we see that understanding the rate of PCPs is very important not just for theoretical purposes.

More broadly, the study of limitations of locally testable codes can be viewed as a branch of the study of classical tradeoffs for error correcting codes. When new families of codes are discovered (e.g., linear, cyclic, maximal distance separable, algebraic geometry, turbo, etc.) it is of great importance to understand how well they match up with known codes in terms of their basic coding-related parameters. Locally testable codes possess a highly desirable coding-related property, namely, the amount of errors in a received word can be estimated by inspecting only a tiny fraction of the codeword. This leads to the possibility of saving computation time, and the energy consumption required by the decoding algorithm, by getting a quick and roughly accurate estimate of the condition of received words and asking for a "re-transmit" in case the word is estimated to be corrupted beyond repair.

Finally, the concept of "locality of computation" is a theme of great interest in numerous settings of theoretical computer science. This is witnessed by the large body of work on property testing and on locally decodable codes. Understanding the limits of LTCs also touches upon questions related to locality of computation in other settings and one may expect to see more connections between LTC rate bounds and other areas in which "local computation" is studied.

1.4   Summary of Results Appearing in the Survey

In the next section we focus on linear codes and ask what limitations can be obtained from studying the structure of the set of dual codewords of small weight. We shall start with random low density parity check codes and use the expander-structure of the constraint graph associated with these codes to argue in Theorem 2 that they are not locally testable even when the query complexity is allowed to be fairly large. Then we shall generalize this result in Theorem 3 and show that all linear LTCs require that their dual code contain many low-weight words and, in Theorem 4, that these words must be nontrivially related. We conclude this section by showing in Theorem 5 that if an LTC has far too many redundant small-weight dual words then it has bad rate.

In Section 3 we shall investigate the rate limitations of group invariant codes. These codes include all known "base-case" LTCs, such as Hadamard and Reed-Muller codes, which serve as the building blocks in more elaborate LTC constructions (such as PCPP-based LTCs). We shall see in Theorem 7 that affine-invariant codes with small dual weight — the most general class of group-invariant codes known to be locally testable — has bad rate.

Results not covered by the survey. Two lines of work on limits of LTCs are not surveyed here. The first set contains the results of [38] which show that 3-query LTCs arising from PCPP-based constructions cannot obtain close-to-optimal soundness in the list decoding regime without suffering a significant decrease in the code-rate. The second line discusses various kinds of 2-query LTCs — linear [8], near-perfect completeness [39], "unique" [40] and "affine" [41] — and shows that there is at most a finite number of (2-query) LTCs of each kind.

2   Limiting Rate of Linear LTCs via the Structure of the Dual Code

This section focuses on limitations on the rate of families of *linear* LTCs. We shall focus on the linear space that is dual to the (linear) code $\mathcal{C} \subseteq \mathbb{F}^n$, this space is also known as the *dual code* and defined as $\mathcal{C}^\perp = \{u \in \mathbb{F}^n \mid u \perp \mathcal{C}\}$ where $u \perp \mathcal{C}$ if and only if $u \perp w$ for all $w \in \mathcal{C}$ and $u \perp w$ denotes the equality $\sum_{i=1}^n u_i w_i = 0$ (in case of inequality we write $u \not\perp w$). We shall take particular interest in the combinatorial structure of the set of dual codewords of *small weight*. We start by explaining why focusing on this structure is all that matters for local testability of linear codes.

2.1   Linear LTCs Are Testable by Linear Testers

A natural way to test whether a word $r \in \mathbb{F}^n$ belongs to an $[n, k, d]_{\mathbb{F}}$ linear code $\mathcal{C} \subset \mathbb{F}^n$ is to project r onto a set of coordinates $I \subset \{1, \ldots, n\}, |I| \leq q$ and accept r if and only if this projection, denoted by $r|_I$, agrees with a projection $w|_I$ of some codeword $w \in \mathcal{C}$. Writing $\mathcal{C}|_I = \{w|_I \mid w \in \mathcal{C}\}$ we can describe this natural test as the test that accepts r if and only if $r|_I \in \mathcal{C}|_I$. The operator that projects $r \in \mathbb{F}^n$ onto I is a *linear* operator, by which we mean that for every $a, b \in \mathbb{F}^n$ and $\alpha, \beta \in \mathbb{F}$ we have $(\alpha a + \beta b)|_I = \alpha(a|_I) + \beta(b|_I)$ and this implies that our natural tester is fact a *linear test* — its acceptance predicate, defined as the subset of $\mathbb{F}^I$ of query-answer tuples accepted by the test, is a linear space, it is the precisely the linear space $\mathcal{C}|_I$.

Accordingly, a *linear tester* for $\mathcal{C}$ is given by a distribution D on subsets I of size at most q. The following theorem of [42] says that without loss of generality linear codes are q-query LTCs if and only if they are testable by a linear tester.

Theorem 1 (Linear LTCs have linear testers). *If $\mathcal{C} \subseteq \mathbb{F}^n$ is a linear (q, ϵ, s, c)-LTC then $\mathcal{C}$ is a $(q, \epsilon, s + (1 - c), 1)$-LTC that can be tested by a linear tester. (Notice the difference between completeness and soundness is maintained when moving from an arbitrary tester to a linear one.)*

Given this theorem we can go one step further and describe the subsets $I \subset \{1, \ldots, n\}$ which correspond to nontrivial linear tests. If I is such that $\mathcal{C}|_I = \mathbb{F}^I$ then the (linear) test associated with I is meaningless — all words must be accepted by it. On the other hand if $\mathcal{C}|_I$ is a subspace strictly contained in $\mathbb{F}^I$ we do get a nontrivial test, meaning that some words $r \in \mathbb{F}^n \setminus \mathcal{C}$ will be rejected by it. In this case, the space that is dual to $\mathcal{C}|_I$, denoted $(\mathcal{C}|_I)^{\perp}$, has positive dimension, so it contains some nonzero words. Any word $u \in (\mathcal{C}|_I)^{\perp}$ can be extended to a word in $\mathbb{F}^n$ that is dual to $\mathcal{C}$ and has its nonzero entries contained in I — set all entries in $\{1, \ldots, n\} \setminus I$ to 0 and notice the word thus obtained is dual to $\mathcal{C}$.

Assuming $(\mathcal{C}|_I)^{\perp}$ is nontrivial we can think of another way to test wether $r|_I \in \mathcal{C}|_I$. Instead of querying all entries in I, pick a uniformly random $u \in (\mathcal{C}|_I)^{\perp}$ and accept r if and only if $u \perp r$. It is easy to see that this test retains perfect completeness, and we now argue that soundness goes down by a factor of at most $(1 - \frac{1}{\mathbb{F}})$. To see this, suppose $r \notin \mathcal{C}|_I$. The set $\{u \in (\mathcal{C}|_I)^{\perp} \mid r \perp u\}$ is a strict subspace of $(\mathcal{C}|_I)^{\perp}$, hence it contains at most a $(1/|\mathbb{F}|)$-fraction of $(\mathcal{C}|_I)^{\perp}$, so a random $u \in (\mathcal{C}|_I)^{\perp}$ will "reject" r (i.e., $u \not\perp r$) with probability at least $(1 - 1/|\mathbb{F}|)$ times the probability that $r|_I \notin \mathcal{C}|_I$. To sum up, if we don't care too much about the exact soundness constant then we may assume without loss of generality that a linear LTC is tested by a tester that is defined by a distribution over $\mathcal{C}^{\perp}_{\leq q}$, the set of words in the dual code $\mathcal{C}^{\perp}$ that have weight at most q. We record this by the following corollary of Theorem 2 (cf. [10, Section 2]). In what follows we use $u \sim D$ to denote that u is sampled according to the distribution D.

Corollary 1 (Linear codes are testable by a distribution over dual words of small weight). *If $\mathcal{C} \subseteq \mathbb{F}^n$ is a linear (q, ϵ, s, c)-LTC then there exists a distribution D over $\mathcal{C}^{\perp}_{\leq q}$ such that for every r that is ϵ-far from $\mathcal{C}$ we have $\Pr_{u \sim D}[u \not\perp r] \geq s + (1 - c)(1 - 1/|\mathbb{F}|)$. (Notice the soundness is $(1 - 1/|\mathbb{F}|)$ times the soundness stated of Theorem 2.)*

All this leads us to consider the *constraint graph* of a tester, a concept that will play a pivotal role in our analysis. Given $U \subseteq \mathcal{C}^{\perp}_{\leq q}$ (U may be a strict subset of $\mathcal{C}^{\perp}_{\leq q}$) we define the constraint graph induced by U to be the bipartite graph $G(\{1, \ldots, n\}, U, E)$ with left vertex set $\{1, \ldots, n\}$, right vertex set U and an edge between i and u if and only if $u_i \neq 0$. Given a distribution D as in the corollary above let $\mathrm{supp}(D) = \{u \in \mathcal{C}^{\perp} \mid D(u) > 0\}$ denote the support of the tester, it is the set of dual words, or linear tests, actually used by the tester. The constraint graph induced by a linear tester associated with D is the constraint graph induced by $\mathrm{supp}(D)$.

2.2   Random Low Density Parity Check Codes

Roughly speaking, a linear code whose dual contains many small-weight words should be hard to construct as the existence of many small-weight words may reduce other parameters of the code, like its rate. Thus, a good starting point is to examine the local testability of the family of *random low density parity check* (LDPC) codes which are known to be asymptotically good [43]. We shall show that testers achieving constant soundness for these codes require linear query complexity, and along the way we shall try to explain the way how this negative result about local testability is related to the structure of the constraint graphs associated with random LDPC codes.

To define our codes we need to describe the concept of a random regular bipartite graph. A bipartite graph is said to be (t, q)-*regular* if all vertices on the left side have degree at most t and all vertices on the right side have degree at most q. A *random (t, q)-regular graph* with n left-hand vertices and $m = \lceil tn/q \rceil$ right-hand ones is obtained as follows. Start with a four-layered graph, the leftmost layer is V, the second and third have tn vertices each, numbered $1, \ldots, tn$, and the rightmost layer is U. Connect $i \in V$ to the t vertices in the second layer numbered $t(i-1)+1, \ldots, ti$. Similarly connect vertex number j in U to the q vertices numbered $q(j-1)+1, \ldots, qj$ in the third layer. (The mth vertex may have less than q neighbors, in case tn/q is not an integer.) To obtain a *random* graph, pick a random permutation on tn elements and use it to construct a matching between the second and third layers. Finally, collapse each 3-edge-long path between $v \in V$ and $u \in U$ to obtain a single edge (collapsing parallel edges when needed), to obtain a random (t, q)-regular graph with n left vertices.

Definition 2 (Random low density parity check code). *The family of (t, q)-regular random LDPC codes is the distribution on families of linear codes obtained by picking the nth member in the family according to the following process. For integers $t < q$ let $G = (V, U, E)$ be a random (t, q)-regular bipartite graph over n left vertices and $m = \lceil tn/q \rceil$ right vertices (notice $m < n$ because $t < q$). Associate each right-hand side vertex $\hat{u} \in U$ with the vector $u = (u_1, \ldots, u_n) \in \mathbb{F}_2^n$ defined by*

$$u_i = \begin{cases} 1 & (i, \hat{u}) \in E \\ 0 & \text{otherwise.} \end{cases}$$

The LDPC code based on G is the code $\mathcal{C} = U^\perp$.

The rate of $\mathcal{C}$ is at least $\frac{n-m}{n} \approx 1 - \frac{t}{q}$ because $\dim(\mathcal{C}^\perp) \leq m$. It is well-known since the work of [43] that a family of random LDPC codes is, with high probability, asymptotically good (cf. [44]). At first glance it may seem that such a family is locally testable. The set of q-query words U *characterizes* $\mathcal{C}$ by which we mean that $w \in \mathcal{C}$ if and only if $w \perp U$. And the random graph G is with high probability an expander which implies that for any set $S \subset \{1, \ldots, n\}, |S| = \epsilon n$ — think of S as indicating the minimal size set of bits that need be flipped in r to obtain a codeword — the set of indices of nonzero entries of a random $u \in U$ hits S with probability proportional to ϵ. In spite of all this $\mathcal{C}$ is not q-testable. This much was conjectured already in [45]. Moreover, $\mathcal{C}$ is not even testable with any sublinear query complexity, i.e., a constant fraction of the received word must be queried in order to distinguish between completely uncorrupted, and severely corrupted, words. This is shown by the following theorem of [42].

Theorem 2 (Random LDPC codes require linear query complexity). *For integers $t < q$ and constants $1/2 > \epsilon > 0, s > 0$ there exists $\mu > 0$ such that for sufficiently large n, a random (t, q)-LDPC code is, with high probability, not $(\mu n, \epsilon, s)$-locally testable.*

Proof (Sketch). Consider a random LDPC code $\mathcal{C}$ based on a random (t, q)-regular graph G and assume that the constraints U that define it are linearly independent, which they are, with high probability. This linear independence implies that for every $u \in U$ there exists a word $r(u) \in \mathbb{F}_2^n$ such that

$$r(u) \not\perp u \quad \text{and} \quad r(u) \perp U \setminus \{u\}. \tag{1}$$

Appealing to the expansion properties of the graph G — which were used in the first place to argue that $\mathcal{C}$ has constant relative distance — we conclude that the code $\mathcal{C}_{-u} = (U \setminus \{u\})^\perp = \{w \mid w \perp (U \setminus \{u\})\}$ has good distance because the constraint graph induced by $U \setminus \{u\}$ is still a good expander. This implies that any word $r(u) \in \mathcal{C}_{-u} \setminus \mathcal{C}$ is ϵ-far from $\mathcal{C}$ for some constant $\epsilon > 0$.

What is the probability with which $r(u)$ is rejected by a q'-query tester? Recall that a linear q'-tester T is defined by a distribution D over $\mathcal{C}_{q'}^\perp$. Expressing a potential linear test $v \in \mathcal{C}_{q'}^\perp$ as a linear combination of elements from U and letting $U(v) \subseteq U$ denote the set of elements that have nonzero coefficients in this expression, we see from Equation (1) that $r(u) \not\perp v$ if and only if $u \in U(v)$. The answer to our question is then

$$\Pr_R[T^{r(u)}[R] = \mathsf{reject}] = \Pr_{v \sim D}[v \perp r(u)] = \Pr_{v \sim D}[u \in U(v)].$$

Taking one step further, for the tester defined by the distribution D to reject each $r(u)$ for $u \in U$, it better be the case that $U(v) \ni u$ for a random $v \sim D$ and uniformly random $u \in U$. This implies that a constant fraction of tests in $\mathrm{supp}(D)$ are, each, a linear combination of a constant fraction of U. Alas, with high probability, all words in $\mathrm{span}(U)$ that are a linear combination of a constant fraction of U must have large weight. This should sound reasonable because U is random, so summing up a constant fraction of its elements should result in a word with pretty large weight. We conclude that any tester that achieves constant soundness must be a distribution over words that have weight $\Omega(n)$, and this completes the proof (sketch).

2.3   LTCs Require Redundant Testers

Our next result rules out the existence of asymptotically good families of LTCs that lack sufficient *redundancy*, a concept we define next. This result can be seen as a generalization of the previous section to the case of codes that have "too few" dual words of weight q so let us explain how we quantify the number of such words and define what we mean by "too few" words.

If $\mathcal{C}_{\leq q}^\perp$ does not span all of $\mathcal{C}^\perp$ then $\mathcal{C}$ cannot be a q-query strong LTC because some non-codeword will be accepted with probability 1. This by itself does not yet mean that $\mathcal{C}$ is not locally testable, as it could be the case that all $r \notin \mathcal{C}$ that are accepted with probability 1 are, say, $(\epsilon/2)$-close to $\mathcal{C}$. A far more interesting case is when $\mathcal{C}_{\leq q}^\perp$ is a basis for $\mathcal{C}^\perp$ but contains no more words. Random (t, q)-regular codes give one example of such codes because it can be verified that the only words of weight at most q are those belonging to the linearly independent set U. We have already seen that such codes are not locally testable but perhaps other codes are? Before we continue let us formally define the redundancy of a code, which is the way we measure how many dual words are out there.

Definition 3 (Redundancy). *Given a set $U \subset \mathbb{F}^n$ let the* redundancy *of U be* $\mathrm{redun}(U) = |U| - \dim(\mathrm{span}(U))$. *It is the number of elements of U that can be removed from U without increasing the linear space that is dual to U (which we think of as a code $\mathcal{C}$). Notice* $\mathrm{redun}(U) = 0$ *if and only if U is linearly independent.*

Let $\mathcal{C}$ be a $[n,k,d]_{\mathbb{F}}$-linear code. For D a distribution over $\mathcal{C}^\perp$ (think of D as a tester for $\mathcal{C}$) let $\mathrm{redun}(D) = \mathrm{redun}(\mathrm{supp}(D))$. *$D$ is said to be a* linearly independent tester *if $\mathrm{redun}(D) = 0$ and if moreover $\mathrm{supp}(D)$ spans $\mathcal{C}^\perp$ we call D a* basis tester *for $\mathcal{C}$. Finally, the q-redundancy of $\mathcal{C}$ is* $\mathrm{redun}_q(\mathcal{C}) = \mathrm{redun}(\mathcal{C}^\perp_{\leq q})$.

The following theorem of [10] shows that any locally testable code with sufficiently large rate must be tested by redundant testers.

Theorem 3 (Linear LTCs require redundant testers). *Let $\mathcal{C}$ be an $[n,k,d= \delta_0 n]_{\mathbb{F}}$-code that is a $(q, \delta_0/3, \epsilon)$-LTC. Then*

$$\mathrm{redun}_q(\mathcal{C}) \geq \frac{\epsilon k}{q} - 1.$$

Moreover, if D, the tester's distribution, is uniformly distributed over $\mathrm{supp}(D)$, then

$$\mathrm{redun}(D) \geq \frac{\epsilon - q/k}{1 - \epsilon} \cdot (n - k).$$

The first equation above implies that every asymptotically good family of q-query LTCs must have linear q-redundancy, to see this set $k = \rho n$ where ρ is the rate of the family of codes. The second equation implies that q-query LTCs with super-constant size that are testable by a *uniform tester*, i.e., a tester whose distribution is uniform over a subset of $\mathcal{C}^\perp_{\leq q}$, must have linear redundancy. All algebraic and affine-invariant codes are testable by uniformly distributed testers, and so are sparse random unbiased codes but it should be stressed that the LTCs obtained by using composition techniques, such as PCPP-based and tensor-product ones, are not necessarily uniform. We point out that both inequalities are known to be nearly tight (cf. [34]).

It may seem that the limitation placed by Theorem 3 on the minimal redundancy of an LTC can be easily overcome. Even if there are precisely $n-k$ linearly independent words in $\mathcal{C}^\perp_{\leq q}$ (this is what happens, for example, with random (t,q)-regular LDPC codes), there are $\binom{n-k}{2}$ words in $\mathcal{C}^\perp_{\leq 2q}$ — take the sumset of $\mathcal{C}^\perp_{\leq q}$ — so clearly this set has superlinear redundancy and for all we know $\mathcal{C}$ may be $2q$-testable without contradicting our theorem. The following stronger version of Theorem 3 is immune to the "sumset" trick and seems to say something deeper about the structure of small weight words of the dual code. To state this theorem we need a more refined definition of redundancy.

Definition 4 (Expected redundancy). *For $U \subset \mathbb{F}^n$, $B = \{b_1, \ldots, b_t\}$ a linearly independent set spanning* $\mathrm{span}(U)$ *(B is not necessarily a subset of U), and $u \in U$ let $B(u)$ be the set of elements of B used to represent u. If $u = \sum_{i=1}^{t} \beta_i b_i$ then this set is*

$$B(u) = \{b_i \in B \mid \beta_i \neq 0\}.$$

For D a distribution on $\mathcal{C}_{\leq q}^{\perp}$ (which we view as a q-query tester for C) let its expected q-redundancy be

$$\mathrm{Eredun}_q(D) = \min_B \mathbb{E}_{u \sim D}[|B(u)|]$$

where the minimum is taken over all bases $B \subset \mathcal{C}_{\leq q}^{\perp}$ which span $\mathcal{C}^{\perp}$. (Notice B is not necessarily a subset of $\mathrm{supp}(D)$*.) The expected q-redundancy of C, denoted as $\mathrm{Eredun}_q(\mathcal{C})$, is the minimal expected q-redundancy of a distribution D on $\mathcal{C}_{\leq q}^{\perp}$.*

The following is the main theorem of [10].

Theorem 4 (LTCs require testers with large expected redundancy). *Let $\mathcal{C}$ be an $[n, k, d = \delta_0 n]_{\mathbb{F}}$-code that is a $(q, \delta_0/3, s)$-LTC. Then*

$$\mathrm{Eredun}_q(\mathcal{C}) \geq \frac{sk}{q}.$$

Returning to the example discussed above, the example which assumed $\mathcal{C}_{\leq q}^{\perp}$ is linearly independent and suggested to use a $2q$-tester distributed over the sumset of $\mathcal{C}_{\leq q}^{\perp}$, it is not hard to see that its expected redundancy is 2 and to see this set $B = \mathcal{C}_{\leq q}^{\perp}$. Theorem 4 thus rules out this case, as well as that of taking as our tester any distribution over the $\Omega(k)$-wise sum of $\mathcal{C}_{\leq q}^{\perp}$.

Informally, this theorem says is that in order for a linear code to be q-query testable it must be the case that for any basis $B \subset \mathcal{C}_{\leq q}^{\perp}$ there exists a linear number of words in $\mathcal{C}_{\leq q}^{\perp} \setminus B$ that are each a linear combination of a constant fraction of B. This means that some nontrivial cancelation is going on by which many small-weight words — a linear number of them — are each a sum of many words from B.

2.4   Dense LTCs Have Small Rate

In the previous section we saw that linear codes with too few dual words of small weight are not locally testable. In this section we discuss the opposite extreme, of codes with too many dual words of small weight. The following definition will be used to capture the notion of "too many" dual words.

Definition 5 (Dense codes). *An $[n, k, d]_{\mathbb{F}}$ linear code $\mathcal{C}$ is said to be (γ, q)-dense if for every $i \in \{1, \ldots, n\}$ there are at least γn^{q-2} dual words u of weight q such that $u_i \neq 0$.*

For instance, the Hadamard code is $(\frac{1}{2}, 3)$-dense because every selection of $j \in \{1, \ldots, n\}$ participates in a dual word of weight 3 that touches i.

Remark 5. A different definition for dense codes can be suggested, one that uses the *total number* of weight-q dual words. For instance, we may decide to call a code $\mathcal{C}$ (γ, q)-dense' if $|\mathcal{C}^{\perp}_{\leq q}| \geq \gamma n^{q-1}$. This definition is problematic, as seen by taking the direct product of the Hadamard code with blocklength n, denoted H_n, with, say, a $[n, k = n/\text{poly} \log n, d]_{\mathbb{F}_2}$-code $\mathcal{C}_0$ that is a $(3, \epsilon, s)$-LTC (codes with these parameters are known to exist). The resulting code

$$\mathcal{C} = \mathcal{C}_0 \times H_n = \{(c, c') \mid c \in \mathcal{C}_0, c' \in H_n\}$$

is a linear 3-query LTC of blocklength $2n$ and can easily be seen to be $(1/4, 3)$-dense' because H_2 is $(1/2, 3)$-dense' but the rate of $\mathcal{C}$ is at least $k/2n$. In other words, we can artificially increase the density' of an LTC at the price of decreasing its rate by a constant factor.

It turns out that it is sufficient to consider the density of weight-3 and weight 4 words, due to the following claim because (γ, q)-density for $q \geq 3$ implies either $(3, \gamma')$- or $(4, \gamma')$-density for $\gamma' > 0$ depending only on γ. The main theorem of [46] shows that dense codes have small rate:

Theorem 5 (Dense codes have small rate). *For every $\gamma > 0$ and integer q there exists $\ell > 0$ depending only on γ and q such that the following holds. If $\mathcal{C}$ is a linear $[n, k, d]_{\mathbb{F}_2}$ code that is (γ, q)-dense, then the dimension k of $\mathcal{C}$ is at most $\log^{\ell} n$.*

The proof relies on results from additive combinatorics and we give a sketch of it next.

Proof (Sketch). Take a generating matrix $A \in \mathbb{F}_2^{n \times k}$ for $\mathcal{C}$, a matrix satisfying $\mathcal{C} = \{Ax \mid x \in \mathbb{F}_2^k\}$. Let $\mathcal{A} = \{A_i \mid i \in \{1, \ldots, n\}\} \subset \mathbb{F}_2^k$ denote the set of rows of the matrix. The density assumption implies

$$\Pr_{a, a' \in \mathcal{A}}[a + a' \in \mathcal{A}] \geq \gamma.$$

The Balog-Szemerédi-Gowers theorem [47,48], together with the Freiman-Ruzsa theorem [49,50], imply that $\mathcal{A}$ contains a subset $\mathcal{A}'$ of size at least $\eta|\mathcal{A}|$ that is an η-fraction of some linear subspace of $\mathbb{F}_2^k$, where $\eta = \gamma^{\text{poly}(1/\gamma)}$. In other words, the set of rows $\mathcal{A}'$ can be viewed, after an appropriate change of basis, as resulting from taking a constant fraction of the rows of a generating matrix of the Hadamard code, which is known to have very bad rate. Consider the residual set $\mathcal{A}'' = \mathcal{A} \setminus \mathcal{A}'$. The assumption that each $i \in \{1, \ldots, n\}$ touches many weight-3 words is now used to argue that $\mathcal{A}''$ is also $(\gamma', 3)$-dense, for $\gamma' > 0$ that depends only on γ, so our argument can be repeated. Continuing in this manner we reach the conclusion that the generating matrix A can be written, after a proper change of basis, as a block-diagonal matrix where each block is a constant fraction of a Hadamard code and Hadamard codes are known to have bad rate. Consequently, $\mathcal{C}$ has small rate and this completes the proof sketch.

2.5   Question: Narrow the Gap between Redundant and Dense LTC Limitations

The rate limitations we have showed regarding both redundant, and dense, LTCs, suggest an interesting avenue for future research — to narrow the gap between these two cases. For simplicity consider the case of an asymptotically good family of *smooth* 3-query LTCs, i.e., LTCs that have a tester which queries each codeword entry with the same probability. The results on redundancy show that each member of the family should have at least a linear number of redundant weight-3 dual words. The result on dense codes shows that the overall number of such words is $o(n^2)$. Here is a seemingly simpler question that is currently open:

Question 2 (Number of small weight dual words of a linear LTC). Prove or refute the following conjecture. Suppose $\{\mathcal{C}_n \subset \{0,1\}^n \mid n \in \mathbb{Z}\}$ is an asymptotically good family of linear $(3, \delta/3, s > 0)$-LTCs of relative distance at least $\delta > 0$. Suppose furthermore that $\mathcal{C}_n$ is testable by a tester associated with the uniform distribution on $(\mathcal{C}_n)_{\leq 3}^{\perp}$, the set of weight-3 words in $\mathcal{C}^{\perp}$. Then $|(\mathcal{C}_n)_{\leq 3}^{\perp}| = \omega(n)$.

3   Limitations on Group-Invariant Codes

We have seen in Section 2.3 that linear LTCs must have dual codes whose small-weight words show a large degree of nontrivial redundancy. Constructing codes that have large rate and such a level of redundancy seems like a hard problem, and one way to get around it is to use codes that are invariant under a "sufficiently rich" group (a concept we explain next), for which the existence of even a single small-weight dual word immediately implies a large number of such words.

A code $\mathcal{C}$ of blocklength n induces a group of automorphisms $\mathrm{aut}(\mathcal{C})$, this is the group of permutations $\pi : \{1, \ldots, n\} \rightarrow \{1, \ldots, n\}$ under which the code is invariant, by which we mean that for every $w = (w_1, \ldots, w_n) \in \mathcal{C}$ the π-permuted word $\pi(w) = (w_{\pi(1)}, \ldots, w_{\pi(n)})$ also belongs to $\mathcal{C}$. It is not hard to verify that $\mathrm{aut}(\mathcal{C})$ is indeed a group and that $\mathrm{aut}(\mathcal{C}^{\perp}) = \mathrm{aut}(\mathcal{C})$. Consequently, if $\mathcal{C}^{\perp}$ contains a word u of weight q then $\mathcal{C}_{\leq q}^{\perp}$ contains $\{\pi(u) \mid \pi \in \mathrm{aut}(\mathcal{C})\}$. Thus, if $\mathrm{aut}(\mathcal{C})$ is sufficiently rich we can hope for $\mathcal{C}_{\leq q}^{\perp}$ to be large and redundant and, if all stars align properly, $\mathcal{C}$ will be a q-query LTC and moreover have large rate and relative distance.

Two notable families of groups that should be mentioned in this context are doubly transitive and affine-invariant ones. A group of permutations G over n elements is said to be *doubly transitive*, or *2-wise transitive*, if for every $i \neq j$ and $i' \neq j' \in \{1, \ldots, n\}$ there exists $\pi \in G$ such that $\pi(i) = i'$ and $\pi(j) = j'$. A conjecture attributed[2] to [18] is that all codes which are invariant under a doubly transitive group (call them doubly transitive codes) are testable with query complexity q' that depends only on the smallest q for which $\mathcal{C}_{\leq q}^{\perp}$ spans $\mathcal{C}^{\perp}$. In particular, this query complexity is conjectured to be independent of the

---

[2] We use the term "attributed" because in [18, Section 5] it appears as an open question.

blocklength of $\mathcal{C}$. (The requirement that $\mathcal{C}_{\leq q}^{\perp}$ span $\mathcal{C}^{\perp}$, cannot be replaced by the weaker assumption that q is the minimal distance of $\mathcal{C}^{\perp}$. [51] showed that if one opts for the weaker assumption then the conjecture is false.) It is shown by [52] that doubly transitive codes with small dual distance are so-called locally correctable codes. These codes are a stronger analog of locally decodable codes (cf. [5,6]), and this implies a polynomial upper bound on their rate of the form of the form $\rho(\mathcal{C}) = O\left(\log n \left(\frac{\log n}{n}\right)^{\frac{2}{q+1}}\right)$, as shown in [53]. This raises the following open problem:

Question 3 (Polynomial rate doubly transitive LTCs). Does there exist a family $\{\mathcal{C}_n \subseteq \mathbb{F}^n \mid n \in \mathbb{Z}\}$ of doubly transitive $(q, \epsilon > 0, s, 1)$-LTCs that has inverse polynomial rate, i.e., $\rho(\mathcal{C}_n) \geq 1/n^{O(1)}$?

A group is said to be *affine-invariant* if $\{1, \ldots, n\}$ can be identified with a vector space $\mathbb{K}^m$ over a finite field $\mathbb{K}$ and G is then isomorphic to the group of invertible affine transformations[3] over $\mathbb{K}^m$. The family of affine-invariant codes includes the Hadamard and Reed-Muller codes as well as dual-BCH codes. [21] showed that, when $|\mathbb{K}|$ is small, every affine-invariant family of codes over $\mathbb{K}^m$, whose dual contains a small-weight word, is locally testable. Since every affine group is doubly transitive, the work of [21] shows that the double-transitivity conjecture does hold in certain interesting special cases. Later on we shall see that affine-invariant codes have small rate, and this answers negatively the question above for this special case.

   A third and final family of group invariant codes considered in the literature is that of *cyclic* codes, i.e., codes invariant under a cyclic group. All affine invariant codes (including Hadamard and Reed-Muller) are, in particular, cyclic. [9] showed that a family of cyclic LTCs cannot be asymptotically good, either its rate or its distance must be less than $1/\sqrt{\log n \log \log n}$. A long-standing open problem in coding theory is whether there exists an asymptotically good family of cyclic codes (cf. [7, Open Problem 9.2]). The result above shows that when local testability is thrown in as a requirement, then indeed asymptotically good codes do not exist.

3.1   Affine Invariant LTCs Have Small Rate

In this section we discuss rate limitations of affine-invariant locally testable codes. More information on this topic can be found in the survey [25]. Recall that if $\mathcal{C}$ is an $[n, k, d]_{\mathbb{F}}$-code affine-invariant code it means we can identify $\{1, \ldots, n\}$ with $\mathbb{K}^m$ for some field $\mathbb{K}$ which is a finite extension[4] of $\mathbb{F}$ and such that the

---

[3] The work of [21] actually talks about the semi-group of all affine transformations, including the non-invertible ones.

[4] The more general case of $\mathbb{K}$ being an arbitrary field, not necessarily extending $\mathbb{F}$, has not been addressed so far. However, it seems reasonable to expect that such codes should not have good rate, regardless of their local testability properties. This is because $\mathbb{K}^m$-affine invariance and $\mathbb{F}$-linearity do not mix well when $\mathbb{K}$ is not an extension of $\mathbb{F}$.

automorphism group of $\mathcal{C}$ contains the affine (semi-)group over $\mathbb{K}^m$. The study of affine invariant LTCs was initiated by [21], as a first step towards characterizing the class of "algebraic" properties which are testable. This class is also an interesting special case of the doubly transitive conjecture of [18]. Indeed, for such codes [21] showed that local testability exists as long as the field $\mathbb{K}$ is sufficiently small and the dual code has constant distance, as seen from their main theorem:

Theorem 6 (Affine invariant codes over small fields with constant dual distance are locally testable). *For fields* $\mathbb{F} \subseteq \mathbb{K}$ *let* $\mathcal{C}$ *be an* $[n = |\mathbb{K}^m|, k, d]_{\mathbb{F}}$ *affine-invariant code such that* $\mathcal{C}^{\perp}$ *contains a word of weight* q_0. *Then* $\mathcal{C}$ *is*

$$\left(q = (|\mathbb{K}|^2 q_0)^{|\mathbb{K}|^2}, s = \frac{1}{2(2q+1)(q+1)} \right) \text{-strongly locally testable}$$

by which we mean that there exists a q-*query linear tester that rejects noncodewords* $r \notin \mathcal{C}$ *with probability at least* $s \cdot \delta_{\mathcal{C}}(r)$.

Now we discuss the rate of such codes. Since affine invariant codes are cyclic, one could get an inverse logarithmic bound on the rate of affine invariant LTCs from what is known on cyclic LTCs. A tighter, inverse polynomial, bound on the rate follows from the result of [52] which says that such codes are locally decodable (and locally correctable) and the result of [53] which bounds the rate of locally decodable codes by $O\left(\log n \left(\frac{\log n}{n} \right)^{\frac{2}{q+1}} \right)$. The following result of [24] gives a stronger bound, showing that the dimension of affine-invariant codes is merely polylogarithmic in the blocklength of the code.

Theorem 7 (Affine invariant LTCs have small rate). *Let* p *be a prime and* r, n, m *be positive integers and let* $\mathbb{F}$ *be the field of size* p^r *and* $\mathbb{K}$ *be its degree* ℓ-*extension, which is of size* $p^{r\ell}$. *Any affine invariant* $[n = |\mathbb{K}^m|, k, d]_{\mathbb{F}}$-*code* $\mathcal{C}$ *such that* $\mathcal{C}^{\perp}$ *contains a word of weight* $q > 0$ *satisfies*

$$k \le (\log_p n)^{q-1}.$$

Notice the theorem shows exponential rate even for large fields $\mathbb{K}$, which are not known to be locally testable. We point out that the theorem as stated in [24] gives more information on affine-invariant codes with small dual distance, showing they are subcodes of low-degree polynomials (Reed-Muller codes). We shall not describe this result, nor shall we go into details of the proof because quite a lot of algebra is needed to describe it. Instead, we point the interested reader to the survey [25] and the relevant papers [21,24].

We end this section by pointing out the following interesting question which addresses the rate of a natural family of codes invariant under a linear group (in particular, Theorem 7 does not apply to such codes):

Question 4 (Rate of linear invariant codes with small dual distance). Let $\mathbb{K}$ *be a finite extension of a finite field* $\mathbb{F}$. *Let* $GL(m, \mathbb{K})$ *denote the general linear group over* $\mathbb{K}$, *containing all invertible* m-*dimensional linear transformations over* $\mathbb{K}$.

Let $\mathcal{C}$ be an $[n = |\mathbb{K}|^m, k, d]_\mathbb{F}$-linear code that is invariant under $GL(m, \mathbb{K})$ and suppose $\mathcal{C}^\perp$ contains a word of weight $q > 0$. How large can k be as a function of the field size $|\mathbb{K}|$ and code distance d?

Acknowledgement

Thanks to Michael Viderman for helpful comments on an earlier draft. The research leading to some of the results surveyed here has received funding from the European Community's Seventh Framework Programme (FP7/2007-2013) under grant agreement number 240258 and from the US-Israel Binational Science Foundation under grant number 2006104.

References

1. Arora, S., Safra, S.: Probabilistic checking of proofs: A new characterization of NP. Journal of the ACM 45(1), 70–122 (1998)
2. Arora, S., Lund, C., Motwani, R., Sudan, M., Szegedy, M.: Proof verification and the hardness of approximation problems. Journal of the ACM 45(3), 501–555 (1998)
3. Dinur, I.: The PCP theorem by gap amplification. Journal of the ACM 54(3), 12:1–12:44 (2007)
4. Goldreich, O., Goldwasser, S., Ron, D.: Property testing and its connection to learning and approximation. J. ACM 45(4), 653–750 (1998)
5. Goldreich, O.: Short locally testable codes and proofs (survey). Electronic Colloquium on Computational Complexity (ECCC) (014) (2005)
6. Trevisan, L.: Some applications of coding theory in computational complexity. Quaderni di Matematica 13, 347–424 (2004)
7. MacWilliams, F., Sloane, N.: The theory of error-correcting codes. North-Holland, Amsterdam (1978)
8. Ben-Sasson, E., Goldreich, O., Sudan, M.: Bounds on 2-query codeword testing. In: Arora, S., Jansen, K., Rolim, J.D.P., Sahai, A. (eds.) RANDOM 2003 and APPROX 2003. LNCS, vol. 2764, pp. 216–227. Springer, Heidelberg (2003)
9. Babai, L., Shpilka, A., Stefankovic, D.: Locally testable cyclic codes. IEEE Transactions on Information Theory 51(8), 2849–2858 (2005)
10. Ben-Sasson, E., Guruswami, V., Kaufman, T., Sudan, M., Viderman, M.: Locally testable codes require redundant testers. In: CCC 2009: Proceedings of the 2009 24th Annual IEEE Conference on Computational Complexity, Washington, DC, USA, pp. 52–61. IEEE Computer Society, Los Alamitos (2009)
11. Bellare, M., Goldreich, O., Sudan, M.: Free bits, PCPs, and nonapproximability—towards tight results. SIAM Journal on Computing 27(3), 804–915 (1998)
12. Meir, O.: On the efficiency of non-uniform pcpp verifiers. Electronic Colloquium on Computational Complexity (ECCC) 15(064) (2008)
13. Blum, M., Luby, M., Rubinfeld, R.: Self-testing/correcting with applications to numerical problems. In: STOC, pp. 73–83. ACM, New York (1990)
14. Bellare, M., Coppersmith, D., Hastad, J., Kiwi, M., Sudan, M.: Linearity testing in characteristic two. IEEE Transactions on Information Theory 42(6), 1781–1795 (1996)

15. Babai, L., Fortnow, L., Levin, L., Szegedy, M.: Checking computations in polylogarithmic time. In: Proceedings of the Twenty-third Annual ACM Symposium on Theory of Computing, pp. 21–32. ACM, New York (1991)
16. Raz, R., Safra, S.: A sub-constant error-probability low-degree test, and a sub-constant error-probability PCP characterization of NP. In: Proceedings of the Twenty-ninth Annual ACM Symposium on Theory of Computing, pp. 475–484. ACM, New York (1997)
17. Arora, S., Sudan, M.: Improved low-degree testing and its applications. Combinatorica 23(3), 365–426 (2003)
18. Alon, N., Kaufman, T., Krivelevich, M., Litsyn, S., Ron, D.: Testing reed-muller codes. IEEE Transactions on Information Theory 51(11), 4032–4039 (2005)
19. Kaufman, T., Ron, D.: Testing polynomials over general fields. SIAM J. Comput. 36(3), 779–802 (2006)
20. Samorodnitsky, A.: Low-degree tests at large distances. In: Johnson, D.S., Feige, U. (eds.) STOC, pp. 506–515. ACM, New York (2007)
21. Kaufman, T., Sudan, M.: Algebraic property testing: the role of invariance. In: Ladner, R.E., Dwork, C. (eds.) STOC, pp. 403–412. ACM, New York (2008)
22. Grigorescu, E., Kaufman, T., Sudan, M.: 2-transitivity is insufficient for local testability. In: IEEE Conference on Computational Complexity, pp. 259–267. IEEE Computer Society, Los Alamitos (2008)
23. Grigorescu, E., Kaufman, T., Sudan, M.: Succinct representation of codes with applications to testing. In: Dinur, I., Jansen, K., Naor, J., Rolim, J. (eds.) APPROX–RANDOM 2009. LNCS, vol. 5687, pp. 534–547. Springer, Heidelberg (2009)
24. Ben-Sasson, E., Sudan, M.: Limits on the rate of locally testable affine-invariant codes. Electronic Colloquium on Computational Complexity (ECCC) (108) (2010)
25. Sudan, M.: Invariance in Property Testing. ECCC, TR10-051 (2010)
26. Ben-Sasson, E., Goldreich, O., Harsha, P., Sudan, M., Vadhan, S.P.: Robust PCPs of proximity, shorter PCPs, and applications to coding. SIAM J. Comput. 36(4), 889–974 (2006)
27. Dinur, I., Reingold, O.: Assignment testers: Towards a combinatorial proof of the PCP theorem. SIAM J. Comput. 36(4), 975–1024 (2006)
28. Meir, O.: Combinatorial construction of locally testable codes. SIAM J. Comput. 39(2), 491–544 (2009)
29. Ben-Sasson, E., Sudan, M.: Robust locally testable codes and products of codes. Random Struct. Algorithms 28(4), 387–402 (2006)
30. Moshkovitz, D., Raz, R.: Two-query pcp with subconstant error. J. ACM 57(5) (2010)
31. Dinur, I., Harsha, P.: Composition of low-error 2-query pcps using decodable pcps. In: FOCS, pp. 472–481. IEEE Computer Society, Los Alamitos (2009)
32. Kaufman, T., Sudan, M.: Sparse random linear codes are locally decodable and testable. In: FOCS, pp. 590–600. IEEE Computer Society, Los Alamitos (2007)
33. Kopparty, S., Saraf, S.: Local list-decoding and testing of random linear codes from high error. In: Schulman, L.J. (ed.) STOC, pp. 417–426. ACM, New York (2010)
34. Ben-Sasson, E., Viderman, M.: Low rate is insufficient for local testability. In: Shaltiel, R. (ed.) Proc. 14th Intl. Workshop on Randomization and Computation - RANDOM 2010 (September 2010)
35. Kilian, J.: A note on efficient zero-knowledge proofs and arguments (extended abstract). In: STOC, pp. 723–732. ACM, New York (1992)
36. Micali, S.: Computationally sound proofs. SIAM J. Comput. 30(4), 1253–1298 (2000)

37. Barak, B., Goldreich, O.: Universal arguments and their applications. SIAM J. Comput. 38(5), 1661–1694 (2008)
38. Ben-Sasson, E., Harsha, P., Lachish, O., Matsliah, A.: Sound 3-query PCPPs are long. In: Aceto, L., Damgård, I., Goldberg, L.A., Halldórsson, M.M., Ingólfsdóttir, A., Walukiewicz, I. (eds.) ICALP 2008, Part I. LNCS, vol. 5125, pp. 686–697. Springer, Heidelberg (2008)
39. Guruswami, V.: On 2-query codeword testing with near-perfect completeness. In: Asano, T. (ed.) ISAAC 2006. LNCS, vol. 4288, pp. 267–276. Springer, Heidelberg (2006)
40. Kol, G., Raz, R.: Bounds on 2-Query Locally Testable Codes with Affine Tests. ECCC Report TR09-138 (2009)
41. Kol, G., Raz, R.: Locally testable codes analogues to the unique games conjecture do not exist. ECCC Report TR09-128 (2009)
42. Ben-Sasson, E., Harsha, P., Raskhodnikova, S.: Some 3CNF properties are hard to test. SIAM J. Comput. 35(1), 1–21 (2005)
43. Gallager, R.: Low-density parity-check codes. IRE Transactions on Information Theory 8(1), 21–28 (1962)
44. Sipser, M., Spielman, D.: Expander codes. IEEE Transactions on Information Theory 42(6), 1710–1722 (1996)
45. Spielman, D.: Computationally efficient error-correcting codes and holographic proofs. PhD thesis, MIT (1995)
46. Ben-Sasson, E., Viderman, M.: Dense locally testable codes have bad rate (2010) (unpublished manuscript)
47. Balog, A., Szemerédi, E.: A statistical theorem of set addition. Combinatorica 14(3), 263–268 (1994)
48. Gowers, W.T.: A new proof of szemerèdi's theorem for arithmetic progressions of length four. Geom. Funct. Anal. 8(3), 529–551 (1998)
49. Freiman, G.A.: Foundations of a structural theory of set addition, vol. 37. American Mathematical Society, Providence (1973)
50. Ruzsa, I.Z.: An analog of freiman's theorem in groups. Astèrique 258, 323–326 (1999)
51. Grigorescu, E., Kaufman, T., Sudan, M.: Succinct representation of codes with applications to testing. In: Approximation, Randomization, and Combinatorial Optimization. Algorithms and Techniques, pp. 534–547 (2009)
52. Kaufman, T., Viderman, M.: Locally testable vs. locally decodable codes. In: Shaltiel, R. (ed.) Proc. 14th Intl. Workshop on Randomization and Computation - RANDOM 2010 (2010)
53. Woodruff, D.: New lower bounds for general locally decodable codes. Electronic Colloquium on Computational Complexity (ECCC) 14(006) (2007)

Testing Juntas: A Brief Survey

Eric Blais

School of Computer Science, Carnegie Mellon University,
Pittsburgh PA 15213, USA
eblais@cs.cmu.edu

Abstract. A function on n variables is called a *k-junta* if it depends on at most k of its variables. In this survey, we review three recent algorithms for testing k-juntas with few queries.

1 Introduction

A function $f : \{0,1\}^n \to \{0,1\}$ is said to be a k-junta if it depends on at most k variables. Juntas provide a clean model for studying learning in the presence of many irrelevant features [7,10] and have consequently been of particular interest to the computational learning theory community [7,8,9,10,22,23,25].

As is typical in the machine learning setting, all learning results on k-juntas assume that the unknown function is a k-junta. In practice, however, it is often not known *a priori* whether a function being learned is a k-junta or not. It is therefore desirable to have an efficient algorithm for *testing* whether a function is a k-junta or "far" from being a k-junta before attempting to run any k-junta learning algorithm.

We consider the problem of testing k-juntas in the standard property testing framework originally defined by Rubinfeld and Sudan [27]. In this framework, we say that a function f is ϵ-*far* from being a k-junta if for every k-junta g, the functions f and g disagree on at least an ϵ fraction of all inputs.

An ϵ-*tester for k-juntas* is an algorithm $\mathcal{A}$ that queries an unknown function f on q inputs of its choosing, and then (1) accepts f with probability at least $2/3$ when f is a k-junta, and (2) rejects f with probability at least $2/3$ when f is ϵ-far from being a k-junta. When the algorithm $\mathcal{A}$ chooses all its queries in advance (i.e., before observing the values of the function on any of its previous queries), it is *non-adaptive*; otherwise it is *adaptive*. The main parameter of interest for our purposes is the number q of queries required by testers for k-juntas. In particular, the question we study is the following:

What is the minimum number of queries required to ϵ-test k-juntas?

A simple way to test k-juntas is to *learn* a target hypothesis k-junta using membership queries, and to then use a separate set of randomly-chosen queries to test this hypothesis [18,22]. Such an approach yields a valid tester but requires $O(k \log n/\epsilon)$ queries. In the rest of this survey, we will examine three algorithms that improve dramatically on this bound by requiring a number of queries that is *independent* of n.

O. Goldreich (Ed.): Property Testing, LNCS 6390, pp. 32–40, 2010.

2   Boolean Functions: Preliminaries

2.1   Basic Definitions

Throughout this survey, $f : \{0,1\}^n \rightarrow \{0,1\}$ represents a (generic) boolean function. The *complement* of f is the function $\bar{f} : \{0,1\}^n \rightarrow \{0,1\}$ defined by $\bar{f}(x) = 1 - f(x)$.

Given $x = (x_1, \ldots, x_n)$ and $y = (y_1, \ldots, y_n)$ from $\{0,1\}^n$, addition and multiplication are defined componentwise: $x + y = (x_1 + y_1, \ldots, x_n + y_n)$ and $x \cdot y = (x_1 y_1, \ldots, x_n y_n)$. We also define a *hybridization* operation: for a set $S \subseteq [n]$, the element $z = x_{\bar{S}} y_S \in \{0,1\}^n$ is formed by setting $z_i = x_i$ for every $i \in \bar{S} = [n] \setminus S$ and setting $z_i = y_i$ for every $i \in S$.

2.2   Notable Boolean Functions

The function that maps all inputs to 0 is the *constant zero* (or just *zero*) function; its complement is the *constant one* function that maps all inputs to 1. When there is an index $i \in [n]$ such that f is defined by $f(x) = x_i$, then we say that f is a *dictator* function. A function is an *anti-dictator* function if its complement is a dictator function.

For a set $S = \{i_1, i_2, \ldots, i_k\} \subseteq [n]$, the *linear* function corresponding to S is the function χ_S defined by $\chi_S(x) = x_{i_1} + x_{i_2} + \cdots + x_{i_k}$. By convention, we define $\chi_\emptyset$ to be the constant zero function. An alternative characterization of linear functions is provided by the following proposition.

Proposition 1. *A function $f : \{0,1\}^n \rightarrow \{0,1\}$ is linear if and only if for every $x, y \in \{0,1\}^n$, $f(x) + f(y) = f(x + y)$.*

For a set $S = \{i_1, i_2, \ldots, i_k\} \subseteq [n]$, the (monotone) *monomial* function corresponding to S is the function ξ_S defined by $\xi_S(x) = x_{i_1} x_{i_2} \cdots x_{i_k}$. (I.e., $\xi_S(x) = 1$ iff $x_{i_1} = \cdots = x_{i_k} = 1$.) As with linear functions, monomials have a useful alternative characterization.

Proposition 2. *A non-constant function $f : \{0,1\}^n \rightarrow \{0,1\}$ is a monomial if and only if for every $x, y \in \{0,1\}^n$, $f(x) \cdot f(y) = f(x \cdot y)$.*

2.3   Influence

For an index $i \in [n]$, the *influence* of the ith variable in the function f is

$$\mathrm{Inf}_f(i) = \Pr_x[f(x) \neq f(x^{(i)})],$$

where the probability is over the uniform distribution on $\{0,1\}^n$ and the input $x^{(i)} \in \{0,1\}^n$ is obtained by flipping the value of the ith variable of x.

The *influence* of the set $S \subseteq [n]$ in the function f is

$$\mathrm{Inf}_f(S) = 2 \Pr_{x,y}[f(x) \neq f(x_{\bar{S}} y_S)].$$

A k-*junta* is a function f for which there are at most k indices $i \in [n]$ such that $\mathrm{Inf}_f(i) > 0$. Alternatively, f is a k-junta if there exists a set $S \subseteq [n]$ of size $|S| \leq k$ such that $\mathrm{Inf}_f(\bar{S}) = 0$.

3 Testing 1-Juntas

We begin with the simplest case: testing 1-juntas. The family of 1-junta functions is small. It contains only the constant functions, the dictator functions, and the anti-dictator functions. Furthermore, dictator functions have the useful distinction of being the only non-constant functions that are both linear functions and monomials. This distinction lies at the heart of the 1-junta tester that we will examine in this section.

3.1 The Algorithm

As suggested above, our main building block for testing 1-juntas is an algorithm that accepts functions that are both linear and monomials. The characterizations of linear functions and of monomials from Propositions 1 and 2 suggest the following simple algorithm for this task:

LINEAR MONOMIAL TEST(f, ϵ)

1. For $O(1/\epsilon)$ randomly selected pairs $x, y \in \{0,1\}^n$,
   1.1. Verify that $f(x) + f(y) = f(x + y)$.
   1.2. Verify that $f(x) \cdot f(y) = f(x \cdot y)$.
2. **Accept** iff all verifications pass.

Clearly, the LINEAR MONOMIAL TEST always accepts the zero function and dictator functions. To accept all 1-juntas, it suffices to test f and its complement $\bar{f}$ for the property of being a linear monomial:

1-JUNTA TEST(f, ϵ)

1. Call LINEAR MONOMIAL TEST(f, ϵ).
2. Call LINEAR MONOMIAL TEST($\bar{f}$, ϵ).
3. **Accept** iff one of the above tests accepts.

The 1-JUNTA TEST algorithm always accepts 1-juntas. To establish that it is a valid tester for 1-juntas, we need to show that it rejects functions that are ϵ-far from 1-juntas with high probability. We do so in two steps.

First, we show that the 1-JUNTA TEST rejects functions that are far from linear with high probability. This statement follows from the *robustness* of the linearity characterization in Proposition 1: when a function f is ϵ-far from linear and x, y are generated uniformly at random, then $f(x) + f(y) \neq f(x + y)$ with probability at least ϵ [3,11].

Lemma 3 (Blum et al. [11], Bellare et al. [3]). *Let f be ϵ-far from linear. Then the* 1-JUNTA TEST *rejects with probability at least $2/3$.*

Second, we show that functions that are ϵ-close to a linear function χ_S for some set S of size $|S| \geq 2$ are rejected by the monomial test in Line 1.2 of the *Linear Monomial Test* with high probability. This is indeed the case, as an elementary counting argument shows.

Lemma 4 (Bellare et al. [4]). *Fix* $0 < \epsilon < \frac{1}{8}$ *and let* f *be* ϵ*-close to* χ_S *for some set* $S \subseteq [n]$ *of size* $|S| \geq 2$. *Then the* 1-JUNTA TEST *rejects with probability at least* $2/3$.

Together, Lemmas 3 and 4 show that the 1-JUNTA TEST rejects functions that are ϵ-far from 1-juntas with probability at least $2/3$. This completes the proof of correctness of the algorithm. We can also easily verify that the tester makes only $O(1/\epsilon)$ queries to the input function; this is optimal.

3.2   History

The problem of testing dictator functions was first studied by Bellare, Goldreich, and Sudan [4] in the context of testing the Long Code for constructing probabilistically-checkable proof (PCP) systems. As pointed out in [26], testing the Long Code is equivalent to testing dictator functions, and their test for dictator functions is roughly equivalent to the 1-JUNTA TEST algorithm above.[1] The analysis of the dictator test was further generalized and extended by Parnas, Ron, and Samorodnitsky [26].

Due to the key role of dictator functions in PCP systems, many other variants of the dictatorship testing problem have been studied – see [13] in this volume and the references therein for more information on this topic.

4   Testing k-Juntas

We now turn our attention to the general problem of testing k-juntas for any value of $k \geq 1$. In contrast to the case of 1-juntas, when $k \geq 2$ the class of k-juntas does not have a simple characterization that directly suggests a testing algorithm. Nonetheless, as we will see in this section it is still possible to test k-juntas with a small number of queries.

4.1   The Algorithm

The algorithm for testing k-juntas relies on two basic components: the INDEPENDENCE TEST, and the idea of *randomly partitioning* the coordinates.

The INDEPENDENCE TEST is a simple algorithm for verifying whether a given function f is independent of a set $S \subseteq [n]$ of coordinates:

INDEPENDENCE TEST(f, S)

1. Generate $x, y \in \{0, 1\}^n$ uniformly at random.
2. **Accept** iff $f(x) = f(x_{\bar{S}} y_S)$.

---

[1] There is one difference: when testing dictator functions, constant functions must be rejected. In our case we want to accept them; this simplifies the algorithm slightly.

By our definition of influence, the probability that the INDEPENDENCE TEST rejects is exactly $\frac{1}{2}\text{Inf}_f(S)$. In particular, when f is independent of the variables in S, then $\text{Inf}_f(S) = 0$ and the test always accepts.

A naïve way to use the INDEPENDENCE TEST for testing k-juntas is to run the test (sufficiently many times) on each singleton set $S = \{1\}, \{2\}, \ldots, \{n\}$ and to accept iff at most k of the sets are rejected. This proposed algorithm is indeed a valid tester for k-juntas, but it requires $\Omega(n)$ queries. A simple trick, however, can dramatically reduce the number of queries required: take a sufficiently fine partition of the coordinates $[n]$ and run the INDEPENDENCE TEST on each part.

---

k-JUNTA TEST(f, ϵ)

1. Randomly partition the coordinates into $O(k^2)$ buckets.
2. Run INDEPENDENCE TEST $\tilde{O}(k^2/\epsilon)$ times.
3. **Accept** iff at most k buckets fail the independence test.

Clearly, the k-JUNTA TEST always accepts k-juntas: if there are only k indices $i \in [n]$ for which $\text{Inf}_f(i) > 0$, then at most k parts in the random partition will have influence $\text{Inf}_f(S) > 0$. Conversely, when f is ϵ-far from being a k-junta, Fischer et al. [17] showed that with high probability over the choice of the random partition, at least $k+1$ parts have large influence.

Lemma 5 (Fischer et al. [17]). *Let $f : \{0,1\}^n \to \{0,1\}$ be ϵ-far from being a k-junta and $s = \Theta(k^2)$. Then with high probability a random partition $S_1 \dot\cup S_2 \dot\cup \cdots \dot\cup S_s$ of $[n]$ will have at least $k+1$ parts with influence $\text{Inf}_f(S_j) > \epsilon/k^2$.*

The proof of Lemma 5 uses Fourier analysis. The rest of the proof of correctness of the k-JUNTA TEST follows almost immediately. The k-JUNTA TEST uses $\tilde{O}(k^4/\epsilon)$ queries. This bound is significant in that it is independent of n; as we discuss below, however, variants on this algorithm can test k-juntas with fewer queries.

4.2 History

Fischer, Kindler, Ron, Safra, and Samorodnitsky [17] first studied the problem of testing juntas and introduced the algorithm presented in this section. They also designed multiple other testing algorithms that improve on the query complexity of the k-JUNTA TEST. In particular, by using the INDEPENDENCE TEST on carefully chosen *sets* of parts in a random partition, they showed that $\tilde{O}(k^2/\epsilon)$ queries are sufficient to test k-juntas.

Fischer et al. [17] also introduced the first non-trivial lower bound on the query complexity of junta testing problem: they showed that for $k = o(\sqrt{n})$, non-adaptive testing algorithms for testing k-juntas must make at least $\tilde{\Omega}(\sqrt{k})$ queries. This lower bound implies a lower bound of $\Omega(\log k)$ queries for all adaptive k-junta testers as well. The lower bound was improved shortly afterwards by Chockler and Gutfreund [14], who showed that $\Omega(k)$ queries are required to test k-juntas (adaptively or non-adaptively).

The gap between the $\Omega(k)$ and $\tilde{O}(k^2/\epsilon)$ bounds on the query complexity of the junta testing problem remained unchanged until recently, when a new algorithm was introduced to test k-juntas with $\tilde{O}(k^{1.5}/\epsilon)$ queries [5]. This was followed by the introduction of another algorithm for testing k-juntas with $O(k \log k + k/\epsilon)$ queries [6]; we examine this algorithm in the next section.

5   Testing k-Juntas Nearly Optimally

The algorithm we saw in the last section relied on the INDEPENDENCE TEST. To improve the query complexity, the algorithm we present in this section relies on a slightly stronger building block.

5.1   The Algorithm

The starting point for the algorithm is an observation due to Blum, Hellerstein, and Littlestone [9]: if we have two inputs $x, y \in \{0, 1\}^n$ such that $f(x) \neq f(y)$, then the set of coordinates $i \in [n]$ for which $x_i \neq y_i$ contains a coordinate that is relevant in f. Furthermore, by performing a binary search over the hybrid inputs formed from x and y, we can identify a relevant coordinate with $O(\log n)$ queries.

Even more interestingly, if we have a partition $\mathcal{I}$ of $[n]$ and we have a pair of inputs x, y such that $f(x) \neq f(y)$, we can use the same binary search idea to identify a part that contains a relevant coordinate with only $O(\log |\mathcal{I}|)$ queries. We use this idea to create an algorithm that attempts to find a part with a relevant coordinate as follows:

FIND RELEVANT PART$(f, \mathcal{I}, S)$

1. Generate $x, y \in \{0, 1\}^n$ uniformly at random.
2. If $f(x) \neq f(x_{\bar{S}} y_S)$ then
    2.1. Use a binary search to identify a part $I \in \mathcal{I}$ that contains a relevant variable;
    2.2. **Return I.**
3. Otherwise, **Return $\emptyset$.**

Note that by the test in Line 2, if the algorithm finds a part with a relevant variable, that relevant variable is guaranteed to be in S. Also, the probability that FIND RELEVANT PART succeeds in identifying a relevant part is the probability that $f(x) \neq f(x_{\bar{S}} y_S)$, which as we have seen previously is exactly $\frac{1}{2}\text{Inf}_f(S)$.

The algorithm we now consider for testing k-juntas uses the FIND RELEVANT PART in the obvious way: after taking a random partition of the coordinates, the algorithm calls this routine a large number of times and rejects the input if it identifies $k + 1$ distinct parts that contain relevant coordinates.

NEARLY OPTIMAL k-JUNTA TEST(f, ϵ)

1. Randomly partition $[n]$ into a partition $\mathcal{I}$ with $\text{poly}(k/\epsilon)$ parts and initialize $J \leftarrow \emptyset$.
2. For each of $O(k/\epsilon)$ rounds,
   2.1. $J \leftarrow J \cup$ FIND RELEVANT PART$(f, \mathcal{I}, \bar{J})$.
   2.2. If J contains $> k$ parts, quit and **Reject**.
3. **Accept**.

As with the algorithms in the previous sections, it is easy to check that this algorithm always accepts k-juntas. Once again, the non-trivial part of the proof of correctness involves showing that functions ϵ-far from k-juntas are rejected with high probability. The key to proving that statement is the following lemma:

Lemma 6 ([6]). *Let $f : \{0,1\}^n \to \{0,1\}$ be ϵ-far from being a k-junta, and let $\mathcal{I}$ be a sufficiently fine partition of $[n]$. Then with high probability every set J formed by taking the union of at most k parts of $\mathcal{I}$ satisfies $\text{Inf}_f(\bar{J}) \geq \epsilon/2$.*

The proof of Lemma 6 can be completed with Fourier analysis. Alternatively, and more generally, it can also be completed using the *Efron-Stein* decomposition of functions [16]. This is the approach taken in [6], and it enables the analysis of the algorithm to hold even in the more general setting where the algorithm is testing functions with any finite product domain and any finite ranges for the property of being k-juntas.[2]

6   Open Problems and Future Directions

There are many possible directions for future research on testing k-juntas. We highlight three particularly intriguing open problems.

6.1   Classical vs. Quantum Property Testing

The field of property testing can be extended to allow the tester to use the quantum oracle model of Beals et al. [2]. The resulting model is called *quantum property testing* and was first studied by Buhrman, Fortnow, Newman, and Röhrig [12]. They showed that there are properties that can be tested with significantly fewer queries in the quantum model than in the classical model and that for some other properties, the extra power of the quantum oracle does not improve the query complexity of the associated testing problem.

The first open problem asks if quantum oracles help when testing juntas: *Is there a gap between the quantum and classical query complexities for testing k-juntas?*

Atıcı and Servedio [1] studied the problem of testing juntas in the quantum model. They showed that in this model, $O(k/\epsilon)$ queries are sufficient and $\Omega(\sqrt{k})$

---

[2] We note that the result in [6] was not the first one to generalize the analysis of a junta testing algorithm to non-boolean functions; Diakonikolas et al. [15] did so as well with a more technically intricate argument.

queries are necessary to ϵ-test k-juntas. At the time that this algorithm was introduced, it provided a quadratic improvement over the query complexity of the best classical k-junta tester. Of course, the algorithm presented in Section 5 reduces the gap to be only logarithmic in k, and in fact our strongest lower bounds in the classical model are not strong enough to guarantee the existence of a gap in the query complexities.

6.2 Adaptive vs. Non-Adaptive Testing

Gonen and Ron [21], and Goldreich and Ron [19] (see also [20] in this volume) recently began a systematic study of the benefits of adaptivity for testing properties in the dense-graph model. They showed that for some properties, there is a gap between the query complexity of the best adaptive and non-adaptive testing algorithms, while for other properties no such gap exists.

The current gap between query complexity of the best adaptive and non-adaptive algorithms for testing k-juntas — $O(k \log k + k/\epsilon)$ and $\tilde{O}(k^{3/2}/\epsilon)$, respectively — leaves the following basic problem open: *Does adaptivity help when testing k-juntas?*

6.3 Improved Testers for Other Properties

Following the work of Fischer et al. [17], junta testers have been used as a basic building block to design testers for many other properties of boolean functions, including function isomorphism [17], halfspaces [24], and many concise representation properties (e.g., being computable by a small decision tree or by a small circuit, having low Fourier degree) [15] (see also [28] in this volume).

All of the above testing algorithms use one of the k-junta testers presented in Section 4. The last open problem that we wish to mention is the following: *Can the NEARLY OPTIMAL k-JUNTA TEST be used (or extended) to obtain improved testing algorithms for function isomorphism, halfspaces, or concise representation properties?*

References

1. Atıcı, A., Servedio, R.A.: Quantum algorithms for learning and testing juntas. Quantum Information Processing 6(5), 323–348 (2007)
2. Beals, R., Buhrman, H., Cleve, R., Mosca, M., de Wolf, R.: Quantum lower bounds by polynomials. J. of the ACM 48(4), 778–797 (2001)
3. Bellare, M., Coppersmith, D., Håstad, J., Kiwi, M., Sudan, M.: Linearity testing in characteristic two. IEEE Transactions on Information Theory 42(6), 1781–1795 (1996)
4. Bellare, M., Goldreich, O., Sudan, M.: Free bits, PCPs and non-approximability – towards tight results. SIAM J. Comput. 27(3), 804–915 (1998)
5. Blais, E.: Improved bounds for testing juntas. In: Goel, A., Jansen, K., Rolim, J.D.P., Rubinfeld, R. (eds.) APPROX and RANDOM 2008. LNCS, vol. 5171, pp. 317–330. Springer, Heidelberg (2008)
6. Blais, E.: Testing juntas nearly optimally. In: Proc. 41st Symposium on Theory of Computing, pp. 151–158 (2009)

7. Blum, A.: Relevant examples and relevant features: thoughts from computational learning theory. In: AAAI Fall Symposium on 'Relevance' (1994)
8. Blum, A.: Learning a function of r relevant variables. In: Proc. 16th Conference on Computational Learning Theory, pp. 731–733 (2003)
9. Blum, A., Hellerstein, L., Littlestone, N.: Learning in the presence of finitely or infinitely many irrelevant attributes. J. Comp. Syst. Sci. 50(1), 32–40 (1995)
10. Blum, A., Langley, P.: Selection of relevant features and examples in machine learning. Artificial Intelligence 97(2), 245–271 (1997)
11. Blum, M., Luby, M., Rubinfeld, R.: Self-testing/correcting with applications to numerical problems. J. Comput. Syst. Sci. 47(3), 549–595 (1993)
12. Buhrman, H., Fortnow, L., Newman, I., Röhrig, H.: Quantum property testing. In: Proc. 14th Symp. on Discrete Algorithms, pp. 480–488 (2003)
13. Chen, V.: Query-Efficient dictatorship testing with perfect completeness. In: Goldreich, O. (ed.) Property Testing. LNCS, vol. 6390, pp. 276–279. Springer, Heidelberg (2010)
14. Chockler, H., Gutfreund, D.: A lower bound for testing juntas. Information Processing Letters 90(6), 301–305 (2004)
15. Diakonikolas, I., Lee, H.K., Matulef, K., Onak, K., Rubinfeld, R., Servedio, R.A., Wan, A.: Testing for concise representations. In: Proc. 48th Symposium on Foundations of Computer Science, pp. 549–558 (2007)
16. Efron, B., Stein, C.: The jackknife estimate of variance. Ann. of Stat. 9(3), 586–596 (1981)
17. Fischer, E., Kindler, G., Ron, D., Safra, S., Samorodnitsky, A.: Testing juntas. J. Comput. Syst. Sci. 68(4), 753–787 (2004)
18. Goldreich, O., Goldwasser, S., Ron, D.: Property testing and its connection to learning and approximation. J. of the ACM 45(4), 653–750 (1998)
19. Goldreich, O., Ron, D.: Algorithmic aspects of property testing in the dense graphs model. In: Dinur, I., Jansen, K., Naor, J., Rolim, J. (eds.) APPROX–RANDOM 2009. LNCS, vol. 5687, pp. 520–533. Springer, Heidelberg (2009)
20. Goldreich, O., Ron, D.: Algorithmic aspects of property testing in the dense graphs model. In: Goldreich, O. (ed.) Property Testing. LNCS, vol. 6390, pp. 295–305. Springer, Heidelberg (2010)
21. Gonen, M., Ron, D.: On the benefits of adaptivity in property testing of dense graphs. In: Charikar, M., Jansen, K., Reingold, O., Rolim, J.D.P. (eds.) RANDOM 2007 and APPROX 2007. LNCS, vol. 4627, pp. 525–539. Springer, Heidelberg (2007)
22. Guijarro, D., Tarui, J., Tsukiji, T.: Finding relevant variables in PAC model with membership queries. In: Watanabe, O., Yokomori, T. (eds.) ALT 1999. LNCS (LNAI), vol. 1720, pp. 313–322. Springer, Heidelberg (1999)
23. Lipton, R.J., Markakis, E., Mehta, A., Vishnoi, N.K.: On the Fourier spectrum of symmetric boolean functions with applications to learning symmetric juntas. In: Proc. 20th Conference on Computational Complexity, pp. 112–119 (2005)
24. Matulef, K., O'Donnell, R., Rubinfeld, R., Servedio, R.A.: Testing halfspaces. In: Proc. 19th Symp. on Discrete Algorithms, pp. 256–264 (2009)
25. Mossel, E., O'Donnell, R., Servedio, R.A.: Learning functions of k relevant variables. J. Comput. Syst. Sci. 69(3), 421–434 (2004)
26. Parnas, M., Ron, D., Samorodnitsky, A.: Testing basic boolean formulae. SIAM J. Discret. Math. 16(1), 20–46 (2003)
27. Rubinfeld, R., Sudan, M.: Self-testing polynomial functions efficiently and over rational domains. In: Proc. 3rd Symp. on Discrete Algorithms, pp. 23–32 (1992)
28. Servedio, R.: Testing by implicit learning: a brief survey. In: Goldreich, O. (ed.) Property Testing. LNCS, vol. 6390, pp. 197–210. Springer, Heidelberg (2010)

Sublinear-time Algorithms[*]

Artur Czumaj[**] and Christian Sohler[***]

[1] Department of Computer Science and Centre for Discrete Mathematics and its
Applications (DIMAP), University of Warwick
A.Czumaj@warwick.ac.uk
[2] Department of Computer Science, TU Dortmund,
christian.sohler@tu-dortmund.de

Abstract. In this paper we survey recent advances in the area of sublinear-time algorithms.

Keywords: Sublinear time algorithms, sublinear approximation algorithms.

1 Introduction

The area of *sublinear-time algorithms* is a new rapidly emerging area of computer science. It has its roots in the study of massive data sets that occur more and more frequently in various applications. Financial transactions with billions of input data and Internet traffic analyses (Internet traffic logs, clickstreams, web data) are examples of modern data sets that show unprecedented scale. Managing and analyzing such data sets forces us to reconsider the traditional notions of efficient algorithms: processing such massive data sets in more than linear time is by far too expensive and often even linear time algorithms may be too slow. Hence, there is the desire to develop algorithms whose running times are not only polynomial, but in fact are *sublinear* in n.

Constructing a sublinear time algorithm may seem to be an impossible task since it allows one to read only a small fraction of the input. However, in recent years, we have seen development of sublinear time algorithms for optimization problems arising in such diverse areas as graph theory, geometry, algebraic computations, and computer graphics. Initially, the main research focus has been on designing efficient algorithms in the framework of *property testing* (for excellent surveys, see [28,32,33,43,53]), which is an alternative notion of approximation for decision problems. But more recently, we have seen some major progress in sublinear-time algorithms in the classical model of randomized and approximation algorithms. In this paper, we survey some of the recent advances in this area. Our main focus is on sublinear-time algorithms for combinatorial problems, especially for graph problems and optimization problems in metric spaces.

---

[*] This survey is a slightly updated version of a survey that appeared in *Bulletin of the EATCS*, 89: 23–47, June 2006.

[**] Research supported by EPSRC award EP/G064679/1 and by the Centre for Discrete Mathematics and its Applications (DIMAP), EPSRC award EP/D063191/1.

[***] Supported by DFG grant SO 514/3-1.

O. Goldreich (Ed.): Property Testing, LNCS 6390, pp. 41–64, 2010.

Our goal is to give a flavor of the area of sublinear-time algorithms. We focus on in our opinion the most representative results in the area and we aim to illustrate main techniques used to design sublinear-time algorithms. Still, many of the details of the presented results are omitted and we recommend the readers to follow the original works. We also do not aim to cover the entire area of sublinear-time algorithms, and in particular, we do not discuss property testing algorithms [28,32,33,43,53], even though this area is very closely related to the research presented in this survey.

Organization. We begin with an introduction to the area and then we give some sublinear-time algorithms for a basic problem in computational geometry [15]. Next, we present recent sublinear-time algorithms for basic graph problems: approximating the average degree in a graph [27,37], estimating the cost of a minimum spanning tree [16] and approximating the size of a maximum matching [51,56]. Then, we discuss sublinear-time algorithms for optimization problems in metric spaces. We present the main ideas behind recent algorithms for estimating the cost of minimum spanning tree [21] and facility location [10], and then we discuss the quality of random sampling to obtain sublinear-time algorithms for clustering problems [22,49]. We finish with some conclusions.

2   Basic Sublinear Algorithms

The concept of sublinear-time algorithms has been known for a very long time, but initially it has been used to denote "pseudo-sublinear-time" algorithms, where after an appropriate *preprocessing*, an algorithm solves the problem in sublinear-time. For example, if we have a set of n numbers, then after an $\mathcal{O}(n \log n)$ preprocessing (sorting), we can trivially solve a number of problems involving the input elements. And so, if the after the preprocessing the elements are put in a sorted array, then in $\mathcal{O}(1)$ time we can find the kth smallest element, in $\mathcal{O}(\log n)$ time we can test if the input contains a given element x, and also in $\mathcal{O}(\log n)$ time we can return the number of elements equal to a given element x. Even though all these results are folklore, this is not what we call nowadays a sublinear-time algorithm.

In this survey, our goal is to study algorithms for which the input is taken to be in any standard representation and with no extra assumptions. Then, an algorithm does not have to read the entire input but it may determine the output by checking only a subset of the input elements. It is easy to see that for many natural problems it is impossible to give any reasonable answer if not all or almost all input elements are checked. But still, for some number of problems we can obtain good algorithms that do not have to look at the entire input. Typically, these algorithms are *randomized* (because most of the problems have a trivial linear-time deterministic lower bound) and they return only an *approximate* solution rather than the exact one (because usually, without looking at the whole input we cannot determine the exact solution). In this survey, we present recently developed sublinear-time algorithm for some combinatorial optimization problems.

Searching in a sorted list. It is well-known that if we can store the input in a sorted array, then we can solve various problems on the input very efficiently. However, the assumption that the input array is sorted is not natural in typical applications. Let us now consider a variant of this problem, where our goal is to *search* for an element x in a linked sorted list containing n *distinct* elements[1]. Here, we assume that the n elements are stored in a doubly-linked list, each list element has access to the next and preceding element in the list, and the list is sorted (that is, if x follows y in the list, then $y < x$). We also assume that we have access to all elements in the list, which for example, can correspond to the situation that all n list elements are stored in an array (but the array is not sorted and we do not impose any order for the array elements). How can we find whether a given number x is in our input or is not?

On the first glace, it seems that since we do not have direct access to the rank of any element in the list, this problem requires $\Omega(n)$ time. And indeed, if our goal is to design a deterministic algorithm, then it is impossible to do the search in $o(n)$ time. However, if we allow randomization, then we can complete the search in $\mathcal{O}(\sqrt{n})$ expected time (and this bound is asymptotically tight).

Let us first sample uniformly at random a set S of $\Theta(\sqrt{n})$ elements from the input. Since we have access to all elements in the list, we can select the set S in $\mathcal{O}(\sqrt{n})$ time. Next, we scan all the elements in S and in $\mathcal{O}(\sqrt{n})$ time we can find two elements in S, p and q, such that $p \leq x < q$, and there is no element in S that is between p and q. Observe that since the input consist of n distinct numbers, p and q are uniquely defined. Next, we traverse the input list containing all the input elements starting at p until we find either the sought key x or we find element q.

Lemma 1. *The algorithm above completes the search in expected $\mathcal{O}(\sqrt{n})$ time. Moreover, no algorithm can solve this problem in $o(\sqrt{n})$ expected time.*

Proof. The running time of the algorithm if equal to $\mathcal{O}(\sqrt{n})$ plus the number of the input elements between p and q. Since S contains $\Theta(\sqrt{n})$ elements, the expected number of input elements between p and q is $\mathcal{O}(n/|S|) = \mathcal{O}(\sqrt{n})$. This implies that the expected running time of the algorithm is $\mathcal{O}(\sqrt{n})$.

For a proof of a lower bound of $\Omega(\sqrt{n})$ expected time, see, e.g., [15].

2.1   Geometry: Intersection of Two Polygons

Let us consider a related problem but this time in a geometric setting. Given two convex polygons A and B in $\mathbb{R}^2$, each with n vertices, determine if they intersect, and if so, then find a point in their intersection.

---

[1] The assumption that the input elements are *distinct* is important. If we allow multiple elements to have the same key, then the search problem requires $\Omega(n)$ time. To see this, consider the input in which about a half of the elements has key 1, another half has key 3, and there is a single element with key 2. Then, searching for 2 requires $\Omega(n)$ time.

It is well known that this problem can be solved in $\mathcal{O}(n)$ time, for example, by observing that it can be described as a linear programming instance in two dimensions, a problem which is known to have a linear-time algorithm (cf. [26]). In fact, within the same time one can either find a point that is in the intersection of A and B, or find a line $\mathcal{L}$ that separates A from B (actually, one can even find a bitangent separating line $\mathcal{L}$, i.e., a line separating A and B which intersects with each of A and B in exactly one point). The question is whether we can obtain a better running time.

The complexity of this problem depends on the input representation. In the most powerful model, if the vertices of both polygons are stored in an array in cyclic order, Chazelle and Dobkin [14] showed that the intersection of the polygons can be determined in logarithmic time. However, a standard geometric representation assumes that the input is not stored in an array but rather A and B are given by their doubly-linked lists of vertices such that each vertex has as its successor the next vertex of the polygon in the clockwise order. Can we then test if A and B intersect?

Chazelle et al. [15] gave an $\mathcal{O}(\sqrt{n})$-time algorithm that uses the approach discussed above for searching in a sorted list. Let us first sample uniformly at random $\Theta(\sqrt{n})$ vertices from each A and B, and let C_A and C_B be the convex hulls of the sample point sets for the polygons A and B, respectively. Using the linear-time algorithm mentioned above, in $\mathcal{O}(\sqrt{n})$ time we can check if C_A and C_B intersects. If they do, then the algorithm will get us a point that lies in the intersection of C_A and C_B, and hence, this point lies also in the intersection of A and B. Otherwise, let $\mathcal{L}$ be the bitangent separating line returned by the algorithm (see Figure 1 (a)).

Let a and b be the points in $\mathcal{L}$ that belong to A and B, respectively. Let a_1 and a_2 be the two vertices adjacent to a in A. We will define now a new polygon P_A. If none of a_1 and a_2 is on the side C_A of $\mathcal{L}$ then we define P_A to be empty. Otherwise, exactly one of a_1 and a_2 is on the side C_A of $\mathcal{L}$; let it be a_1. We define polygon P_A by walking from a to a_1 and then continue walking along the boundary of A until we cross $\mathcal{L}$ again (see Figure 1 (b)). In a similar way we define polygon P_B. Observe that the expected size of each of P_A and P_B is at most $\mathcal{O}(\sqrt{n})$.

It is easy to see that A and B intersect if and only if either A intersects P_B or B intersects P_A. We only consider the case of checking if A intersects P_B.

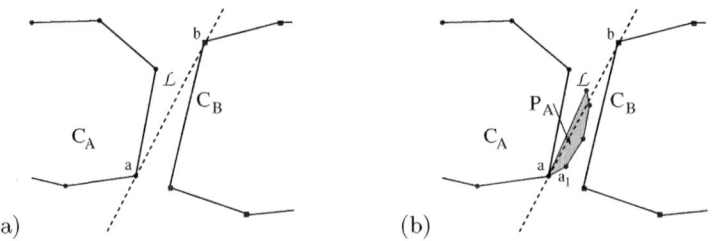

Fig. 1. (a) Bitangent line $\mathcal{L}$ separating C_A and C_B, and (b) the polygon P_A

We first determine if C_A intersects P_B. If yes, then we are done. Otherwise, let $\mathcal{L}_A$ be a bitangent separating line that separates C_A from P_B. We use the same construction as above to determine a subpolygon Q_A of A that lies on the P_B side of $\mathcal{L}_A$. Then, A intersects P_B if and only if Q_A intersects P_B. Since Q_A has expected size $\mathcal{O}(\sqrt{n})$ and so does P_B, testing the intersection of these two polygons can be done in $\mathcal{O}(\sqrt{n})$ expected time. Therefore, by our construction above, we have solved the problem of determining if two polygons of size n intersect by reducing it to a constant number of problem instances of determining if two polygons of expected size $\mathcal{O}(\sqrt{n})$ intersect. This leads to the following lemma.

Lemma 2. [15] *The problem of determining whether two convex n-gons intersect can be solved in $\mathcal{O}(\sqrt{n})$ expected time, which is asymptotically optimal.*

Chazelle et al. [15] gave not only this result, but they also showed how to apply a similar approach to design a number of sublinear-time algorithms for some basic geometric problems. For example, one can extend the result discussed above to test the intersection of two convex polyhedra in $\mathbb{R}^3$ with n vertices in $\mathcal{O}(\sqrt{n})$ expected time. One can also approximate the volume of an n-vertex convex polytope to within a relative error $\varepsilon > 0$ in expected time $\mathcal{O}(\sqrt{n}/\varepsilon)$. Or even, for a pair of two points on the boundary of a convex polytope P with n vertices, one can estimate the length of an optimal shortest path outside P between the given points in $\mathcal{O}(\sqrt{n})$ expected time.

In all the results mentioned above, the input objects have been represented by a linked structure: either every point has access to its adjacent vertices in the polygon in $\mathbb{R}^2$, or the polytope is defined by a doubly-connected edge list, or so. These input representations are standard in computational geometry, but a natural question is whether this is necessary to achieve sublinear-time algorithms — what can we do if the input polygon/polytop is represented by a set of points and no additional structure is provided to the algorithm? In such a scenario, it is easy to see that no $o(n)$-time algorithm can solve exactly any of the problems discussed above. That is, for example, to determine if two polygons with n vertices intersect one needs $\Omega(n)$ time. However, still, we can obtain some approximation to this problem, one which is described in the framework of *property testing*.

Suppose that we relax our task and instead of determining if two (convex) polytopes A and B in $\mathbb{R}^d$ intersects, we just want to distinguish between two cases: either A and B are intersection-free, or one has to "significantly modify" A and B to make them intersection-free. The definition of the notion of "significantly modify" may depend on the application at hand, but the most natural characterization would be to remove at least εn points in A and B, for an appropriate parameter ε (see [20] for a discussion about other geometric characterization). Czumaj et al. [25] gave a simple algorithm that for any $\varepsilon > 0$, can distinguish between the case when A and B do not intersect, and the case when at least εn points has to be removed from A and B to make them intersection-free: the algorithm returns the outcome of a test if a random sample of $\mathcal{O}((d/\varepsilon) \log(d/\varepsilon))$ points from A intersects with a random sample of $\mathcal{O}((d/\varepsilon) \log(d/\varepsilon))$ points from B.

Sublinear-time algorithms: perspective. The algorithms presented in this section should give a flavor of the area and give us the first impression of what do we mean by sublinear-time and what kind of results one can expect. In the following sections, we will present more elaborate algorithms for various combinatorial problems for graphs and for metric spaces.

3   Sublinear Time Algorithms for Graphs Problems

In the previous section, we introduced the concept of sublinear-time algorithms and we presented two basic sublinear-time algorithms for geometric problems. In this section, we will discuss sublinear-time algorithms for graph problems. Our main focus is on sublinear-time algorithms for graphs, with special emphasizes on sparse graphs represented by adjacency lists where combinatorial algorithms are sought.

3.1   Approximating the Average Degree

Assume we have access to the degree distribution of the vertices of an undirected connected graph $G = (V, E)$, i.e., for any vertex $v \in V$ we can query for its degree. Can we achieve a good approximation of the average degree in G by looking at a sublinear number of vertices? At first sight, this seems to be an impossible task. It seems that approximating the average degree is equivalent to approximating the average of a set of n numbers with values between 1 and $n-1$, which is not possible in sublinear time. However, Feige [27] proved that one can approximate the average degree in $\mathcal{O}(\sqrt{n}/\varepsilon)$ time within a factor of $2 + \varepsilon$.

The difficulty with approximating the average of a set of n numbers can be illustrated with the following example. Assume that almost all numbers in the input set are 1 and a few of them are $n - 1$. To approximate the average we need to approximate how many occurrences of $n - 1$ exist. If there is only a constant number of them, we can do this only by looking at $\Omega(n)$ numbers in the set. So, the problem is that these large numbers can "hide" in the set and we cannot give a good approximation, unless we can "find" at least some of them.

Why is the problem less difficult, if, instead of an arbitrary set of numbers, we have a set of numbers that are the vertex degrees of a graph? For example, we could still have a few vertices of degree $n - 1$. The point is that in this case any edge incident to such a vertex can be seen at another vertex. Thus, even if we do not sample a vertex with high degree we will see all incident edges at other vertices in the graph. Hence, vertices with a large degree cannot "hide."

We will sketch a proof of a slightly weaker result than that originally proven by Feige [27]. Let d denote the average degree in $G = (V, E)$ and let d_S denote the random variable for the average degree of a set S of s vertices chosen uniformly at random from V. We will show that if we set $s \geq \beta \sqrt{n}/\varepsilon^{\mathcal{O}(1)}$ for an appropriate constant β, then $d_S \geq (\frac{1}{2} - \varepsilon) \cdot d$ with probability at least $1 - \varepsilon/64$. Additionally, we observe that Markov inequality immediately implies that $d_S \leq (1 + \varepsilon) \cdot d$ with probability at least $1 - 1/(1 + \varepsilon) \geq \varepsilon/2$. Therefore, our algorithm will pick $8/\varepsilon$ sets S_i, each of size s, and output the set with the smallest average degree. Hence, the probability that all of the sets S_i have too high average degree is at most $(1 - \varepsilon/2)^{\varepsilon/8} \leq 1/8$. The probability that one of them has too small average degree is at most $\frac{8}{\varepsilon} \cdot \frac{\varepsilon}{64} = 1/8$. Hence, the output value will satisfy both inequalities with probability at least $3/4$. By replacing ε with $\varepsilon/2$, this will yield a $(2 + \varepsilon)$-approximation algorithm.

Now, our goal is to show that with high probability one does not underestimate the average degree too much. Let H be the set of the $\sqrt{\varepsilon n}$ vertices with highest degree in G and let $L = V \setminus H$ be the set of the remaining vertices. We first argue that the sum of the degrees of the vertices in L is at least $(\frac{1}{2} - \varepsilon)$ times the sum of the degrees of all vertices. This can be easily seen by distinguishing between edges incident to a vertex from L and edges within H. Edges incident to a vertex from L contribute with at least 1 to the sum of degrees of vertices in L, which is fine as this is at least $1/2$ of their full contribution. So the only edges that may cause problems are edges within H. However, since $|H| = \sqrt{\varepsilon n}$, there can be at most εn such edges, which is small compared to the overall number of edges (which is at least $n - 1$, since the graph is connected).

Now, let d_H be the degree of a vertex with the smallest degree in H. Since we aim at giving a lower bound on the average degree of the sampled vertices, we can safely assume that all sampled vertices come from the set L. We know that each vertex in L has a degree between 1 and d_H. Let X_i, $1 \leq i \leq s$, be the random variable for the degree of the ith vertex from S. Then, it follows from Hoeffding bounds that

$$\mathbf{Pr}[\sum_{i=1}^{s} X_i \leq (1 - \varepsilon) \cdot \mathbf{E}[\sum_{i=1}^{s} X_i]] \leq e^{-\frac{\mathbf{E}[\sum_{i=1}^{r} X_i] \cdot \varepsilon^2}{d_H}} \ .$$

We know that the average degree is at least $d_H \cdot |H|/n$, because any vertex in H has at least degree d_H. Hence, the average degree of a vertex in L is at least $(\frac{1}{2} - \varepsilon) \cdot d_H \cdot |H|/n$. This just means $\mathbf{E}[X_i] \geq (\frac{1}{2} - \varepsilon) \cdot d_H \cdot |H|/n$. By linearity of expectation we get $\mathbf{E}[\sum_{i=1}^{s} X_i] \geq s \cdot (\frac{1}{2} - \varepsilon) \cdot d_H \cdot |H|/n$. This implies that, for our choice of s, with high probability we have $d_S \geq (\frac{1}{2} - \varepsilon) \cdot d$.

Feige showed the following result, which is stronger with respect to the dependence on ε.

Theorem 1. [27] *Using $\mathcal{O}(\varepsilon^{-1} \cdot \sqrt{n/d_0})$ queries, one can estimate the average degree of a graph within a ratio of $(2 + \varepsilon)$, provided that $d \geq d_0$.*

Feige also proved that $\Omega(\varepsilon^{-1} \cdot \sqrt{n/d})$ queries are required, where d is the average degree in the input graph. Finally, any algorithm that uses only degree queries and estimates the average degree within a ratio $2 - \delta$ for some constant δ requires $\Omega(n)$ queries.

Interestingly, if one can also use neighborhood queries, then it is possible to approximate the average degree using $\widetilde{\mathcal{O}}(\sqrt{n}/\varepsilon^{\mathcal{O}(1)})$ queries with a ratio of $(1+\varepsilon)$, as shown by Goldreich and Ron [37]. The model for neighborhood queries is as follows. We assume we are given a graph and we can query for the ith neighbor of vertex v. If v has at least i neighbors we get the corresponding neighbor; otherwise we are told that v has less than i neighbors. We remark that one can simulate degree queries in this model with $\mathcal{O}(\log n)$ queries. Therefore, the algorithm from [37] uses only neighbor queries.

For a sketch of a proof, let us assume that we know the set H. Then we can use the following approach. We only consider vertices from L. If our sample contains a vertex from H we ignore it. By our analysis above, we know that there are only few edges within H and that we make only a small error in estimating the number of edges within L. We loose the factor of two, because we "see" edges from L to H only from one side. The idea behind the algorithm from [37] is to approximate the fraction of edges from L to H and add it to the final estimate. This has the effect that we count any edge between L and H twice, canceling the effect that we see it only from one side. This is done as follows. For each vertex v we sample from L we take a random set of incident edges to estimate the fraction $\lambda(v)$ of its neighbors that is in H. Let $\hat{\lambda}(v)$ denote the estimate we obtain. Then our estimate for the average degree will be $\sum_{v \in S \cap L} (1 + \hat{\lambda}(v)) \cdot d(v) / |S \cap L|$, where $d(v)$ denotes the degree of v. If for all vertices we estimate $\lambda(v)$ within an additive error of ε, the overall error induced by the $\hat{\lambda}$ will be small. This can be achieved with high probability querying $\mathcal{O}(\log n / \varepsilon^2)$ random neighbors. Then the output value will be a $(1 + \varepsilon)$-approximation of the average degree. The assumption that we know H can be dropped by taking a set of $\mathcal{O}(\sqrt{n}/\varepsilon)$ vertices and setting H to be the set of vertices with larger degree than all vertices in this set (breaking ties by the vertex number).

(We remark that the outline of a proof given above is different from the proof in [37].)

Theorem 2. [37] *Given the ability to make neighbor queries to the input graph G, there exists an algorithm that makes $\mathcal{O}(\sqrt{n/d_0} \cdot \varepsilon^{-\mathcal{O}(1)})$ queries and approximates the average degree in G to within a ratio of $(1 + \varepsilon)$.*

In their paper Goldreich and Ron also discuss the more general question of approximating average parameters in graphs. They point out that an algorithm's ability to approximate a graph parameter is closely related to the type of query the algorithm may ask. This raises the question, which types of queries are natural and which graph parameters can be approximated with such natural queries. Besides the result on the average degree discussed, they also prove that one can approximate the average distance in an unweighted graph with $O(\sqrt{n}/\varepsilon^{O(1)})$ time, which can be further improved as a function of the average degree. Their algorithm is allowed to perform distance queries that output the distance between any two query vertices in constant time. Their model can be viewed as a special case of the distance oracle model for metric spaces (see Section 4) as it considers shortest path metrics of undirected graphs. This restriction allows to achieve query times sublinear in n, which is impossible for most problems in the more general model.

3.2 Minimum Spanning Trees

One of the most fundamental graph problems is to compute a minimum spanning tree. Since the minimum spanning tree is of size linear in the number of vertices, no sublinear algorithm for sparse graphs can exists. It is also know that no constant factor approximation algorithm with $o(n^2)$ query complexity in dense graphs (even in metric spaces) exists [40]. Given these facts, it is somewhat surprising that it is possible to approximate the cost of a minimum spanning tree in sparse graphs [16] as well as in metric spaces [21] to within a factor of $(1 + \varepsilon)$.

In the following we will explain the algorithm for sparse graphs by Chazelle et al. [16]. We will prove a slightly weaker result than in [16]. Let $G = (V, E)$ be an undirected connected weighted graph with maximum degree D and integer edge weights from $\{1, \ldots, W\}$. We assume that the graph is given in adjacency list representation, i.e., for every vertex v there is a list of its at most D neighbors, which can be accessed from v. Furthermore, we assume that the vertices are stored in an array such that it is possible to select a vertex uniformly at random. We assume also that the values of D and W are known to the algorithm.

The main idea behind the algorithm is to express the cost of a minimum spanning tree as the number of connected components in certain auxiliary subgraphs of G. Then, one runs a randomized algorithm to estimate the number of connected components in each of these subgraphs. The algorithm to estimate the number of connected components is based on a property tester for connectivity in the bounded degree graph model by Goldreich and Ron [35].

To start with basic intuitions, let us assume that $W = 2$, i.e., the graph has only edges of weight 1 or 2. Let $G^{(1)} = (V, E^{(1)})$ denote the subgraph that contains all edges of weight (at most) 1 and let $c^{(1)}$ be the number of connected components in $G^{(1)}$. It is easy to see that the minimum spanning tree has to link these connected components by edges of weight 2. Since any connected component in $G^{(1)}$ can be spanned by edges of weight 1, any minimum spanning tree of G has $c^{(1)} - 1$ edges of weight 2 and $n - 1 - (c^{(1)} - 1)$ edges of weight 1. Thus, the weight of a minimum spanning tree is

$$n - 1 - (c^{(1)} - 1) + 2 \cdot (c^{(1)} - 1) = n - 2 + c^{(1)} = n - W + c^{(1)} \ .$$

Next, let us consider an arbitrary integer value for W. Defining $G^{(i)} = (V, E^{(i)})$, where $E^{(i)}$ is the set of edges in G with weight at most i, one can generalize the formula above to obtain that the cost MST of a minimum spanning tree can be expressed as

$$MST = n - W + \sum_{i=1}^{W-1} c^{(i)} \ .$$

This gives the following simple algorithm.

$$\boxed{\begin{array}{l} \textsc{ApproxMSTWeight}(G, \varepsilon) \\ \quad \textbf{for } i = 1 \textbf{ to } W - 1 \\ \qquad \text{Compute estimator } \widehat{c}^{(i)} \text{ for } c^{(i)} \\ \quad \textbf{output } \widetilde{MST} = n - W + \sum_{i=1}^{W-1} \widehat{c}^{(i)} \end{array}}$$

Thus, the key question that remains is how to estimate the number of connected components. This is done by the following algorithm.

$$\boxed{\begin{array}{l} \textsc{ApproxConnectedComps}(G, s) \\ \quad \{ \textit{ Input: an arbitrary undirected graph } G \ \} \\ \quad \{ \textit{ Output: } \hat{c}\text{: an estimation of the number of connected components of } G \ \} \\ \qquad \text{choose } s \text{ vertices } u_1, \ldots, u_s \text{ uniformly at random} \\ \quad \textbf{for } i = 1 \textbf{ to } s \textbf{ do} \\ \qquad \text{choose } X \text{ according to } \mathbf{Pr}[X \geq k] = 1/k \\ \qquad \text{run breadth-fist-search (BFS) starting at } u_i \text{ until either} \\ \qquad\qquad (1) \text{ the whole connected component containing } u_i \\ \qquad\qquad\quad \text{has been explored, or} \\ \qquad\qquad (2) \ X \text{ vertices have been explored} \\ \qquad \textbf{if } \text{BFS stopped in case (1) } \textbf{then } b_i = 1 \\ \qquad \textbf{else } b_i = 0 \\ \quad \textbf{output } \hat{c} = \frac{n}{s} \sum_{i=1}^{s} b_i \end{array}}$$

To analyze this algorithm let us fix an arbitrary connected component C and let $|C|$ denote the number of vertices in the connected component. Let c denote the number of connected components in G. We can write

$$\mathbf{E}[b_i] = \sum_{\text{connected component } C} \mathbf{Pr}[u_i \in C] \cdot \mathbf{Pr}[X \geq |C|] = \sum_{\text{connected component } C} \frac{|C|}{n} \cdot \frac{1}{|C|} = \frac{c}{n} \ .$$

By linearity of expectation we obtain $\mathbf{E}[\hat{c}] = c$.

To show that $\hat{c}$ is concentrated around its expectation, we apply Chebyshev inequality. Since b_i is an indicator random variable, we have

$$\mathbf{Var}[b_i] = \mathbf{E}[b_i^2] - \mathbf{E}[b_i]^2 \leq \mathbf{E}[b_i^2] = \mathbf{E}[b_i] = c/n \ .$$

The b_i are mutually independent and so we have

$$\mathbf{Var}[\hat{c}] = \mathbf{Var}\left[\frac{n}{s} \cdot \sum_{i=1}^{s} b_i\right] = \frac{n^2}{s^2} \cdot \sum_{i=1}^{s} \mathbf{Var}[b_i] \leq \frac{n \cdot c}{s} \ .$$

With this bound for $\mathbf{Var}[\hat{c}]$, we can use Chebyshev inequality to obtain

$$\mathbf{Pr}[|\hat{c} - \mathbf{E}[\hat{c}]| \geq \lambda n] \leq \frac{n \cdot c}{s \cdot \lambda^2 \cdot n^2} \leq \frac{1}{\lambda^2 \cdot s} \ .$$

From this it follows that one can approximate the number of connected components within additive error of λn in a graph with maximum degree D in

$\mathcal{O}(\frac{D \cdot \log n}{\lambda^2 \cdot \varrho})$ time and with probability $1 - \varrho$. The following somewhat stronger result has been obtained in [16]. Notice that the obtained running time is *independent of the input size* n.

Theorem 3. [16] *The number of connected components in a graph with maximum degree D can be approximated with additive error at most $\pm \lambda n$ in $\mathcal{O}(\frac{D}{\lambda^2} \log(D/\lambda))$ time and with probability $3/4$.*

Now, we can use this procedure with parameters $\lambda = \varepsilon/(2W)$ and $\varrho = \frac{1}{4W}$ in algorithm APPROXMSTWEIGHT. The probability that at least one call to APPROXCONNECTEDCOMPS is not within an additive error $\pm \lambda n$ is at most $1/4$. The overall additive error is at most $\pm \varepsilon n/2$. Since the cost of the minimum spanning tree is at least $n - 1 \geq n/2$, it follows that the algorithms computes in $\mathcal{O}(D \cdot W^3 \cdot \log n/\varepsilon^2)$ time a $(1 \pm \varepsilon)$-approximation of the weight of the minimum spanning tree with probability at least $3/4$. In [16], Chazelle et al. proved a slightly stronger result which has running time *independent of the input size*.

Theorem 4. [16] *Algorithm APPROXMSTWEIGHT computes a value $\widetilde{MST}$ that with probability at least $3/4$ satisfies*

$$(1 - \varepsilon) \cdot MST \leq \widetilde{MST} \leq (1 + \varepsilon) \cdot MST .$$

The algorithm runs in $\widetilde{\mathcal{O}}(D \cdot W/\varepsilon^2)$ time.

The same result also holds when D is only the average degree of the graph (rather than the maximum degree) and the edge weights are reals from the interval $[1, W]$ (rather than integers) [16]. Observe that, in particular, for sparse graphs for which the ratio between the maximum and the minimum weight is constant, the algorithm from [16] *runs in constant time!*

It was also proved in [16] that any algorithm estimating MST requires $\Omega(D \cdot W/\varepsilon^2)$ time.

3.3   Constant Time Approximation Algorithms for Maximum Matching

The next result we will explain here is an elegant technique to construct constant time approximation algorithms for graphs with bounded degree, as introduced by Nguyen and Onak [51].

Let $G = (V, E)$ be an undirected graph with maximum degree D. Define a randomized (α, β)-*approximation algorithm* to be an algorithm that returns with probability at least $2/3$ a solution with cost at most $\alpha Opt + \beta n$, where n is the size of the input and Opt denotes the cost of an optimal solution. For a graph we will define the input size to be the cardinality of its vertex set. We will consider the problem of computing *the size of maximum matching*, i.e., the size of a maximum size set $M \subseteq E$ such that no two edges are incident to the same vertex of G. It is known that the following simple greedy algorithm (that returns a *maximal matching*) provides a 2-approximation to this problem.

GREEDYMATCHING(G)
{ *Input: an undirected graph $G = (V, E)$* }
{ *Output: a matching $M \subseteq E$* }
    $M \leftarrow \emptyset$
        for each edge $(u, v) \in E$ **do**
            Let $V(M)$ be the set of vertices of edges in M
            if $u, v \notin V(M)$ **then** $M \leftarrow M \cup \{e\}$
    return M

An important property of GREEDYMATCHING is that in the **for**-loop of the algorithm the edges are considered in an arbitrary ordering. We further observe that at any stage of the algorithm, the set M is a subset of the edges that have already been processed. Furthermore, if we consider an edge e then we know that neighboring edges can only be in M if they appear in the ordering before e. Now assume that the edges are inserted in a random order and let us try to determine for some fixed edge e whether it is contained in the constructed greedy matching. We could, of course, simply run the algorithm to do so by exploring the entire graph. However, our goal is to solve it using local computations that consider only the subgraph of the input graph close to e. In order to determine whether e is in the matching it suffices to determine for all its neighboring edges whether they are in M at the time e is considered by the algorithm. If e appears earlier than all of its neighbors in the random ordering, then we know that e is in the matching. Otherwise, we have to recursively solve the problem for all neighbors of e that appear before e in the random ordering. It may seem in the first place that this reasoning does not help because we now have to determine for a bigger set of edges whether they are in the matching. However, we also gained something: all edges we have to consider recursively are known to appear before e in the random ordering. This makes it less likely that some of their neighbors again appear even earlier in the sequence, which in turn means that we have to recurse for fewer of their neighbors. Thus, typically, this process stops after a constant number of steps.

Let us now try to formalize our findings. We obtain a random ordering of the edges by picking a priority $p(e)$ for each edge uniformly at random from $[0, 1]$. The random order we consider is now defined by increasing priorities. The benefit of this approach is that we do not have to compute a random ordering for the whole vertex set to run the local algorithm. Instead we can draw $p(e)$ at random whenever we consider an edge e for the first time. If we now want to determine whether an edge e is in the matching we only have to recurse with edges having a smaller priority than e. Thus, we have to follow all paths of decreasing priority starting at the endpoints of e.

For a fixed path of length k in the graph, the probability that the priorities along the path are decreasing is $1/k!$ (this can be seen by the fact that for any sequence of k distinct priorities just one of them is decreasing; the case that probabilities are equal occurs with probability 0). Since the input graph has maximum degree D, the number of paths of length k starting from a vertex v is at most D^k. Hence, there are at most $2D^k$ paths starting at the endpoints of an edge e. For a large enough constant c this implies that for $k \geq 2^{cD}$, with (large) constant probability there is no path of length k starting from an endpoint of e that has decreasing priorities. This implies that we can determine whether e is in the matching by looking at all vertices with distance at most 2^{cD} from the endpoints of e.

Once we have an oracle to determine whether $e \in M$, we can sample edges to determine whether a given edge e is in M or not. Using a sample of size $\Theta(D/\varepsilon^2)$ we can approximate the number of edges in the matching up to additive error εn. This gives a constant-time $(2, \varepsilon)$-approximation algorithm for estimating the size of maximum matching, assuming D and ε are constant. The algorithm can be further improved to an $(1, \varepsilon)$-approximation using a more complicated approximation algorithm that greedily improves the matching using short augmenting paths. The query complexity of the improved algorithm is $2^{D^{\mathcal{O}(1/\varepsilon)}}$.

A further improvement has been done in a subsequent work by Yoshida et al. [56]. In that paper, the authors reduce the query complexity to $D^{\mathcal{O}(1/\varepsilon^2)} + \mathcal{O}(1/\varepsilon)^{\mathcal{O}(1/\varepsilon)}$ time. The source of improvement is here the idea to consider the edge with lowest priority first. If this edge turns out to be in the matching then we are already done and do not have to perform the remaining recursive calls.

Theorem 5. [51,56] *For any integer $1 \leq k < \frac{n}{2}$, there is a $(1 + \frac{1}{k}, \varepsilon n)$-approximation algorithm with query complexity $D^{\mathcal{O}(k^2)} k^{\mathcal{O}(k)} \varepsilon^{-2}$ for the size of the maximum matching for graphs with n vertices and degree bound D.*

3.4   Other Sublinear-time Results for Graphs

In this section, our main focus was on combinatorial algorithms for sparse graphs. In particular, we did not discuss a large body of algorithms for dense graphs represented in the adjacency matrix model. Still, we mention the results of approximating the size of the maximum cut in *constant time* for dense graphs [30,34], and the more general results about approximating all dense problems in Max-SNP in *constant time* [2,8,30]. Similarly, we also would like to mention about the existence of a large body of property testing algorithms for graphs, which in many situations can lead to sublinear-time algorithms for graph problems. To give representative references, in addition to the excellent survey expositions [28,32,33,43,53], we would like to mention the recent results on testability of graph properties, as described, e.g., in [3,4,5,6,11,12,19,23,36,46].

4   Sublinear Time Approximation Algorithms for Problems in Metric Spaces

One of the most widely considered models in the area of sublinear time approximation algorithms is the *distance oracle model* for metric spaces. In this model, the input of an algorithm is a set P of n points in a metric space (P, d). We assume that it is possible to compute the distance $d(p, q)$ between any pair of points p, q in constant time. Equivalently, one could assume that the algorithm is given access to the $n \times n$ distance matrix of the metric space, i.e., we have oracle access to the matrix of a weighted undirected complete graph. Since the full description size of this matrix is $\Theta(n^2)$, we will call any algorithm with $o(n^2)$ running time a *sublinear algorithm*.

Which problems can and cannot be approximated in sublinear time in the distance oracle model? One of the most basic problems is to find (an approximation) of the shortest or the longest pairwise distance in the metric space. It turns out that the shortest distance cannot be approximated. The counterexample is a uniform metric (all distances are 1) with one distance being set to some very small value ε. Obviously, it requires $\Omega(n^2)$ time to find this single short distance. Hence, no sublinear time approximation algorithm for the shortest distance problem exists. What about the longest distance? In this case, there is a very simple $\frac{1}{2}$-approximation algorithm, which was first observed by Indyk [40]. The algorithm chooses an arbitrary point p and returns its furthest neighbor q. Let r, s be the furthest pair in the metric space. We claim that $d(p, q) \geq \frac{1}{2} d(r, s)$. By the triangle inequality, we have $d(r, p) + d(p, s) \geq d(r, s)$. This immediately implies that either $d(p, r) \geq \frac{1}{2} d(r, s)$ or $d(p, s) \geq \frac{1}{2} d(r, s)$. This shows the approximation guarantee.

In the following, we present some recent sublinear-time algorithms for a few optimization problems in metric spaces.

4.1   Minimum Spanning Trees

We can view a metric space as a weighted complete graph G. A natural question is whether we can find out anything about the minimum spanning tree of that graph. As already mentioned in the previous section, it is not possible to find in $o(n^2)$ time a spanning tree in the distance oracle model that approximates the minimum spanning tree within a constant factor [40]. However, it is possible to *approximate the weight* of a minimum spanning tree within a factor of $(1 + \varepsilon)$ in $\widetilde{O}(n/\varepsilon^{O(1)})$ time [21].

The algorithm builds upon the ideas used to approximate the weight of the minimum spanning tree in graphs described in Section 3.2 [16]. Let us first observe that for the metric space problem we can assume that the maximum distance is $O(n/\varepsilon)$ and the shortest distance is 1. This can be achieved by first approximating the longest distance in $O(n)$ time and then scaling the problem appropriately. Since by the triangle inequality the longest distance also provides a lower bound on the minimum spanning tree, we can round up to 1 all edge weights that are smaller than 1. Clearly, this does not significantly change the weight of the minimum spanning tree. Now we could apply the algorithm APPROXMSTWEIGHT from Section 3.2, but this would not give us an $o(n^2)$ algorithm. The reason is that in the metric case we have a complete graph, i.e., the average degree is $D = n - 1$, and the edge weights are in the interval $[1, W]$, where $W = O(n/\varepsilon)$. So, we need a different approach. In the following we will outline an idea how to achieve a randomized $o(n^2)$ algorithm. To get a near linear time algorithm as in [21] further ideas have to be applied.

The first difference to the algorithm from Section 3.2 is that when we develop a formula for the minimum spanning tree weight, we use geometric progression instead of arithmetic progression. Assuming that all edge weights are powers of $(1 + \varepsilon)$, we define $G^{(i)}$ to be the subgraph of G that contains all edges of length at most $(1 + \varepsilon)^i$. We denote by $c^{(i)}$ the number of connected components in $G^{(i)}$. Then we can write

$$MST = n - W + \varepsilon \cdot \sum_{i=0}^{r-1} (1 + \varepsilon)^i \cdot c^{(i)} \; , \tag{1}$$

where $r = \log_{1+\varepsilon} W - 1$.

Once we have (1), our approach will be to approximate the number of connected components $c^{(i)}$ and use formula (1) as an estimator. Although geometric progression has the advantage that we only need to estimate the connected components in $r = \mathcal{O}(\log n / \varepsilon)$ subgraphs, the problem is that the estimator is multiplied by $(1 + \varepsilon)^i$. Hence, if we use the procedure from Section 3.2, we would get an additive error of $\varepsilon \, n \cdot (1 + \varepsilon)^i$, which, in general, may be much larger than the weight of the minimum spanning tree.

The basic idea how to deal with this problem is as follows. We will use a different graph traversal than BFS. Our graph traversal runs only on a subset of the vertices, which are called *representative vertices*. Every pair of representative vertices are at distance at least $\varepsilon \cdot (1 + \varepsilon)^i$ from each other. Now, assume there are m representative vertices and consider the graph induced by these vertices (there is a problem with this assumption, which will be discussed later). Running algorithm APPROXCONNECTEDCOMPS on this induced graph makes an error of $\pm \lambda m$, which must be multiplied by $(1 + \varepsilon)^i$ resulting in an additive error of $\pm \lambda \cdot (1 + \varepsilon)^i \cdot m$. Since the m representative vertices have pairwise distance $\varepsilon \cdot (1 + \varepsilon)^i$, we have a lower bound $MST \geq m \cdot \varepsilon \cdot (1 + \varepsilon)^i$. Choosing $\lambda = \varepsilon^2 / r$ would result in a $(1 + \varepsilon)$-approximation algorithm.

Unfortunately, this simple approach does not work. One problem is that we cannot choose a random representative point. This is because we have no a priori knowledge of the set of representative points. In fact, in the algorithm the points are chosen greedily during the graph traversal. As a consequence, the decision whether a vertex is a representative vertex or not, depends on the starting point of the graph traversal. This may also mean that the number of representative vertices in a connected component also depends on the starting point of the graph traversal. However, it is still possible to cope with these problems and use the approach outlined above to get the following result.

Theorem 6. [21] *The weight of a minimum spanning tree of an n-point metric space can be approximated in $\widetilde{\mathcal{O}}(n/\varepsilon^{\mathcal{O}(1)})$ time to within a $(1+\varepsilon)$ factor and with confidence probability at least $\frac{3}{4}$.*

Extensions: Sublinear-time $(2 + \varepsilon)$-approximation of metric TSP and Steiner trees. Let us remark here one direct corollary of Theorem 6. By the well known relationship (see, e.g., [55]) between minimum spanning trees, travelling salesman tours, and minimum Steiner trees, the algorithm for estimating the weight of the minimum spanning tree from Theorem 6 immediately yields $\tilde{\mathcal{O}}(n/\varepsilon^{\mathcal{O}(1)})$ time $(2 + \varepsilon)$-approximation algorithms for two other classical problems in metric spaces (or in graphs satisfying the triangle inequality): estimating the weight of the *travelling salesman tour* and the *minimum Steiner tree*.

4.2   Uniform Facility Location

Similarly to the minimum spanning tree problem, one can estimate the cost of the *metric uniform facility location* problem in $\tilde{\mathcal{O}}(n/\varepsilon^{\mathcal{O}(1)})$ time [10]. This problem is defined as follows. We are given an n-point metric space (P, d). We want to find a subset $F \subseteq P$ of open facilities such that

$$|F| + \sum_{p \in P} d(p, F)$$

is minimized. Here, $d(p, F)$ denotes the distance from p to the nearest point in F. It is known that one cannot find a solution that approximates the optimal solution within a constant factor in $o(n^2)$ time [54]. However, it is possible to approximate the *cost* of an optimal solution within a constant factor.

The main idea is as follows. Let us denote by $B(p, r)$ the set of points from P with distance at most r from p. For each $p \in P$ let r_p be the unique value that satisfies

$$\sum_{q \in B(p, r_p)} (r_p - d(p, q)) = 1 \ .$$

Then one can show that

Lemma 3. [10]

$$\frac{1}{4} \cdot Opt \ \leq \ \sum_{p \in P} r_p \ \leq \ 6 \cdot Opt \ ,$$

where Opt denotes the cost of an optimal solution to the metric uniform facility location problem.

Now, the algorithm is based on a randomized algorithm that for a given point p, estimates r_p to within a constant factor in time $\mathcal{O}(r_p \cdot n \cdot \log n)$ (recall that $r_p \leq 1$). Thus, the smaller r_p, the faster the algorithm. Now, let p be chosen uniformly at random from P. Then the expected running time to estimate r_p is $\mathcal{O}(n \log n \cdot \sum_{p \in P} r_p/n) = \mathcal{O}(n \log n \cdot \mathbf{E}[r_p])$. We pick a random sample set S of $s = 100 \log n/\mathbf{E}[r_p]$ points uniformly at random from P. (The fact that we do not know $\mathbf{E}[r_p]$ can be dealt with by using a logarithmic number of guesses.) Then we use our algorithm to compute for each $p \in S$ a value $\hat{r}_p$ that approximates r_p within a constant factor. Our algorithm outputs $\frac{n}{s} \cdot \sum_{p \in S} \hat{r}_p$ as an estimate for the cost of the facility location problem. Using Hoeffding bounds it is easy to prove that $\frac{n}{s} \cdot \sum_{p \in S} r_p$ approximates $\sum_{p \in P} r_p = Opt$ within a constant factor and with high probability. Clearly, the same statement is true, when we replace the r_p values by their constant approximations $\hat{r}_p$. Finally, we observe that expected running time of our algorithm will be $\tilde{\mathcal{O}}(n/\varepsilon^{\mathcal{O}(1)})$. This allows us to conclude with the following.

Theorem 7. **[10]** *There exists an algorithm that computes a constant factor approximation to the cost of the metric uniform facility location problem in* $\mathcal{O}(n \log^2 n)$ *time and with high probability.*

4.3 Clustering via Random Sampling

The problems of clustering large data sets into subsets (clusters) of similar characteristics are one of the most fundamental problems in computer science, operations research, and related fields. Clustering problems arise naturally in various massive datasets applications, including data mining, bioinformatics, pattern classification, etc. In this section, we will discuss *uniform random sampling* for clustering problems in metric spaces, as analyzed in two recent papers [22,49].

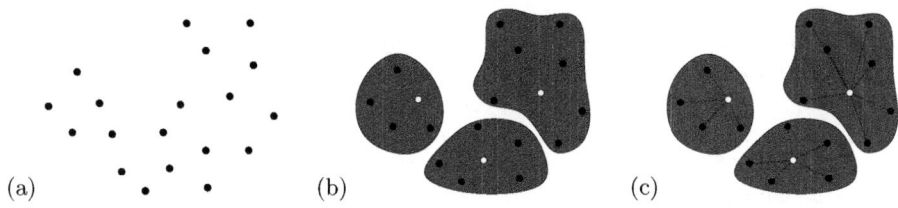

(a)   (b)   (c)

Fig. 2. (a) A set of points in a metric space, (b) its 3-clustering (white points correspond to the center points), and (c) the distances used in the cost for the 3-median

Let us consider a classical clustering problem known as the *k-median problem*. Given a finite metric space (P, d), the goal is to find a set $C \subseteq P$ of k centers (points in P) that minimizes $\sum_{p \in P} d(p, C)$, where $d(p, C)$ denotes the distance from p to the nearest point in C. The k-median problem has been studied in numerous research papers. It is known to be $\mathcal{NP}$-hard and there exist constant-factor approximation algorithms running in $\tilde{\mathcal{O}}(n\,k)$ time. In two recent papers [22,49], the authors asked the question about the quality of the uniformly random sampling approach to k-median, that is, what is the quality of the following generic scheme:

(1) choose a multiset $S \subseteq P$ of size s i.u.r. (with repetitions),
(2) run an α-approximation algorithm $\mathbb{A}_\alpha$ on input S to compute a solution C^*, and
(3) **return** set C^* (the clustering induced by the solution for the sample).

The goal is to show that already a sublinear-size sample set S will suffice to obtain a good approximation guarantee. Furthermore, as observed in [49] (see also [48]), in order to have any approximation guarantee, one has to consider the quality of approximation as a function of the diameter of the metric space. Therefore, we consider a model with the diameter of the metric space Δ given, that is, with $d : P \times P \to [0, \Delta]$.

Using techniques from statistics and computational learning theory, Mishra et al. [49] proved that if we sample a set S of $s = \widetilde{O}\left(\left(\frac{\alpha \Delta}{\varepsilon}\right)^2 (k \ln n + \ln(1/\delta))\right)$ points from P i.u.r. (*independently and uniformly at random*) and run α-approximation algorithm $\mathbb{A}_\alpha$ to find an approximation of the k-median for S, then with probability at least $1 - \delta$, the output set of k centers has *average distance* to the nearest center of at most $2 \cdot \alpha \cdot \overline{\mathrm{med}}(P, k) + \varepsilon$, where $\overline{\mathrm{med}}(P, k)$ denotes the *average distance* to the k-median C, that is, $\overline{\mathrm{med}}(P, k) = \frac{\sum_{v \in P} d(v,C)}{n}$. We will now briefly sketch the analysis due to Czumaj and Sohler [22] of a similar approximation guarantee but with a smaller bound for s.

Let C_{opt} denote an optimal set of centers for P and let $\overline{\mathrm{cost}}(X, C)$ be the average cost of the clustering of set X with center set C, that is, $\overline{\mathrm{cost}}(X, C) = \frac{\sum_{x \in X} d(x,C)}{|X|}$. Notice that $\overline{\mathrm{cost}}(P, C_{opt}) = \overline{\mathrm{med}}(P, k)$. The analysis of Czumaj and Sohler [22] is performed in two steps.

(i) We first show that there is a set of k centers $C \subseteq S$ such that $\overline{\mathrm{cost}}(S, C)$ is a good approximation of $\overline{\mathrm{med}}(P, k)$ with high probability.

(ii) Next we show that with high probability, every solution C for P with cost much bigger than $\overline{\mathrm{med}}(P, k)$ is either not a feasible solution for S (i.e., $C \not\subseteq S$) or $\overline{\mathrm{cost}}(S, C) \gg \alpha \cdot \overline{\mathrm{med}}(P, k)$ (that is, the cost of C for the sample set S is large with high probability).

Since S contains a solution with cost at most $c \cdot \overline{\mathrm{med}}(P, k)$ for some small c, $\mathbb{A}_\alpha$ will compute a solution C^* with cost at most $\alpha \cdot c \cdot \overline{\mathrm{med}}(P, k)$. Now we have to prove that no solution C for P with cost much bigger than $\overline{\mathrm{med}}(P, k)$ will be returned, or in other words, that if C is feasible for S then its cost is larger than $\alpha \cdot c \cdot \overline{\mathrm{med}}(P, k)$. But this is implied by (ii). Therefore, the algorithm will not return a solution with too large cost, and the sampling is a $(c \cdot \alpha)$-approximation algorithm.

Theorem 8. [22] *Let $0 < \delta < 1$, $\alpha \geq 1$, $0 < \beta \leq 1$ and $\varepsilon > 0$ be approximation parameters. If $s \geq \frac{c \cdot \alpha}{\beta} \cdot \left(k + \frac{\Delta}{\varepsilon \cdot \beta} \cdot \left(\alpha \cdot \ln(1/\delta) + k \cdot \ln\left(\frac{k \Delta \alpha}{\varepsilon \beta^2}\right)\right)\right)$ for an appropriate constant c, then for the solution set of centers C^*, with probability at least $1 - \delta$ it holds the following*

$$\overline{\mathrm{cost}}(V, C^*) \leq 2(\alpha + \beta) \cdot \overline{\mathrm{med}}(P, k) + \varepsilon .$$

To give the flavor of the analysis, we will sketch (a simpler) part (i) of the analysis:

Lemma 4. *If $s \geq \frac{3\Delta\alpha(1+\alpha/\beta)\ln(1/\delta)}{\beta \cdot \overline{\mathrm{med}}(P,k)}$ then $\mathbf{Pr}\left[\overline{\mathrm{cost}}(S, C^*) \leq 2(\alpha+\beta) \cdot \overline{\mathrm{med}}(P, k)\right] \geq 1 - \delta$.*

Proof. We first show that if we consider the clustering of S with the optimal set of centers C_{opt} for P, then $\overline{\mathrm{cost}}(S, C_{opt})$ is a good approximation of $\overline{\mathrm{med}}(P, k)$. The problem with this bound is that in general, we cannot expect C_{opt} to be contained in the sample set S. Therefore, we have to show also that the optimal set of centers for S cannot have cost much worse than $\overline{\mathrm{cost}}(S, C_{opt})$.

Let X_i be the random variable for the distance of the ith point in S to the nearest center of C_{opt}. Then, $\overline{\mathrm{cost}}(S, C_{opt}) = \frac{1}{s} \sum_{1 \leq i \leq s} X_i$, and, since $\mathbf{E}[X_i] = \overline{\mathrm{med}}(P, k)$, we also have $\overline{\mathrm{med}}(P, k) = \frac{1}{s} \cdot \mathbf{E}\left[\sum X_i\right]$. Hence,

$$\mathbf{Pr}\left[\overline{\mathrm{cost}}(S, C_{opt}) > \left(1 + \tfrac{\beta}{\alpha}\right) \cdot \overline{\mathrm{med}}(P, k)\right] = \mathbf{Pr}\left[\sum_{1 \leq i \leq s} X_i > \left(1 + \tfrac{\beta}{\alpha}\right) \cdot \mathbf{E}\left[\sum_{1 \leq i \leq s} X_i\right]\right].$$

Observe that each X_i satisfies $0 \leq X_i \leq \Delta$. Therefore, by Chernoff-Hoeffding bound we obtain:

$$\mathbf{Pr}\left[\sum_{1 \leq i \leq s} X_i > (1 + \beta/\alpha) \cdot \mathbf{E}\left[\sum_{1 \leq i \leq s} X_i\right]\right] \leq e^{-\frac{s \cdot \overline{\mathrm{med}}(P,k) \cdot \min\{(\beta/\alpha), (\beta/\alpha)^2\}}{3\,\Delta}} \leq \delta .$$

$$(2)$$

This gives us a good bound for the cost of $\overline{\mathrm{cost}}(S, C_{opt})$ and now our goal is to get a similar bound for the cost of the optimal set of centers for S. Let C be the set of k centers in S obtained by replacing each $c \in C_{opt}$ by its nearest neighbor in S. By the triangle inequality, $\overline{\mathrm{cost}}(S, C) \leq 2 \cdot \overline{\mathrm{cost}}(S, C_{opt})$. Hence, multiset S contains a set of k centers whose cost is at most $2 \cdot (1 + \beta/\alpha) \cdot \overline{\mathrm{med}}(P, k)$ with probability at least $1 - \delta$. Therefore, the lemma follows because $\mathbb{A}_\alpha$ returns an α-approximation C^* of the k-median for S.

Next, we only state the other lemma that describes part (ii) of the analysis of Theorem 8.

Lemma 5. *Let $s \geq \frac{c \cdot \alpha}{\beta} \cdot \left(k + \frac{\Delta}{\varepsilon \cdot \beta} \cdot \left(\alpha \cdot \ln(1/\delta) + k \cdot \ln\left(\frac{k \Delta \alpha}{\varepsilon \beta^2}\right)\right)\right)$ for an appropriate constant c. Let $\mathbb{C}$ be the set of all sets of k centers C of P with $\overline{\mathrm{cost}}(P, C) > (2\,\alpha + 6\,\beta) \cdot \overline{\mathrm{med}}(P, k)$. Then,*

$$\mathbf{Pr}\left[\exists C_b \in \mathbb{C} : C_b \subseteq S \text{ and } \overline{\mathrm{cost}}(S, C_b) \leq 2\,(\alpha + \beta)\,\overline{\mathrm{med}}(P, k)\right] \leq \delta . \qquad \square$$

Observe that comparing the result from [49] to the result in Theorem 8, Theorem 8 improves the sample complexity by a factor of $\Delta \cdot \log n/\varepsilon$ while obtaining a slightly worse approximation ratio of $2\,(\alpha + \beta)\,\overline{\mathrm{med}}(P, k) + \varepsilon$, instead of $2\,\alpha\,\overline{\mathrm{med}}(P, k) + \varepsilon$ as in [49]. However, since the polynomial-time algorithm with the best known approximation guarantee has $\alpha = 3 + \frac{1}{c}$ for the running time of $\mathcal{O}(n^c)$ time [9], this significantly improves the running time of [49] for all realistic choices of the input parameters while achieving the same approximation guarantee. As a highlight, Theorem 8 yields a sublinear-time algorithm that in time $\widetilde{\mathcal{O}}((\frac{\Delta}{\varepsilon} \cdot (k + \log(1/\delta)))^2)$ — *fully independent of n* — returns a set of k centers for which the average distance to the nearest median is at most $\mathcal{O}(\overline{\mathrm{med}}(P, k)) + \varepsilon$ with probability at least $1 - \delta$.

Extensions. The result in Theorem 8 can be significantly improved if we assume the input points are in *Euclidean space* $\mathbb{R}^d$. In this case the approximation guarantee can be improved to $(\alpha + \beta)\,\overline{\mathrm{med}}(P, k) + \varepsilon$ at the cost of increasing the sample size to $\widetilde{\mathcal{O}}(\frac{\Delta \cdot \alpha}{\varepsilon \cdot \beta^2} \cdot (k\,d + \log(1/\delta)))$.

Furthermore, a similar approach as that sketched above can be applied to study similar generic sample schemes for other clustering problems. As it is shown in [22], almost identical analysis lead to sublinear (independent on n) sample complexity for the classical *k-means problem*. Also, a more complex analysis can be applied to study the sample complexity for the *min-sum k-clustering problem* [22].

4.4   Other Results

Indyk [40] was the first who observed that some optimization problems in metric spaces can be solved in sublinear-time, that is, in $o(n^2)$ time. He presented $(\frac{1}{2} - \varepsilon)$-approximation algorithms for MaxTSP and the maximum spanning tree problems that run in $O(n/\varepsilon)$ time [40]. He also gave a $(2 + \varepsilon)$-approximation algorithm for the minimum routing cost spanning tree problem and a $(1 + \varepsilon)$ approximation algorithm for the average distance problem; both algorithms run in $O(n/\varepsilon^{O(1)})$ time.

There is also a number of sublinear-time algorithms for various clustering problems in either Euclidean spaces or metric spaces, when the number of clusters is small. For radius (*k-center*) and *diameter clustering* in Euclidean spaces, sublinear-time property testing algorithms [1,23] and tolerant testing algorithms [52] have been developed. The first sublinear algorithm for the *k-median* problem was a bicriteria approximation algorithm [40]. This algorithm computes in $\widetilde{O}(n\,k)$ time a set of $O(k)$ centers that are a constant factor approximation to the k-median objective function. Later, standard constant factor approximation algorithms were given that run in time $\widetilde{O}(n\,k)$ (see, e.g., [47,54]). These sublinear-time results have been extended in many different ways, e.g., to efficient data streaming algorithms and very fast algorithms for Euclidean k-median and also to k-means, see, e.g., [9,13,17,29,38,39,44,45,48]. For another clustering problem, the *min-sum k-clustering problem* (which is complement to the Max-k-Cut), for the basic case of $k = 2$, Indyk [42] (see also [41]) gave a $(1 + \varepsilon)$-approximation algorithm that runs in time $O(2^{1/\varepsilon^{O(1)}} n\,(\log n)^{O(1)})$, which is sublinear in the full input description size. No such results are known for $k \geq 3$, but recently, [24] gave a constant-factor approximation algorithm for min-sum k-clustering that runs in time $\mathcal{O}(n\,k\,(k\,\log n)^{\mathcal{O}(k)})$ and a polylogarithmic approximation algorithm running in time $\widetilde{\mathcal{O}}(n\,k^{\mathcal{O}(1)})$.

4.5   Limitations: What Cannot Be Done in Sublinear-Time

The algorithms discussed in the previous sections may suggest that many optimization problems in metric spaces have sublinear-time algorithms. However, it turns out that the problems listed in the previous sections are more like exceptions than a norm. Indeed, most of the problems have a trivial lower bound that exclude sublinear-time algorithms. We have already mentioned in Section 4 that the problem of approximating the cost of the lightest edge in a finite metric space (P, d) requires $\Omega(n^2)$, even if randomization is allowed. The other problems for which no sublinear-time algorithms are possible include estimation of the cost of minimum-cost matching, the cost of minimum-cost bi-chromatic matching, the cost of minimum *non-uniform* facility location, the cost of k-median for $k = n/2$; all these problems require $\Omega(n^2)$ (randomized) time to estimate the cost of their optimal solution to within any constant factor [10].

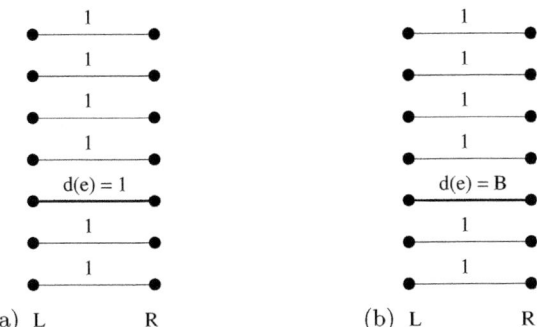

Fig. 3. Two instance of the metric matching which are indistinguishable in $o(n^2)$ time and whose cost differ by a factor greater than λ. The perfect matching connecting L with R is selected at random and the edge e is selected as a random edge from the matching. We set $B = n(\lambda - 1) + 2$. The distances not shown are all equal to $n^3\lambda$.

To illustrate the lower bounds, we give two instances of the metric spaces which are indistinguishable by any $o(n^2)$-time algorithm for which the cost of the minimum-cost matching in one instance is greater than λ times the one in the other instance (see Figure 3). Consider a metric space (P, d) with $2n$ points, n points in L and n points in R. Take a random perfect matching $\mathbb{M}$ between the points in L and R, and then choose an edge $e \in \mathbb{M}$ at random. Next, define the distance in (P, d) as follows:

- $d(e)$ is either 1 or B, where we set $B = n(\lambda - 1) + 2$,
- for any $e^* \mathbb{M} \setminus \{e\}$ set $d(e^*) = 1$, and
- for any other pair of points $p, q \in P$ not connected by an edge from $\mathbb{M}$, $d(p, q) = n^3\lambda$.

It is easy to see that both instances define properly a metric space (P, d). For such problem instances, the cost of the minimum-cost matching problem will depend on the choice of $d(e)$: if $d(e) = B$ then the cost will be $n - 1 + B > n\lambda$, and if $d(e) = 1$, then the cost will be n. Hence, any λ-factor approximation algorithm for the matching problem must distinguish between these two problem instances. However, this requires to find if there is an edge of length B, and this is known to require time $\Omega(n^2)$, even if a randomized algorithm is used.

5   Conclusions

It would be impossible to present a complete picture of the large body of research known in the area of sublinear-time algorithms in such a short paper. In this survey, our main goal was to give some flavor of the area and of the types of the results achieved and the techniques used. For more details, we refer to the original works listed in the references.

We did not discuss two important areas that are closely related to sublinear-time algorithms: property testing and data streaming algorithms. For interested readers, we recommend the surveys in [7,28,32,33,43,53] and [50], respectively.

References

1. Alon, N., Dar, S., Parnas, M., Ron, D.: Testing of clustering. SIAM Journal on Discrete Mathematics 16(3), 393–417 (2003)
2. Alon, N., Fernandez de la Vega, W., Kannan, R., Karpinski, M.: Random sampling and approximation of MAX-CSPs. Journal of Computer and System Sciences 67(2), 212–243 (2003)
3. Alon, N., Fischer, E., Krivelevich, M., Szegedy, M.: Efficient testing of large graphs. Combinatorica 20(4), 451–476 (2000)
4. Alon, N., Fischer, E., Newman, I., Shapira, A.: A combinatorial characterization of the testable graph properties: it's all about regularity. SIAM Journal on Computing 39(1), 143–167 (2009)
5. Alon, N., Shapira, A.: Every monotone graph property is testable. SIAM Journal on Computing 38(2), 505–522 (2008)
6. Alon, N., Shapira, A.: A characterization of the (natural) graph properties testable with one-sided error. SIAM Journal on Computing 37(6), 1703–1727 (2008)
7. Alon, N., Shapira, A.: Homomorphisms in graph property testing - A survey. In: Klazar, M., Kratochvil, J., Loebl, M., Matousek, J., Thomas, R., Valtr, P. (eds.) Topics in Discrete Mathematics, dedicated to Jarik Nesetril on the occasion of his 60th Birthday, pp. 281–313
8. Arora, S., Karger, D.R., Karpinski, M.: Polynomial time approximation schemes for dense instances of $\mathcal{NP}$-hard problems. Journal of Computer and System Sciences 58(1), 193–210 (1999)
9. Arya, V., Garg, N., Khandekar, R., Meyerson, A., Munagala, K., Pandit, V.: Local search heuristics for k-median and facility location problems. SIAM Journal on Computing 33(3), 544–562 (2004)
10. Bădoiu, M., Czumaj, A., Indyk, P., Sohler, C.: Facility location in sublinear time. In: Caires, L., Italiano, G.F., Monteiro, L., Palamidessi, C., Yung, M. (eds.) ICALP 2005. LNCS, vol. 3580, pp. 866–877. Springer, Heidelberg (2005)
11. Benjamini, I., Schramm, O., Shapira, A.: Every minor-closed property of sparse graphs is testable. In: Proceedings of the 40th Annual ACM Symposium on Theory of Computing (STOC), pp. 393–402 (2008)
12. Borgs, C., Chayes, J., Lovász, L., Sos, V.T., Szegedy, B., Vesztergombi, K.: Graph limits and parameter testing. In: Proceedings of the 38th Annual ACM Symposium on Theory of Computing (STOC) (2006)
13. Charikar, M., O'Callaghan, L., Panigrahy, R.: Better streaming algorithms for clustering problems. In: Proceedings of the 35th Annual ACM Symposium on Theory of Computing (STOC), pp. 30–39 (2003)
14. Chazelle, B., Dobkin, D.P.: Intersection of convex objects in two and three dimensions. Journal of the ACM 34(1), 1–27 (1987)
15. Chazelle, B., Liu, D., Magen, A.: Sublinear geometric algorithms. SIAM Journal on Computing 35(3), 627–646 (2006)
16. Chazelle, B., Rubinfeld, R., Trevisan, L.: Approximating the minimum spanning tree weight in sublinear time. SIAM Journal on Computing 34(6), 1370–1379 (2005)
17. Chen, K.: On k-median clustering in high dimensions. In: Proceedings of the 17th Annual ACM-SIAM Symposium on Discrete Algorithms (SODA), pp. 1177–1185 (2006)
18. Czumaj, A., Ergün, F., Fortnow, L., Magen, A., Newman, I., Rubinfeld, R., Sohler, C.: Sublinear-time approximation of Euclidean minimum spanning tree. SIAM Journal on Computing 35(1), 91–109 (2005)

19. Czumaj, A., Shapira, A., Sohler, C.: Testing hereditary properties of non-expanding bounded-degree graphs. SIAM Journal on Computing 38(6), 2499–2510 (2009)
20. Czumaj, A., Sohler, C.: Property testing with geometric queries. In: Meyer auf der Heide, F. (ed.) ESA 2001. LNCS, vol. 2161, pp. 266–277. Springer, Heidelberg (2001)
21. Czumaj, A., Sohler, C.: Estimating the weight of metric minimum spanning trees in sublinear-time. SIAM Journal on Computing 39(3), 904–922 (2009)
22. Czumaj, A., Sohler, C.: Sublinear-time approximation for clustering via random sampling. Random Structures and Algorithms 30(1-2), 226–256 (2007)
23. Czumaj, A., Sohler, C.: Abstract combinatorial programs and efficient property testers. SIAM Journal on Computing, 34(3), 580–615 (2005)
24. Czumaj, A., Sohler, C.: Small space representations for metric min-sum k-clustering and their applications. Theory of Computing Systems 46(3), 416–442 (2010)
25. Czumaj, A., Sohler, C., Ziegler, M.: Property testing in computational geometry. In: Paterson, M. (ed.) ESA 2000. LNCS, vol. 1879, pp. 155–166. Springer, Heidelberg (2000)
26. Dyer, M., Megiddo, N., Welzl, E.: Linear programming. In: Goodman, J.E., O'Rourke, J. (eds.) Handbook of Discrete and Computational Geometry, 2nd edn., pp. 999–1014. CRC Press, Boca Raton (2004)
27. Feige, U.: On sums of independent random variables with unbounded variance and estimating the average degree in a graph. SIAM Journal on Computing 35(4), 964–984 (2006)
28. Fischer, E.: The art of uninformed decisions: A primer to property testing. Bulletin of the EATCS 75, 97–126 (2001)
29. Frahling, G., Sohler, C.: Coresets in dynamic geometric data streams. In: Proceedings of the 37th Annual ACM Symposium on Theory of Computing (STOC), pp. 209–217 (2005)
30. Frieze, A., Kannan, R.: Quick approximation to matrices and applications. Combinatorica 19(2), 175–220 (1999)
31. Frieze, A., Kannan, R., Vempala, S.: Fast Monte-Carlo algorithms for finding low-rank approximations. Journal of the ACM 51(6), 1025–1041 (2004)
32. Goldreich, O.: Combinatorial property testing (a survey). In: Pardalos, P., Rajasekaran, S., Rolim, J. (eds.) Proc. DIMACS Workshop on Randomization Methods in Algorithm Design. DIMACS, Series in Discrete Mathetaics and Theoretical Computer Science, vol. 43, pp. 45–59. American Mathematical Society, Providence (1997)
33. Goldreich, O.: Property testing in massive graphs. In: Abello, J., Pardalos, P.M., Resende, M.G.C. (eds.) Handbook of massive data sets, pp. 123–147. Kluwer Academic Publishers, Dordrecht (2002)
34. Goldreich, O., Goldwasser, S., Ron, D.: Property testing and its connection to learning and approximation. Journal of the ACM 45(4), 653–750 (1998)
35. Goldreich, O., Ron, D.: Property Testing in Bounded Degree Graphs. Algorithmica 32(2), 302–343 (2002)
36. Goldreich, O., Ron, D.: A sublinear bipartiteness tester for bounded degree graphs. Combinatorica 19(3), 335–373 (1999)
37. Goldreich, O., Ron, D.: Approximating average parameters of graphs. Random Structures and Algorithms 32(4), 473–493 (2008)
38. Har-Peled, S., Mazumdar, S.: Coresets for k-means and k-medians and their applications. In: Proceedings of the 36th Annual ACM Symposium on Theory of Computing (STOC), pp. 291–300 (2004)

39. Har-Peled, S., Kushal, A.: Smaller coresets for k-median and k-means clustering. Discrete & Computational Geometry 37(1), 3–19 (2007)
40. Indyk, P.: Sublinear time algorithms for metric space problems. In: Proceedings of the 31st Annual ACM Symposium on Theory of Computing (STOC), pp. 428–434 (1999)
41. Indyk, P.: A sublinear time approximation scheme for clustering in metric spaces. In: Proceedings of the 40th IEEE Symposium on Foundations of Computer Science (FOCS), pp. 154–159 (1999)
42. Indyk, P.: High-Dimensional Computational Geometry. PhD thesis, Stanford University (2000)
43. Kumar, R., Rubinfeld, R.: Sublinear time algorithms. SIGACT News 34, 57–67 (2003)
44. Kumar, A., Sabharwal, Y., Sen, S.: A simple linear time $(1+\varepsilon)$-approximation algorithm for k-means clustering in any dimensions. In: Proceedings of the 45th IEEE Symposium on Foundations of Computer Science (FOCS), pp. 454–462 (2004)
45. Kumar, A., Sabharwal, Y., Sen, S.: Linear time algorithms for clustering problems in any dimensions. In: Caires, L., Italiano, G.F., Monteiro, L., Palamidessi, C., Yung, M. (eds.) ICALP 2005. LNCS, vol. 3580, pp. 1374–1385. Springer, Heidelberg (2005)
46. Lovász, L., Szegedy, B.: Graph limits and testing hereditary graph properties. Technical Report, MSR-TR-2005-110, Microsoft Research (August 2005)
47. Mettu, R., Plaxton, G.: Optimal time bounds for approximate clustering. Machine Learning 56(1-3), 35–60 (2004)
48. Meyerson, A., O'Callaghan, L., Plotkin, S.: A k-median algorithm with running time independent of data size. Machine Learning 56(1-3), 61–87 (2004)
49. Mishra, N., Oblinger, D., Pitt, L.: Sublinear time approximate clustering. In: Proceedings of the 12th Annual ACM-SIAM Symposium on Discrete Algorithms (SODA), pp. 439–447 (2001)
50. Muthukrishnan, S.: Data streams: Algorithms and applications. Foundations and Trends in Theoretical Computer Science 1(2) (August 2005)
51. Nguyen, H., Onak, K.: Constant-time approximation algorithms via local improvements. In: Proceedings of the 49th IEEE Symposium on Foundations of Computer Science (FOCS), pp. 489–498 (2008)
52. Parnas, M., Ron, D., Rubinfeld, R.: Tolerant property testing and distance approximation. Journal of Computer and System Sciences 72(6), 1012–1042 (2006)
53. Ron, D.: Property testing. In: Pardalos, P.M., Rajasekaran, S., Reif, J., Rolim, J.D.P. (eds.) Handobook of Randomized Algorithms, vol. II, pp. 597–649. Kluwer Academic Publishers, Dordrecht (2001)
54. Thorup, M.: Quick k-median, k-center, and facility location for sparse graphs. SIAM Journal on Computing 34(2), 405–432 (2005)
55. Vazirani, V.V.: Approximation Algorithms. Springer, New York (2004)
56. Yoshida, Y., Yamamoto, M., Ito, H.: Improved constant-time approximation algorithms for maximum independent sets and maximum matchings. In: Proceedings of the 41st Annual ACM Symposium on Theory of Computing (STOC), pp. 225–234 (2009)

Short Locally Testable Codes and Proofs: A Survey in Two Parts

Oded Goldreich

Department of Computer Science, Weizmann Institute of Science, Rehovot, Israel
oded.goldreich@weizmann.ac.il

Abstract. We survey known results regarding locally testable codes and locally testable proofs (known as PCPs), with emphasis on the length of these constructs. Local testability refers to approximately testing large objects based on a very small number of probes, each retrieving a single bit in the representation of the object. This yields super-fast approximate-testing of the corresponding property (i.e., be a codeword or a valid proof). We also review the related concept of local decodable codes.

The survey consists of two independent (i.e., self-contained) parts that cover the same material at different levels of rigor and detail. Still, in spite of the repetitions, there may be a benefit in reading both parts.

Keywords: Error Correcting Codes, Probabilistically Checkable Proofs (PCP), Locally Testable Codes, Locally Decodable Codes, Self-Correction, Low-Degree Tests, Derandomization, Private Information Retrieval.

This is a revised version of [36].

PART I: A HIGH-LEVEL OVERVIEW

The title of this survey refers to two types of objects (i.e., codes and proofs) and two adjectives (i.e., *local testability* and *short*). A clarification of these terms is in place.

Codes, proofs and their length. Codes are sets of strings (of equal length), typically, having a large pairwise distance. Equivalently, codes are viewed as mappings from short (k-bit) strings to longer (n-bit) strings, called codewords, such that the codewords are distant from one another. We will focus on *codes with relative constant distance*; that is, every two n-bit codewords are at distance $\Omega(n)$ apart. The length of the code is measured in terms of the length of the pre-image (i.e., we are interested in the growth of n as a function of k). Turning to proofs, these are defined with respect to a verification procedure for assertions of a certain length, and their length is measured in terms of the length of the assertion. The verification procedure must satisfy the natural completeness and soundness properties: For valid assertions there should be strings, called proofs, that are accepted (in conjunction with the assertion) by the verification procedures, whereas for false assertions no such strings may exist. The reader may envision proof systems for the set of satisfiable propositional formulae (i.e., assertions of satisfiability of given formulae).

O. Goldreich (Ed.): Property Testing, LNCS 6390, pp. 65–104, 2010.

Local testability. By local testability we mean that the object can be tested for the natural property (i.e., being a codeword or a valid proof) using a small (typically constant)[1] number of probes, each recovering individual bits in a standard representation of the object. Thus, local testability allows for super-fast testing of the corresponding objects. The tests are probabilistic and hence the result is correct only with high probability.[2] Furthermore, correctness refers to a *relaxed notion of deciding* (which was formulated, in general terms, in the context of property testing [57,38]): It is required that valid objects be accepted with high probability, whereas objects that are "far" from being valid should be rejected with high probability. Specifically, in the case of codes, codewords should be accepted (with high probability), whereas strings that are "far" from the code should be rejected (with high probability). In the case of proofs, valid proofs (which exist for correct assertions) should be accepted (with high probability), whereas strings that are "far" from being valid proofs (and, in particular, all strings in case no valid proofs exist) should be rejected (with high probability).[3]

Our notion of locally testable proofs is closely related to the notion of a PCP (i.e., probabilistically checkable proof), and we will ignore the difference in the sequel. The difference is that in the definition of locally testable proofs we required rejection of strings that are far from any valid proof, also in the case that valid proofs exists (i.e., the assertion is valid). In contrast, the standard rejection criteria of PCPs refers only to false assertions. Still, all known PCP constructions actually satisfy the stronger definition.

The very possibility of local testability. Indeed, local testability of either codes or proofs is quite challenging, regardless of the issue of length:

- For codes, the simplest example of a locally testable code (of constant relative distance) is the Hadamard code and testing it amounts to linearity testing. However, the exact analysis of the natural linearity tester (of Blum, Luby and Rubinfeld [22]) turned out to be highly complex (cf. [22,6,31,12,13,10,46]).
- For proofs, the simplest example of a locally testable proof is the "inner verifier" of the PCP construction of Arora, Lund, Motwani, Sudan and Szegedy [4], which in turn is based on the Hadamard code.

In both cases, the constructed object has exponential length in terms of the relevant parameter (i.e., the amount of information being encoded in the code or the length of the assertion being proved).

Local testability at a polynomial blow-up. Achieving local testability by codes and proofs that have polynomial length turns out to be even more challenging.

---

[1] In this part, we associate local testability with tests that perform a constant number of probes.

[2] It is easy to see that deterministic tests will perform very poorly, and the same holds with respect to probabilistic tests that make no error.

[3] Indeed, in the case the assertion is false, there exist no valid proofs. In this case all strings are defined to be far from a valid proof.

- In the case of codes, a direct interpretation of *low-degree tests* (cf. [6,7,35,57,34]), proposed in [34,57], yields a locally testable code of quadratic length over a *sufficiently large alphabet*. Similar (and actually better) results for *binary* codes required additional ideas, and have appeared only later (cf. [41]).
- The case of proofs is far more complex: Achieving locally testable proof of polynomial length is essentially the contents of the celebrated PCP Theorem of Arora, Lund, Motwani, Safra, Sudan and Szegedy [5,4].

We focus on even *shorter* codes and proofs; specifically, codes and proofs of *nearly linear length*. The latter term has been given quite different interpretations, and here we adopt the most strict interpretation by which nearly linear means linear up to polylogarithmic factors.

Local testability with a polylogarithmic (length) overhead: The ultimate goal is to obtain locally testable codes and proofs of minimal length. The currently known results get very close to obtaining this goal.

Theorem 1. (Dinur [26], building on [20]): *There exist locally testable codes and proofs of length that is only a polylogarithmic factor larger than the relevant parameter. That is, the length function $\ell : \mathbb{N} \to \mathbb{N}$ satisfies $\ell(k) = \widetilde{O}(k) = k \cdot \mathrm{poly}(\log k)$.*

One may wonder whether or not a polylogarithmic overhead in inherent to local testability of codes and proofs. This is indeed a fundamental open problem.

Open Problem 2. *Do there exist locally testable codes and proofs of linear length?*

In the rest of this part of the survey, we motivate the study of short locally testable objects, comment on the relation between such codes and proofs, and discuss a somewhat related coding problem.

Motivation for the Study of Short Locally Testable Codes and Proofs

Local testability offers an extremely strong notion of efficient testing: The tester makes only a constant number of bit probes, and determining the probed locations (as well as the final decision) is typically done in time that is polylogarithmic in the length of the probed object.

The length of an error-correcting code is widely recognized as one of the two most fundamental parameters of the code (the second one being its distance). In particular, the length of the code is of major importance in applications, because it determines the overhead involved in encoding information.

The same considerations apply also to proofs. However, in the case of proofs, this obvious point was blurred by the indirect, unexpected and highly influential applications of locally testable proofs (known as PCPs) to the theory of approximation algorithms. In our view, the significance of locally testable proofs (i.e., PCPs) extends far beyond their applicability to deriving non-approximability

results. The mere fact that proofs can be transformed into a format that supports super-fast probabilistic verification is remarkable. From this perspective,
the question of how much redundancy is introduced by such a transformation is
a fundamental one. Furthermore, locally testable proofs (i.e., PCPs) have been
used not only to derive non-approximability results but also for obtaining positive results (e.g., CS-proofs [48,53] and their applications [8,24]), and the length
of the PCP affects the complexity of those applications.

Turning back to the celebrated application of PCP to the study of approximation algorithms, we note that the length of PCPs is also relevant to non-
approximability results; specifically, the length of PCPs affects the *tightness*
with respect to the running time of the non-approximability results derived. For
example, suppose (exact) SAT has complexity $2^{\Omega(n)}$. The original PCP Theorem [5,4] only implies that approximating MaxSAT requires time $2^{n^{\alpha}}$, for some
(small) $\alpha > 0$. The work of [55] makes α arbitrarily close to 1, whereas the results
of [41,21] further improve the lower bound to $2^{n^{1-o(1)}}$ and the results of [20,26]
yields a lower bound of $2^{n/\mathrm{poly}(\log n)}$.[4]

On the Relation between Locally Testable Codes and Proofs

Locally testable codes seem related to locally testable proofs (PCPs). In fact,
the use of codes with some "local testability" features is implicit in known PCP
constructions. Furthermore, the known constructions of locally testable proofs
(PCPs) provides a transformation of *standard proofs* (for say SAT) to *locally
testable proofs* (i.e., PCP-oracles) such that transformed strings are accepted
with probability one by the PCP verifier. Moreover, starting from different standard proofs, one obtains locally testable proofs that are far apart, and hence
constitute a good code. It is tempting to think that the PCP verifier yields a
codeword tester, but this is not really the case. Note that our definition of a locally testable proof requires rejection of strings that are far from any valid proof,
but it is not clear that the only valid proofs (w.r.t the constructed PCP verifier)
are those that are obtained by the aforementioned transformation of standard
proofs to locally testable ones.[5] In fact, the standard PCP constructions accept
also valid proofs that are not in the range of the corresponding transformation.

In spite of the above, locally testable codes and proofs are related, and the
feeling is that locally testable codes are the combinatorial counterparts of locally
testable proofs (PCPs), which are complexity theoretic in nature. From that
perspective, one should expect (or hope) that it would be easier to construct
locally testable codes than it is to construct PCPs. This feeling was among
the main motivations of Goldreich and Sudan, and indeed their first result was
along this vein: They showed a relatively simple construction (i.e., simple in

---

[4] Using [54] (or [27]) allows to achieve the lower bound of $2^{n^{1-o(1)}}$ simultaneously
with optimal approximation ratios, but this is currently unknown for the better
lower bound of $2^{n/\mathrm{poly}(\log n)}$.

[5] Let alone that the standard definition of PCP refers only to the case of false assertions, in which case all strings are far from a valid proof (which does not exist).

comparison to PCP constructions) of a locally testable code of length $\ell(k) = k^c$ for any constant $c > 1$ [41, Sec. 3]. Unfortunately, their stronger result, providing a locally testable code of shorter length (i.e., length $\ell(k) = k^{1+o(1)}$) is obtained by constructing and using a corresponding locally testable proof (i.e., PCP). Subsequent works have mostly followed this route, with the notable exception of Meir's work [51].

Locally Decodable Codes

Locally *decodable* codes are in some sense complimentary to local *testable* codes. Here, one is given a slightly corrupted codeword (i.e., a string close to some unique codeword), and is required to recover individual bits of the encoded information based on a constant number of probes (per recovered bit). That is, a code is said to be locally decodable if whenever relatively few location are corrupted, the decoder is able to recover each information-bit, with high probability, based on a constant number of probes to the (corrupted) codeword.

The best known locally decodable codes are of strictly sub-exponential length. Specifically, k information bits can be encoded by codewords of length $n = \exp(k^{o(1)})$ that are locally decodable using three bit-probes (cf. [29], building over [61]). The problem is related to the construction of (information theoretic secure) Private Information Retrieval schemes, introduced in [25].

A natural relaxation of the definition of locally decodable codes requires that, whenever few location are corrupted, the decoder should be able to recover most of the individual information-bits (based on a constant number of queries), and for the rest of the locations the decoder may output a fail symbol (but not the wrong value). That is, the decoder must still avoid errors (with high probability), but on a few bit-locations it is allowed to sometimes say "don't know". This relaxed notion of local decodability can be supported by codes that have length $\ell(k) = k^c$ for any constant $c > 1$ (cf. [15]).

An obvious open problem is to separate locally decodable codes from relaxed locally decodable codes. This may follow by either improving the $\Omega(k^{1+\frac{1}{q-1}})$ lower bound on the length of q-query locally decodable codes (of [45]), or by providing relaxed locally decodable codes of length $\ell(k) = k^{1+o(1)}$.

PART II: A MORE DETAILED AND RIGOROUS ACCOUNT

In this part we provide a general treatment of local testability. In contrast to Part I, here we allow the tester to use a number of queries that is a (typically small) predetermined function of the length parameter, rather than insisting on a constant number of queries. The latter special case is indeed an important one.

1   Introduction

Codes (i.e., error correcting codes) and proofs (i.e., automatically verifiable proofs) are fundamental to computer science as well as to related disciplines such as mathematics and computer engineering. Redundancy is inherent to error-correcting codes, whereas testing validity is inherent to proofs. In this survey we also consider less traditional combinations such as testing validity of codewords and the use of proofs that contain redundancy. The reader may wonder why we explore these non-traditional possibilities, and the answer is that they offer various advantages (as will be elaborated next).

Testing the validity of codewords is natural in settings in which one may want to take an action in case the codeword is corrupted. For example, when storing data in an error correcting format, one may want to recover the data and re-encode it whenever one finds that the current encoding is corrupted. Doing so may allow to maintain the data integrity over eternity, although the encoded bits may all get corrupted in the course of time. Of course, one can use the error-correcting decoding procedure associated with the code in order to check whether the current encoding is corrupted, but the question is whether one can check (or just approximately check) this property *much faster*.

Loosely speaking, locally testable codes are error correcting codes that allow for a super-fast testing of whether or not a give string is a valid codeword. In particular, the tester works in sub-linear time and reads very few of the bits of the tested object. Needless to say, the answer provided by such a tester can only be approximately correct, but this would suffice in many applications (including the one outlined above).

Similarly, locally testable proofs are proofs that allow for a super-fast probabilistic verification. Again, the tester works in sub-linear time and reads very few of the bits of the tested object. The tester's (a.k.a. verifier's) verdict is only correct with high probability, but this may suffice for many applications, where the assertion is rather mundane but of great practical importance. In particular, it suffices in applications in which proofs are used for establishing the correctness of *specific* computations of practical interest. Lastly, we comment that such *locally testable proofs must be redundant* (or else there would be no chance for verifying them based on inspecting only a small portion of them).

Our focus is on relatively *short* locally testable codes and proofs, which is not surprising in view of the fact that *we envision such objects being actually used in practice*. Of course, we do not mean to suggest that one may use in practice any of the constructions surveyed here (especially not the ones that provide the stronger bounds). We rather argue that this direction of research may find applications in practice. Furthermore, it may even be the case that some of the current concepts and techniques may lead to such applications.

Organization: In Section 2 we provide a quite comprehensive definitional treatment of locally testable codes and proofs, while relating them to PCPs, PCPs of proximity, and property testing. In Section 3, we survey the main results regarding locally testable codes and proofs as well as many of the underlying ideas. In Section 4 we consider locally decodable codes, which are somewhat complementary to locally testable codes.

Caveat: Our exposition of locally testable/decodable codes is aimed at achieving the best possible length, regardless of whether or not the code is popular (i.e., used in practice). Thus, we do not survey here results that refer to the testing (and decoding) features of various popular codes, unless these features are instructive for our aim.

2   Definitions

Local testability is formulated by considering oracle machines. That is, the tester is an oracle machine, and the object that it tests is viewed as an oracle. For simplicity, we confine ourselves to *non-adaptive* probabilistic oracle machines; that is, machines that determine their queries based on their explicit input (which in case of codes is merely a length parameter) and their internal coin tosses (but not depending on previous oracle answers). When talking about oracle access to a string $w \in \{0,1\}^n$ we viewed w as a function $w : \{1, ..., n\} \to \{0,1\}$.

2.1   Codeword Testers

We consider codes mapping sequences of k (input) bits into sequences of $n \geq k$ (output) bits. Such a generic code is denoted by $C : \{0,1\}^k \to \{0,1\}^n$, and the elements of $\{C(x) : x \in \{0,1\}^k\} \subseteq \{0,1\}^n$ are called codewords (of C).

The distance of a code $C : \{0,1\}^k \to \{0,1\}^n$ is the minimum (Hamming) distance between its codewords; that is, $\min_{x \neq y}\{\Delta(C(x), C(y))\}$, where $\Delta(u, v)$ denotes the number of bit-locations on which u and v differ. Throughout this work, *we focus on codes of linear distance*; that is, codes $C : \{0,1\}^k \to \{0,1\}^n$ of distance $\Omega(n)$.

The distance of $w \in \{0,1\}^n$ from a code $C : \{0,1\}^k \to \{0,1\}^n$, denoted $\Delta_C(w)$, is the minimum distance between w and the codewords; that is, $\Delta_C(w) \overset{\text{def}}{=} \min_x\{\Delta(w, C(x))\}$. For $\delta \in [0,1]$, the n-bit long strings u and v are said to be δ-far (resp., δ-close) if $\Delta(u, v) > \delta \cdot n$ (resp., $\Delta(u, v) \leq \delta \cdot n$). Similarly, w is δ-far from C (resp., δ-close to C) if $\Delta_C(w) > \delta \cdot n$ (resp., $\Delta_C(w) \leq \delta \cdot n$).

Definition 2.1 (codeword tests, basic version): *Let* $C : \{0,1\}^k \to \{0,1\}^n$ *be a code of distance* d, *and let* $q \in \mathbb{N}$ *and* $\delta \in (0,1)$. *A* q-*local (codeword)* δ-*tester for* C *is a probabilistic (non-adaptive) oracle machine* M *that makes at most* q *queries and satisfies the following two conditions:*

Accepting codewords (a.k.a. completeness): *For any* $x \in \{0,1\}^k$, *given oracle access to* $w = C(x)$, *machine* M *accepts with probability 1. That is,* $\Pr[M^{C(x)}(1^k)=1] = 1$, *for any* $x \in \{0,1\}^k$.
Rejection of non-codeword (a.k.a. soundness): *For any* $w \in \{0,1\}^n$ *that is* δ-*far from* C, *given oracle access to* w, *machine* M *rejects with probability at least* $1/2$. *That is,* $\Pr[M^w(1^k)=1] \leq 1/2$, *for any* $w \in \{0,1\}^n$ *that is* δ-*far from* C.

We call q *the* query complexity *of* M, *and* δ *the* proximity parameter.

The above definition is interesting only in case δn is smaller than the covering radius of C (i.e., the smallest r such that for every $w \in \{0,1\}^n$ it holds that

$\Delta_C(w) \leq r$). Clearly, $r \geq d/2$, and so the definition is certainly interesting in the case that $\delta < d/2n$, and indeed we will focus on this case. On the other hand, observe that $q = \Omega(1/\delta)$ must hold, which means that we focus on the case that $d = \Omega(n/q)$.

We next consider families of codes $C = \{C_k : \{0,1\}^k \to \{0,1\}^{n(k)}\}_{k \in K}$, where $n, d : \mathbb{N} \to \mathbb{N}$ and $K \subseteq \mathbb{N}$, such that C_k has distance $d(k)$. In accordance with the above, our main interest is in the case that $\delta(k) < d(k)/2n(k)$. Furthermore, seeking constant query complexity, we focus on the case $d = \Omega(n)$.

Definition 2.2 (codeword tests, asymptotic version): *For functions $n, d : \mathbb{N} \to \mathbb{N}$, let $C = \{C_k : \{0,1\}^k \to \{0,1\}^{n(k)}\}_{k \in K}$ be such that C_k is a code of distance $d(k)$. For functions $q : \mathbb{N} \to \mathbb{N}$ and $\delta : \mathbb{N} \to (0,1)$, we say that a machine M is a q-local (codeword) δ-tester for $C = \{C_k\}_{k \in K}$ if, for every $k \in K$, machine M is a $q(k)$-local $\delta(k)$-tester for C_k. Again, q is called the* query complexity of M, *and δ the* proximity parameter.

Recall that being particularly interested in constant query complexity (and recalling that $d(k)/n(k) \geq 2\delta(k) = \Omega(1/q(k))$), we focus on the case that $d = \Omega(n)$ and δ is a constant smaller than $d/2n$. In this case, we may consider a stronger definition.

Definition 2.3 (locally testable codes): *Let n, d and C be as in Definition 2.2 and suppose that $d = \Omega(n)$. We say that C is* locally testable *if for every constant $\delta > 0$ there exists a constant q and a probabilistic polynomial-time oracle machine M such that M is a q-local δ-tester for C.*

We will be concerned of the growth rate of n as a function of k, for locally testable codes $C = \{C_k : \{0,1\}^k \to \{0,1\}^{n(k)}\}_{k \in K}$ of distance $d = \Omega(n)$. More generally, for $d = \Omega(n)$, we will be interested in the trade-off between n, the proximity parameter δ, and the query complexity q.

2.2   Proof Testers

We start by recalling the standard definition of PCP. (For an introduction to the subject as well as a wider perspective, see [37, Chap. 9]).

Definition 2.4 (PCP, standard definition): *A* probabilistically checkable proof (PCP) system *for a set S is a probabilistic (non-adaptive) polynomial-time oracle machine (called a* verifier), *denoted V, satisfying*

Completeness: *For every $x \in S$ there exists an oracle π_x such that V, on input x and access to oracle π_x, always accepts x; that is, $\Pr[V^{\pi_x}(x) = 1] = 1$.*

Soundness: *For every $x \notin S$ and every oracle π, machine V, on input x and access to oracle π, rejects x with probability at least $\frac{1}{2}$; that is, $\Pr[V^\pi(x) = 1] \leq 1/2$,*

Let $Q_x(r)$ denote the set of oracle positions inspected by V on input x and random-tape $r \in \{0,1\}^{\text{poly}(|x|)}$. The query complexity *of V is defined as $q(n) \stackrel{\text{def}}{=}$*

$\max_{x \in \{0,1\}^n, r \in \{0,1\}^{\mathrm{poly}(n)}} \{|Q_x(r)|\}$. *The* proof complexity *of V is defined as* $p(n) \overset{\text{def}}{=}$ $\max_{x \in \{0,1\}^n} \{|\bigcup_{r \in \{0,1\}^{\mathrm{poly}(n)}} Q_x(r)|\}$.

Note that in the case that the verifier V uses a logarithmic number of coin tosses, its proof complexity is polynomial. In general, the proof complexity is upper-bounded by $2^r \cdot q$, where r and q are the randomness complexity and the query complexity of the proof tester. Thus, the trade-off between the query complexity and the proof complexity is typically captured by the trade-off between the query complexity and the randomness complexity. Furthermore, focusing on the randomness complexity allows for better bounds when composing proofs (cf. §3.2.2).

All known PCP constructions can be easily modified such that the oracle locations accessed by V are a prefix of the oracle (i.e., $\bigcup_{r \in \{0,1\}^{\mathrm{poly}(|x|)}} Q_x(r) \subseteq$ $\{1, ..., p(|x|)\}$, for every x).[6] (For simplicity, the reader may assume that this is the case throughout the rest of this exposition.) More importantly, all known PCP constructions can be easily modified to satisfy the following definition, which is closer in spirit to the definition of locally testable codes.

Definition 2.5 (PCP, augmented): *For functions $q : \mathbb{N} \to \mathbb{N}$ and $\delta : \mathbb{N} \to (0,1)$, we say that a PCP system V for a set S is a q-locally δ-testable* proof system *if it has query complexity q and satisfies the following condition, which augments the standard soundness condition.[7]*

Rejecting invalid proofs: *For every $x \in \{0,1\}^*$ and every oracle π that is δ-far from $\Pi_x \overset{\text{def}}{=} \{w : \Pr[V^w(x) = 1] = 1\}$, machine V, on input x and access to oracle π, rejects x with probability at least $\frac{1}{2}$.*

The proof complexity of V is defined as in Definition 2.4.

Note that Definition 2.5 uses the tester V itself in order to define the set (denoted Π_x) of valid proofs (for $x \in S$). That is, V is used both to define the set of valid proofs and to test for the proximity of a given oracle to this set. A more general definition (presented next), refers to an arbitrary proof system, and lets Π_x equal the set of valid proofs (in that system) for $x \in S$. Obviously, it must hold that $\Pi_x \neq \emptyset$ if and only if $x \in S$. Typically, one also requires the existence of a polynomial-time

---

[6] Recall that p denotes the proof complexity of the system. In fact, for every $x \in \{0,1\}^n$, it holds that $\bigcup_{r \in \{0,1\}^{\mathrm{poly}(n)}} Q_x(r) = \{1, ..., p(n)\}$.

[7] Definition 2.5 relies on two natural conventions:

1. All strings in Π_x are of the same length, which equals $|\bigcup_{r \in \{0,1\}^{\mathrm{poly}(n)}} Q_x(r)|$, where $Q_x(r)$ is as in Definition 2.4. Furthermore, we consider only π's of this length.
2. If $\Pi_x = \emptyset$ (which happens if and only if $x \notin S$), then every π is considered δ-far from Π_x.

These conventions will also be used in Definition 2.6.

procedure that, on input a pair (x, π), determines whether or not $\pi \in \Pi_x$.[8] For simplicity we assume that, for some function $p : \mathbb{N} \to \mathbb{N}$ and every $x \in \{0, 1\}^*$, it holds that $\Pi_x \subseteq \{0, 1\}^{p(|x|)}$. The resulting definition follows.

Definition 2.6 (locally testable proofs): *Suppose that, for some function $p :$ $\mathbb{N} \to \mathbb{N}$ and every $x \in \{0, 1\}^*$, it holds that $\Pi_x \subseteq \{0, 1\}^{p(|x|)}$. For functions $q :$ $\mathbb{N} \to \mathbb{N}$ and $\delta : \mathbb{N} \to (0, 1)$, we say that a probabilistic* (non-adaptive) *polynomial-time oracle machine V is a q-locally δ-tester for proofs in $\{\Pi_x\}_{x \in \{0,1\}^*}$ if V has query complexity q and satisfies the following conditions:*

Technical condition: *On input x, machine V issues queries in $\{1, ..., p(|x|)\}$.*

Accepting valid proofs: *For every $x \in \{0, 1\}^*$ and every oracle $\pi \in \Pi_x$, machine V, on input x and access to oracle π, accepts x with probability 1.*

Rejecting invalid proofs: *For every $x \in \{0, 1\}^*$ and every oracle π that is δ-far from Π_x, machine V, on input x and access to oracle π, rejects x with probability at least $\frac{1}{2}$.*

The proof complexity *of V is defined as p,[9] and δ is called the* proximity parameter. *In such a case, we say that $\Pi = \{\Pi_x\}_{x \in \{0,1\}^*}$ is q-locally δ-testable, and that $S = \{x \in \{0, 1\}^* : \Pi_x \neq \emptyset\}$ has q-locally δ-testable proofs of length p.*

We say that Π is locally testable *if for every constant $\delta > 0$ there exists a constant q such that Π is q-locally δ-testable. In such a case, we say that S has* locally testable proofs of length p.

This notion of locally testable proofs is closely related to the notion of probabilistically checkable proofs (i.e., PCPs). The difference is that in the definition of locally testable proofs we required rejection of strings that are far from any valid proof, also in the case that valid proofs exists (i.e., the assertion is valid). In contrast, the standard rejection criteria of PCPs refers only to false assertions. Still, all known PCP constructions actually satisfy the stronger definition.[10]

Needless to say, the new term "locally testable proof" was introduced to match the term "locally testable codes". In retrospect, "locally testable proofs" seems a more fitting term than "probabilistically checkable proofs", because it stresses the positive aspect (of locality) rather than the negative aspect (of being probabilistic). The latter perspective has been frequently advocated by Leonid Levin.

2.3   Discussion

We first comment about a few definitional choices made above. Firstly, we chose to present testers that always accept valid objects (i.e., accept valid codewords

---

[8] Recall that in the case that the verifier V uses a logarithmic number of coin tosses, its proof complexity is polynomial (and so the "effective length" of the strings in Π_x must be polynomial in $|x|$). Furthermore, if in addition it holds that $\Pi_x = \{w : \Pr[V^w(x) = 1] = 1\}$, then (scanning all possible coin tosses of) V yields a polynomial-time procedure for determining whether a given pair (x, π) satisfies $\pi \in \Pi_x$.

[9] Note that by the technical condition, the current definition of the proof complexity of V is lower-bounded by the definition used in Definition 2.4.

[10] In some cases this holds only under a weighted version of the Hamming distance, rather under the standard Hamming distance. Alternatively, these constructions can be easily modified to work under the standard Hamming distance.

(resp., valid proofs) with probability 1). This is more appealing than allowing two-sided error, but the latter weaker notion is meaningful too. A second choice was to fix the error probability (i.e., probability of accepting far from valid objects), rather than introducing yet another parameter. Needless to say, the error probability can be reduced by sequential applications of the tester.

In the rest of this section, we consider an array of definitional issues. First, we consider two natural strengthenings of the definition of local testability (cf. §2.3.1). We next discuss the relation of local testability to property testing (cf. §2.3.2), and the relation of locally testable proofs to PCP of proximity (as defined in [15], cf. §2.3.3). Finally, we discuss the relation between locally testable codes and proofs (cf. §2.3.4), and the motivation for the study of *short* local testable codes and proofs (cf. §2.3.5).[11] Finally (in §2.3.6), we mention a weaker definition, which seem natural only in the context of codes.

2.3.1   Stronger Definitions

The definitions of testers presented so far, allow for the construction of a different tester for each relevant value of the proximity parameter. However, whenever such testers are actually constructed, they tend to be "uniform" over all relevant values of the proximity parameter. Thus, it is natural to present a single tester for all relevant values of the proximity parameter, provide this tester with the said parameter, allow it to behave accordingly, and measure its query complexity as a function of that parameter. For example, we may strengthen Definition 2.3, by requiring the existence of a function $q : (0, 1) \to \mathbb{N}$ and an oracle machine M such that, for every constant $\delta > 0$, all (sufficiently large) k and all $w \in \{0, 1\}^{n(k)}$, the following conditions hold:

1. On input $(1^k, \delta)$, machine M makes $q(\delta)$ queries.
2. If w is a codeword of C then $\Pr[M^w(1^k, \delta) = 1] = 1$.
3. If w is δ-far from $\{C(x) : x \in \{0, 1\}^k\}$ then $\Pr[M^w(1^k, \delta) = 1] \leq 1/2$.

An analogous strengthening applies to Definition 2.6. A special case of interest is when $q(\delta) = O(1/\delta)$. In this case, it makes sense to ask whether or not an even stronger "uniformity" condition may hold. Like in Definitions 2.1 and 2.2 (resp., Definitions 2.5 and 2.6), the tester M is not given the proximity parameter (and so its query complexity cannot depend on it), but we only require it to reject with probability proportional to the distance of the oracle from the relevant set. For example, we may strengthen Definition 2.3, by requiring the existence of an oracle machine M and a *constant* q such that, for every constant $\delta > 0$, every (sufficiently large) k and $w \in \{0, 1\}^{n(k)}$, the following conditions hold:

1. On input 1^k, machine M makes q queries.
2. If w is a codeword of C then $\Pr[M^w(1^k, \delta) = 1] = 1$.
3. If w is δ-far from $\{C(x) : x \in \{0, 1\}^k\}$ then $\Pr[M^w(1^k, \delta) = 1] < 1 - \Omega(\delta)$.

---

[11] The text of §2.3.5 is almost identical to a corresponding motivational text that appears in Part I.

2.3.2   Relation to Property Testing

Locally testable codes (and their corresponding testers) are essentially special cases of property testing algorithms, as defined in [57,38]. Specifically, the property being tested is membership in a predetermined code. The only difference between the definitions presented in Section 2.1 and the formulation that is standard in the property testing literature is that in the latter the tester is given the proximity parameter as input and determines its behavior (and in particular the number of queries) accordingly. This difference is eliminated in the first strengthening outlined in §2.3.1, while the second strengthening is related to the notion of proximity oblivious testing (cf. [39]). We note, however, that most of the property testing literature is concerned with "natural" objects (e.g., graphs, sets of points, functions) presented in a "natural" form rather than with objects designed artificially to withstand errors (i.e., codewords of error correcting codes).

Our general formulation of proof testing (i.e., Definition 2.6) can be viewed as a generalization of property testing. That is, we view the set Π_x as a set of objects having a certain x-dependent property (rather than as a set of valid proofs for some property of x). In other words, Definition 2.6 allows to consider properties that are parameterized by auxiliary information (i.e., x), whereas traditional property testing may be viewed as referring to the case that x only determines the length of strings in Π_x (e.g., $\Pi_x = \emptyset$ for every $x \notin \{1\}^*$ or, equivalently, $\Pi_x = \Pi_y$ for every $|x| = |y|$).[12]

2.3.3   Relation to PCPs of Proximity

Our definition of a locally testable proof is related but different from the definition of a PCP of proximity (appearing in [15]).[13] We start by reviewing the definition of a PCP of proximity.

Definition 2.7 (PCPs of Proximity): *A PCP of proximity for a set S with proximity parameter δ is a probabilistic* (non-adaptive) *polynomial-time oracle machine, denoted V, satisfying*

Completeness: *For every $x \in S$ there exists a string π_x such that V always accepts when given access to the oracle (x, π_x); that is, $\Pr[V^{x,\pi_x}(1^{|x|}) = 1] = 1$.*
Soundness: *For every x that is δ-far from $S \cap \{0,1\}^{|x|}$ and for every string π, machine V rejects with probability at least $\frac{1}{2}$ when given access to the oracle (x, π); that is, $\Pr[M^{x,\pi}(1^{|x|}) = 1] \leq 1/2$.*

The query complexity of V is defined as in case of PCP, but here also queries to the x-part are counted.

The oracle (x, π) is actually a concatenation of two oracles: the input-oracle x (which replaces an explicitly given input in the definitions of PCPs and locally testable proofs), and a proof-oracle π (exactly as in the prior definitions). Note

---

[12] In fact, in the context of property testing, the length of the oracle must always be given to the tester (although some sources neglect to state this fact).

[13] We mention that PCPs of proximity are almost identical to Assignment Testers, defined independently by Dinur and Reingold [28]. Both notions are (important) special cases of the general definition of a "PCP spot-checker" formulated before in [30].

that Definition 2.7 refers to the distance of the input-oracle to S, whereas locally testable proofs refer to the distance of the proof-oracle from the set Π_x of valid proofs of membership of $x \in S$.

Still, PCPs of proximity can be defined within the framework of locally testable proofs. Specifically, consider an extension of Definition 2.6, where (relative) distances are measured according to a weighted Hamming distance; that is, for a weight function $\omega : \{1, ..., n\} \rightarrow [0, 1]$ and $u, v \in \{0, 1\}^n$, we let $\delta_\omega(u, v) = \sum_{i=1}^n \omega(i) \cdot \Delta(u_i, v_i)$. (Indeed, the standard notion of relative distance between $u, v \in \{0, 1\}^n$ is obtained by $\delta_\omega(u, v)$ when using the uniform weighting function (i.e., $\omega(i) = 1/n$ for every $i \in \{1, ..., n\}$).) Now, Definition 2.7 can be viewed as a special case of (the extended) Definition 2.6 when applied to the (rather artificial) set of proofs $\Pi_{1^n} = \{(x, \pi) : x \in S \cap \{0, 1\}^n \wedge \pi \in \Pi'_x\}$, where $\Pi'_x = \{\pi : \Pr[V^{x,\pi}(1^{|x|}) = 1] = 1\}$, by using the weighted Hamming distance δ_ω for ω that is uniform on the input-part of the oracle; that is, for $(x, \pi), (x', \pi') \in \{0, 1\}^{n+p}$, we use $\delta_\omega((x, \pi), (x', \pi')) \stackrel{\text{def}}{=} \Delta(x, x')/n$, which corresponds to $\omega(i) = 1/n$ if $i \in \{1, ..., n\}$ and $\omega(i) = 0$ otherwise. Alternatively, weights can be approximately replaced by repetitions (provided that the tester checks the consistency of the repetitions).[14]

We mention that PCPs of proximity (of constant query complexity) yield a simple way of obtaining locally testable codes. More generally, we can combine any code C_0 with any PCP of proximity V, and obtain a q-locally testable code with distance essentially determined by C_0 and rate determined by V, where q is the query complexity of V. Specifically, x will be encoded by appending $c = C_0(x)$ by a proof that c is a codeword of C_0, and distances will be determined by the weighted Hamming distance that assigns uniform weights to the first part of the new code. As in the previous paragraph, these weights can be implemented by making suitable repetitions.

Finally, we comment that the definition of a PCP of proximity can be extended by providing the verifier with part of the input in an explicit form. That is, referring to Definition 2.7, we let $x = (x', x'')$, and provide V with explicit input $(x', 1^{|x|})$ and input-oracle x'' (rather than with explicit input $1^{|x|}$ and input-oracle x). Clearly, the extended formulation implies PCP as a special case (i.e., $x'' = \lambda$). More interestingly, an extended PCP of proximity for a set of pairs R (e.g., the witness relation of an NP-set), yields a PCP for the set $S \stackrel{\text{def}}{=} \{x' : \exists x'' \text{ s.t. } (x', x'') \in R\}$.

---

[14] That is, given a verifier V as in Definition 2.7, and denoting by n and $p = p(n)$ the sizes of the two parts of its oracle, we consider proofs of length $t \cdot n + p$, where $t = p/o(n)$ (e.g., $t = (p/n) \cdot \log n$). We consider a verifier V' with syntax as in Definition 2.6 that, on input 1^n and oracle access to $w = (u_1, ..., u_t, v) \in \{0, 1\}^{t \cdot n + p}$, where $u_i \in \{0, 1\}^n$ and $v \in \{0, 1\}^p$, selects uniformly $i \in \{1, ..., t\}$ and invokes $V^{u_i, v}(1^n)$. In addition, V' performs a number of repetition tests that is inversely proportional to the proximity parameter, where in each test V' selects uniformly $i, i' \in \{1, ..., t\}$ and $j \in \{1, ..., n\}$ and checks that u_i and $u_{i'}$ agree on their j-th bit. Thus, V' essentially emulates the PCP of proximity V, and the fact that V satisfies Definition 2.7 can be captured by saying that V' satisfies Definition 2.6.

2.3.4   Relating Locally Testable Codes and Proofs

Locally testable codes can be thought of as the combinatorial counterparts of the complexity theoretic notion of locally testable proofs (PCPs). This perspective raises the question of whether one of these notions implies (or is useful towards the understanding of) the other.

Do PCPs imply locally testable codes?. The use of codes with features related to local testability is implicit in known PCP constructions. Furthermore, the known constructions of locally testable proofs (PCPs) provides a transformation of *standard proofs* (for say SAT) to *locally testable proofs* (i.e., PCP-oracles), such that transformed strings are accepted with probability one by the PCP verifier. Specifically, denoting by S_x the set of standard proofs referring to an assertion x, there exists a polynomial-time mapping f_x of S_x to $R_x \stackrel{\text{def}}{=} \{f_x(y) : y \in S_x\}$ such that for every $\pi \in R_x$ it holds that $\Pr[V^\pi(x) = 1] = 1$, where V is the PCP verifier. Moreover, starting from different standard proofs, one obtains locally testable proofs that are far apart, and hence constitute a good code (i.e., for every x and every $y \neq y' \in S_x$, it holds that $\Delta(f_x(y), f_x(y')) \geq \Omega(|f_x(y)|)$). It is tempting to think that the PCP verifier yields a codeword tester, but this is not really the case. Note that Definition 2.5 requires rejection of strings that are far from any valid proof (i.e., any string far from Π_x), but it is not clear that the only valid proofs (w.r.t V) are those in R_x (i.e., the proofs obtained by the transformation f_x of standard proofs (in S_x) to locally testable ones).[15] In fact, the standard PCP constructions accept also valid proofs that are not in the range of the corresponding transformation (i.e., f_x); that is, Π_x as in Definition 2.5 is a strict subset of R_x (rather than $\Pi_x = R_x$). We comment that most known PCP constructions can be (non-trivially)[16] modified to yield $\Pi_x = R_x$, and thus to yield a locally testable code (but this is not necessarily the best way to design locally testable codes, see one alternative in §2.3.3).

Do locally testable codes imply PCPs?. Saying that locally testable codes are the combinatorial counterparts of locally testable proofs (PCPs), raises the expectation (or hope) that it would be easier to construct locally testable codes than it is to construct PCPs. The reason being that combinatorial objects (e.g., codes) should be easier to understand than complexity theoretic ones (e.g., PCPs). Indeed, this feeling was among the main motivations of Goldreich and Sudan, and their first result (cf. [41, Sec. 3]) was along this vein: They showed a relatively simple construction (i.e., simple in comparison to PCP constructions) of a locally testable code of length $\ell(k) = k^c$ for any constant $c > 1$. Unfortunately, their stronger result, providing a locally testable code of shorter length (i.e., length $\ell(k) = k^{1+o(1)}$) is obtained by constructing (cf. [41, Sec. 4]) and using

---

[15] Let alone that Definition 2.4 refers only to the case of false assertions, in which case all strings are far from a valid proof (which does not exist).

[16] The interested reader is referred to [41, Sec. 5.2] for a discussion of typical problems that arise.

(cf. [41, Sec. 5]) a corresponding locally testable proof (i.e., PCP). Subsequent works have mostly followed this route, with the notable exception of Meir's work [51], which provides a combinatorial construction of a locally testable code that does not seem to yield a corresponding locally testable proof.[17]

2.3.5 Motivation for the Study of Short Locally Testable Codes and Proofs

Local testability offers an extremely strong notion of efficient testing: The tester makes only a constant number of bit probes, and determining the probed locations (as well as the final decision) is typically done in time that is polylogarithmic in the length of the probed object. Recall that the tested object is supposed to be related to some primal object; in the case of codes, the probed object is supposed to encode the primal object, whereas in the case of proofs the probed object is supposed to help verify some property of the primal object. In both cases, the length of the secondary (probed) object is of natural concern, and this length is stated in terms of the length of the primary object.

The length of codewords in an error-correcting code is widely recognized as one of the two most fundamental parameters of the code (the second one being the code's distance). In particular, the length of the code is of major importance in applications, because it determines the overhead involved in encoding information.

As argued in Section 1, the same considerations apply also to proofs. However, in the case of proofs, this obvious point was blurred by the indirect, unexpected and highly influential applications of PCPs to the theory of approximation algorithms. In our view, the significance of locally testable proofs (or PCPs) extends far beyond their applicability to deriving non-approximability results. The mere fact that proofs can be transformed into a format that supports super-fast probabilistic verification is remarkable. From this perspective, the question of how much redundancy is introduced by such a transformation is a fundamental one. Furthermore, locally testable proofs (i.e., PCPs) have been used not only to derive non-approximability results but also for obtaining positive results (e.g., CS-proofs [48,53] and their applications [8,24]), and the length of the PCP affects the complexity of those applications.

Turning back to the celebrated application of PCP to the study of approximation algorithms, we note that the length of PCPs is also relevant to non-approximability results; specifically, the length of PCPs affects the *tightness with respect to the running time* of the non-approximability results derived from these PCPs. For example, suppose (exact) SAT has complexity $2^{\Omega(n)}$. The original PCP Theorem [5,4] only implies that approximating MaxSAT requires time $2^{n^{\alpha}}$, for some (small) $\alpha > 0$. The work of [55] makes α arbitrarily close to 1, whereas

---

[17] We mention that the prior work of Ben-Sasson and Sudan [20] also shows some deviation from this route (i.e., it reversed the course to the "right one"): First codes are constructed, and next they are used towards the construction of proofs (rather than the other way around).

the results of [41,21] further improve the lower bound to $2^{n^{1-o(1)}}$ and the results of [20,26] yields a lower bound of $2^{n/\text{poly}(\log n)}$. We mention that the result of [54] (cf. [27]) allows to achieve the lower bound of $2^{n^{1-o(1)}}$ simultaneously with optimal approximation ratios, but this is currently unknown for the better lower bound of $2^{n/\text{poly}(\log n)}$.

2.3.6   A Weaker Definition

One of the concrete motivations for local testable codes refers to settings in which one may want to re-encode the information when discovering that the codeword is corrupted. In such a case, assuming that re-encoding is based solely on the corrupted codeword, one may assume (or rather needs to assume) that the corrupted codeword is not too far from the code. Thus, the following version of Definition 2.1 may suffice for various applications.

Definition 2.8 (weak codeword tests): *Let* $C : \{0,1\}^k \to \{0,1\}^n$ *be a code of distance* d, *and let* $q \in \mathbb{N}$ *and* $\delta_1, \delta_2 \in (0,1)$ *be such that* $\delta_1 < \delta_2$. *A* weak q-local (codeword) (δ_1, δ_2)-tester *for* C *is a probabilistic* (non-adaptive) *oracle machine* M *that makes at most* q *queries, accepts any codeword, and rejects non-codewords that are both* δ_1-far *and* δ_2-close *to* C. *That is, the rejection condition of Definition 2.1 is modified as follows.*

Rejection of non-codeword (weak version): *For any* $w \in \{0,1\}^n$ *such that* $\Delta_C(w) \in [\delta_1 n, \delta_2 n]$, *given oracle access to* w, *machine* M *rejects with probability at least* $1/2$.

Needless to say, there is something highly non-intuitive in this definition: It requires rejection of non-codewords that are somewhat far from the code, but not the rejection of codewords that are very far from the code. Still, such weak codeword testers may suffice in some applications. Interestingly, such weak codeword testers do exist and even achieve linear length (cf. [58, Chap. 5]). We note that the non-monotonicity of the rejection probability of testers has been observed before, the most famous example being linearity testing (cf. [22] and [10]).

2.4   A Confused History

There is a fair amount of confusion regarding credits for some of the definitions presented in this section.[18] We refer mainly to the definition of locally testable codes. This definition (or at least a related notion)[19] is arguably implicit in [7]

---

[18] Some confusion exists also with respect to some of the results and constructions described in Section 3, but in comparison to what will be discussed here the latter confusion is minor.

[19] The related notion refers to the following relaxed notion of codeword testing: For two fixed good codes $C_1 \subseteq C_2 \subset \{0,1\}^n$, one has to accept (with high probability) every codeword of C_1, but reject (with high probability) every string that is far from being a codeword of C_2. Indeed, our definitions refer to the special (natural) case that $C_2 = C_1$, but the more general case suffices for the construction of PCPs (and is implicitly achieved in most of them).

as well as in subsequent works on PCP (see §2.3.4). Furthermore, the definition of locally testable codes has appeared independently in the works of Friedl and Sudan [34] and Rubinfeld and Sudan [57] as well as in the PhD Thesis of Arora [3].

3   Results and Ideas

We review the known constructions of locally testable codes and proofs, starting from codes and proofs of exponential length and concluding with codes and proofs of nearly linear length. We mention that random linear codes (of linear length) require any codeword tester to read a linear number of bits of the codeword [18], providing an indication to the non-triviality of local testability.

3.1   The Mere Existence of Locally Testable Codes and Proofs

The mere existence of locally testable codes and proofs, regardless of their length, is non-obvious. Thus, we start by recalling the simplest constructions known.

3.1.1   The Hadamard Code Is Locally Testable

The simplest example of a locally testable code (of constant relative distance) is the Hadamard code. This code, denoted C_{Had}, maps $x \in \{0,1\}^k$ to a string, of length $n = 2^k$, that provides the evaluation of all GF(2)-linear functions at x; that is, the coordinates of the codeword are associated with linear functions $\ell(z) = \sum_{i=1}^k \ell_i z_i$ and so $C_{Had}(x)_\ell = \ell(x) = \sum_{i=1}^k \ell_i x_i$. Testing whether a string $w \in \{0,1\}^{2^k}$ is a codeword amounts to linearity testing. This is the case because w is a codeword of C_{Had} if and only if, when viewed as a function $w : \{0,1\}^k \to \{0,1\}$, it is linear (i.e., $w(z) = \sum_{i=1}^k c_i z_i$ for some c_i's, or equivalently $w(y+z) = w(y) + w(z)$ for all y, z). Specifically, local testability is achieved by uniformly selecting $y, z \in \{0,1\}^k$ and checking whether $w(y+z) = w(y) + w(z)$. The exact analysis of this natural tester, due to Blum, Luby and Rubinfeld [22], turned out to be highly complex (cf. [22,6,31,12,13,10,46]). Denoting by $\mathrm{rej}(w)$ the probability that the test rejects the string w and by $R(\delta)$ be the minimum of $\mathrm{rej}(w)$ taken over all strings that are at distance $\delta \cdot |w|$ from C_{Had}, it is known that $R(\delta) \geq \Gamma(\delta)$, where the function $\Gamma : [0, 0.5] \to [0,1]$ is defined as follows:

$$\Gamma(x) \overset{\text{def}}{=} \begin{cases} 3x - 6x^2 & 0 \leq x \leq 5/16 \\ 45/128 & 5/16 \leq x \leq \tau_2 \text{ where } \tau_2 \approx 44.9962/128 \\ x + \delta(x) & \tau_2 \leq x \leq 1/2, \\ \quad \text{where } \delta(x) \overset{\text{def}}{=} 1376x^3(1 - 2x)^{12}. \end{cases} \tag{1}$$

The lower bound Γ is composed of three different bounds with "phase transitions" at $x = \frac{5}{16}$ and at $x = \tau_2$ (where $\tau_2 \approx \frac{44,9962}{128}$ is the solution to $x + \delta(x) = 45/128$).[20] It was shown in [10] that the first segment of this bound

---

[20] The third segment is due to [46], which improves over the prior bound of [10] that asserted $R(x) \geq \max(45/128, x)$ for every $x \in [5/16, 1/2]$.

(i.e., for $x \in [0, 5/16]$) is the best possible, and that the first "phase transitions" (i.e., at $x = \frac{5}{16}$) is indeed a reality; in other words, $R = \Gamma$ in the interval $[0, 5/16]$.[21] We highlight the fact that the detection probability of the aforementioned test does not increase monotonically with the distance (of the string from the code), since Γ decreases in the interval $[1/4, 5/16]$ (while equaling R in this interval).

Other codes. We mention that Reed-Muller Codes of constant order are also locally testable [1]. These codes have sub-exponential length, but are quite popular in practice. The Long Code is also locally testable [11], but this code has double-exponential length (and was introduced merely for the design of PCPs).[22]

3.1.2   The Hadamard-Based PCP of [4]

The simplest example of a locally testable proof (for a set not known to be in $\mathcal{BPP}$) is the "inner verifier" of the PCP construction of Arora, Lund, Motwani, Sudan and Szegedy [4], which in turn is based on the Hadamard code. Specifically, proofs of the satisfiability of a given system of quadratic equations over $GF(2)$ are presented by providing a Hadamard encoding of the outer-product of a satisfying assignment with itself (i.e., a satisfying assignment $\alpha \in \{0,1\}^n$ is presented by $C_{Had}(\beta)$, where $\beta = (\beta_{i,j})_{i,j \in [n]}$ and $\beta_{i,j} = \alpha_i \alpha_j$). Given an alleged proof $\pi \in \{0,1\}^{2^{n^2}}$, the proof-tester proceeds as follows:

1. Tests that π is indeed a codeword of the Hadamard Code. If the test passes then w is close to some $C_{Had}(\beta)$, for an arbitrary $\beta = (\beta_{i,j})_{i,j \in [n]}$.
2. Tests that the aforementioned β is indeed an outer-product of some $\alpha \in \{0,1\}^n$ with itself. Note that the Hadamard encoding of α is supposed to be part of the Hadamard encoding of β (because $\sum_{i=1}^{n} c_i \alpha_i = \sum_{i=1}^{n} c_i \alpha_i^2$ is supposed to equal $\sum_{i=1}^{n} c_i \beta_{i,i}$). So we would like to test that the latter codeword matches the former one. Specifically, we wish to test whether $(\beta_{i,j})_{i,j \in [n]}$ equals $(\alpha_i \alpha_j)_{i,j \in [n]}$ (i.e., the equality of two matrices). This can be done by uniformly selecting $(r_1, ..., r_n), (s_1, ..., s_n) \in \{0,1\}^n$, and comparing $\sum_{i,j} r_i s_j \beta_{i,j}$ and $\sum_{i,j} r_i s_j \alpha_i \alpha_j = (\sum_i r_i \alpha_i)(\sum_j s_j \alpha_j)$.

   The above would have been fine if $w = C_{Had}(\beta)$, but we only know that w is close to $C_{Had}(\beta)$. The Hadamard encoding of α is a tiny part of the latter, and so we should not try to retrieve the latter directly (because this tiny part may be totally corrupted). Instead, we use the paradigm of self-correction (cf. [22]): In general, for any fixed $c = (c_{i,j})_{i,j \in [n]}$, whenever we wish to retrieve $\sum_{i=1}^{n} c_{i,j} \beta_{i,j}$, we uniformly select $r = (r_{i,j})_{i,j \in [n]}$ and retrieve both $w(r)$ and $w(r + c)$. Thus, we obtain a self-corrected value of $w(c)$; that is, if w is δ-close to $C_{Had}(\beta)$ then $w(r + c) - w(r) = \sum_{i=1}^{n} c_{i,j} \beta_{i,j}$ with probability at least $1 - 2\delta$ (over the choice of r).

---

[21] In contrast, the lower bound provided by the other two segments (i.e., for $x \in [5/16, 1/2]$) is unlikely to be tight, and in particular it is unlikely that the "phase transitions" at $x = \tau_2$ represents the behavior of R itself. Also note that $\delta(x) > 59(1 - 2x)^{12}$ for every $x > \tau_2$, but $\delta(x) < 0.0001$ for every $x < 1/2$.

[22] Interestingly, the best results are obtained by using a relaxed notion of local testability [43,44].

Using self-correction, we indirectly obtain bits in $C_{Had}(\alpha)$, for $\alpha = (\alpha_i)_{i \in [n]} = (\beta_{i,i})_{i \in [n]}$. Similarly, we can obtain any other desired bit in $C_{Had}(\beta)$, which in turn allows us to test whether $(\beta_{i,j})_{i,j \in [n]} = (\alpha_i \alpha_j)_{i,j \in [n]}$. In fact, we are checking whether $(\beta_{i,j})_{i,j \in [n]} = (\beta_{i,i} \beta_{j,j})_{i,j \in [n]}$, by comparing $\sum_{i,j} r_i s_j \beta_{i,j}$ and $(\sum_i r_i \beta_{i,i})(\sum_j s_j \beta_{j,j})$, for randomly selected $(r_1, ..., r_n)$, $(s_1, ..., s_n) \in \{0,1\}^n$.

3. Finally, we need to check whether the aforementioned α satisfies the given system of equations. Towards this end, we uniformly selects a linear combination of the equations, and check whether α satisfies the resulting (single) equation. Note that the value of the corresponding linear expression (in quadratic (and linear) forms) appears as a bit of the Hadamard encoding of β, but again we retrieve it from w by using self correction.

One key observation underlying the analysis of Steps 2 and 3 is that for $(u_1, ..., u_n) \neq (v_1, ..., v_n) \in \{0,1\}^n$, if we uniformly select $(r_1, ..., r_n) \in \{0,1\}^n$ then $\Pr[\sum_i r_i u_i = \sum_i r_i v_i] = 1/2$. Similarly, for n-by-n matrices $A \neq B$, when $r, s \in \{0,1\}^n$ are uniformly selected (vectors), it holds that $\Pr[As = Bs] = 2^{-\text{rank}(A-B)}$ and it follows that $\Pr[rAs = rBs] \leq 3/4$.

3.2 Locally Testable Codes and Proofs of Polynomial Length

The constructions presented in Section 3.1 have exponential length in terms of the relevant parameter (i.e., the amount of information being encoded in the code or the length of the assertion being proved). Achieving local testability by codes and proofs that have polynomial length turns out to be more challenging.

3.2.1 Locally Testable Codes of Quadratic Length

A direct interpretation of *low-degree tests* (cf. [6,7,35,57,34]), proposed by Friedl and Sudan [34] and Rubinfeld and Sudan [57], yields a locally testable code of quadratic length over a *sufficiently large alphabet*. Similar (and actually better) results for *binary* codes required additional ideas, and have appeared only later (cf. [41]). We sketch both constructions below, starting with locally testable codes over very large alphabets (which are defined analogously to the binary case).

We will consider a code $C : \Sigma^k \to \Sigma^n$ of linear distance, with $|\Sigma| \gg k$ and $n > k^2$. For parameters $m \ll d < \log k$ (such that $k < d^m$), consider a finite field F of size $O(d)$ and an alphabet $\Sigma = F^{d+1}$ (see below).[23] Viewing the information as an m-variant polynomial p of total degree d over F, we encode it by providing its value on all possible lines over F^m, where each such line is defined by two points in F^m. Actually, the value of p on such a line can be represented by a univariate polynomial of degree d. Thus, the code maps $\log_2 |F|^{\binom{m+d}{d}} > (d/m)^m \log |F|$ bits of information (which may be viewed as $k \stackrel{\text{def}}{=} (d/m)^m/(d+1) \approx d^{m-1}/m^m$ long sequences over $\Sigma = F^{d+1}$) to sequences

---

[23] Indeed, it would have been more natural to present the code as a mapping from sequences over F to sequences over $\Sigma = F^{d+1}$. Following the convention of using the same alphabet for both the information and the codeword, we just pack every $d+1$ elements of F as an element of Σ.

of length $n \stackrel{\text{def}}{=} |F|^{2m} = O(d)^{2m}$ over Σ. Note that the smaller m, the better the rate (i.e., relation of n to k) is, but this comes at the expense of using a larger alphabet. In particular, we consider two instantiations:

1. Using $d = m^m$, we get $k \approx m^{m^2 - 2m}$ and $n = m^{2m^2 + o(m)}$, which yields $n \approx \exp(\sqrt{\log k}) \cdot k^2$ and $\log |\Sigma| = \log |F|^{d+1} \approx d \log d \approx \exp(\sqrt{\log k})$.
2. Letting $d = m^c$ for any constant $c > 1$, we get $k \approx m^{(c-1)m}$ and $n = m^{2cm + o(m)}$, which yields $n \approx k^{2c/(c-1)}$ and $\log |\Sigma| \approx d \log d \approx (\log k)^c$.

As for the codeword tester, it uniformly selects two intersecting lines and checks that the corresponding univariate polynomials agree on the point of intersection. Thus, this tester makes two queries (to an oracle over the alphabet Σ). The analysis of this tester reduces to the analysis of the corresponding low degree test, undertaken in [4,55].

The above tester uses only two queries, but the entire description (which refers to codes over a large alphabet) deviates from the bulk of our treatment, which has focused on a binary alphabet. We comment that 2-query locally testable *binary* codes are essentially impossible (cf., [14]), but we have already seem that 3-query tests are possible. A natural way of reducing the alphabet size of codes is via the well-known paradigm of *concatenated codes* [32].[24] However, local testability can be maintained only in special cases. In particular, observe that, for each of the two queries made by the tester of C, the tester does not need the entire polynomial represented in $\Sigma = F^{d+1}$, but rather only its value at a specific point. Thus, encoding Σ by an error correcting code that supports recovery of the said value while using a constant number of probes will do.[25] In particular, for integers h, e such that $d + 1 = h^e$, Goldreich and Sudan used an encoding of $F^{d+1} = F^{h^e}$ by sequences of length $|F|^{eh}$ over F, and provided a testing and recovery procedure that makes $O(e)$ queries [41, Sec. 3.3]. We mention that the case of $e = 1$ and $|F| = 2$ corresponds to the Hadamard code, and that a bigger constant e allow for shorter codes. The resulting concatenated code, C′, is a locally testable code over F, and has length $n \cdot O(d)^{eh} = n \cdot \exp((e \log d) \cdot d^{1/e})$. Using constant $e = 2c$ and setting $d = m^c \approx (\log k)^c$, we get $n \approx k^{2c/(c-1)} \cdot \exp(\widetilde{O}(\log k)^{1/2})$ and $|F| = \text{poly}(\log k)$. Finally, a *binary* locally testable code is obtained by concatenating C′ with the Hadamard code, while noting that the latter supports a "local recovery" property that suffices to emulate the tester for C′. In particular, the tester of C′ merely checks a linear (over F) equation

---

[24] A concatenated code is obtained by encoding the symbols of an "outer code" (using the coding method of the "inner code"). Specifically, let $C_1 : \Sigma_1^{k_1} \to \Sigma_1^{n_1}$ be the outer code and $C_2 : \Sigma_2^{k_2} \to \Sigma_2^{n_2}$ be the inner code, where $\Sigma_1 \equiv \Sigma_2^{k_2}$. Then, the concatenated code $C : \Sigma_2^{k_1 k_2} \to \Sigma_2^{n_1 n_2}$ is obtained by $C(x_1, ..., x_{k_1}) = (C_2(y_1), ..., C_2(y_{n_1}))$, where $x_i \in \Sigma_2^{k_2} \equiv \Sigma_1$ and $(y_1, ..., y_{n_1}) = C_1(x_1, ..., x_{k_1})$. Using a good inner code for relatively short sequences, allows to transform good codes for a large alphabet into good codes for a smaller alphabet.

[25] Indeed, this property is related to locally decodable codes, to be discussed in Section 4. Here we need to recover one out of $|F|$ specific linear combinations of the encoded $(d + 1)$-long sequence of F-symbols. In contrast, locally decodable refers to recovering one out of the original F-symbols of the $(d + 1)$-long sequence.

referring to a constant number of F-elements, and for $F = GF(2^\ell)$, this can be emulated by checking *related* random linear combinations of the bits representing these elements, which in turn can be locally recovered (or rather self-corrected) from the Hadamard code. The final result is a locally testable (binary) code of nearly quadratic length.[26]

3.2.2   Locally Testable Proofs of Polynomial Length: The PCP Theorem

The case of proofs is far more complex: Achieving locally testable proofs of polynomial length is essentially the contents of the celebrated PCP Theorem of Arora, Lund, Motwani, Safra, Sudan and Szegedy [5,4]. The construction is analogous to (but far more complex than) the one presented in the case of codes:[27] First one constructs proofs over a large alphabet, and next one composes such proofs with corresponding "inner" proofs (over a smaller alphabet, and finally a binary one). Our exposition focuses on the construction of these proof systems and blurs the issues involved in their composition.[28]

The first step is to introduce the following NP-complete problem. The input to the problem consists of a finite field F, a subset $H \subset F$ of size $\lfloor |F|^{1/15} \rfloor$, an integer $m < |H|$, and a $(3m + 4)$-variant polynomial $P : F^{3m+4} \to F$ of total degree $3m|H| + O(1)$. The problem is to determine whether there exists an m-variant ("assignment") polynomial $A : F^m \to F$ of total degree $m|H|$ such that $P(x, z, y, \tau, A(x), A(y), A(z)) = 0$ for every $x, y, z \in H^m$ and $\tau \in \{0,1\}^3 \subset H$. Note that the problem-instance can be explicitly described by a sequence of $|F|^{3m+4} \log_2 |F|$ bits, whereas the solution sought can be explicitly described by a sequence of $|F|^m \log_2 |F|$ bits. We comment that the NP-completeness of the aforementioned problem can be proved via a reduction from 3SAT, by identifying the variables of the formula with H^m and essentially letting P be a low-degree extension of a function $f : H^{3m} \times \{0,1\}^3 \to \{0,1\}$ that encodes the structure of the formula (by considering all possible 3-clauses). In fact, the resulting P has degree $|H|$ in each of the first $3m$ variables and constant degree in each of the other variables, and this fact can be used to improve the parameters below (but not in a fundamental way).

The proof that a given input P satisfies the aforementioned condition consists of an m-variant polynomial $A : F^m \to F$ (which is supposed to be of total degree $m|H|$) as well as $3m + 4$ auxiliary polynomials $A_i : F^{3m+1} \to F$, for $i = 1, ..., 3m + 1$ (each supposedly of degree $(3m|H| + O(1)) \cdot m|H|$). The polynomial A is supposed to satisfy the conditions of the problem, and in particular $P(x, z, y, \tau, A(x), A(y), A(z)) = 0$ should hold for every $x, y, z \in H^m$ and

---

[26] Actually, the aforementioned result is only implicit in [41], because Goldreich and Sudan apply these ideas directly to a truncated version of the low-degree based code.

[27] Our presentation reverses the historical order in which the corresponding results (for codes and proofs) were achieved. That is, the constructions of locally testable proofs of polynomial length predated the coding counterparts.

[28] This section is significantly more complex than the rest of this article, and some readers may prefer to skip it and proceed directly to Section 3.3. For further details regarding the proof composition paradigm, the reader is referred to [37, Sec. 9.3.2].

$\tau \in \{0,1\}^3 \subset H$. Furthermore, $A_0(x,z,z,\tau) \stackrel{\text{def}}{=} P(x,z,y,\tau,A(x),A(y),A(z))$ should vanish on H^{3m+1}. The auxiliary polynomials are given to assist the verification of the latter condition. In particular, it should be the case that A_i vanishes on $F^i H^{3m+1-i}$, a condition that is easy to test for A_{3m+1} (assuming that A_{3m+1} is a low degree polynomial). Checking that A_{i-1} agrees with A_i on $F^{i-1} H^{3m+1-(i-1)}$, for $i = 1, ..., 3m+1$, and that all A_i's are low degree polynomials, establishes the claim for A_0. Thus, testing an alleged proof $(A, A_1, ..., A_{3m+1})$ is performed as follows:

1. Testing that A is a polynomial of total degree $m|H|$. This is done by selecting a random line through F^m, and testing whether A restricted to this line agrees with a degree $m|H|$ univariate polynomial.

2. Testing that, for $i = 1, ..., 3m+1$, the polynomial A_i is of total degree $d \stackrel{\text{def}}{=} (3m|H|+O(1)) \cdot m|H|$. Here we select a random line through F^{3m+1}, and test whether A_i restricted to this line agrees with a degree d univariate polynomial.

3. Testing that, for $i = 1, ..., 3m+1$, the polynomial A_i agrees with A_{i-1} on $F^{i-1} H^{3m+1-(i-1)}$. This is done by uniformly selecting $r' = (r_1, ..., r_{i-1}) \in F^{i-1}$ and $r'' = (r_{i+1}, ..., r_{3m+1}) \in F^{3m+1-i}$, and comparing $A_{i-1}(r', e, r'')$ to $A_i(r', e, r'')$, for every $e \in H$. In addition, we check that both functions when restricted to the axis-parallel line $(r', \cdot, r'')$ agree with a univariate polynomial of degree d.[29] We stress that the values of A_0 are computed according to the given polynomial P by accessing A at the appropriate locations (i.e., by definition $A_0(x,z,z,\tau) = P(x,z,y,\tau,A(x),A(y),A(z))$).

4. Testing that A_{3m+1} vanishes on F^{3m+1}. This is done by uniformly selecting $r \in F^{3m+1}$, and testing whether $F(r) = 0$.

The above description (which follows [59, Apdx. C]) is somewhat different than the original presentation in [4], which in turn follows [6,7,31].[30] The above tester may be viewed as making $O(m|F|)$ queries to an oracle over the alphabet F, or alternatively, as making $O(m|F| \log |F|)$ binary queries.[31] Note that we have already obtained a highly non-trivial tester. It makes $O(m|F| \log |F|)$ queries in order to verify a claim regarding an input of length $n \stackrel{\text{def}}{=} |F|^{3m+4} \log_2 |F|$. Using $m = \log n / \log \log n$, $|H| = \log n$ and $|F| = \text{poly}(\log n)$, we have obtained a tester of poly-logarithmic query complexity.

To further reduce the query complexity, one invokes the "proof composition" paradigm, introduced by Arora and Safra [5]. Specifically, one composes an "outer" tester (as described above) with an "inner" tester that checks the residual condition that the "outer" tester determines for the answers it obtains. This composition is more problematic than one suspects, because we wish the "inner"

---

[29] Thus, effectively, we are self-correcting the values at H (on the said line), based on the values at F (on that line).

[30] The point is that the sum-check, which originates in [50], is replaced by an analogous process (which happens to be non-adaptive).

[31] Another alternative perspective is obtained by applying so-called parallelization (cf. [49,4]). The result is a test making a constant number of queries that are each answered by strings of length $\text{poly}(|F|)$.

tester to perform its task without reading its entire input (i.e., the answers to the "outer" tester). This seems quite paradoxical, since it is not clear how the "inner" tester can operate without reading its entire input. The problem can be resolved by using a "proximity tester" (i.e., a PCP of proximity) as an "inner" tester, provided that it suffices to have such a proximity test (for the answers to the "outer" tester). Thus, the challenge is to reach a situation in which the "outer" tester is robust in the sense that, when the assertion is false, the answers obtained by this tester are far from being convincing (i.e., they are far from any sequence of answers that is accepted by this tester). Two approaches towards obtaining such robust testers are known.

- One approach, introduced in [4], is to convert the "outer" tester into one that makes a constant number of queries over some larger alphabet, and furthermore have the answer be presented in an error correcting format. Thus, robustness is guaranteed by the fact that the answers correspond to a constant-length sequence of codewords, and so any two (properly formatted) sequences are at constant relative distance of one another.

    The implementation of this approach consists of two steps (and is based on some specifics). The first step is to convert the "outer" tester into one that makes a constant number of queries over some larger alphabet. This step uses the so-called parallelization technique (cf. [49,4]). Next, one applies an error correcting code to these $O(1)$ longer answers, and assumes that the "proximity tester" can handle inputs presented in this format (i.e., that it can test an input that is presented by an encoding of a constant number of its parts).[32]

- An alternative approach, pursued and advocated in [15], is to take advantage of the specific structure of the queries, "bundle" the answers together and furthermore show that the "bundled" answers are "robust" in a sense that fits proximity testing. In particular, the (generic) parallelization step is avoided, and is replaced by a closer analysis of the specific (outer) tester. We will demonstrate this approach next.

First, we show how the queries of the aforementioned tester can be "bundled" (into a constant number of bundles). In particular, we consider the following "bundling" that accommodates all types of tests (and in particular the $m + 1$ different sub-tests performed in Steps 2 and 3). Consider

$$B(x_1, ..., x_{3m+1}) = (A_1(x_1, x_2, ..., x_{3m+1}), A_2(x_2, ..., x_{3m+1}, x_1), ..., A_{3m+1}$$
$$(x_{3m+1}, x_1, ..., x_{3m}))$$

and perform all $3m + 1$ tests of Step (3) by selecting uniformly $(r_2, ..., r_{3m+1}) \in F^{3m}$ and querying B at $(e, r_2, ..., r_{3m+1})$ and $(r_{3m+1}, e, ..., r_{3m})$ for all $e \in F$. Thus, all $3m + 1$ tests of Step (3) can be performed by retrieving the values of

---

[32] The aforementioned assumption holds trivially in case one uses a generic "proximity tester" (i.e., a PCP of proximity or an Assignment Tester) as done in [28]. But the aforementioned approach can be (and was in fact originally) applied with a specific "proximity tester" that can only handle inputs presented in one specific format (cf. [4]).

B on a single *axis parallel* random line through F^{3m+1}. Furthermore, note that all $3m + 1$ tests of Step (2) can be performed by retrieving the values of B on a single (arbitrary) random line through F^{3m+1}. Finally, observe that these tests are "robust" in the sense that if, for some i, the function A_i is (say) 0.01-far from satisfying the condition (i.e., being low-degree or agreeing with A_{i-1}) then with constant probability many of the values of A_i on an appropriate random line will not fit to what is needed. This robustness property is inherited by B, as well as by B' (resp., A') that is obtained by applying a good binary error-correcting code on B (resp., on A). Thus, we may replace A and the A_i's by A' and B', and conduct all all tests by making $O(m^2|F|\log|F|)$ queries to $A' : F^m \times [O(\log|F|)] \to \{0,1\}$ and $B' : F^{3m+1} \times [O(\log|F|^{3m+1})] \to \{0,1\}$. The *robustness property* asserts that if the original polynomial P had no solution (i.e., an A as above) then the answers obtained by the tester will be far from satisfying the residual decision predicate of the tester.

Once the robustness property of the resulting ("outer") tester fits the proximity testing feature of the "inner tester", composition is possible. Indeed, we compose the "outer" tester with an "inner tester" that checks whether the residual decision predicate of the "outer tester" is satisfies. The benefit of this composition is that the query complexity is reduced from poly-logarithmic to polynomial in a double-logarithm. At this point we can afford the Hadamard-Based proof tester (because the overhead in the proof complexity will only be exponential in a polynomial in a double-logarithmic function), and obtain a locally testable proof of polynomial length. That is, we compose the poly(log log)-query tester (acting as an outer tester) with the Hadamard-Based tester (acting as an inner tester), and obtain a locally testable proof of polynomial length (as asserted by the PCP Theorem).

Digest: the proof composition paradigm. The PCP Theorem asserts a PCP system that obtains simultaneously the minimal possible randomness and query complexity (up to a multiplicative factor, assuming that $\mathcal{P} \neq \mathcal{NP}$). The foregoing construction obtains this remarkable result by combining two different PCPs: the first PCP obtains logarithmic randomness but uses poly-logarithmically many queries, whereas the second PCP uses a constant number of queries but has polynomial randomness complexity. We stress that *each of these two PCP systems is highly non-trivial and very interesting by itself.* We also highlight the fact that these PCPs are combined using a very simple composition method (which refers to auxiliary properties such as robustness and proximity testing). Details follow.[33]

Loosely speaking, the proof composition paradigm refers to composing two proof systems such that the "inner" verifier is used for probabilistically verifying the acceptance criteria of the "outer" verifier. That is, the combined verifier selects coins for the "outer" verifier, determines the corresponding locations that the "outer" verifier wishes to inspect (in the proof), and verifies that the "outer" verifier would have accepted the values that reside in these locations. The latter verification is performed by invoking the "inner" verifier, *without reading the values residing in all the aforementioned locations.* Indeed, the aim is to conduct this

---

[33] Our presentation of the composition paradigm follows [15], rather than the original presentation of [5,4].

("composed") verification while using much fewer queries than the query complexity of the "outer" proof system. In particular, the inner verifier cannot afford to read its input, which makes the composition more subtle than the term suggests.

In order for the proof composition to work, the combined verifiers should satisfy some auxiliary conditions. Specifically, the *outer* verifier should be robust in the sense that its soundness condition guarantee that, with high probability, the oracle answers are "far" from satisfying the residual decision predicate (rather than merely not satisfying it).[34] The *inner* verifier is given oracle access to its input and is charged for each query made to it, but is only required to reject (with high probability) inputs that are far from being valid (and, as usual, accept inputs that are valid). That is, the inner verifier is actually a verifier of proximity.

Composing two such PCPs yields a new PCP, where the new proof oracle consists of the proof oracle of the "outer" system and a sequence of proof oracles for the "inner" system (one "inner" proof per each possible random-tape of the "outer" verifier). The resulting verifier selects coins for the outer-verifier and uses the corresponding "inner" proof in order to verify that the outer-verifier would have accepted under this choice of coins. Note that such a choice of coins determines locations in the "outer" proof that the outer-verifier would have inspected, and the combined verifier provides the inner-verifier with oracle access to these locations (which the inner-verifier considers as its input) as well as with oracle access to the corresponding "inner" proof (which the inner-verifier considers as its proof-oracle).

The quantitative effect of such a composition is easy to analyze. Specifically, composing an outer-verifier of randomness-complexity r' and query-complexity q' with an inner-verifier of randomness-complexity r'' and query-complexity q'' yields a PCP of randomness-complexity $r(n) = r'(n) + r''(q'(n))$ and query-complexity $q(n) = q''(q'(n))$, because $q'(n)$ represents the length of the input (oracle) that is accessed by the inner-verifier. Thus, assuming $q''(m) \ll m$, the query complexity is significantly decreased (from $q'(n)$ to $q''(q'(n))$), while the increase in the randomness complexity is moderate provided that $r''(q'(n)) \ll r'(n)$. Furthermore, the verifier resulting from the composition inherits the robustness features of the composed verifier, which is important in case we wish to compose the resulting verifier with another inner-verifier.

3.3   Locally Testable Codes and Proofs of Nearly Linear Length

We now move on to even *shorter* codes and proofs; specifically, codes and proofs of *nearly linear length*. The latter term has been given quite different interpretations, and we start by sorting these out. Currently, this taxonomy is relevant mainly for second-level discussions and review of some past works.[35]

---

[34] Furthermore, the latter predicate, which is well-defined by the non-adaptive nature of the outer verifier, must have a circuit of size bounded by a polynomial in the number of queries.

[35] Things were different when the original version of this text [36] was written. At that time, only T2-nearly linear length was know for $O(1)$-local testability, and the T3-nearly linear result achieved by Dinur [26] seemed a daring conjecture (which was, nevertheless, stated in [36, Conj. 3.3]).

3.3.1 Types of Nearly Linear Functions

A few common interpretations of this term are listed below (going from the most liberal to the most strict one).

T1-nearly linear: A very liberal notion, which seems at the verge of an abuse of the term, refers to a sequence of functions $f_\epsilon : \mathbb{N} \to \mathbb{N}$ such that, for every $\epsilon > 0$, it holds that $f_\epsilon(n) \leq n^{1+\epsilon}$. That is, each function is actually of the form $n \mapsto n^c$, for some constant $c > 1$, but the sequence as a whole can be viewed as approaching linearity.

   The PCP of Polishchuk and Spielman [55] and the simpler locally testable code of Goldreich and Sudan [41, Thm. 2.4] have nearly linear length in this sense.

T2-nearly linear: A more reasonable notion of nearly linear functions refers to individual functions f such that $f(n) = n^{1+o(1)}$. Specifically, for some function $\epsilon : \mathbb{N} \to [0,1]$ that goes to zero, it holds that $f(n) \leq n^{1+\epsilon(n)}$. Common sub-types include the following:

1. $\epsilon(n) = 1/\log\log n$.
2. $\epsilon(n) = 1/(\log n)^c$ for some constant $c \in (0,1)$.
   The locally testable codes and proofs of [41,21,15] have nearly linear length in this sense. Specifically, in [41, Sec. 4-5] and [21] any $c > 1/2$ will do, whereas in [15] any $c > 0$ will do.
3. $\epsilon(n) = \frac{\exp((\log\log n)^c)}{\log n}$ for some constant $c \in (0,1)$.
   Note that $\mathrm{poly}(\log\log n) < \exp((\log\log n)^c) < (\log n)^{o(1)}$, for any constant $c \in (0,1)$.

   Indeed, the case in which $\epsilon(n) = \frac{O(\log\log n)}{\log n}$ (or so) deserves a special category, presented next.

T3-nearly linear: The strongest notion interprets near-linearity as linearity up to a poly-logarithmic (or quasi-poly-logarithmic) factor. In the former case $f(n) = \widetilde{O}(n) \overset{\text{def}}{=} \mathrm{poly}(\log n) \cdot n$, which corresponds to the case of $f(n) \leq n^{1+\epsilon(n)}$ with $\epsilon(n) = O(\log\log n)/\log n$, whereas the latter case corresponds to $\epsilon(n) = \mathrm{poly}(\log\log n)/\log n$ (i.e., in which case $f(n) \leq (\log n)^{\mathrm{poly}(\log\log n)} \cdot n$). The recent results of [20,26] refer to this notion.

We note that while [20,26] achieve T3-nearly linear length, the low-error results of [54,27] only achieve T2-nearly linear length.

3.3.2 Local Testability with Nearly Linear Length

The celebrated gap amplification technique of Dinur [26] is best known for providing an alternative proof of the PCP Theorem. However, applying this technique to a PCP that was (previously) provided by Ben-Sasson and Sudan [20] yields locally testable codes and proofs of T3-nearly linear length. In particular, the overhead in the code and proof length is only polylogarithmic in the length of the primal object (which establishes [36, Conj. 3.3]).

Theorem 3.1 (Dinur [26], building on [20]): *There exists a constant q and a poly-logarithmic function* $f : \mathbb{N} \to \mathbb{N}$ *such that there exist q-locally testable codes and proofs of length* $f(k) \cdot k$, *where k denotes the length of the actual information* (i.e., the assertion in case of proofs and the encoded information in case of codes).

The proof of Theorem 3.1 combines the PCP system of Ben-Sasson and Sudan [20] with the gap amplification method of Dinur [26]. The latter is reviewed in §3.3.3. We mention that the PCP system of [20] is based on the NP-completeness of a certain code (of length $n = \tilde{O}(k)$), and on a randomized reduction of testing whether a given n-bit long string is a codeword to a constant number of similar tests that refer to $\sqrt{n}$-bit long strings. Applying this reduction $\log\log n$ times yields a PCP of query complexity poly$(\log n)$ and length $\tilde{O}(n)$, which in turn yields a 3-query "PCP with soundness error $1 - 1/\text{poly}(\log n)$".

We mention that in the original version of this survey [36], we conjectured that a polylogarithmic (length) overhead is inherent to local testability (or, at least, that linear length $O(1)$-local testability is impossible). We currently have mixed feelings with respect to this conjecture (even when confined to proofs), and thus rephrase it as an open problem.

Open Problem 3.2 *Determine whether there exist locally testable codes and proofs of linear length.*

3.3.3   The Gap Amplification Method

Essentially, Theorem 3.1 is proved by applying the gap amplification method (of Dinur [26]) to the (weak) PCP system constructed by Ben-Sasson and Sudan [20]. The latter PCP system has length $\ell(k) = \tilde{O}(k)$, but its soundness error is $1 - 1/\text{poly}(\log k)$ (i.e., its rejection probability is at least $1/\text{poly}(\log k)$). Each application of the gap amplification step *doubles the rejection probability while essentially maintaining the initial complexities.* That is, in each step, the constant query complexity of the verifier is preserved and its randomness complexity is increased only by a constant term (and so the length of the PCP oracle is increased only by a constant factor). Thus, starting from the system of [20] and applying $O(\log\log k)$ amplification steps, we essentially obtain Theorem 3.1. (Note that a PCP system of polynomial length can be obtained by starting from a trivial "PCP" system that has rejection probability $1/\text{poly}(k)$, and applying $O(\log k)$ amplification steps.)

In order to describe the aforementioned process we need to *redefine PCP systems so as to allow arbitrary soundness error*. In fact, for technical reasons, it is more convenient to describe the process as an iterated reduction of a "constraint satisfaction" problem to itself. Specifically, we refer to systems of 2-variable constraints, which are readily represented by (labeled) graphs such that the vertices correspond to (non-Boolean) variables and the edges are associated with constraints.

Definition 3.3 (CSP with 2-variable constraints): *For a fixed finite set* Σ, *an instance of* CSP *consists of a graph* $G = (V, E)$ *(which may have parallel edges and self-loops) and a sequence of 2-variable constraints* $\Phi = (\phi_e)_{e \in E}$ *associated*

with the edges, where each constraint has the form $\phi_e : \Sigma^2 \to \{0, 1\}$. The value
of an assignment $\alpha : V \to \Sigma$ is the number of constraints satisfied by α; that is,
the value of α is $|\{(u, v) \in E : \phi_{(u,v)}(\alpha(u), \alpha(v)) = 1\}|$. We denote by $\mathtt{vlt}(G, \Phi)$
(standing for violation) the fraction of unsatisfied constraints under the best
possible assignment; that is,

$$\mathtt{vlt}(G, \Phi) = \min_{\alpha:V \to \Sigma} \left\{ \frac{|\{(u, v) \in E : \phi_{(u,v)}(\alpha(u), \alpha(v)) = 0\}|}{|E|} \right\}. \tag{2}$$

For various functions $\tau : \mathbb{N} \to (0, 1]$, we will consider the promise problem
$\mathtt{gapCSP}_\tau^\Sigma$, having instances as above, such that the YES-instances are fully satisfiable instances (i.e., $\mathtt{vlt} = 0$) and the NO-instances are pairs (G, Φ) for which
$\mathtt{vlt}(G, \Phi) \geq \tau(|G|)$ holds, where $|G|$ denotes the number of edges in G.

Note that 3SAT is reducible to $\mathtt{gapCSP}_{\tau_0}^{\Sigma_0}$ for $\Sigma_0 = \{\mathtt{F}, \mathtt{T}\}^3$ and $\tau_0(m) = 1/m$
(e.g., replace each clause by a vertex, and use edge constraints that enforce
mutually consistent and satisfying assignments to each pair of clauses). Furthermore, the PCP system of [20] yields a reduction of 3SAT to $\mathtt{gapCSP}_{\tau_1}^{\Sigma_0}$ for
$\tau_1(m) = 1/\mathrm{poly}(\log m)$ where the size of the graph is nearly linear in the length
of the input formula. Our goal is to reduce $\mathtt{gapCSP}_{\tau_0}^{\Sigma_0}$ (or rather $\mathtt{gapCSP}_{\tau_1}^{\Sigma_0}$) to
$\mathtt{gapCSP}_c^\Sigma$, for some fixed finite Σ and constant $c > 0$, where in the case of
$\mathtt{gapCSP}_{\tau_1}^{\Sigma_0}$ we wish the reduction to preserve the length of the instance up to a
polylogarithmic factor. The PCP Theorem (resp., a PCP of nearly linear length)
follows by showing a simple PCP system for $\mathtt{gapCSP}_c^\Sigma$. As noted above, the reduction is obtained by repeated applications of an amplification step that is
captured by the following lemma.

Lemma 3.4 (amplifying reduction of $\mathtt{gapCSP}$ to itself): *For some finite Σ and
constant $c > 0$, there exists a polynomial-time computable function f such that,
for every instance (G, Φ) of $\mathtt{gapCSP}^\Sigma$, it holds that $(G', \Phi') = f(G, \Phi)$ is an
instance of $\mathtt{gapCSP}^\Sigma$ and the two instances are related as follows:*

1. *If $\mathtt{vlt}(G, \Phi) = 0$ then $\mathtt{vlt}(G', \Phi') = 0$.*
2. *$\mathtt{vlt}(G', \Phi') \geq \min(2 \cdot \mathtt{vlt}(G, \Phi), c)$.*
3. *$|G'| = O(|G|)$.*

That is, satisfiable instances are mapped to satisfiable instances, whereas instances that violate a ν fraction of the constraints are mapped to instances that
violate at least a $\min(2\nu, c)$ fraction of the constraints. Furthermore, the mapping
increases the number of edges (in the instance) by at most a constant factor. We
stress that both Φ and Φ' consists of Boolean constraints defined over Σ^2. Thus,
by iteratively applying Lemma 3.4 for a logarithmic (resp., double-logarithmic)
number of times, we reduce $\mathtt{gapCSP}_{\tau_0}^\Sigma$ (resp., $\mathtt{gapCSP}_{\tau_1}^\Sigma$) to $\mathtt{gapCSP}_c^\Sigma$.

Outline of the proof of Lemma 3.4: Before turning to the proof, let us
highlight the difficulty that it needs to address. Specifically, the lemma asserts a
"violation amplifying effect" (i.e., Items 1 and 2), while maintaining the alphabet
Σ and allowing only a moderate increase in the size of the graph (i.e., Item 3).

Waiving the latter requirements allows a relatively simple proof that mimics (an augmented version of) the "parallel repetition" of the corresponding PCP. Thus, the challenge is significantly decreasing the "size blow-up" that arises from parallel repetition and maintaining a fixed alphabet. The first goal (i.e., Item 3) calls for a suitable derandomization, and indeed we shall use a "pseudorandom" generator based on random walks on expander graphs. The second goal (i.e., fixed alphabet) can be handled by using the proof composition paradigm, which was outlined in §3.2.2.

The lemma is proved by presenting a three-step reduction. The first step is a pre-processing step that makes the underlying graph suitable for further analysis (e.g., the resulting graph will be an expander). The value of vlt may decrease during this step by a constant factor. The heart of the reduction is the second step in which we increase vlt by any desired constant factor. This is done by a construction that corresponds to taking a random walk of constant length on the current graph. The latter step also increases the alphabet Σ, and thus a post-processing step is employed to regain the original alphabet (by using any inner PCP systems; e.g., the one presented in §3.1.2). Details follow.

We first stress that the aforementioned Σ and c, as well as the auxiliary parameters d and t (to be introduced in the following two paragraphs), are fixed constants that will be determined such that various conditions (which arise in the course of our argument) are satisfied. Specifically, t will be the last parameter to be determined (and it will be made greater than a constant that is determined by all the other parameters).

We start with the pre-processing step. Our aim in this step is to reduce the input (G, Φ) of gapCSP$^\Sigma$ to an instance (G_1, Φ_1) such that G_1 is a d-regular expander graph.[36] Furthermore, each vertex in G_1 will have at least $d/2$ self-loops, the number of edges will be preserved up to a constant factor (i.e., $|G_1| = O(|G|)$), and vlt$(G_1, \Phi_1) = \Theta(\text{vlt}(G, \Phi))$. This step is quite simple: essentially, the original vertices are replaced by expanders of size proportional to their degree, and a big (dummy) expander is "superimposed" on the resulting graph.

The main step is aimed at increasing the fraction of violated constraints by a sufficiently large constant factor. The intuition underlying this step is that the probability that a random (t-edge long) walk on the expander G_1 intersects a fixed set of edges is closely related to the probability that a random sample of (t) edges intersects this set. Thus, we may expect such walks to hit a violated edge with probability that is $\min(\Theta(t \cdot \nu), c)$, where ν is the fraction of violated edges. Indeed, the current step consists of reducing the instance (G_1, Φ_1) of gapCSP$^\Sigma$ to an instance (G_2, Φ_2) of gapCSP$^{\Sigma'}$ such that $\Sigma' = \Sigma^{d^t}$ and the following holds:

---

[36] A d-regular graph is a graph in which each vertex is incident to exactly d edges. Loosely speaking, an expander graph has the property that each moderately balanced cut (i.e., partition of its vertex set) has relatively many edges crossing it. An equivalent definition, also used in the actual analysis, is that, except for the largest eigenvalue (which equals d), all the eigenvalues of the corresponding adjacency matrix have absolute value that is bounded away from d.

1. The vertex set of G_2 is identical to the vertex set of G_1, and each t-edge long path in G_1 is replaced by a corresponding edge in G_2, which is thus a d^t-regular graph.
2. The constraints in Φ_2 refer to each element of Σ' as a Σ-labeling of the ("distance $\leq t$") neighborhood of a vertex, and mandates that the two corresponding labelings (of the endpoints of the G_2-edge) are consistent as well as satisfy Φ_1. That is, the following two types of conditions are enforced by the constraints of Φ_2:

   (consistency): If vertices u and w are connected in G_1 by a path of length at most t and vertex v resides on this path, then the Φ_2-constraint associated with the G_2-edge between u and w mandates the equality of the entries corresponding to vertex v in the Σ'-labeling of vertices u and w.

   (satisfying Φ_1): If the G_1-edge (v, v') is on a path of length at most t starting at u, then the Φ_2-constraint associated with the G_2-edge that corresponds to this path enforces the Φ_1-constraint that is associated with (v, v').

Clearly, $|G_2| = d^{t-1} \cdot |G_1| = O(|G_1|)$, because d is a constant and t will be set to a constant. (Indeed, the relatively moderate increase in the size of the graph corresponds to the low randomness-complexity of selecting a random walk of length t in G_1.)

Turning to the analysis of this step, we note that $\mathtt{vlt}(G_1, \Phi_1) = 0$ implies $\mathtt{vlt}(G_2, \Phi_2) = 0$. The interesting fact is that the fraction of violated constraints increases by a factor of $\Omega(\sqrt{t})$; that is, $\mathtt{vlt}(G_2, \Phi_2) \geq \min(\Omega(\sqrt{t} \cdot \mathtt{vlt}(G_1, \Phi_1)), c)$. Here we merely provide a rough intuition and refer the interested reader to [26]. We may focus on any Σ'-labeling of the vertices of G_2 that is consistent with some Σ-labeling of G_1, because relatively few inconsistencies (among the Σ-values assigned to a vertex by the Σ'-labeling of other vertices) can be ignored, while relatively many such inconsistencies yield violation of the "equality constraints" of many edges in G_2. Intuitively, relying on the hypothesis that G_1 is an expander, it follows that the set of violated edge-constraints (of Φ_1) with respect to the aforementioned Σ-labeling causes many more edge-constraints of Φ_2 to be violated (because each edge-constraint of Φ_1 is enforced by many edge-constraints of Φ_2). The point is that *any set F of edges of G_1 is likely to appear on a $\min(\Omega(t) \cdot |F|/|G_1|, \Omega(1))$ fraction of the edges of G_2* (i.e., t-paths of G_1). (Note that the claim would have been obvious if G_1 were a complete graph, but it also holds for an expander.)[37]

The factor of $\Omega(\sqrt{t})$ gained in the second step makes up for the constant factor lost in the first step (as well as the constant factor to be lost in the last step). Furthermore, for a suitable choice of the constant t, the aforementioned gain yields an overall constant factor amplification (of $\mathtt{vlt}$). However, so far we obtained an instance of $\mathtt{gapCSP}^{\Sigma'}$ rather than an instance of $\mathtt{gapCSP}^{\Sigma}$, where $\Sigma' = \Sigma^{d^t}$. The purpose of the last step is to reduce the latter instance to an instance of

---

[37] We mention that, due to a technical difficulty, it is easier to establish the claimed bound of $\Omega(\sqrt{t} \cdot \mathtt{vlt}(G_1, \Phi_1))$ rather than $\Omega(t \cdot \mathtt{vlt}(G_1, \Phi_1))$.

gapCSP$^\Sigma$. This is done by viewing the instance of gapCSP$^{\Sigma'}$ as a PCP-system,[38] and composing it with an inner-verifier using the proof composition paradigm outlined in §3.2.2. We stress that the inner-verifier used here needs only handle instances of constant size (i.e., having description length $O(d^t \log |\Sigma|)$), and so the verifier presented in §3.1.2 will do. The resulting PCP-system uses randomness $r \stackrel{\text{def}}{=} \log_2 |G_2| + O(d^t \log |\Sigma|)^2$ and a constant number of binary queries, and has rejection probability $\Omega(\text{vlt}(G_2, \Phi_2))$, which is independent of the choice of the constant t. For $\Sigma = \{0,1\}^{O(1)}$, we can obtain an instance of gapCSP$^\Sigma$ that has a $\Omega(\text{vlt}(G_2, \Phi_2))$ fraction of violated constraints. Furthermore, the size of the resulting instance (which is used as the output (G', Φ') of the three-step reduction) is $O(2^r) = O(|G_2|)$, where the equality uses the fact that d and t are constants. Recalling that $\text{vlt}(G_2, \Phi_2) \geq \min(\Omega(\sqrt{t} \cdot \text{vlt}(G_1, \Phi_1)), c)$ and $\text{vlt}(G_1, \Phi_1) = \Omega(\text{vlt}(G, \Phi))$, this completes the (outline of the) proof of the entire lemma.   □

Reflection. In contrast to the proof outlined in §3.2.2. which combines two remarkable constructs by using a simple composition method, the current proof of the PCP Theorem is based on developing a powerful "combining method" that improves the quality of the main system to which it is applied. This new method, captured by the amplification step (Lemma 3.4), does not merely obtain the best of the combined systems, but rather obtains a better system than the one given. However, the quality-amplification offered by Lemma 3.4 is rather moderate, and thus many applications are required in order to derive the desired result. Taking the opposite perspective, one may say that remarkable results are obtained by a gradual process of many moderate amplification steps.

3.4 Additional Considerations

Our motivation for studying locally testable codes and proofs referred to super-fast testing, but our actual definitions have focused on the query complexity of these testers. While the query complexity of testing has a natural appeal, the hope is that low query complexity testers would also yield super-fast testing. Indeed, in the case of codes, it is typically the case that the testing time is related to the query complexity. However, in the case of proofs there is a seemingly unavoidable (linear) dependence of the verification time on the input length. This (linear) dependence can be avoided if one considers PCP-of-Proximity (see Section 2.3.3) rather than standard PCP. But even in this case, additional work is needed in order to derive testers that work is sub-linear time. The interested reader is referred to [16,52].

4 Locally Decodable Codes

Locally *decodable* codes are complimentary to local *testable* codes. Recall that the latter are required to allow for super-fast rejection of strings that are far from

---

[38] The PCP-system referred to here has arbitrary soundness error (i.e., it rejects the instance (G_2, Φ_2) with probability $\text{vlt}(G_2, \Phi_2) \in [0, 1]$).

being codewords (while accepting all codewords). In contrast, in case of locally decodable codes, we are guaranteed that the input is close to a codeword, and are required to recover individual bits of the encoded information based on a *small number of probes* (per recovered bit). As in case of local testability, the case when the operation (in this case decoding) is performed based on a *constant number of probes* is of special interest.

Local decodability is of natural practical appeal, which in turn provides additional motivation for local testability. The point is that it makes little sense to try to recover part of the data when the codeword is too corrupt. Thus, one should first apply local testability to check that the received codeword is not too corrupt, and apply local decodability only in case the codeword test passes.

4.1   Definitions

We follow the conventions of Section 2.1, but extend the treatment to codes over any finite alphabet Σ (rather than insisting on $\Sigma = \{0, 1\}$).

Definition 4.1 (locally decodable codes, basic version): *Let* $C : \Sigma^k \to \Sigma^n$ *be a code, and let* $q \in \mathbb{N}$ *and* $\delta \in (0, 1)$. *A* q-local δ-decoder for C *is a probabilistic (non-adaptive) oracle machine* M *that makes at most* q *queries and satisfies the following condition:*

Local recovery from somewhat corrupted codewords: *For every* $i \in [k]$ *and* $x = (x_1, ..., x_k) \in \Sigma^k$, *and any* $w \in \Sigma^n$ *that is* δ-close *to* $C(x)$, *on input* i *and oracle access to* w, *machine* M *outputs* x_i *with probability at least* $2/3$. *That is,* $\Pr[M^w(1^k, i) = x_i] > 2/3$, *for any* $w \in \Sigma^n$ *that is* δ-far *from* $C(x)$.

We call q *the* query complexity *of* M, *and* δ *the* proximity parameter.

Note that the proximity parameter must be smaller than the covering radius of the code (as otherwise the definition cannot possibly be satisfied (at least for some w and i)). One may strengthen Definition 4.1 by requiring that the bits of an uncorrupted codeword be always recovered correctly (rather than with high probability); that is, for every $i \in [k]$ and $x = (x_1, ..., x_k) \in \Sigma^k$, it must hold that $\Pr[M^{C(x)}(1^k, i) = x_i] = 1$. Turning to families of codes, we present the following definition (which potentially allows the alphabet to grow with k).

Definition 4.2 (locally decodable codes, asymptotic version): *For functions* $n, \sigma : \mathbb{N} \to \mathbb{N}$, *let* $C = \{C_k : [\sigma(k)]^k \to [\sigma(k)]^{n(k)}\}_{k \in K}$. *We say that* C *is a* local decodable code *if there exist constants* $\delta > 0$ *and* q *and a machine* M *that is a* q-local δ-decoder for C_k, *for every* $k \in K$.

We mention that locally decodable codes are related to (information theoretic secure) Private Information Retrieval (PIR) schemes, introduced in [25]. In the latter a user wishes to recover a bit of data from a k-bit long database, copies of which are held by s servers, without revealing any information to any single server. To that end, the user (secretly) communicates with each of the servers, and the issue is to minimize the total amount of communication. As we shall see, certain s-server PIR schemes yield $2s$-locally decodable codes of length exponential in the communication complexity of the PIR.

Related notions of local recovery. The notion of local decodability is a special case of a general notion of local recovery, where one may be required to recover an arbitrary function $f : \Sigma^k \to \{0,1\}^*$ of the original information based on a constant number of probes to the (corrupted) codeword. The function f must be restricted in two ways: Firstly, it should have a small range (e.g., its range may be Σ), and secondly it should come from a small predetermined set $\mathcal{F}$ of functions. Definition 4.1 may be recast in these terms, by considering the set of projection functions (i.e., $\{f_i : \Sigma^k \to \Sigma\}$ where $f_i(x_1, ..., x_k) = x_i$). We believe that this is the most natural special case of the general notion of local recovery. In §3.2.1 we referred to another special case, where the alphabet is associated with a finite field F and the recovery function $f_e : F^k \to F$ is one out of $|F|$ possible linear functions (specifically, $f_e(x_1, ..., x_k) = \sum_{i=1}^{k} e^{i-1} x_i$, for $e \in F$).[39] Another natural case (also used in §3.2.1) is that of the recovery of (correct) symbols of the codeword, which may be viewed as self-correction. (In this case each admissible function determines one codeword symbol as a function of the encoded message.)

4.2   Results

The best known locally decodable codes are of strictly sub-exponential length; that is, k information bits can be encoded by codewords of length $n = \exp(k^{o(1)})$ that are locally decodable (cf. [29], building on [61]). This result disproves [36, Conj. 4.4],

Theorem 4.3 (Efremenko [29], building on Yekhanin [61]): *For some $\delta > 0$ there exists a code C : $\{0,1\}^k \to \{0,1\}^n$ that has a 3-local δ-decoder such that $n = \exp(2^{\tilde{O}(\sqrt{\log k})}) = \exp(k^{o(1)})$. Furthermore, 2^d-local decodability can be obtained with $n = \exp(2^{\tilde{O}(\sqrt[d]{\log k})})$.*

In this section we only outline a couple of codes of lesser performance. Specifically, we will present longer codes that are $O(1)$-locally decodable as well as shorter codes that are poly$(\log k)$-locally decodable.

4.2.1   Locally Decodable Codes of Sub-exponential Length
For any $d \geq 1$, there is a simple construction of a 2^d-locally 2^{-d-2}-decodable binary code of length $n = 2^{d \cdot k^{1/d}}$. For $h = k^{1/d}$, we identify $[k]$ with $[h]^d$, and view $x \in \{0,1\}^k$ as $(x_{i_1,...,i_d})_{i_1,...,i_d \in [h]}$. We encode x by providing the parity of all $x_{i_1,...,i_d}$ residing in each of the $(2^h)^d$ sub-cubes of $[h]^d$; that is, for every $(S_1, ..., S_d) \in 2^{[h]} \times \cdots \times 2^{[h]}$, we provide

$$\mathrm{C}(x)_{S_1,...,S_d} = \oplus_{i_1 \in S_1,...,i_d \in S_d} x_{i_1,...,i_d}. \tag{3}$$

Indeed, the Hadamard code is the special case in which $d = 1$. To recover the value of $x_{i_1,...,i_d}$, at any desired $(i_1, ..., i_d) \in [h]^d$, the decoder uniformly

---

[39] Indeed, the value $f_e(x_1, ..., x_k)$ is the evaluation at e of the polynomial $p(\zeta) = \sum_{i=1}^{k} x_i \zeta^{i-1}$ represented by the coefficients $(x_1, ..., x_k)$.

selects $(R_1, ..., R_d) \in 2^{[h]} \times \cdots \times 2^{[h]}$, and recovers the (possibly corrupted) values $C(x)_{S_1,...,S_d}$, where each S_j either equals R_j or equals $R_j \triangle \{i_j\}$, where $R \triangle \{i\} = R \setminus \{i\}$ if $i \in R$ and $R \triangle \{i\} = R \cup \{i\}$ otherwise. The key observation is that each of the decoder's queries is uniformly distributed. Thus, with probability at least $3/4$, XORing the 2^d answers, yields the desired result (because $\bigoplus_{S_1 \in \{R_1, R_1 \triangle \{i_1\}\}, ..., S_d \in \{R_d, R_d \triangle \{i_d\}\}} C(x)_{S_1,...,S_d}$ equals $C(x)_{\{i_1\},...,\{i_d\}} = x_{i_1,...,i_d}$).

We comment that a related code (of length $n = 2^{d^d \cdot k^{1/d}}$) allows for recovery based on $d + 1$ (rather 2^d) queries. The original presentation, due to [2] (building on [25]), is in terms of PIR schemes (with $s = (d + 1)/2$ servers and overall communication $d^d \cdot k^{1/d} = \exp(\widetilde{O}(s)) \cdot k^{1/(2s-1)}$). In particular, in the case that $d = 2$, we use two servers, sending (R_1, R_2, R_3) to one server and $(R_1 \triangle \{i_1\}, R_2 \triangle \{i_2\}, R_3 \triangle \{i_3\})$ to the other server. Upon receiving (S_1, S_2, S_3), each server replies with the bit $C(x)_{S_1,S_2,S_3}$ as well as the three $k^{1/3}$-bit long sequences $(C(x)_{S_1 \triangle \{i\}, S_2, S_3})_{i \in [k^{1/3}]}$, $(C(x)_{S_1, S_2 \triangle \{i\}, S_3})_{i \in [k^{1/3}]}$, and $(C(x)_{S_1, S_2, S_3 \triangle \{i\}})_{i \in [k^{1/3}]}$, which contain the bits $C(x)_{S_1 \triangle \{i_1\}, S_2, S_3}$, $C(x)_{S_1, S_2 \triangle \{i_2\}, S_3}$, and $C(x)_{S_1, S_2, S_3 \triangle \{i_3\}}$. Thus, the user obtains the bits $C(x)_{R_1,R_2,R_3}$, $C(x)_{R_1 \triangle \{i_1\}, R_2, R_3}$, $C(x)_{R_1, R_2 \triangle \{i_2\}, R_3}$, and $C(x)_{R_1, R_2, R_3 \triangle \{i_3\}}$ from the first server, and the bits $C_{R_1 \triangle \{i_1\}, R_2 \triangle \{i_2\}, R_3 \triangle \{i_3\}}$, $C_{R_1, R_2 \triangle \{i_2\}, R_3 \triangle \{i_3\}}$, $C_{R_1 \triangle \{i_1\}, R_2, R_3 \triangle \{i_3\}}$, $C_{R_1 \triangle \{i_1\}, R_2 \triangle \{i_2\}, R_3}$ from the second server.

The corresponding locally decodable code is obtained by a generic transformation that applies to any PIR scheme with s servers, in which the user makes uniformly distributed queries of length $\mathsf{qst}(k)$, gets answers of length $\mathsf{ans}(k)$, and recovers the desired value by XORing some predetermined bits contained in the answers. In this case, the resulting code will contain the Hadamard encoding of each of the possible answers provided by each of the servers; that is, if the j-th server answers according to $A_j(x, q) \in \{0,1\}^{\mathsf{ans}(k)}$, where $x \in \{0,1\}^k$ and $q \in \{0,1\}^{\mathsf{qst}(k)}$, then $C(x)_{j,q,\ell} = C_{\mathrm{Had}}(A_j(x,q))_\ell$, for every $\ell \in \{0,1\}^{\mathsf{ans}(k)}$. Thus, the length of the code is $s \cdot 2^{\mathsf{qst}(k)} \cdot 2^{\mathsf{ans}(k)}$. Now, on input $i \in [k]$, the decoder emulates the PIR user, obtaining the query sequence $(q_1, ..., q_s)$ and the desired linear combinations $(\ell_1, ..., \ell_s)$. It uniformly selects $r_1, ..., r_s \in \{0,1\}^{\mathsf{ans}(k)}$, queries the (possibly corrupted) codeword at locations $(1, q_1, r_1), (1, q_1, r_1 \oplus \ell_1), ..., (s, q_s, r_s), (s, q_s, r_s \oplus \ell_s)$, and XORs the corresponding $2s$ answers. Note that each of these queries is uniformly distributed in $\{j\} \times \{0,1\}^{\mathsf{qst}(k)} \times \{0,1\}^{\mathsf{ans}(k)}$, for some $j \in [s]$, and that $C(x)_{j,q_j,r_j} \oplus C(x)_{j,q_j,r_j \oplus \ell_j} = C_{\mathrm{Had}}(A_j(x, q_j))_{\ell_j}$.

4.2.2 Polylog-local Decoding for Codes of Nearly Linear Length

We will consider a code $C : \Sigma^k \to \Sigma^n$ of linear distance, while identifying Σ with a finite field (denoted F). For parameters h and $m = \log_h k$, consider a finite field F of size $O(m \cdot h)$, and a subset $H \subset F$ of size h. Viewing the information as a function $f : H^m \to F$, we encode it by providing the values of its low-degree extension $\widehat{f} : F^m \to F$ on all points in F, where $\widehat{f}$ is an m-variant polynomial of degree $|H| - 1$ in each variable. Thus, the code maps $k = h^m$ long sequences over F (which may be viewed as $h^m \log |F|$ bits of information) to sequences of length $n \stackrel{\text{def}}{=} |F|^m = O(mh)^m = O(m)^m \cdot k$ over F. This code has relative distance $mh/|F|$. Note that the smaller m, the better the rate (i.e., relation of n to k) is, but this comes at the expense of using a larger alphabet F (as well as larger query complexity of the decoder presented below).

The decoder works by applying the self-correction paradigm. Given a point $x \in H^m$ and access to an oracle $w : F^m \to F$ that is $1/2$-close to $\hat{f}$, the value of $f(x)$ is recovered by uniformly selecting a line through x, querying for the $|F|$ values of w along the line, finding the degree mh univariate polynomial with the greatest agreement with these values, and evaluating it at the appropriate point. Thus, we obtain an $|F|$-local decoder.

Using a constant m, we obtain an $O(k^{1/m})$-locally decodable code of constant rate (i.e., $n = O(k)$), over an alphabet of size $O(k^{1/m})$. On the other hand, using $m = \epsilon \log k / \log \log k$ (for any constant $\epsilon > 0$), we obtain a poly$(\log k)$-locally decodable code of length $n = k^{1+\epsilon}$, over an alphabet of size poly$(\log k)$. Concatenation with any reasonable[40] binary code (coupled with a trivial decoder that reads the entire codeword), yields a binary poly$(\log k)$-locally decodable code of length $n = k^{1+\epsilon}$.

4.2.3 Lower Bounds

It is known that locally decodable codes cannot be T2-nearly linear:[41] Specifically, any q-locally decodable code $C : \Sigma^k \to \Sigma^n$ must satisfy $n = \Omega(k^{1+\frac{1}{q-1}})$ (cf. [45]). For $q = 2$ and $\Sigma = \{0,1\}$, an exponential lower bound is known (cf. [47], following [40]).

We mention that our past conjectures regarding lower bounds for locally decodable (binary) codes were disproved twice. Our conjectured lower bound of $n > \exp(k^{\Omega(1/q)})$ for q-locally decodable codes was disproved by [9], and our conjectured lower bound of $n > \exp(k^{\Omega(1)})$ for any locally decodable code was disproved by [29] (after being vastly shaken by [61]). Given this history, we dare not make any further conjectures, but instead pose the following open problem.

Open Problem 4.4 *Determine whether there exist locally decodable codes of polynomial length.*

Recall that we know, for a fact, that T2-nearly linear length is impossible, and it is very tempting to conjecture that T1-nearly linear length is impossible too (i.e., any locally decodable code $C : \Sigma^k \to \Sigma^n$ requires $n > k^{1+\Omega(1)}$). Still, let us pose this too as an open problem.

4.3 Relaxations

In light of the fact that locally decodable codes cannot be T2-nearly linear, it is natural to seek relaxations to the notion of locally decodable codes. One natural

---

[40] Indeed, we may use any good code (i.e., linear length and linear distance), as such can be easily constructed for block length $O(\log \log k)$. But we can even use the Hadamard code, because the length overhead caused by it in this setting is negligible.

[41] See terminology in §3.3.1.

relaxation requires local recovery of most individual information-bits, allowing for recovery-failure (but not error) on the rest [15]: That is, it is requires that, whenever few location are corrupted, the decoder should be able to recover most of the individual information-bits, based on a constant number of queries, and for the rest of the locations the decoder may output a fail symbol (but not the wrong value). Augmenting these requirements by the requirement that whenever the codeword is not corrupted – all bits are recovered correctly (with high probability), yields the following definition.

Definition 4.5 (locally decodable codes, relaxed): *For functions $n, \sigma : \mathbb{N} \to \mathbb{N}$, let $\mathrm{C} = \{C_k : \{0,1\}^k \to \{0,1\}^{n(k)}\}_{k \in K}$. For $q \in \mathbb{N}$ and $\delta, \rho \in (0,1)$, a q-local relaxed (δ, ρ)-decoder for C is a probabilistic* (non-adaptive) *oracle machine M that makes at most q queries and satisfies the following conditions:*

Local recovery from uncorrupted codewords: *For every $i \in [k]$ and $x = (x_1, ..., x_k) \in \Sigma^k$, it holds that* $\Pr[M^{C(x)}(1^k, i) = x_i] > 2/3$,

Relaxed local recovery from somewhat corrupted codewords: *For every $x = (x_1, ..., x_k) \in \Sigma^k$, and any $w \in \Sigma^n$ that is δ-close to $\mathrm{C}(x)$, the following two conditions hold:*

  1. *For every $i \in [k]$, it holds that* $\Pr[M^{C(x)}(1^k, i) \in \{x_i, \bot\}] > 2/3$, *where $\bot$ is a special* ("failure") *symbol.*
  2. *There exists a set $I_w \subseteq [k]$ of size at least ρk such that, for every $i \in I_w$, it holds that* $\Pr[M^{C(x)}(1^k, i) = x_i] > 2/3$.[42]

In such a case, C is said to be locally relaxed-decodable.

It turns out (cf. [15]) that Condition 2, in the relaxed recovery requirement, essentially follows from the other requirements. That is, codes satisfying the other requirements can be transformed into locally relaxed-decodable codes, while essentially preserving their rate (and distance). Furthermore, the resulting codes satisfy the following stronger form of Condition 2: *There exists a set $I_w \subseteq [k]$ of density at least $1 - O(\Delta(w, \mathrm{C}(x))/n)$ such that for every $i \in I_w$ it holds that* $\Pr[M^{C(x)}(1^k, i) = x_i] > 2/3$.

Theorem 4.6 [15]: *There exist locally relaxed-decodable codes of T1-nearly linear length. Specifically, for every $\epsilon > 0$, there exists codes of length $n = k^{1+\epsilon}$ that have a $O(1/\epsilon^2)$-local relaxed $(\Omega(\epsilon), 1 - O(\epsilon))$-decoder.*

An obvious open problem is to separate locally decodable codes from relaxed ones. This may follow by either improving the aforementioned lower bound on the length of locally decodable codes or by providing relaxed locally decodable codes of T2-nearly linear length.

---

[42] We stress that it is not required that $\Pr[M^{C(x)}(1^k, i) = \bot] > 2/3$ for $i \in [k] \setminus I_w$. Adding this requirement collapses the notion of relaxed-decodability to ordinary decodability (cf. [23]).

Acknowledgments

We are grateful to Madhu Sudan, Luca Trevisan and Salil Vadhan for related discussions. We are also grateful to Omer Tamuz for useful comments and suggestions regarding this article.

References

1. Alon, N., Krivelevich, M., Kaufman, T., Litsyn, S., Ron, D.: Testing low-degree polynomials over GF(2). In: Arora, S., Jansen, K., Rolim, J.D.P., Sahai, A. (eds.) RANDOM 2003 and APPROX 2003. LNCS, vol. 2764, pp. 188–199. Springer, Heidelberg (2003)
2. Ambainis, A.: An upper bound on the communication complexity of private information retrieval. In: Degano, P., Gorrieri, R., Marchetti-Spaccamela, A. (eds.) ICALP 1997. LNCS, vol. 1256, pp. 401–407. Springer, Heidelberg (1997)
3. Arora, S.: Probabilistic checking of proofs and the hardness of approximation problems. PhD thesis, UC Berkeley (1994)
4. Arora, S., Lund, C., Motwani, R., Sudan, M., Szegedy, M.: Proof verification and the hardness of approximation problems. Journal of the ACM 45(3), 501–555 (1998); Preliminary Version in 33rd FOCS (1992)
5. Arora, S., Safra, S.: Probabilistic checking of proofs: A new characterization of NP. Journal of the ACM 45(1), 70–122 (1998); Preliminary Version in 33rd FOCS (1992)
6. Babai, L., Fortnow, L., Lund, C.: Non-deterministic exponential time has two-prover interactive protocols. Computational Complexity 1(1), 3–40 (1991)
7. Babai, L., Fortnow, L., Levin, L.A., Szegedy, M.: Checking computations in poly-logarithmic time. In: Proc. 23rd ACM Symposium on the Theory of Computing, pp. 21–31 (May 1991)
8. Barak, B.: How to go beyond the black-box simulation barrier. In: Proc. 42nd IEEE Symposium on Foundations of Computer Science, pp. 106–115 (October 2001)
9. Beimel, A., Ishai, Y., Kushilevitz, E., Raymond, J.F.: Breaking the $O(n^{1/(2k-1)})$ barrier for information-theoretic private information retrieval. In: Proc. 43rd IEEE Symposium on Foundations of Computer Science, pp. 261–270 (November 2002)
10. Bellare, M., Coppersmith, D., Håstad, J., Kiwi, M., Sudan, M.: Linearity testing in characteristic two. In: Proceedings of the 36th IEEE Symposium on Foundations of Computer Science, pp. 432–441 (1995)
11. Bellare, M., Goldreich, O., Sudan, M.: Free bits, PCPs, and nonapproximability— towards tight results. SIAM Journal on Computing 27(3), 804–915 (1998); Preliminary Version in 36th FOCS (1995)
12. Bellare, M., Goldwasser, S., Lund, C., Russell, A.: Efficient probabilistically checkable proofs and applications to approximation. In: Proc. 25th ACM Symposium on the Theory of Computing, pp. 294–304 (May 1993)
13. Bellare, M., Sudan, M.: Improved non-approximability results. In: Proceedings of the 26th Annual ACM Symposium on the Theory of Computing, pp. 184–193 (1994)
14. Ben-Sasson, E., Goldreich, O., Sudan, M.: Bounds on 2-query codeword testing. In: Arora, S., Jansen, K., Rolim, J.D.P., Sahai, A. (eds.) RANDOM 2003 and APPROX 2003. LNCS, vol. 2764, pp. 216–227. Springer, Heidelberg (2003)

15. Ben-Sasson, E., Goldreich, O., Harsha, P., Sudan, M., Vadhan, S.: Robust PCPs of proximity, shorter PCPs and applications to coding. In: Proc. 36th ACM Symposium on the Theory of Computing, pp. 1–10 (June 2004); See ECCC Technical Report TR04-021 (March 2004)

16. Ben-Sasson, E., Goldreich, O., Harsha, P., Sudan, M., Vadhan, S.: Short PCPs verifiable in polylogarithmic time. In: 20th IEEE Conference on Computational Complexity, pp. 120–134 (2005)

17. Ben-Sasson, E., Guruswami, V., Kaufman, T., Sudan, M., Viderman, M.: Locally testable codes require redundant testers. In: 24th IEEE Conference on Computational Complexity, pp. 52–61 (2009)

18. Ben-Sasson, E., Harsha, P., Raskhodnikova, S.: Some 3CNF properties are hard to test. In: Proc. 35th ACM Symposium on the Theory of Computing, pp. 345–354 (June 2003)

19. Ben-Sasson, E., Sudan, M.: Robust locally testable codes and products of codes. In: Jansen, K., Khanna, S., Rolim, J.D.P., Ron, D. (eds.) RANDOM 2004 and APPROX 2004. LNCS, vol. 3122, pp. 286–297. Springer, Heidelberg (2004); See ECCC TR04-046 (2004)

20. Ben-Sasson, E., Sudan, M.: Short PCPs with polylog query complexity. SIAM Journal on Computing 38(2), 551–607 (2008); Preliminary Version in 37th STOC (2005)

21. Ben-Sasson, E., Sudan, M., Vadhan, S., Wigderson, A.: Randomness-efficient low degree tests and short PCPs via epsilon-biased sets. In: Proc. 35th ACM Symposium on the Theory of Computing, pp. 612–621 (June 2003)

22. Blum, M., Luby, M., Rubinfeld, R.: Self-testing/correcting with applications to numerical problems. Journal of Computer and System Science 47(3), 549–595 (1993); Preliminary Version in 22nd STOC (1990)

23. Buhrman, H., de Wolf, R.: On relaxed locally decodable codes (July 2004) (unpublished manuscript)

24. Canetti, R., Goldreich, O., Halevi, S.: The random oracle methodology, revisited. In: Proc. 30th ACM Symposium on the Theory of Computing, pp. 209–218 (May 1998)

25. Chor, B., Goldreich, O., Kushilevitz, E., Sudan, M.: Private Information Retrieval. Journal of the ACM 45(6), 965–982 (1998)

26. Dinur, I.: The PCP theorem by gap amplification. Journal of the ACM 54(3), Art. 12 (2007); Extended abstract in 38th STOC (2006)

27. Dinur, I., Harsha, P.: Composition of low-error 2-query PCPs using decodable PCPs. In: Goldreich, O. (ed.) Property Testing. LNCS, vol. 6390, pp. 280–288. Springer, Heidelberg (2010)

28. Dinur, I., Reingold, O.: Assignment-testers: Towards a combinatorial proof of the PCP-Theorem. SIAM Journal on Computing 36(4), 975–1024 (2006); Extended abstract in 45th FOCS (2004)

29. Efremenko, K.: 3-query locally decodable codes of subexponential length. In: 41st ACM Symposium on the Theory of Computing, pp. 39–44 (2009)

30. Ergün, F., Kumar, R., Rubinfeld, R.: Fast approximate PCPs. In: Proc. 31st ACM Symposium on the Theory of Computing, pp. 41–50 (May 1999)

31. Feige, U., Goldwasser, S., Lovász, L., Safra, S., Szegedy, M.: Interactive proofs and the hardness of approximating cliques. Journal of the ACM 43(2), 268–292 (1996); Preliminary version in 32nd FOCS (1991)

32. Forney, G.D.: Concatenated Codes. MIT Press, Cambridge (1966)

33. Fortnow, L., Rompel, J., Sipser, M.: On the power of multi-prover interactive protocols. Theoretical Computer Science 134(2), 545–557 (1994)

34. Friedl, K., Sudan, M.: Some improvements to total degree tests. In: Proc. 3rd Israel Symposium on Theoretical and Computing Systems, Tel Aviv, Israel, January 4-6, pp. 190–198 (1995)
35. Gemmell, P., Lipton, R., Rubinfeld, R., Sudan, M., Wigderson, A.: Self-testing/correcting for polynomials and for approximate functions. In: Proc. 23rd ACM Symposium on the Theory of Computing, pp. 32–42 (1991)
36. Goldreich, O.: Short locally testable codes and proofs (survey). ECCC Technical Report TR05-014 (January 2005)
37. Goldreich, O.: Computational Complexity: A Conceptual Perspective. Cambridge University Press, Cambridge (2008)
38. Goldreich, O., Goldwasser, S., Ron, D.: Property testing and its connection to learning and approximation. Journal of the ACM 45(4), 653–750 (1998); Preliminary Version in 37th FOCS (1996)
39. Goldreich, O., Ron, D.: On proximity oblivious testing. ECCC, TR08-041 (2008); Also in the proceedings of the 41st STOC (2009)
40. Goldreich, O., Karloff, H., Schulman, L., Trevisan, L.: Lower bounds for linear locally decodable codes and private information retrieval. In: Proc. 17th Conference on Computational Complexity, Montréal, Québec, Canada, May 21-24, pp. 175–183 (2002)
41. Goldreich, O., Sudan, M.: Locally testable codes and PCPs of almost linear length. In: Proc. 43rd IEEE Symposium on Foundations of Computer Science, pp. 13–22 (November 2002); See ECCC Report TR02-050 (2002)
42. Harsha, P., Sudan, M.: Small PCPs with low query complexity. Computational Complexity 9(3-4), 157–201 (2000); Preliminary Version in 18th STACS (2001)
43. Håstad, J.: Clique is hard to approximate within $n^{1-\epsilon}$. Acta Mathematica 182, 105–142 (1999); Preliminary Versions in 28th STOC (1996), and 37th FOCS (1997)
44. Håstad, J.: Some optimal inapproximability results. Journal of the ACM 48(4), 798–859 (2001); Preliminary Version in 29th STOC (1997)
45. Katz, J., Trevisan, L.: On the efficiency of local decoding procedures for error-correcting codes. In: Proc. 32nd ACM Symposium on the Theory of Computing, pp. 80–86 (2000)
46. Kaufman, T., Litsyn, S., Xie, N.: Breaking the ϵ-soundness bound of the linearity test over GF(2). SIAM Journal on Computing 39(5), 1988–2003 (2009/2010)
47. Kerenidis, I., de Wolf, R.: Exponential lower bound for 2-query locally decodable codes via a quantum argument. In: Proc. 35th ACM Symposium on the Theory of Computing, pp. 106–115 (June 2003)
48. Kilian, J.: A note on efficient zero-knowledge proofs and arguments (extended abstract). In: Proc. 24th ACM Symposium on the Theory of Computing, pp. 723–732 (May 1992)
49. Lapidot, D., Shamir, A.: Fully parallelized multi prover protocols for NEXP-time. In: Proc. 32nd IEEE Symposium on Foundations of Computer Science, pp. 13–18 (October 1991) (extended abstract)
50. Lund, C., Fortnow, L., Karloff, H., Nisan, N.: Algebraic methods for interactive proof systems. Journal of the ACM 39(4), 859–868 (1992)
51. Meir, O.: Combinatorial construction of locally testable codes. SIAM Journal on Computing 39(2), 491–544 (2009); Extended abstrat in 40th STOC (2008)
52. Meir, O.: Combinatorial PCPs with efficient verifiers. In: 50th IEEE Symposium on Foundations of Computer Science, pp. 463–471 (2009)
53. Micali, S.: Computationally sound proofs. SIAM Journal on Computing 30(4), 1253–1298 (2000); Preliminary Version in 35th FOCS (1994)

54. Moshkovitz, D., Raz, R.: Two query PCP with sub-constant error. In: 49th IEEE Symposium on Foundations of Computer Science, pp. 314–323 (2008)
55. Polishchuk, A., Spielman, D.A.: Nearly-linear size holographic proofs. In: Proc. 26th ACM Symposium on the Theory of Computing, pp. 194–203 (May 1994)
56. Raz, R.: A parallel repetition theorem. SIAM Journal of Computing 27(3), 763–803 (1998); Preliminary Version in 27th STOC (1995)
57. Rubinfeld, R., Sudan, M.: Robust characterizations of polynomials with applications to program testing. SIAM Journal on Computing 25(2), 252–271 (1996); Preliminary Version in 3rd SODA (1992)
58. Spielman, D.: Computationally efficient error-correcting codes and holographic proofs. PhD thesis, Massachusetts Institute of Technology (June 1995)
59. Sudan, M.: Efficient checking of polynomials and proofs and the hardness of approximation problems. Ph.D. Thesis, Computer Science Division, University of California at Berkeley (1992); Also appears as Lecture Notes in Computer Science, Vol. 1001, Springer (1996)
60. Szegedy, M.: Many-valued logics and holographic proofs. In: Wiedermann, J., Van Emde Boas, P., Nielsen, M. (eds.) ICALP 1999. LNCS, vol. 1644, pp. 676–686. Springer, Heidelberg (1999)
61. Yekhanin, S.: Towards 3-Query locally decodable codes of subexponential length. In: 39th ACM Symposium on the Theory of Computing, pp. 266–274 (2007)

Introduction to Testing Graph Properties

Oded Goldreich

Department of Computer Science and Applied Mathematics,
Weizmann Institute of Science, Rehovot, Israel
`oded@wisdom.weizmann.ac.il`

Abstract. The aim of this article is to introduce the reader to the study of testing graph properties, while focusing on the main models and issues involved. No attempt is made to provide a comprehensive survey of this study, and specific results are often mentioned merely as illustrations of general themes.

Keywords: Graph Properties, randomized algorithms, approximation problems.

1   The General Context

In general, property testing is concerned with super-fast (probabilistic) algorithms for deciding whether a given object has a predetermined property or is *far* from any object having this property. Such algorithms, called testers, obtain local views of the object by making adequate queries; that is, the object is seen as a function and the tester gets oracle access to this function, and thus may be expected to work in time that is sub-linear in the size of the object.

Looking at the foregoing formulation, we first note that property testing is concerned with promise problems (cf. [26,30]), rather than with standard decision problems. Specifically, objects that neither have the property nor are far from having the property are discarded. The exact formulation of these promise problems refers to a *distance measure* defined on the set of all relevant objects (i.e., this distance measure coupled with a distance parameter determine the set of objects that are far from the property). Thus, the choice of natural distance measures is crucial to the study of property testing. Secondly, we note that the requirement that the algorithms operate in sub-linear time (i.e., without reading their entire input) calls for a specification of the *type of queries* that these algorithms can make to their input. Thus, the choice of natural query types is also crucial to the study of property testing. These two general considerations will become concrete once we delve into the actual subject matter (i.e., testing graph properties).

1.1   Why Graphs?

Let us start with an empirical observation, taken from Shimon Even's book *Graph Algorithms* [25] (published in 1979):

O. Goldreich (Ed.): Property Testing, LNCS 6390, pp. 105–141, 2010.
ⓒ Springer-Verlag Berlin Heidelberg 2010

Graph theory has long become recognized as one of the more useful mathematical subjects for the computer science student to master. The approach which is natural in computer science is the algorithmic one; our interest is not so much in existence proofs or enumeration techniques, as it is in finding efficient algorithms for solving relevant problems, or alternatively showing evidence that no such algorithms exist. Although algorithmic graph theory was started by Euler, if not earlier, its development in the last ten years has been dramatic and revolutionary.

Meditating on these facts, one may ask what is the source of this ubiquitous use of graphs in computer science. The most common answer is that graphs arise naturally as a model (or an abstraction) of numerous natural and artificial objects. Another answer is that graphs help visualize binary relations over finite sets. These two different answers correspond to two types of models of testing graph properties that will be discussed below.

1.2   Why Testing?

Suppose we are given a huge graph representing some binary relation over a huge data-set (see below), and we need to determine whether the graph (equivalently, the relation) has some predetermined property. Since the graph is huge, we cannot or do not want to even scan all of it (let alone process all of it). The question is whether it is possible to make meaningful statements about the entire graph based only on a "small portion" of it. Of course, such statements will at best be approximations. But in many settings approximations are good enough.

   As a motivation, let us consider a well-known example in which fast approximations are possible and useful. Suppose that some cost function is defined over a huge set, and that one wants to obtain the average cost of an element in the set. To be more specific, let $\mu : S \to [0, 1]$ be a cost function, and suppose we want to estimate $\overline{\mu} \stackrel{\text{def}}{=} \frac{1}{|S|} \sum_{x \in S} \mu(x)$. Then, uniformly (and independently) selecting $m \stackrel{\text{def}}{=} O(\epsilon^{-2} \log(1/\delta))$ sample points, $x_1, ..., x_m$, in S we obtain with probability at least $1 - \delta$ an estimate of $\overline{\mu}$ within $\pm \epsilon$. That is,

$$\mathbf{Pr}_{x_1,...,x_m \in S} \left[\left| \frac{1}{m} \sum_{i=1}^{m} \mu(x_i) - \overline{\mu} \right| > \epsilon \right] < \delta. \tag{1}$$

Turning back to graphs, we note that they capture more complex features of data sets; that is, graphs capture relations among pairs of elements (rather then functions of single elements). Specifically, a symmetric binary relation $R \subseteq S \times S$ is represented by a graph $G = (S, R)$, where the elements of S are viewed as vertices and the elements in R are viewed as edges.

   The study of testing graph properties reveals that sampling a huge data set may be useful not only towards approximating various statistics regarding a function defined over the set, but also towards approximating various properties regarding a binary relation defined on this set. As we shall see, in many cases, the sampling method used (or at least its analysis) is significantly more sophisticated than the one employed in gathering statistics of the former type. But before doing so, we wish to further discuss the potential benefit in the notion of approximation underlining the definition of property testing.

Firstly, being close to a graph that has the property is a notion of approximation that, in certain applications, may be of direct value. Furthermore, in some cases, being close to a graph having the property translates to a standard notion of approximation (see Section 2.2). In other cases, it translates to a notion of "dual approximation" (see, again, Section 2.2).

Secondly, in some cases, we may be forced to take action without having the time to run a decision procedure, while given the option of modifying the graph in the future, at a cost proportional to the number of added/omitted edges. For example, suppose we are given a graph that represents some suggested design, where bipartite graphs correspond to good designs and changes in the design correspond to edge additions/omissions. Using a `Bipartiteness` tester, we may (with high probability) accept any good design, while rejecting designs that will cost a lot to modify. That is, we may still accept designs that are not good, but only such that are close to being good and thus will not cost too much to modify later.

Thirdly, we may use the property tester as a preliminary stage before running a slower exact decision procedure. In case the graph is far from having the property, with high probability, we obtain an indication towards this fact, and save the time we might have used running the decision procedure. Furthermore, if the tester has one-sided error (i.e., it always accepts a graph having the property) and the tester has rejected, then we have obtained an absolutely correct answer without running the slower decision procedure at all. The saving provided by using a property tester as a preliminary stage may be very substantial in many natural settings where *typical* graphs either have the property or are very far from having the property. Furthermore, *if* it is *guaranteed* that graphs either have the property or are very far from having it *then* we may not even need to run the (exact) decision procedure at all.

1.3   Three Models of Testing Graph Properties

A graph property is a set of graphs closed under graph isomorphism (renaming of vertices).[1] Let Π be such a property. A Π-tester is a *randomized* algorithm that is given oracle access to a graph, $G = (V, E)$, and has to determine whether the graph is in Π or is far from being in Π. The type of oracle (equiv., the type of queries allowed) and distance-measure depend on the model, and we focus on three such models:

1. The adjacency predicate model [32]: Here the Π-tester is given oracle access to a symmetric function $g : V \times V \to \{0, 1\}$ that represents the adjacency predicate of the graph G; that is $g(u, v) = 1$ if and only if $(u, v) \in E$. In

---

[1] That is, Π is a graph property if, for every graph $G = (V, E)$ and every permutation π over V, it holds that $G \in \Pi$ if and only if $\pi(G) \in \Pi$, where $\pi(G) \stackrel{\text{def}}{=} (V, \{\{\pi(u), \pi(v)\} : \{u, v\} \in E\})$.

this model distances between graphs are measured according to their representation; that is, if the graphs G and G' are represented by the functions g and g', then their relative distance is the fraction of pairs (u, v) such that $g(u, v) \neq g'(u, v)$.

Note that saying that $G = ([N], E)$ is ϵ-far from the graph property Π means that for every $G' \in \Pi$ it holds that G is ϵ-far from G'. Since Π is closed under graph isomorphism, this means that G is ϵ-far from any isomorphic copy of $G' = ([N], E')$; that is, for every permutation π over $[N]$, it holds that $|\{(u, v) : g(u, v) \neq g'(\pi(u), \pi(v))\}| > \epsilon N^2$, where g and g' are as above.

Finally, note that this notion of distance between graphs is most meaningful in the case that the graph is dense (since in this case fractions of the number of possible vertex pairs are closely related to fractions of the actual number of edges). Thus, this model is often called the dense graph model.

2. The incidence function model [34]: Here, for some fixed upper bound d (on the degrees of vertices in G), the Π-tester is given oracle access to a function $g : V \times [d] \to V \cup \{\bot\}$ that represents the graph $G = (V, E)$ such that $g(u, i) = v$ if v is the ith vertex incident at u and $g(u, i) = \bot$ if u has less than i neighbors. That is, $E = \{(u, v) : \exists i \ f(u, i) = v\}$, where we always assume that $g(u, i) = v$ if and only if there exists a $j \in [d]$ such that $g(v, j) = u$.

Indeed, only graphs of degree at most d can be represented in this model, which is called the bounded-degree graph model.

In this model too, distances between graphs are measured according to their representation, but here the representation is different and so the distances are different. Specifically, if the graphs G and G' are represented by the functions g and g', then their relative distance is the fraction of pairs (u, i) such that $g(u, i) \neq g'(u, i)$. Again, saying that $G = ([N], E)$ is ϵ-far from the graph property Π means that for every $G' \in \Pi$ it holds that G is ϵ-far from G'. Since Π is closed under graph isomorphism (and the ordering of the vertices incident at each vertex is arbitrary), this means that for every permutation π over $[N]$, it holds that

$$\sum_{u \in V} |\{v : \exists i \ g(u, i) = v\} \triangle \{v : \exists i \ g'(\pi(u), i) = \pi(v)\}| > \epsilon dN \,,$$

where g and g' are as above, and $\triangle$ denotes the symmetric difference (i.e., $A \triangle B = (A \cup B) \setminus (A \cap B)$).

3. The general graph model [52,46]: In contrast to the foregoing two models in which the oracle queries and the distances between graphs are linked to the representation of graphs as functions, in the following model the representation is blurred and the query types and distance measure are decoupled.

The relative distance between the graphs $G = ([N], E)$ and $G' = ([N], E)$ is usually defined as $\frac{|E \triangle E'|}{\max(|E|, |E'|)}$; that is, the absolute distance is normalized by the actual number of edges rather than by an absolute upper bound (on the number of edges) such as $N^2/2$ or $dN/2$.

The types of queries typically considered are the two types of queries considered in the previous two models. That is, the algorithm may ask whether two vertices are adjacent in the graph and may also ask for a specific neighbor of a specific vertex.

Needless to say, the general graph model is the most general one, and it is indeed closest to actual algorithmic applications.[2] The fact that this model has so far received relatively little attention merely reflects the fact that its study is overly complex. Given that current studies of the other models still face formidable difficulties (and that these models offer a host of interesting open problems), it is natural that researchers shy away from yet another level of complication.

The current focus on query complexity. Although property testing is motivated by referring to super-fast algorithms, research in the area tends to focus on the *query complexity* of testing various properties. This focus should be viewed as providing an initial estimate to the actual complexity of the testing problems involved; certainly, query complexity lower bounds imply corresponding bounds on the time complexity, whereas the latter is typically at most exponential in the query complexity. Furthermore, in many cases, the time complexity is polynomial in the query complexity and this fact is typically stated. Thus, we will follow the practice of focusing on the query complexity of testing, but also mention time complexity upper bounds whenever they are of interest.

1.4   Organization

The following three sections are devoted to the three models discussed above: We start with the dense graph model (Section 2), then move to the bounded-degree model (Section 3), and finally get to the general graph model (Section 4). In each model we review the definition of testing (when specialized to that model), provide a taste of the known results, and demonstrate some of the ideas involved (by focusing on testing Bipartiteness, which seems a good benchmark).

We conclude this article with a discussion of a few issues that are relevant to all models; these include the treatment of directed graphs (Section 5.1), the related notions of tolerant testing and distance approximation (Section 5.2), and the notion of proximity oblivious testing (Section 5.3).

The appendix presents three observations that occurred to us in the process of writing this article. These refer to testing (degree) regularity in the dense graph model (Appendix A.1), non-adaptive testers in the bounded-degree graph model (Appendix A.2), and testing strong connectivity of directed graphs by only using forward queries (Appendix A.3).

---

[2] In other words, this model is relevant for most applications, since these seem to refer to general graphs (which model various natural and artificial objects). In contrast, the dense graph model is relevant to applications that refer to (dense) binary relations over finite graphs.

2   The Dense Graph Model

In the adjacency matrix model (a.k.a the dense graph model), an N-vertex graph $G = ([N], E)$ is represented by the Boolean function $g : [N] \times [N] \to \{0, 1\}$ such that $g(u, v) = 1$ if and only if u and v are adjacent in G (i.e., $\{u, v\} \in E$). Distance between graphs is measured in terms of their aforementioned representation (i.e., as the fraction of (the number of) different matrix entries (over N^2)), but occasionally one uses the more intuitive notion of the fraction of (the number of) unordered vertex pairs over $\binom{N}{2}$.[3] Recall that we are interested in *graph properties*, which are sets of graphs that are closed under isomorphism; that is, Π is a graph property if for every graph $G = ([N], E)$ and every permutation π of $[N]$ it holds that $G \in \Pi$ if and only if $\pi(G) \in \Pi$, where $\pi(G) \stackrel{\text{def}}{=} ([N], \{\{\pi(u), \pi(v)\} : \{u, v\} \in E\})$. We now spell out the meaning of property testing in this model.

Definition 2.1 (testing graph properties in the adjacency matrix model): *A tester for a graph property Π is a probabilistic oracle machine that, on input parameters N and ϵ and access to (the adjacency predicate of) an N-vertex graph $G = ([N], E)$, outputs a binary verdict that satisfies the following two conditions.*

1. *If $G \in \Pi$ then the tester accepts with probability at least $2/3$.*
2. *If G is ϵ-far from Π then the tester accepts with probability at most $1/3$, where G is ϵ-far from Π if for every N-vertex graph $G' = ([N], E') \in \Pi$ it holds that the symmetric difference between E and E' has cardinality that is greater than $\epsilon \cdot \binom{N}{2}$.*

If the tester accepts every graph in Π with probability 1, then we say that it has one-sided error. *A tester is called* non-adaptive *if it determines all its queries based solely on its internal coin tosses (and the parameters N and ϵ); otherwise it is called* adaptive.

The query complexity of a tester is the number of queries it makes to any N-vertex graph, as a function of the parameters N and ϵ. We say that a tester is efficient if it runs in time that is polynomial in its query complexity, where basic operations on elements of $[N]$ (and in particular, uniformly selecting an element in $[N]$) are counted at unit cost.

We stress that testers are defined as (uniform)[4] algorithms that are given the size parameter N and the distance (or proximity) parameter ϵ as explicit inputs. This uniformity (over the values of the distance parameter) makes the positive

---

[3] Indeed, there is a tiny discrepancy between these two measures, but it is immaterial in all discussions.

[4] That is, we refer to the standard (uniform) model of computation (cf., e.g., [31, Sec. 1.2.3]), which does not allow for hard-wiring some parameters (e.g., input length) into the computing device (as done in the case of non-uniform circuit families).

results stronger and more appealing (especially in light of a separation result shown in [10]). In contrast, negative results typically refer to a fixed value of the distance parameter.

The study of property testing in the dense graph model was initiated by Goldreich, Goldwasser, and Ron [32], as a concrete and yet general framework for the study of property testing at large. From that perspective, it was most natural to represent graphs as Boolean functions, and the adjacency matrix representation was the obvious choice. This dictated the choice of the type of queries as well as the distance measure. In retrospect, the dense graph model seems most natural when graphs are viewed as representing generic (symmetric) binary relations (cf. the second motivation to the study of graphs mentioned in Section 1.1 as well as the discussion of sampling in Section 1.2).

2.1 A Taste of the Known Results

We first mention that graph properties of arbitrary query complexity are known: Specifically, in this model, graph properties (even those in $\mathcal{P}$) may have query complexity ranging from $O(1/\epsilon)$ to $\Omega(N^2)$, and the same holds also for monotone graph properties in $\mathcal{NP}$ (cf. [33]).[5] In this overview, we focus on properties that can be tested within *query complexity that only depends on the proximity parameter* (i.e., ϵ); that is, *the query complexity does not depend on the size of the graph being tested.* Interestingly, there is much to say about this class of properties. Let us start with a brief summary, and provide more details later.

1. A celebrated result of Alon, Fischer, Newman, and Shapira [3] provides a combinatorial characterization of the class of properties that can be tested within query complexity that only depends on the proximity parameter. This class contains natural properties that are not testable in query complexity poly$(1/\epsilon)$; see [1].
2. The prior work of Goldreich, Goldwasser, and Ron [32] provides a natural class of properties that can be tested within query complexity poly$(1/\epsilon)$. This class consists of so-called "partition problems" and includes sets such as k-colorability, for any fixed $k \geq 2$, and graphs containing a clique for density ρ, for any fixed $\rho > 0$.
3. A relatively recent work of Goldreich and Ron [38] initiates a study of the class of properties that can be tested within query complexity $\widetilde{O}(1/\epsilon)$.

Before providing more details on the foregoing results, we mention that, when disregarding a possible quadratic blow-up in the query complexity, we may assume that the tester in canonical in the following sense.

---

[5] We mention that a full query complexity hierarchy is established in [33] by using unnatural graph properties, starting from the $\Omega(N^2)$ lower bound of [32], which also uses an unnatural graph property. In contrast, the $\Omega(N)$ lower bound established in [27] (following [2]) refers to the natural property of testing whether an N-vertex graph consists of two isomorphic copies of some $N/2$-vertex graph.

Theorem 2.2 (canonical testers [40, Thm 2]):[6] *Let Π be any graph property. If there exists a tester with query complexity $q(N, \epsilon)$ for Π, then there exists a tester for Π that uniformly selects a set of $O(q(N, \epsilon))$ vertices and accepts iff the induced subgraph has property Π', where Π' is a graph property that may depend on N as well as on Π. Furthermore, if the original tester has one-sided error, then so does the new tester, and a sample of $2q(N, \epsilon)$ vertices suffices.*

Indeed, the resulting tester is called canonical. We warn that Π' need not equal Π (let alone that Π' may depend on N), and that the time complexity of the canonical tester may be significantly larger than the time complexity of the original tester. Still, in many natural cases (e.g., k-colorability), $\Pi' = \Pi$.

2.1.1 Testability in $q(\epsilon)$ Queries, for Any Function q

As stated above, a celebrated result of Alon *et al.* [3] provides a combinatorial characterization of the class of properties that can be tested within query complexity that only depends on the proximity parameter. This characterization refers to the notion of a *regularity instance*, where regularity is in the sense of Szemerédi's Regularity Lemma [57]. The result essentially asserts that a graph property can be tested in query complexity that only depends on ϵ if and only if it can be characterized in terms of a constant number of regularity instances. The lesson from this characterization is that, when ignoring the specific dependency on ϵ, *testing graph properties in query complexity that only depends on ϵ reduces to graph regularity.* This lesson makes more concrete the feeling already raised by Theorem 2.2 that testing in this model reduces to combinatorics.

The downside of the algorithms that emerge from this characterization is that their query complexity is related to the proximity parameter via a function that grows tremendously fast. Specifically, in the general case, the query complexity is only upper bounded by a tower of a tower of exponents (in a monotonically growing function of $1/\epsilon$, which in turn depends on the property at hand).

Interestingly, it is known that a super-polynomial dependence on the proximity parameter is inherent to the foregoing result. Actually, as shown by Alon [1], such a dependence is essential even for testing *triangle freeness*. Indeed, this fact provides a nice demonstration of the non-triviality of testing graph properties. *One might have guessed that $O(1/\epsilon)$ or $O(1/\epsilon^3)$ queries would have sufficed to detect a triangle in any graph that is ϵ-far from being triangle free, but Alon's result asserts that this guess is wrong and that* $\mathrm{poly}(1/\epsilon)$ *queries do not suffice.* We mention that the best upper bound known for the query complexity of testing triangle freeness is $\mathtt{tf}(\mathrm{poly}(1/\epsilon))$, where $\mathtt{tf}$ is the tower function defined inductively by $\mathtt{tf}(n) = \exp(\mathtt{tf}(n-1))$ with $\mathtt{tf}(1) = 2$ (cf. [1]).

---

[6] As pointed out in [10], the statement of [40, Thm 2] should be corrected such that the auxiliary property Π' may depend on N and not only on Π. Thus, on input N and ϵ (and oracle access to an N-vertex graph G), the canonical tester checks whether a random induced subgraph of size $s = O(q(N, \epsilon))$ has the property Π', where Π' itself (or rather its intersection with the set of s-vertex graphs) may depend on N. In other words, the tester's decision depends only on the induced subgraph that it sees and on the size parameter N.

Perspective. It is indeed an amazing fact that many properties can be tested within (query) complexity that only depends on the proximity parameter (rather than also on the size of the object being tested). This amazing statement seems to shadow the question of the form of the aforementioned dependence, and blurs the difference between a reasonable dependence (e.g., a polynomial relation) and a prohibiting one (e.g., a tower-function relation). We beg to disagree with this sentiment and claim that, as in the context of standard approximation problems (cf. [44]), *the dependence of the complexity on the approximation* (or proximity) *parameter is a key issue.*

We wish to stress that we do value the impressive results of [2,7,8,29] (let alone [3]), which refer to graph property testers having query complexity that is independent of the graph size but depends prohibitively on the proximity parameter. We view such results as an impressive first step, which called for further investigation directed at determining the actual dependency of the complexity on the proximity parameter.

While it is conceivable that there exist (natural) graph properties that can be tested in $\exp(1/\epsilon)$ queries but not in $\operatorname{poly}(1/\epsilon)$ queries, we are not aware of such a property.[7] We thus move directly from complexities of the form $\mathtt{tf}(1/\epsilon)$ (and larger) to complexities of the form $\operatorname{poly}(1/\epsilon)$.

2.1.2   Testability in $\operatorname{poly}(1/\epsilon)$ Queries

Testers of query complexity $\operatorname{poly}(1/\epsilon)$ are known for several natural graph properties [32].

- k-Colorability, for any fixed $k \geq 2$. The query-complexity is $\operatorname{poly}(k/\epsilon)$. For $k = 2$ the running-time is $\widetilde{O}(1/\epsilon^3)$, whereas for $k > 2$ the running-time is $\exp(\operatorname{poly}(1/\epsilon))$ (and running-time polynomial in $1/\epsilon$ is unlikely, since k-Colorability is NP-complete, for $k \geq 3$).

  The k-Colorability tester has one-sided error; that is, in case the graph is k-colorable, the tester always accepts. Furthermore, when rejecting a graph, this tester always supplies a small counterexample (i.e., a $\operatorname{poly}(1/\epsilon)$-size subgraph that is not k-colorable).

  The 2-Colorability (equivalently, Bipartiteness) Tester is presented in §2.3. An improved analysis has been obtained by Alon and Krivelevich [4].
- ρ-Clique, for any fixed $\rho > 0$, where ρ-Clique is the set of graphs that have a clique of density ρ (i.e., N-vertex graphs having a clique of size ρN).
- ρ-CUT, for any fixed $\rho > 0$, where ρ-CUT is the set of graphs that have a cut of density at least ρ (compared to N^2).

  A generalization to k-way cuts has query-complexity $\operatorname{poly}((\log k)/\epsilon)$.
- ρ-Bisection, for any fixed $\rho > 0$, where ρ-Bisection is the set of graphs that have a bisection of density at most ρ (i.e., an N-vertex graph is in ρ-Bisection if its vertex set can be partitioned into two equal parts with at most ρN^2 edges going between them).

---

[7] Needless to say, demonstrating the existence of such (natural) properties is an interesting open problem.

Except for k-Colorability, all the other testers have two-sided error, and this is unavoidable for any tester of $o(N)$ query complexity for any of these properties.

All the above property testing problems are special cases of the General Graph Partition Testing Problem, which is parameterized by a set of lower and upper bounds. In this problem one needs to determine whether there exists a k-partition of the vertices so that the number of vertices in each part as well as the number of edges between each pair of parts falls between the corresponding lower and upper bounds (in the set of parameters). For example, ρ-clique is expressible as a 2-partition in which one part has ρN vertices, and the number of edges in this part is $\binom{\rho N}{2}$. A tester for the general problem also appears in [32]: The tester uses $\widetilde{O}(k^2/\epsilon)^{2k+O(1)}$ queries, and runs in time exponential in its query-complexity.

From testing to searching. Interestingly, the testers for (all cases of) the General Graph Partition Problem can be modified into algorithms that find an (implicit representation of an) approximately adequate partition whenever it exists. That is, if the graph has the desired (partitioning) property, then the testing algorithm may actually output auxiliary information that allows to reconstruct, in $\mathrm{poly}(1/\epsilon) \cdot N$-time, a partition that approximately obeys the property. For example, for ρ-CUT, we can construct a partition with at least $(\rho - \epsilon) \cdot N^2$ crossing edges. We comment that this notion of an implicit representation of an adequate structure may be relevant for other sets in $\mathcal{NP}$, where this structure corresponds to an NP-witness. (Indeed, an interesting algorithmic application was presented in [28], where an implicit partition of an imaginary hypergraph is used in order to efficiently construct a regular partition (with almost optimal parameters) of a given graph.)

Back to testing graph properties. Although many natural graph properties can be formulated as partition problems, many other properties that can be tested with $\mathrm{poly}(1/\epsilon)$ queries cannot be formulated as such problems. The list include the set of regular graphs, connected graphs, planar graphs, and more. We identify three classes of such natural properties:

1. Properties that only depends on the vertex degree distribution (e.g., degree regularity and average degree). For example, for any fixed $\rho > 0$, the set of N-vertex graphs having ρN^2 edges can be tested using $O(1/\epsilon^2)$ queries, which is the best result possible.[8] The same holds with respect to testing degree regularity, where the $\Omega(1/\epsilon^2)$ queries lower bound follows by reduction to estimating the average value of Boolean functions and a corresponding upper bound can be obtained by building on the $\widetilde{O}(1/\epsilon^3)$-query algorithm presented in the proof of [32, Prop. 10.2.1.3].[9]

2. Properties that are satisfied only by sparse graphs (i.e., N-vertex graphs having $O(N)$ edges)[10] such as Cycle-freeness and Planarity. These

---

[8] Both upper and lower bounds can be proved by reduction to the problem of estimating the average value of Boolean functions (cf. [22]).

[9] For the lower bound, consider the problem of distinguishing between a random N-vertex graph in which each vertex has degree either $(0.5 + \epsilon)N$ or $(0.5 - \epsilon)N$ and a random $(N/2)$-regular N-vertex graph. For the upper bound, see Appendix A.1.

properties can be tested by rejecting any graph that is not sufficiently sparse (see [32, Prop. 10.2.1.2]).

3. Properties that are almost trivial in the sense that, for some constant $c > 0$ and every $\epsilon > N^{-c}$, all N-vertex graphs are ϵ-close to the property. For example, every N-vertex graph is N^{-1}-close to being connected (or being Hamiltonian or Eulerian). These properties can be tested by accepting any N-vertex graph if $\epsilon > N^{-c}$ (without making any query), and inspecting the entire graph otherwise (where, in this case $\binom{N}{2} = \text{poly}(1/\epsilon)$). (See [32, Prop. 10.2.1.1].)

In view of all of the foregoing, we believe that characterizing the class of graph properties that can be tested in $\text{poly}(1/\epsilon)$ queries may be very challenging. We mention that the special case of induced subgraph freeness properties was resolved in [9].

2.1.3   Testability in $\widetilde{O}(1/\epsilon)$ Queries

While Theorem 2.2 may be interpreted as suggesting that testing in the dense graph model leaves no room for algorithmic design, this conclusion is valid only if one ignores a possible quadratic blow-up in the query complexity (and also disregards the time complexity). As advocated by Goldreich and Ron [38], a finer examination of the model, which takes into account the exact query complexity (i.e., cares about a quadratic blow-up), reveals the role of algorithmic design. In particular, the results in [38] distinguish adaptive testers from non-adaptive ones, and distinguish the latter from canonical testers. These results refer to testability in $\widetilde{O}(1/\epsilon)$ queries. In particular, it is shown that:

– Testing every "non-trivial for testing" graph property requires $\Omega(1/\epsilon)$ queries, even when adaptive testers are allowed. Furthermore, any canonical tester for such a property requires $\Omega(1/\epsilon^2)$ queries.
– There exists a natural graph property that can be tested by $\widetilde{O}(1/\epsilon)$ adaptive queries, requires $\Omega(\epsilon^{-4/3})$ non-adaptive queries, and is actually testable by $O(\epsilon^{-4/3})$ non-adaptive queries.
– There exists a natural graph property that can be tested by $\widetilde{O}(1/\epsilon)$ adaptive queries but requires $\Omega(\epsilon^{-3/2})$ non-adaptive queries.
– There exist an infinite class of natural graph properties that can be tested by $\widetilde{O}(1/\epsilon)$ non-adaptive queries.

All the above testers have one-sided error probability and are efficient, whereas the lower bounds hold also for two-sided error testers (regardless of efficiency).

The foregoing results seem to indicate that even at this low complexity level (i.e., testing in $\widetilde{O}(1/\epsilon)$ adaptive queries) there is a lot of structure and much to be understood. In particular, it is conjectured in [38] that, for every $t \geq 4$, there exists graph properties that can be tested by $\widetilde{O}(1/\epsilon)$ adaptive queries and have non-adaptive query complexity $\Theta(\epsilon^{-2+\frac{2}{t}})$.

---

[10] Actually, this class can be extended by considering a more relaxed notion of sparseness that includes N-vertex graphs having $O(N^{2-\Omega(1)})$ edges.

2.1.4 Reflections

Let us reflect about some issues that arise from the foregoing exposition.

Adaptive testers versus non-adaptive ones. Recall that Theorem 2.2 asserts that canonical testers (which are in particular non-adaptive) have query complexity that is at most quadratic in the query complexity of general (possibly adaptive) testers. Still the results surveyed in §2.1.3 indicate that such a gap may exist. An interesting question, raised by Michael Krivelevich, is whether such a gap exists also for properties having query complexity that is significantly larger than $\widetilde{O}(1/\epsilon)$. In particular, we mention that testing Bipartiteness, which has non-adaptive query complexity $\widetilde{\Theta}(\epsilon^{-2})$ (cf. [4,21])[11] and requires $\Omega(\epsilon^{-3/2})$ adaptive queries [21], may be testable in $o(\epsilon^{-2})$ adaptive queries (cf. [41]).

One-sided versus two-sided error probability. As noted above, for many natural properties there is a significant gap between the complexity of one-sided and two-sided error testers. For example, ρ-CUT has a two-sided error tester of query complexity poly$(1/\epsilon)$, but no one-sided error tester of query complexity $o(N^2)$. In general, the interested reader may contrast the characterization of two-sided error testers in [3] with the results in [8].

A contrast to recognizing graph properties. The notion of testing a graph property Π is a *relaxation* of the classical notion of *recognizing the graph property Π*, which has received much attention since the early 1970's (cf. [47]). In the classical (recognition) problem there are no margins of error; that is, one is required to accept all graphs having property Π and reject all graphs that lack property Π. In 1975, Rivest and Vuillemin resolved the Aanderaa–Rosenberg Conjecture, showing that any deterministic procedure for deciding any non-trivial monotone N-vertex graph property must examine $\Omega(N^2)$ entries in the adjacency matrix representing the graph. The query complexity of randomized decision procedures was conjectured by Yao to be $\Omega(N^2)$, and the currently best lower bound is $\Omega(N^{4/3})$. This stands in striking contrast to the aforementioned results regarding testing graph properties that establish that many natural (non-trivial) monotone graph properties can be *tested* by examining a constant number of locations in the matrix (where this constant depends on the constant value of the proximity parameter).

Graph properties are poor codes. We note that with the exception of two properties, which each contains a single N-vertex graph, the adjacency matrix representation of any property Π_N of N-vertex graphs yields a code over $\{0,1\}^{\binom{N}{2}}$ with relative distance at most $O(1/N)$. Specifically, if Π_N neither consists of the N-vertex clique nor of the N-vertex independent set, then Π_N contains a graph $G = ([N], E)$ that contains two vertices $u, v \in [N]$ that have different neighborhoods in G. Consider a permutation π that transposes u and v, while leaving the rest of $[N]$ intact, and let $G' = ([N], \{\pi(a), \pi(b) : (a, b) \in E\})$. Then $G' \in \Pi_N$, but G' is $\frac{2N}{\binom{N}{2}}$-close to G.

---

[11] The $\widetilde{O}(\epsilon^{-2})$ upper bound is due to [4], improving over [32], whereas the $\Omega(\epsilon^{-2})$ lower bound is due to [21].

2.2 Testing versus Other Forms of Approximation

We shortly discuss the relation of the notion of approximation underlying the definition of testing graph properties (in the dense graph model)[12] to more traditional notions of approximation. Throughout this section, we refer to randomized algorithms that have a small error probability, which we ignore for simplicity.

Application to the standard notion of approximation: The relation of testing graph properties to standard notions of approximation is best illustrated in the case of Max-CUT. Any tester for the set ρ-CUT, working in time $T(\epsilon, N)$, yields an algorithm for approximating the size of the maximum cut in an N-vertex graph, up to additive error ϵN^2, in time $\frac{1}{\epsilon} \cdot T(\epsilon, N)$. Thus, for any constant $\epsilon > 0$, using the above tester of [32], we can approximate the size of the max-cut to within ϵN^2 in constant time. This yields a constant time approximation scheme (i.e., to within any constant relative error) for dense graphs, which improves over previous work of Arora *et al.* [12] and de la Vega [24] who solved this problem in polynomial-time (i.e., in $O(N^{1/\epsilon^2})$–time and $\exp(\widetilde{O}(1/\epsilon^2)) \cdot N^2$–time, respectively). In the latter works the problem is solved by actually finding approximate max-cuts. Finding an approximate max-cut does not seem to follow from the mere existence of a tester for ρ-cut; yet, as mentioned above, the tester in [32] can be used to find such a cut in time linear in N.

Relation to "dual approximation" (cf. [44, Chap. 3]): To illustrate this relation, we consider the aforementioned ρ-Clique Tester. The traditional notion of approximating Max-Clique corresponds to distinguishing the case in which the max-clique has size at least ρN from, say, the case in which the max-clique has size at most $\rho N/2$. On the other hand, when we talk of testing ρ-Clique, the task is to distinguish the case in which an N-vertex graph has a clique of size ρN from the case in which it is ϵ-far from the class of N-vertex graphs having a clique of size ρN. This is equivalent to the "dual approximation" task of distinguishing the case in which an N-vertex graph has a clique of size ρN from the case in which any ρN subset of the vertices misses at least ϵN^2 edges. To demonstrate that these two tasks are vastly different we mention that whereas the former task is NP-Hard, for $\rho < 1/4$ (see [15,42]), the latter task can be solved in $\exp(O(1/\epsilon^2))$-time, for any $\rho, \epsilon > 0$. We believe that there is no absolute sense in which one of these approximation tasks is more important than the other: Each of these tasks may be relevant in some applications and irrelevant in others.

2.3 A Benchmark: Testing Bipartiteness

The Bipartite tester is extremely simple: It selects a tiny, random set of vertices and checks whether the induced subgraph is bipartite.

Algorithm 2.3 (Bipartite Tester in the Dense Graph Model [32]): *On input N, ϵ and oracle access to an adjacency predicate of an N-vertex graph, $G = (V, E)$:*

---

[12] Analogous relations hold also in the other models of testing graph properties.

1. *Uniformly select a subset of $\widetilde{O}(1/\epsilon^2)$ vertices of* V.
2. *Accept if and only if the subgraph induced by this subset is bipartite.*

Step (2) amounts to querying the predicate on all pairs of vertices in the subset selected at Step (1), and testing whether the induced graph is bipartite (e.g., by running BFS). As will become clear from the analysis, it actually suffice to query only $\widetilde{O}(1/\epsilon^3)$ of these pairs. We comment that a more complex analysis due to Alon and Krivelevich [4] implies that the Algorithm 2.3 is a Bipartite Tester even if one selects only $\widetilde{O}(1/\epsilon)$ vertices (rather than $\widetilde{O}(1/\epsilon^2)$) in Step (1)).

Theorem 2.4 [32]: *Algorithm 2.3 is a* Bipartite *Tester (in the dense graph model). Furthermore, the algorithm always accepts a bipartite graph, and in case of rejection it provides a witness of length* poly$(1/\epsilon)$ *(that the graph is not bipartite).*

Proof: Let R be the subset selected in Step (1), and G_R the subgraph of G induced by R. Clearly, if G is bipartite then so is G_R, for any R. The point is to prove that if G is ϵ-far from bipartite then the probability that G_R is bipartite is at most $1/3$. Thus, from this point on we assume that at least ϵN^2 edges have to be omitted from G to make it bipartite.

We view R as a union of two disjoint sets U and S, where $t \overset{\text{def}}{=} |U| = O(\epsilon^{-1} \cdot \log(1/\epsilon))$ and $m \overset{\text{def}}{=} |S| = O(t/\epsilon)$. We will consider all possible partitions of U, and associate a partial partition of V with each such partition of U. The idea is that in order to be consistent with a given partition, (U_1, U_2), of U, all neighbors of U_1 (respectively, U_2) must be placed opposite to U_1 (respectively, U_2). We will show that, with high probability, most high-degree vertices in V do neighbor U and so are forced by its partition. Since there are relatively few edges incident to vertices that do not neighbor U, it follows that, with very high probability, each such partition of U is detected as illegal by G_R. Details follow, but before we proceed let us stress the key observation: *It suffices to rule out relatively few (partial) partitions of* V (i.e., these induced by partitions of U), rather than all possible partitions of V.

We use the notations $\Gamma(v) \overset{\text{def}}{=} \{u : (u,v) \in E\}$ and $\Gamma(X) \overset{\text{def}}{=} \cup_{v \in X} \Gamma(v)$. Given a partition (U_1, U_2) of U, we define a (possibly partial) partition, (V_1, V_2), of V so that $V_1 \overset{\text{def}}{=} \Gamma(U_2)$ and $V_2 \overset{\text{def}}{=} \Gamma(U_1)$ (assume, for simplicity that $V_1 \cap V_2$ is indeed empty). As suggested above, if one claims that G can be "bi-partitioned" with U_1 and U_2 on different sides, then $V_1 = \Gamma(U_2)$ must be on the opposite side to U_2 (and $\Gamma(U_1)$ opposite to U_1). Note that the partition of U places no restriction on vertices that have no neighbor in U. Thus, we first ensure that *almost all* "influential" (i.e., "high-degree") vertices in V have a neighbor in U.

Technical Definition 2.4.1 (high-degree vertices and good sets): *We say that a vertex $v \in$ V is of* high-degree *if it has degree at least $\frac{\epsilon}{3}N$. We call* U good *if all but at most $\frac{\epsilon}{3}N$ of the high-degree vertices in V have a neighbor in U.*

We comment that NOT insisting that a good set U neighbors *all* high-degree vertices allows us to show that, with high probability, a random U of size unrelated to the size of the graph is good. (In contrast, if we were to insist that a good U neighbors *all* high-degree vertices, then we would have had to use $|U| = \Omega(\log N)$.)

Claim 2.4.2. *With probability at least 5/6, a uniformly chosen set U of size t is good.*

Proof: For any high-degree vertex v, the probability that v does not have any neighbor in a uniformly chosen U is at most $(1 - \epsilon/3)^t < \frac{\epsilon}{18}$ (since $t = \Omega(\epsilon^{-1} \log(1/\epsilon))$). Hence, the expected number of high-degree vertices that do not have a neighbor in a random set U is less than $\frac{\epsilon}{18} \cdot N$, and the claim follows by Markov's Inequality.   □

Technical Definition 2.4.3 (disturbing a partition of U): *We say that an edge disturbs a partition* (U_1, U_2) *of* U *if both its end-points are in the same* $\Gamma(U_i)$, *for some* $i \in \{1, 2\}$.

Claim 2.4.4. *For any good set* U *and any partition of* U, *at least* $\frac{\epsilon}{3} N^2$ *edges disturb the partition.*

Proof: Each partition of V has at least ϵN^2 violating edges (i.e., edges with both end-points on the same side). We upper bound the number of these edges that are not disturbing. Actually, we upper bound the number of edges that have an end-point not in $\Gamma(U)$.

- The number of edges incident to high-degree vertices that do not neighbor U is bounded by $\frac{\epsilon}{3} N \cdot N$ (since there are at most $\frac{\epsilon}{3} N$ such vertices).
- The number of edges incident to vertices that are not of high-degree is bounded by $N \cdot \frac{\epsilon}{3} N$ (since each such vertex has at most $\frac{\epsilon}{3} N$ incident edges).

This leaves us with at least $\frac{\epsilon}{3} N^2$ violating edges connecting vertices in $\Gamma(U)$ (i.e., edges disturbing the partition of U).   □

The theorem follows by observing that G_R is bipartite only if either (1) the set U is not good; or (2) the set U is good and there exists a partition of U so that none of the disturbing edges occurs in G_R. Using Claim 2.4.2 the probability of event (1) is bounded by 1/6, whereas by Claim 2.4.4 the probability of event (2) is bounded by the probability that there exists a partition of U so that none of the corresponding $\geq \frac{\epsilon}{3} N^2$ disturbing edges has both end-points in the second sample S. Actually, we pair the m vertices of S, and consider the probability that none of these pairs is a disturbing edge for a partition of U. Thus the probability of event (2) is bounded by

$$2^{|U|} \cdot \left(1 - \frac{\epsilon}{3}\right)^{m/2} < \frac{1}{6}$$

where the inequality holds since $m = \Omega(t/\epsilon)$. The theorem follows.   ∎

Comment: The procedure employed in the proof yields a randomized poly($1/\epsilon$) $\cdot$ N-time algorithm for 2-partitioning a bipartite graph such that (with high probability) at most ϵN^2 edges lie within the same side. This is done by running the tester, determining a partition of U (defined as in the proof) that is consistent with the bipartite partition of R, and partitioning V as done in the proof (with vertices that do not neighbor U, or neighbor both U_1, U_2, placed arbitrarily). Thus, the placement of each vertex is determined by inspecting at most $\widetilde{O}(1/\epsilon)$ entries of the adjacency matrix. Furthermore, the aforementioned partition of U constitutes a succinct representation of the 2-partition of the entire graph. All this is a typical consequence of the fact that the analysis of the tester follows the "enforce-and-test" paradigm (see [55, Sec. 4]).

3   The Bounded-Degree Graph Model

The bounded-degree model refers to a fixed degree bound, denoted $d \geq 2$. An N-vertex graph $G = ([N], E)$ (of maximum degree d) is represented in this model by a function $g : [N] \times [d] \to \{0, 1, ..., N\}$ such that $g(v, i) = u \in [N]$ if u is the ith neighbor of v and $g(v, i) = 0$ if v has less than i neighbors.[13] Distance between graphs is measured in terms of their aforementioned representation (i.e., as the fraction of (the number of) different array entries (over dN)), but occasionally we shall use the more intuitive notion of the fraction of (the number of) edges over $dN/2$. We now spell out the meaning of property testing in this model.

Definition 3.1 (testing graph properties in the bounded-degree model): *For a fixed d, a* tester *for a graph property Π is a probabilistic oracle machine that, on input parameters N and ϵ and access to (the incidence function of) an N-vertex graph $G = ([N], E)$ of maximum degree d, outputs a binary verdict that satisfies the following two conditions.*

1. *If $G \in \Pi$ then the tester accepts with probability at least 2/3.*
2. *If G is ϵ-far from Π then the tester accepts with probability at most 1/3, where G is ϵ-far from Π if for every N-vertex graph $G' = ([N], E') \in \Pi$ of maximum degree d it holds that the symmetric difference between E and E' has cardinality that is greater than $\epsilon \cdot dN/2$.*

One-sided testers and non-adaptive testers are defined as in Definition 2.1.

The query complexity of a tester is defined as in Section 2; ditto for its efficiency.

The study of property testing in the bounded-degree graph model was initiated by Goldreich and Ron [34], with the aim of allowing the consideration of sparse graphs, which appear in numerous applications (cf. the first motivation to the study of graphs mentioned in Section 1.1). The point was that the

---

[13] For simplicity, we assume here that the neighbors of v appear in an arbitrary order in the sequence $g(v, 1), ..., g(v, \deg(v))$, where $\deg(v) \stackrel{\text{def}}{=} |\{i : g(v, i) \neq 0\}|$. Also, we shall always assume that if $g(v, i) = u \in [N]$ then there exists $j \in [d]$ such that $g(u, j) = v$.

dense graph model seems irrelevant to sparse graphs, both because the distance measure that underlies it deems all sparse graphs as close to one another, and because adjacency queries seems unsuitable for sparse graphs. Sticking to the paradigm of representing graphs as functions, where both the distance measure and the type of queries are determined by the representation, the aforementioned representation seemed the most natural choice. Indeed, a conscious decision was (and is) made not to capture, at this point (and in this model), sparse graphs that do not have constant (or low) maximum degree.

3.1   A Taste of the Known Results

We first mention that, also in this model, graph properties of arbitrary query complexity are known: Specifically, in this model, graph properties (in $\mathcal{NP}$) may have query complexity ranging from $O(1/\epsilon)$ to $\Omega(N)$, and furthermore such properties are monotone and natural (cf. [33], which builds over [20]). In particular, testing 3-Colorability requires $\Omega(N)$ queries, whereas testing 2-Colorability (i.e., Bipartiteness) requires $\Omega(\sqrt{N})$ queries [34] and can be done using $\widetilde{O}(\sqrt{N}) \cdot \mathrm{poly}(1/\epsilon)$ queries [35]. We also mention that many natural properties are testable in query complexity that only depends on the proximity parameter (i.e., ϵ). A partial list includes k-edge connectivity, for every fixed k, and Planarity (cf. [34] and [18], respectively). Details follow.

3.1.1   Testability in $q(\epsilon)$ Queries, for Any Function q
We first mention, that with the exception of properties that only depend on the degree distribution, adaptive testers are essential for obtaining query complexity that only depends on ϵ (cf. [54]).[14] Still, as observed in [39], at the cost of an exponentially blow-up in the query complexity, we may assume that the tester's adaptivity is confined to performing (full, BFS-like) searches of a predetermined depth from several randomly selected vertices. However, the best testing results are typically obtained by testers that either perform more adaptive searchers or perform DFS-like rather than BFS-like searchers. A few examples follow, where all testers are efficient (i.e., their running time is polynomial in their query complexity).

Testing connectivity. Graph connectivity can be tested in $\widetilde{O}(1/\epsilon)$ queries [34]. Essentially, the tester starts a search (e.g., a BFS) from a few randomly selected vertices, but each such search is terminated after a predetermined number of vertices is encountered (rather than after visiting all vertices that are at a predetermined distance from the start vertex). This tester rejects if and only if it detects a small connected component, and thus it has one-sided error. The result essentially extends to k-edge connectivity, for any $k \geq 2$, but the query complexity is $\widetilde{O}(k^3/\epsilon^c)$, where $c = \min(k-1, 3)$ (cf. [34]).

---

[14] Actually, the result extends to query complexity of the form $o(\sqrt{N} \cdot q(\epsilon))$, for any function q. In contrast, note that triangle-freeness can be tested by $O(\sqrt{N/\epsilon})$ non-adaptive queries; see Appendix A.2.

Testing cycle-freeness. Cycle-freeness can be tested in $\widetilde{O}(\epsilon^{-3})$ queries, by a tester having two-sided error [34]. Essentially, the tester compares the number of edges to the number of connected components, while fully exploring any small connected components that it happens to visit. The two-sided error is unavoidable by any tester that has query complexity $o(\sqrt{N})$ (cf. [34, Prop. 4.3]). Viewing cycle-free graphs as graphs that have no K_3-minor, leads us to the following general result of Benjamini, Schramm, and Shapira [18], which refers to graph minors (to be briefly recalled next).

The graph H is a minor of the graph G, if H can be obtained from G by a sequence of edge removal, vertex removal, and edge contraction operations. We say that G is H-minor free if H is not a minor of G. Thus, a graph is cycle-free if and only if it is K_3-minor free, where K_k denotes the k-vertex clique. (The notion of minor freeness extends to sets of graphs; that is, for a set of graphs $\mathcal{H}$, the graph G is $\mathcal{H}$-minor free if no element of $\mathcal{H}$ is a minor of G.) Lastly, a graph property is minor-closed if it is closed under removal of edges, removal of vertices, and edge contraction. Note that, for every finite sets of graphs $\mathcal{H}$, the property of being $\mathcal{H}$-minor free (e.g., Planarity) is minor-closed.

Theorem 3.2 ([43], improving over [18]):[15] *Any minor-closed property can be tested in query complexity* $\exp(\mathrm{poly}(1/\epsilon))$.

We mention that this tester has two-sided error, which is unavoidable for any tester of query complexity $o(\sqrt{N})$, except for the case that the forbidden minors are all cycle-free.

3.1.2   Testability in $\widetilde{O}(N^{1/2}) \cdot \mathrm{poly}(1/\epsilon)$ Queries

The query complexity of testing two natural properties is $\widetilde{\Theta}(N^{1/2}) \cdot \mathrm{poly}(1/\epsilon)$, and in both cases the time complexity has the same form. The properties are Bipartiteness and Expansion. In both cases, the algorithm is based on taking many (i.e., $\widetilde{O}(N^{1/2}) \cdot \mathrm{poly}(1/\epsilon)$) *random walks* from a few randomly selected vertices, where each walk has length $\mathrm{poly}(\epsilon^{-1} \log N)$.

The foregoing algorithmic approach originates in [35], where it was applied to testing Bipartiteness; for further details see §3.2.2. This approach was also suggested for testing Expansion [36], but the analysis was successfully completed only in [45,50]. We mention that the Bipartite tester has one-sided error, and whenever it rejects it may also output a short proof that the graph is not bipartite (i.e., an odd cycle of length $\mathrm{poly}(\epsilon^{-1} \log N)$).

The $\Omega(N^{1/2})$ lower bound on the query complexity of testing each of the aforementioned properties was proved in [34]; for details see §3.2.1. We note that the lower bound for testing Bipartiteness stands in sharp contrast to the situation in the dense graph model, where Bipartite testing is possible in $\mathrm{poly}(1/\epsilon)$-time. This discrepancy is due to the difference between the notions of relative distance employed in the two models.

An application to the study of the dense graph model. We mention that the Bipartiteness tester of the bounded-degree model was used in order to derive

---

[15] The query complexity obtained in [18] is triple-exponential in $1/\epsilon$.

an alternative Bipartite tester for the dense graph model [41]. In the case that almost all vertices in the N-vertex graph have degree $O(\epsilon^{0.99}N)$, this tester improves over the ones presented in [32,4]. Essentially, this dense-graph model tester invokes the bounded-degree model tester on the subgraph induced by a sample S of $\widetilde{O}(1/\epsilon)$ random vertices (and emulates neighbor queries regarding a vertex $v \in S$ by making adjacency queries of the form (v, w) for every $w \in S$).

3.1.3   Reflections

The fact that the bounded-degree model is closer (than the dense graph model) to standard algorithmic research offers greater interaction at the technical level. Indeed, techniques such as local search and random walks are quite basic in both domains, and the relationship becomes even tighter when we shall move to the general graph model (in Section 4). At the current point, we mention that the idea underlying the cycle-freeness tester (outlined in §3.1.1) was employed to the design of an algorithm for approximating the minimum spanning tree weight in sub-linear time [23].

We also mention that the idea underlying the expansion tester has become quite pivotal in the contents of testing distributions, which emerged with [13].

3.2   A Benchmark: Testing Bipartiteness

Both the following lower and upper bounds reflect the fact that being far from Bipartiteness does not require having constant size cycles of odd length. We comment that a simplified version of the upper bound implies that odd cycles of logarithmic length must exist (cf. [35, Prop. 1]).

3.2.1   A Lower Bound

In contrast to Theorem 2.4, under the incidence function representation, there exists no Bipartite tester of complexity that is independent of the graph size.

Theorem 3.3 [34]: *Testing* Bipartiteness *(with constant ϵ and d) requires* $\Omega(\sqrt{N})$ *queries (in the incidence function model).*

Proof Idea: For any (even) N, we consider the following two families of graphs:

1. The first family, denoted $\mathcal{G}_1^N$, consists of all degree-3 graphs that are composed of the union of a Hamiltonian cycle and a perfect matching. That is, there arc N edges connecting the vertices in a cycle, and the other $N/2$ edges are a perfect matching.
2. The second family, denoted $\mathcal{G}_2^N$, is the same as the first *except* that the perfect matchings allowed are restricted as follows: the distance on the cycle between every two vertices that are connected by a perfect matching edge must be odd.

Clearly, all graphs in $\mathcal{G}_2^N$ are bipartite. It can be shown that almost all graphs in $\mathcal{G}_1^N$ are far from being bipartite. On the other hand, one can prove that a testing algorithm that performs $o(\sqrt{N})$ queries cannot distinguish between a graph chosen randomly from $\mathcal{G}_2^N$ (which is always bipartite) and a graph chosen randomly from $\mathcal{G}_1^N$ (which with high probability is far from bipartite). Loosely speaking, this is the case since in both cases the algorithm is unlikely to encounter a cycle (among the vertices that it has inspected).   ■

3.2.2   An Algorithm

The lower bound of Theorem 3.3 is essentially tight. Furthermore, the following natural algorithm constitutes a Bipartite tester of running time $\text{poly}((\log N)/\epsilon)$ · $\sqrt{N}$.

Algorithm 3.4 (Bipartite Tester in the Bounded-Degree Model [35]): *On input N, d, ϵ and oracle access to an incidence function for an N-vertex graph, $G = (V, E)$, of degree bound d, repeat $T \stackrel{\text{def}}{=} \Theta(\frac{1}{\epsilon})$ times:*

1. *Uniformly select s in V.*
2. *(Try to find an odd cycle through vertex s):*
   (a) *Perform $K \stackrel{\text{def}}{=} \text{poly}((\log N)/\epsilon) \cdot \sqrt{N}$ random walks starting from s, each of length $L \stackrel{\text{def}}{=} \text{poly}((\log N)/\epsilon)$.*
   (b) *Let R_0 (respectively, R_1) denote the vertices set reached from s in an even (respectively, odd) number of steps in any of these walks.*
   (c) *If $R_0 \cap R_1$ is not empty then reject.*

If the algorithm did not reject in any of the foregoing T iterations, then it accepts.

Theorem 3.5 [35]: *Algorithm 3.4 is a Bipartite Tester (in the incidence function model). Furthermore, the algorithm always accepts a bipartite graph, and in case of rejection it provides a witness of length $\text{poly}((\log N)/\epsilon)$ (that the graph is not bipartite).*

Motivation – the special case of rapid mixing graphs. The proof of Theorem 3.5 is quite involved. As a motivation, we consider the special case where the graph has a "rapid mixing" feature. It is convenient to modify the random walks so that at each step each neighbor is selected with probability $1/2d$, and otherwise (with probability at least $1/2$) the walk remains in the present vertex. Furthermore, we will consider a single execution of Step (2) starting from an arbitrary vertex, s, which is fixed in the rest of the discussion. The rapid mixing feature we assume is that, for every vertex v, a (modified) random walk of length L starting at s reaches v with probability approximately $1/N$ (say, up-to a factor of 2). Note that if the graph is an expander then this is certainly the case (since $L = \omega(\log N)$).

The key quantities in the analysis are the following probabilities, referring to the parity *of the length of a path obtained from the random walk by omitting the self-loops* (transitions that remain at current vertex). Let $p^0(v)$ (respectively, $p^1(v)$) denote the probability that a (modified) *random walk of length L, starting at s, reaches v while making an even* (respectively, *odd*) *number of real* (i.e., non-self-loop) *steps*. By the rapid mixing assumption (for every $v \in V$), it holds that

$$\frac{1}{2N} < p^0(v) + p^1(v) < \frac{2}{N}. \tag{2}$$

We consider two cases regarding the sum $\sum_{v \in V} p^0(v)p^1(v)$: If the sum is (relatively) "small", we show that V can be 2-partitioned so that there are relatively few edges between vertices that are placed in the same side, which implies that G is close to being bipartite. Otherwise (i.e., when the sum is not "small"), we show that with significant probability, when Step (2) is started at vertex s it is completed by rejecting G. These two cases are analyzed in the following two (corresponding) claims.

Claim 3.5.1. *Suppose $\sum_{v \in V} p^0(v)p^1(v) \leq \epsilon/50N$. Let $V_1 \overset{\text{def}}{=} \{v \in V : p^0(v) < p^1(v)\}$ and $V_2 = V \setminus V_1$. Then, the number of edges with both end-points in the same V_σ is bounded above by ϵdN.*

Proof Sketch: Consider an edge (u, v) where, without loss of generality, both u and v are in V_1. Then, both $p^1(v)$ and $p^1(u)$ are greater than $\frac{1}{2} \cdot \frac{1}{2N}$. However, one can show that $p^0(v) > \frac{1}{3d} \cdot p^1(u)$: Observe that an $(L-1)$-step walk of path-parity 1 ending at u is almost as likely as an L-step walk of path-parity 1 ending at u, and that once an $(L-1)$-step walk reaches u, with probability exactly $1/2d$, it continues to v in the next step. Thus, the edge (u, v) contributes at least $\frac{(1/4N)^2}{3d}$ to the sum $\sum_{w \in V} p^0(w)p^1(w)$. It follows that we can have at most $(\epsilon/50N)/(1/48dN^2)$ such edges, and the claim follows. $\square$

Claim 3.5.2. *Suppose $\sum_{v \in V} p^0(v)p^1(v) \geq \epsilon/50N$, and that Step (2) is started with vertex s. Then, with probability at least $2/3$, the set $R_0 \cap R_1$ is not empty (and rejection follows).*

Proof Sketch: Consider the probability space defined by an execution of Step (2) with start vertex s. For every $i \neq j$ such that $i, j \in [K]$, we define an indicator random variable $\zeta_{i,j}$ representing *the event that the vertex encountered in the Lth step of the ith walk equals the vertex encountered in the Lth step of the jth walk, and that the ith walk corresponds to an even-path whereas the jth to an odd-path.* (That is, $\zeta_{i,j} = 1$ if the foregoing event holds, and $\zeta_{i,j} = 0$ otherwise.) Then

$$\mathbf{E}[|R_0 \cap R_1|] > \sum_{i \neq j} \mathbf{E}[\zeta_{i,j}]$$
$$= K(K-1) \cdot \sum_{v \in V} p^0(v)p^1(v)$$
$$> \frac{500N}{\epsilon} \cdot \sum_{v \in V} p^0(v)p^1(v)$$
$$\geq 10$$

where the second inequality is due to the setting of K, and the third to the claim's hypothesis. Intuitively, with high probability, it should hold that $|R_0 \cap R_1| > 0$. This is indeed the case, but proving it is less straightforward than it seems; the problem being that the $\zeta_{i,j}$'s are not pairwise independent. Yet, since the sum of the covariances of the dependent $\zeta_{i,j}$'s is quite small, Chebyshev's Inequality is still very useful (cf. [11, Sec. 4.3]). Specifically, letting $\mu \overset{\text{def}}{=} \sum_{v \in V} p^0(v)p^1(v)$ $(= \mathbf{E}[\zeta_{i,j}])$, and $\overline{\zeta}_{i,j} \overset{\text{def}}{=} \zeta_{i,j} - \mu$, we get:

$$\mathbf{Pr}\left[\sum_{i \neq j} \zeta_{i,j} = 0\right] < \frac{\mathbf{Var}\left[\sum_{i \neq j} \zeta_{i,j}\right]}{(K^2\mu)^2}$$

$$= \frac{1}{K^4\mu^2} \cdot \left(\sum_{i,j} \mathbf{E}\left[\overline{\zeta}_{i,j}^2\right] + 2\sum_{i,j,k} \mathbf{E}\left[\overline{\zeta}_{i,j}\overline{\zeta}_{i,k}\right]\right)$$

$$< \frac{1}{K^2\mu} + \frac{2}{K\mu^2} \cdot \mathbf{E}[\zeta_{1,2}\zeta_{1,3}]$$

For the second term, we observe that $\mathbf{Pr}[\zeta_{1,2} = \zeta_{1,3} = 1]$ is upper bounded by $\mathbf{Pr}[\zeta_{1,2} = 1] = \mu$ times the probability that the Lth vertex of the first walk appears as the Lth vertex of the third path. Using the rapid mixing hypothesis, we upper bound the latter probability by $2/N$, and obtain

$$\mathbf{Pr}[|R_0 \cap R_1| = 0] < \frac{1}{K^2\mu} + \frac{2}{K\mu^2} \cdot \mu \cdot \frac{2}{N}$$

$$< \frac{1}{3}$$

where the last inequality uses $\mu \geq \epsilon/50N$ and $K^2 \geq 6 \cdot 50N/\epsilon$ (along with $\epsilon > 5000/N$). The claim follows. □

Beyond rapid mixing graphs. The proof in [35] refers to a more general sum of products; that is, $\sum_{u \in U} p_{\text{odd}}(u)p_{\text{even}}(u)$, where $U \subseteq V$ is an appropriate set of vertices, and $p_{\text{odd}}(v)$ (respectively, $p_{\text{even}}(v)$) is essentially the probability that an L-step random walk (starting at s) passes through v after more than $L/2$ steps and the corresponding path to v has odd (respectively, even) parity. Much of the analysis in [35] goes into selecting the appropriate U (and an appropriate starting vertex s), and pasting together many such U's to cover all of V. Loosely speaking, U and s are selected so that there are few edges from U and the rest of the graph, and $p_{\text{odd}}(u) + p_{\text{even}}(u) \approx 1/\sqrt{|V| \cdot |U|}$, for every $u \in U$. The selection is based on the "combinatorial treatment of expansion" of Mihail [49]. Specifically, we use the contrapositive of the standard analysis, which asserts that rapid mixing occurs when all cuts are relatively large, to assert the existence of small cuts which partition the graph so that vertices reached with relatively high probability (in a short random walk) are on one side and the rest of the graph on the other. The first set corresponds to the aforementioned U and the cut is relatively small with respect to U. A start vertex s for which the corresponding sum is big is shown to cause Step (2) to reject (when started with this s), whereas a small corresponding sum enables to 2-partition U while having few violating edges among the vertices in each part of U.

The actual argument of [35] proceeds in iterations. In each iteration a vertex s for which Step (2) accepts with high probability is fixed, and an appropriate set of remaining vertices, U, is found. The set U is then 2-partitioned so that there are few violating edges inside U. Since we want to paste all these partitions together, U may not contain vertices treated in previous iterations. This complicates the analysis, since it must refer to the part of G, denoted H, not treated in previous iterations. We consider walks over an (imaginary) Markov Chain representing the H-part of the walks performed by the algorithm on G. Statements about rapid mixing are made with respect to the Markov Chain, and linked to what happens in random walks performed on G. In particular, a subset U of H is determined so that the vertices in U are reached with probability $\approx 1/\sqrt{|V| \cdot |U|}$ (in the chain) and the cut between U and the rest of H is small. Linking the sum of products defined for the chain with the actual walks performed by the algorithm, we infer that U may be partitioned with few violating edges inside it. Edges to previously treated parts of the graphs are charged to these parts, and edges to the rest of $H \setminus U$ are accounted for by using the fact that this cut is small (relative to the size of U).

4   The General Graph Model

In contrast to the foregoing two models in which the oracle queries and the distances between graphs are linked to the representation of graphs as functions, in the following model the representation is blurred and the query types and distance measure are decoupled. This decoupling makes the current model closer in spirit to standard studies in graph algorithms.

Giving up on the representation as a yardstick for the relative distance between graphs, leaves us with no absolute point of reference. Instead, we just define the relative distance between graphs in relation to the actual number of edges in these graphs; specifically, the relative distance between the graphs $G = ([N], E)$ and $G' = ([N], E)$ may be defined as $\frac{|E \triangle E'|}{\max(|E|,|E'|)}$ (or, alternatively, as $\frac{|E \triangle E'|}{(|E|+|E'|)/2}$).[16]

Turning to the question of query types, we again need to make a choice, which is now free from representation considerations. The most natural choice is to allow both *adjacency queries* and *incidence queries* (i.e., the two types of queries that were each allowed in one of the previous queries).[17] However, other choices has been considered too (cf. [17]). We note that, typically, adjacency queries become more useful as the graph becomes more dense, whereas incidence queries (a.k.a neighbor queries) become more useful as the graph becomes more sparse (cf. [17]).

Definition 4.1 (testing graph properties in the general model): *A tester for a graph property Π is a probabilistic oracle machine that, on input parameters*

---

[16] Needless to say, these two definitions may not yield the same result, but they are related by a factor of at most 2.

[17] Recall that the incidence query (u, i) is answered with 0 if u has less than i neighbors. Thus, the incidence queries allow to emulate degree queries at logarithmic cost.

N and ϵ and access to a function answering adjacency queries and incidence queries regarding an N-vertex graph G = ([N], E), outputs a binary verdict that satisfies the following two conditions.

1. *If G ∈ Π then the tester accepts with probability at least 2/3.*
2. *If G is ϵ-far from Π then the tester accepts with probability at most 1/3, where G is ϵ-far from Π if for every N-vertex graph G' = ([N], E') ∈ Π it holds that the symmetric difference between E and E' has cardinality that is greater than ϵ · max($|E|, |E'|$).*

One-sided testers and non-adaptive testers are defined as in Definition 2.1.

The query complexity of a tester is defined as in Section 2; ditto for its efficiency.

The study of property testing in the general graph model was initiated by Parnas and Ron [52], who only considered incidence queries, and extended by Kaufman, Krivelevich, and Ron [46], who considered both types of queries. Needless to say, the aim of these works was to allow the consideration of arbitrary graphs and so strengthen the relation between property testing and standard algorithmic studies. However, forsaking the paradigm of representing graphs as functions means that the connection to the rest of property testing is a bit weakened (or at least becomes more cumbersome). Still, we believe that the trade-off is worthwhile.

4.1   A Taste of the Known Results

It is natural to attempt to extend testers designed for the bounded-degree model to the general graph model. Such extensions face two potential difficulties, which refer to two ways in which the general graph model extends the bounded-degree model:

1. Firstly, the maximum degree of vertices in the graph may no longer be constant, and the question is how does the performance of the tester depends on the degree bound, d. Formally, one should think of the degree bound d as a variable, and analyze the tester accordingly.

   Note that when d increases, relative distances decrease and so testing may become easier. On the other hand, we can no longer scan all neighbors of a given vertex at constant cost.
2. Treating the maximum degree as a variable, raises the question of what happens when there is a significant discrepancy among the degrees of the various vertices. Such a situation can break the balance between the aforementioned positive and negative effects of increasing the maximum degree. Specifically, the algorithmic operations may becomes more costly when the maximum degree increases, but when using the distance measure of Definition 4.1 the distances no longer vary with the maximum degree (i.e., d) but rather vary with the average degree. Thus, we may be in trouble if the maximum degree is significantly larger than the average degree.

The effect of the foregoing issues is tester-dependent. For example, the operation of the Connectivity tester (outlined in §3.1.1) is not affected by the possible discrepancies in the vertex degrees, and so this tester (as is) applies also to the general graph model (cf. [52]). In contrast, the Bipartiteness tester presented in Algorithm 3.4 should be modified to the current setting. Details follow.

4.2   A Benchmark: Testing Bipartiteness

Firstly, it was shown in [46] that the algorithm's performance does not deteriorate when d increases. Next, an algorithm for the general graph model was obtained by emulating Algorithm 3.4 on an imaginary graph that is obtained by replacing vertices of high degree by adequate gadgets. Specifically, a vertex having degree that is t times larger than the average degree is replaced by a t-by-t bipartite expander graph, while connecting the original neighbors to vertices on one of the sides of the expander (such that no vertex has degree greater than twice the average degree). This replacement preserves the distance to Bipartiteness (up to a constant factor). We warn that implementing the emulation (of Algorithm 3.4 on this imaginary graph) is not straightforward. In particular, it seems to require a procedure for sampling edges in the actual graph such that almost all edges are sampled with probability that is approximately (up to a constant factor) the uniform one.[18] For details, see [46].

As evident from the above description, the extension of a tester from the bounded-degree model to the general graph model may require ideas that are specific to the property at hand. For example, the gadgets used above should preserve Bipartiteness (as well as distance to Bipartiteness).

Another issue that arises is that one may hope to perform better when the degree bound d (whether maximum or average) is large. Indeed, we know that in case of Bipartiteness, dense graphs can be tested with much fewer queries than sparse graphs (recall Algorithm 2.3). Thus, an optimal tester for the general graph model should be able to match the result of the dense graph model whenever the actual graph happens to be dense. Such a result is indeed provided by [46], who show a Bipartiteness tester (for the general graph model) that is optimal for all possible edge densities.

Theorem 4.2 (Testing Bipartiteness in the General Graph Model [46]): *Ignoring factors that are polynomial in $\epsilon^{-1} \log N$, the query* (and time) *complexity of testing* Bipartiteness *is* $\min(\sqrt{N}, N^2/M)$, *where M denotes the number of edges in the input graph.*

Note that dealing with $M \gg N^{3/2}$ requires some deviation from the aforementioned emulation (of Algorithm 3.4). Indeed, in such a case the tester of [46] behaves quite differently. Specifically, it takes $K = \sqrt{N^2/M}$ random walks (rather than N^2/M random walks), from each random start vertex, and checks for collisions among the endpoints these K walks by using $\binom{K}{2}$ adjacency queries. We mention that the use of adjacency queries is necessary for an $o(\sqrt{N})$ query tester of Bipartiteness.

---

[18] A more accurate sampling procedure is implicit in the subsequent work of [37].

An opposite behavior. In contrast to the case of testing Bipartiteness, where the complexity improves with the edge density, in the case of testing triangle-freeness we see the opposite behavior [5].[19] Furthermore, in contrast to testing Bipartiteness, there is a gap between the complexity of testing triangle-freeness in the bounded-degree model and the corresponding complexity in the general graph model even when the graph is sparse (i.e., $M = O(N)$). For example, in the general graph model, the complexity is $\Omega(N^{1/3})$ as long as $M = N^{2-o(1)}$ [5].

4.3   Reflections

The bulk of algorithmic research regarding graphs refers to general graphs. Of special interest are graphs that are neither very dense nor have a bounded degree. In contrast, research in testing properties of graphs started (in [32]) with the study of dense graphs, proceeded to the study of bounded-degree graphs (in [34]), and reached general graphs only in [52,46]. This evolution has historical reasons to be reviewed first.

Testing graph properties was initially conceived (in [32]) as a special case of the framework of testing properties of functions. Thus, graphs had to be represented by functions, and two standard representations of graphs (indeed, the two reviewed in Sections 2 and 3) seemed most fitting in this context. We stress that both models were formulated in a way that identifies the graphs with a specific functional representation, which in turn defines the type of queries allowed to the tester as well as the notion of fractional distance (which underlies the performance guarantee).

The identification of graphs with any specific functional representation was abandoned by Parnas and Ron [52] who developed a more general model by decoupling the type of queries allowed to the tester from the distance measure: Whatever is the mechanism of accessing the graph, the distance between graphs is defined as the number of edges in their symmetric difference (rather than the number of different entries with respect to some specific functional representation). Furthermore, the relative distance may be defined as the size of the symmetric difference divided by the actual (total) number of edges in both graphs (rather than divided by some (possibly non-tight) upper-bound on the latter quantity). Also, as advocated by Kaufman *et al.* [46], it is reasonable to allow the tester to perform both adjacency and neighbor queries (and indeed each type of query may be useful in a different range of edge densities). Needless to say, this model seems adequate for the study of testing properties of arbitrary graphs, and it strictly generalizes the positive aspects of the two prior models (i.e., the models based on the adjacency matrix and bounded-degree incidence list representations).

---

[19] This is to be expected in light of the fact that testing triangle-freeness has complexity $O(d/\epsilon)$ in the bounded-degree model [34], whereas in the dense graph model testing triangle-freeness requires more than $\mathrm{poly}(1/\epsilon)$ queries [1].

We wish to advocate further study of the latter model. We believe that this model, which allows for a meaningful treatment of property testing of general graphs, is the one that is most relevant to computer science applications. Furthermore, it seems that designing testers in this model requires the development of algorithmic techniques that may be applicable also in other areas of algorithmic research. As an example, we mention that techniques in [46] underly the average degree approximation of [37]. (Likewise techniques of [34] underly the minimum spanning tree weight approximation of [23]; indeed, as noted next, the bounded-degree incidence list model is also more algorithmic oriented than the adjacency matrix model.)

Let us focus on the algorithmic contents of property testing of graphs. Recall that, when ignoring a quadratic blow-up in the query complexity, property testing in the adjacency matrix representation reduces to sheer combinatorics (as reflected in the notion of canonical testers, see Theorem 2.2). Indeed, as shown in [38], a finer look (which does not allow for ignoring quadratic blow-ups in complexity) reveals the role of algorithmic design also in this model. But still property testing in the incidence list representation seems to require more sophisticated algorithms. Testers in the general graph models seem to require even more algorithmic ideas (cf. [46]).

To summarize, we advocate further study of the model of [52,46] for two reasons. The first reason is that we believe in the greater relevance of this model to computer science applications. The second reason is that we believe in the greater potential of this model to have cross fertilization with other branches of algorithmic research. Nevertheless, this advocation is not meant to undermine the study of the dense graph and bounded-degree models. The latter have their own merits and also offer a host of interesting open problems, which are potentially relevant to computer science at large.

5   Additional Issues

In this section we discuss three issues that are relevant to each of the three models discussed in the prior corresponding three sections.

5.1   Directed Graphs

So far our discussion was confined to undirected graphs. Nevertheless, the three models extend naturally to the case of directed graphs. Actually, when considering incidence queries, two different sub-models emerge (cf. [16]): In the first model the tester may only query for edges in the forward direction (resp., backward direction), whereas in the second model both forward and backward directions are allowed. That is, in the second model, the directed graph $G = ([N], E)$ is represented by two functions, g_{out} and g_{in}, such that $g_{out}(u, i) = v$ (resp., $g_{in}(u, i) = v$) if the ith out-going edge of u leads to v (resp., the ith in-coming edge of u arrives from v).

The gap between these two query models was demonstrated by Bender and Ron, who initiated the study of testing properties of directed graphs [16]. In particular, they showed that while strong connectivity in bounded-degree directed graphs can be tested by $\widetilde{O}(1/\epsilon)$ forward and backward queries [16, Sec. 5.1], when only forward (resp., backward) queries are allowed no tester can work with $o(\sqrt{N})$ queries (even when allowing two-sided error [16, Sec. 5.2]).[20]

Another task studied in [16] is testing whether a given directed graph is acyclic (i.e., has no directed cycles). They presented an Acyclicity tester of poly$(1/\epsilon)$ complexity in the adjacency predicate model, and showed that in the incidence list model no Acyclicity tester can work with $o(N^{1/3})$ queries (even when both forward and backward queries are allowed). The question of whether Acyclicity can be tested with $o(N)$ queries (in the bounded-degree digraph model) remains open. In general, it seems that the study of this model deserves more attention than it has received so far. (We mention that testing directed graphs in the dense digraph model was further studied in [6,51].)

5.2   Tolerant Testing and Distance Approximation

Recall that property testing calls for distinguishing objects having a predetermined property from object that are far from any objects that has this property (i.e., are far from the property). A more "tolerant" notion requires distinguishing objects that are close to having the property from objects that are far from this property. Such a distinguisher is called a tolerant tester, and is a special case of a distance approximator that given any object is required to approximate its distance to the property. The study of these related notions was initiated by Parnas, Ron, and Rubinfeld [53].

Definition 5.1 (sketch for the generic case): *Let Π be a set of functions over a finite set Ω. A* distance approximator *for Π is a probabilistic oracle machine T that on input an approximation parameter ϵ and access to any function f outputs with probability at least $2/3$ a value that approximates the relative distance of f to Π up to an additive term of ϵ; that is, $\mathbf{Pr}[|T^f - \delta_\Pi(f)| \le \epsilon] \ge 2/3$, where $\delta_\Pi(f) \stackrel{\text{def}}{=} \min_{g \in \Pi}\{\delta(f,g)\}$ and $\delta(f,g) \stackrel{\text{def}}{=} \mathbf{Pr}_{x \in \Omega}[f(x) \ne g(x)]$.*

A simple observation is that any tester that makes uniformly distributed queries offers some level of tolerance. Specifically, if a tester makes $q(\epsilon)$ queries and each query is uniformly distributed, then this tester distinguishes between objects that are ϵ-far from the property and objects that are $(\epsilon/10q(\epsilon))$-close to the property. Needless to say, the challenge is to provide stronger relations between property testing and distance approximators. Such a result was provided by Fischer and Newman [29]: They showed that, *in the dense graph model, testability in a number of queries that only depends on ϵ implies distance approximator in a number of queries that only depends on ϵ.* In the the bounded-degree model, many of the known testers were extended to yield distance approximators (cf. [48]).

---

[20] The lower bound can be strengthened to $\Omega(N)$ when considering only one-sided error testers. In the case of two-sided error, some improvements are possible; see Appendix A.3.

5.3   Proximity Oblivious Testing

Note that in order to satisfy the property testing requirement, any tester (of a reasonable property) must obtain the proximity parameter as auxiliary input and determine its actions accordingly. The question, addressed here, is what does the tester do with this parameter (or how does the parameter affect the actions of the tester). A very minimal effect is exhibited by testers that, based on the value of the proximity parameter, determine the number of times that a basic test is invoked, where the basic test is oblivious of the proximity parameter. For example, the celebrated linearity tester of [19] repeats a basic test that consists of selecting two random points, x and y, and probing the value of the function at the points x, y, and $x + y$. This basic test is repeated for a number of times that is inversely proportional to the proximity parameter.

Our focus here is on such basic tests (i.e., basic tests that are oblivious of the proximity parameter), called proximity oblivious testers. Although proximity oblivious testers were implicit in prior works (see, e.g., [19,2,3]), their general study was initiated by Goldreich and Ron [39].

Definition 5.2 (sketch for the generic case): *Let Π be a set of functions over a finite set Ω. A proximity-oblivious tester for Π is a probabilistic oracle machine T that, when given oracle access to any function f over Ω, satisfies the following two conditions:*

1. *The machine T accepts each function in Π with probability 1.*
2. *For some (monotone) function $\rho : (0,1] \rightarrow (0,1]$, each function $f \notin \Pi$ is rejected by T with probability at least $\rho(\delta_\Pi(f))$, where $\delta_\Pi(f)$ is as in Definition 5.1.*

The function ρ is called the detection probability *of the tester T.*

Indeed, we require that $\rho(\epsilon) > 0$ for every $\epsilon > 0$, whereas extending Item 2 to $f \in \Pi$ (while avoiding contradiction with Item 1) mandates extending ρ so that $\rho(0) = 0$. The requirement that ρ is monotone (i.e., monotonically increasing) does not rule out cases where the tight lower-bound is non-monotone (e.g., [14]), because ρ is not required to be tight.

Indeed, using a proximity-oblivious tester T, we can obtain a standard (one-sided error) tester (of error probability at most $1/3$). Specifically, given the proximity parameter ϵ, the standard tester invokes T for $\Theta(1/\rho(\epsilon))$ times, and accepts if and only if all these invocations accept. Two natural questions regarding proximity oblivious testers are:

1. Which properties have proximity oblivious tests (of small query complexity)?
2. How does the detection probability of such tests grow as a function of the distance of the object from the property, and how does this relate to the query complexity of the best (standard) tester for the corresponding property.

Goldreich and Ron [39] provide a mix of positive and negative results regarding the foregoing questions. In particular, they provide a characterizations of the graph properties that have constant-query proximity-oblivious testers in the two main models discussed in this article (i.e., the dense graphs model and the bounded-degree graph model). It follows that constant-query proximity-oblivious testers do not exist for many easily testable properties (e.g., `Bipartiteness` in the dense graph model). Also, even when proximity-oblivious testers exist, repeating them does not necessarily yield the best standard testers for the corresponding property (e.g., `Clique Collection` in the dense graph model).

Acknowledgments

We are grateful to Tali Kaufman, Michael Krivelevich, Dana Ron, Asaf Shapira, and Omer Tamuz for useful comments and suggestions regarding this article.

References

1. Alon, N.: Testing subgraphs of large graphs. Random Structures and Algorithms 21, 359–370 (2002)
2. Alon, N., Fischer, E., Krivelevich, M., Szegedy, M.: Efficient Testing of Large Graphs. Combinatorica 20, 451–476 (2000)
3. Alon, N., Fischer, E., Newman, I., Shapira, A.: A Combinatorial Characterization of the Testable Graph Properties: It's All About Regularity. In: 38th STOC, pp. 251–260 (2006)
4. Alon, N., Krivelevich, M.: Testing k-Colorability. SIAM Journal on Disc. Math. 15(2), 211–227 (2002)
5. Alon, N., Kaufman, T., Krivelevich, M., Ron, D.: Testing triangle freeness in general graphs. In: 17th SODA, pp. 279–288 (2006)
6. Alon, N., Shapira, A.: Testing subgraphs in directed graphs. JCSS 69, 354–482 (2004)
7. Alon, N., Shapira, A.: Every Monotone Graph Property is Testable. In: 37th STOC, pp. 128–137 (2005)
8. Alon, N., Shapira, A.: A Characterization of the (natural) Graph Properties Testable with One-Sided. In: 46th FOCS, pp. 429–438 (2005)
9. Alon, N., Shapira, A.: A Characterization of Easily Testable Induced Subgraphs. Combinatorics Probability and Computing 15, 791–805 (2006)
10. Alon, N., Shapira, A.: A Separation Theorem in Property Testing. Combinatorica 28(3), 261–281 (2008)
11. Alon, N., Spencer, J.H.: The Probabilistic Method. John Wiley & Sons, Inc., Chichester (1992)
12. Arora, S., Karger, D., Karpinski, M.: Polynomial time approximation schemes for dense instances of NP-hard problems. JCSS 58(1), 193–210 (1999)
13. Batu, T., Fortnow, L., Rubinfeld, R., Smith, W.D., White, P.: Testing that Distributions are Close. In: 41st FOCS, pp. 259–269 (2000)
14. Bellare, M., Coppersmith, D., Håstad, J., Kiwi, M., Sudan, M.: Linearity testing in characteristic two. In: The 36th FOCS, pp. 432–441 (1995)
15. Bellare, M., Goldreich, O., Sudan, M.: Free Bits, PCPs and Non-approximability – Towards Tight Results. SIAM Journal on Computing 27(3), 804–915 (1998)
16. Bender, M., Ron, D.: Testing acyclicity of directed graphs in sublinear time. Random Structures and Algorithms, 184–205 (2002)

17. Ben-Eliezer, I., Kaufman, T., Krivelevich, M., Ron, D.: Comparing the strength of query types in property testing: the case of testing k-colorability. In: 19th SODA (2008)

18. Benjamini, I., Schramm, O., Shapira, A.: Every Minor-Closed Property of Sparse Graphs is Testable. In: 40th STOC, pp. 393–402 (2008)

19. Blum, M., Luby, M., Rubinfeld, R.: Self-Testing/Correcting with Applications to Numerical Problems. JCSS 47(3), 549–595 (1993)

20. Bogdanov, A., Obata, K., Trevisan, L.: A lower bound for testing 3-colorability in bounded-degree graphs. In: 43rd FOCS, pp. 93–102 (2002)

21. Bogdanov, A., Trevisan, L.: Lower Bounds for Testing Bipartiteness in Dense Graphs. In: IEEE Conference on Computational Complexity, pp. 75–81 (2004)

22. Canetti, R., Even, G., Goldreich, O.: Lower Bounds for Sampling Algorithms for Estimating the Average. In: IPL, vol. 53, pp. 17–25 (1995)

23. Chazelle, B., Rubinfeld, R., Trevisan, L.: Approximating the minimum spanning tree weight in sublinear time. In: Orejas, F., Spirakis, P.G., van Leeuwen, J. (eds.) ICALP 2001. LNCS, vol. 2076, pp. 190–200. Springer, Heidelberg (2001)

24. de la Vega, W.F.: MAX-CUT has a randomized approximation scheme in dense graphs. Random Structures and Algorithms 8(3), 187–198 (1996)

25. Even, S.: Graph Algorithms. Computer Science Press, Rockville (1979)

26. Even, S., Selman, A.L., Yacobi, Y.: The Complexity of Promise Problems with Applications to Public-Key Cryptography. Inform. and Control 61, 159–173 (1984)

27. Fischer, E., Matsliah, A.: Testing graph isomorphism. In: 17th SODA, pp. 299–308 (2006)

28. Fischer, E., Matsliah, A., Shapira, A.: Approximate hypergraph partitioning and applications. In: Proceedings of 48th FOCS, pp. 579–589 (2007)

29. Fischer, E., Newman, I.: Testing versus estimation of graph properties. In: 37th STOC, pp. 138–146 (2005)

30. Goldreich, O.: On Promise Problems. In: memory of Shimon Even (1935–2004). ECCC, TR05-018 (January 2005); See also in Theoretical Computer Science: Essays in Memory of Shimon Even, Springer, LNCS Festschrift, Vol. 3895 (March 2006)

31. Goldreich, O.: Computational Complexity: A Conceptual Perspective. Cambridge University Press, Cambridge (2008)

32. Goldreich, O., Goldwasser, S., Ron, D.: Property testing and its connection to learning and approximation. Journal of the ACM, 653–750 (July 1998); Extended abstract in 37th FOCS (1996)

33. Goldreich, O., Krivelevich, M., Newman, I., Rozenberg, E.: Hierarchy Theorems for Property Testing. In: Goldreich, O. (ed.) Property Testing. LNCS, vol. 6390, pp. 295–300. Springer, Heidelberg (2010)

34. Goldreich, O., Ron, D.: Property testing in bounded degree graphs. Algorithmica, 302–343 (2002)

35. Goldreich, O., Ron, D.: A sublinear bipartite tester for bounded degree graphs. Combinatorica 19(3), 335–373 (1999)

36. Goldreich, O., Ron, D.: On Testing Expansion in Bounded-Degree Graphs. ECCC, TR00-020 (March 2000)

37. Goldreich, O., Ron, D.: Approximating Average Parameters of Graphs. Random Structures and Algorithms 32(3), 473–493 (2008)

38. Goldreich, O., Ron, D.: Algorithmic Aspects of Property Testing in the Dense Graphs Model. In: Goldreich, O. (ed.) Property Testing. LNCS, vol. 6390, pp. 301–311. Springer, Heidelberg (2010)

39. Goldreich, O., Ron, D.: On Proximity Oblivious Testing. ECCC, TR08-041 (2008); Also in the proceedings of the 41st STOC (2009)
40. Goldreich, O., Trevisan, L.: Three theorems regarding testing graph properties. Random Structures and Algorithms 23(1), 23–57 (2003)
41. Gonen, M., Ron, D.: On the Benefit of Adaptivity in Property Testing of Dense Graphs. In: Charikar, M., Jansen, K., Reingold, O., Rolim, J.D.P. (eds.) RANDOM 2007 and APPROX 2007. LNCS, vol. 4627, pp. 525–539. Springer, Heidelberg (2007); To appear in Algorithmica (special issue of RANDOM and APPROX 2007)
42. Håstad, J.: Clique is hard to approximate within $n^{1-\epsilon}$. Acta Mathematica 182, 105–142 (1999) (Preliminary Version in 28th STOC, 1996 and 37th FOCS, 1996)
43. Hassidim, A., Kelner, J., Nguyen, H., Onak, K.: Local Graph Partitions for Approximation and Testing. In: 50th FOCS, pp. 22–31 (2009)
44. Hochbaum, D. (ed.): Approximation Algorithms for NP-Hard Problems. PWS (1996)
45. Kale, S., Seshadhri, C.: Testing expansion in bounded degree graphs. In: Aceto, L., Damgård, I., Goldberg, L.A., Halldórsson, M.M., Ingólfsdóttir, A., Walukiewicz, I. (eds.) ICALP 2008, Part I. LNCS, vol. 5125, pp. 527–538. Springer, Heidelberg (2008); Preliminary version appeared as TR07-076, ECCC (2007)
46. Kaufman, T., Krivelevich, M., Ron, D.: Tight Bounds for Testing Bipartiteness in General Graphs. SIAM Journal on Computing 33(6), 1441–1483 (2004)
47. Lovász, L., Young, N.: Lecture notes on evasiveness of graph properties. Technical Report TR-317-91, Princeton University, Computer Science Department (1991)
48. Marko, S., Ron, D.: Distance approximation in bounded-degree and general sparse graphs. Transactions on Algorithms 5(2), Article no. 22 (2009)
49. Mihail, M.: Conductance and convergence of Markov chains– A combinatorial treatment of expanders. In: 30th FOCS, pp. 526–531 (1989)
50. Nachmias, A., Shapira, A.: Testing the expansion of a graph. TR07-118, ECCC (2007)
51. Orenstein, Y.: Testing properties of directed graphs. Master's thesis, School of Electrical Engineering (2010)
52. Parnas, M., Ron, D.: Testing the diameter of graphs. Random Structures and Algorithms 20(2), 165–183 (2002)
53. Parnas, M., Ron, D., Rubinfeld, R.: Tolerant Property Testing and Distance Approximation. Journal of Computer and System Sciences 72(6), 1012–1042 (2006)
54. Raskhodnikova, S., Smith, A.: A note on adaptivity in testing properties of bounded-degree graphs. ECCC, TR06-089 (2006)
55. Ron, D.: Algorithmic and Analysis Techniques in Property Testing. Foundations and Trends in TCS 5(2), 73–205 (2010)
56. Rubinfeld, R., Sudan, M.: Robust characterization of polynomials with applications to program testing. SIAM Journal on Computing 25(2), 252–271 (1996)
57. E. Szemerédi. Regular partitions of graphs. In: Proceedings, Collogue Inter. CNRS, pp. 399–401 (1978)

Appendix: In Passing – Three Unrelated Observations

The following three observations occurred to us in the process of writing this article.

A.1   Testing Degree Regularity in the Dense Graph Model

We improve the $\widetilde{O}(\epsilon^{-3})$ query upper bound of [32, Prop. 10.2.1.3] to an optimal quadratic bound.

Proposition A.1. *In the dense graph model, degree regularity can be tested in $O(\epsilon^{-2})$ non-adaptive queries.*

Proof: We start by reviewing the $\widetilde{O}(\epsilon^{-3})$-query tester presented in the proof of [32, Prop. 10.2.1.3]. This tester selects $O(1/\epsilon)$ random vertices, and estimates the degree of each of them up to $\pm\epsilon N/100$ using a sample of $s = \widetilde{O}(1/\epsilon^2)$ random vertices (and making the corresponding s queries). This tester accepts if and only if all these estimates are at most $\epsilon N/20$ apart. The analysis is based on the observation that if the tester accepts with high probability, then all but $\epsilon' N$ vertices have degree that is within $\pm\epsilon' N$ units of some value, where $\epsilon' = \epsilon/13$. By omitting and adding at most $\epsilon' N^2$ vertices (i.e., from/to the exceptional vertices), we reach a situation in which all vertices have degrees that at most $D \stackrel{\text{def}}{=} 4\epsilon' N$ units apart. At this point, we are done by applying a theorem of Noga Alon (cf. [32, Apdx. D]) that asserts that such a graph is $((3D/N) + o(1))$-close to being regular.

We improve the foregoing upper bound as follows. For a sufficiently large constant c, let $\ell \stackrel{\text{def}}{=} \log_2(c/\epsilon)$, and consider an algorithm that, for every $i \in [\ell]$, proceeds as follows:

1. The algorithm selects uniformly $c \cdot 2^i$ vertices, and estimates the degree of each of these vertices up to $\pm 2^{4i/5}\epsilon \cdot N/c$ units by using a sample of $s_i \stackrel{\text{def}}{=} c^3 \cdot 2^{-3i/2}\epsilon^{-2}$ random vertices.

   Note that with probability at least

   $$1 - c \cdot 2^i \cdot \exp(-2s_i \cdot (2^{4i/5}\epsilon/c)^2) = 1 - c \cdot 2^i \cdot \exp(-2c \cdot 2^{i/10})$$
   $$> 1 - 2^{-i-c}$$

   all these estimates are as desired.
2. If two of these estimates are more than $2^{1+(4i/5)}\epsilon \cdot N$ units apart, then the algorithm rejects.

(The algorithm accepts if and only if it does not reject in any of these ℓ iterations.) The query complexity of this algorithm is $\sum_{i \in [\ell]} c2^i \cdot c^3 2^{-3i/2}\epsilon^{-2} = O(\epsilon^{-2})$, and it accepts each regular graph with high probability (i.e., whenever all the foregoing degree estimates are adequate).

On the other hand, if a graph is accepted with high probability, then, for every $i \in [\ell]$, it holds that all but at most a 2^{-i} fraction of the vertices have degree that is within $2^{1+4i/5}\epsilon \cdot N/c$ of the average degree, denoted ρ. For each value of $i \in [\ell]$, let us denote the set of deviating vertices by B_i; that is, each vertex in $[N] \setminus B_i$ has degree $(\rho \pm 2^{1+4i/5}\epsilon/c) \cdot N$. Thus (dealing separately with each $B_i \setminus B_{i+1}$ as well as with B_ℓ and $[N] \setminus B_1$), we may omit at most $40\epsilon N^2/c$ edges from the graph, and obtain a graph in which every vertex has degree at most $(\rho + 2\epsilon/c)N$. Next, by adding at most $42\epsilon N^2/c$ edges to the graph, we can obtain a graph in which every vertex has degree at least $(\rho - 2\epsilon/c)N$, and if we add these edges uniformly (among the vertices) then each vertex in the resulting graph has degree $(\rho \pm 44\epsilon/c)N$. At this point we can apply the result of aforementioned result of Noga Alon, and be done. ∎

A.2   Non-Adaptive Testers in the Bounded-Degree Graph Model

Recall that, for any function q, if a property can be tested in $o(\sqrt{N} \cdot q(\epsilon))$ non-adaptive queries in the bounded-degree graph model, then it depends only on the vertex degree distribution [54]. In contrast, we show that triangle-freeness can be tested by $O(\sqrt{N/\epsilon})$ non-adaptive queries (in the same model).

The tester selects at random $O(\sqrt{N/\epsilon})$ vertices, queries for the neighbors of each of them, and accepts if and only if the subgraph discovered contains no triangles. Note that if the input graph is ϵ-far from triangle-freeness, then it contains $\Omega(\epsilon N)$ triangles, whereas a random sample of $O(\sqrt{N/\epsilon})$ vertices is likely to hit two vertices of such a triangle.

The argument can be extended to testing H-freeness,[21] for any fixed H, with $O((N/\epsilon)^{1-\frac{1}{\beta(H)}})$ non-adaptive queries, where $\beta(H)$ denotes the minimum vertex cover of H. In this case, if the input graph is ϵ-far from being H-free, then a sample of $O((N/\epsilon)^{1-\frac{1}{\beta(H)}})$ random vertices is likely to hit all vertices in a vertex cover of one of the copies of H. A more general statement, with weaker quantitative bounds, follows.

Proposition A.2. *Let Π be a graph property having a q-query proximity-oblivious tester of detection probability ρ, in the bounded-degree model. Then, in this model, Π can be tested by $O(N^{\frac{q-1}{q}}/\rho(\epsilon))$ non-adaptive queries.*

Actually, Proposition A.2 holds also when q is an upper bound on the number of different vertices that appear in the queries of the proximity-oblivious tester.

Proof: The main observation is that a sample of $O(N^{1-(1/q)})$ vertices (along with the neighbor queries that correspond to each vertex) is likely to allow for the emulation of a random execution of the proximity-oblivious tester (POT). Specifically, given a q-query POT, we consider the following non-adaptive POT:

1. Select a random sample of $O(N^{1-(1/q)})$ vertices, denoted S, and query the neighborhood of each vertex in S. For every $(v, i) \in S \times [d]$, denote the oracle answer by $\Gamma_i(v)$.

    These are all the queries made by the new POT, and the following steps only involve computations (and no actual queries).

---

[21] Here, we refer to subgraph freeness.

2. Select and fix random coins for T, deriving a residual deterministic oracle machine T'.

3. Let $S = \{s_1, ..., s_{|S|}\}$, and $\overline{S} \stackrel{\text{def}}{=} \{(s_{(i-1)q+1}, ..., s_{iq}) : i \in [|S|/2q]\}$; that is, $\overline{S}$ consists of q-sequences of elements in S such that no element appears twice. For every $(v_1, ..., v_q) \in \overline{S}$, try to emulate an execution of T using the information obtained in Step 1. For $j = 1, ..., q$, proceed as follows, where initially the permutation $\pi : [N] \to [N]$ is totally undetermined.

   (a) Obtain the jth query of T', denoted (u_j, i_j).

   If π is undetermined on u_j, then determine $\pi(u_j) = v_j$.

   If π is determined on u_j and $\pi(u_j) \notin S$, then this emulation is terminated.

   Thus, the algorithm proceeds to Step 3b only if $\pi(u_j) \in S$, whereas in this case the value of $\Gamma_{i_j}(\pi(u_j))$ is known.

   (b) Let $a_j = \Gamma_{i_j}(\pi(u_j))$, and suppose that $a_j \in [N]$ (as otherwise we provide a_j as the oracle answer to T', and proceed to the next iteration).[22] If π^{-1} is undetermined on a_j, then select at random a vertex u such that π is undetermined on u, and determine $\pi(u) = a_j$. Provide u as the oracle answer to T', and proceed to the next iteration.

   Note that it is quite likely that $a_j \notin S$, and in this case if T' subsequently issues a query of the form $(u, \cdot)$ then the emulation will be terminated (in the corresponding execution of Step 3a).

   If the current emulation is successfully completed, then halt and output the corresponding verdict of T'. Otherwise, proceed to the next $(v_1, ..., v_q) \in \overline{S}$, while resetting π to be totally undetermined.

4. If no emulation is successfully completed, then halt and output the verdict 1 (i.e., accept).

Each execution of Step 3b may yield a value $a_j \notin S$, with probability at least $1 - (|S|/N)$. However, with probability at least $|S|/2N$, it holds that $a_j \in S$. Thus, for each $(v_1, ..., v_q) \in \overline{S}$, we complete an emulation of T' (in Step 3) with probability at least $(|S|/2N)^{q-1} \gg 1/|\overline{S}|$. Furthermore, such an emulation correspond to the execution of T' on a random isomorphic copy of the input graph.

To see that, with high probability, at least one of the $|\overline{S}|$ emulations is completed, we consider all $|\overline{S}|$ emulations simultaneously. Let $u_1^{(i)}, ..., u_q^{(i)}$ denote the sequence of vertices that occur in the ith emulation, and let $\pi^{(i)}$ denote the corresponding permutation. We partition the $|S|/2$ samples that do not appear in $\overline{S}$ into q equal sets, denoted $S_1, ..., S_q$, and terminate the ith emulation in iteration $j < q$ if $a_j^{(i)} \notin S_j$. (Indeed, this only makes early termination more likely; cf. Step 3b.) Still, on can show by induction on j, that with high probability the number of emulations that are not terminated by iteration j exceeds $|\overline{S}| \cdot (|S|/4qN)^j$. Furthermore, the queries issued in the $j+1^{\text{st}}$ iteration are mostly different, because they are determined based on different sequences in $\overline{S}$. Using $|\overline{S}| \cdot (|S|/4qN)^{q-1} > 1$, we conclude that, with high probability, there exists an emulation that does not terminate before the last iteration.

---

[22] Recall that in this case a_j is a fixed indication that the relevant vertex has less than i_j neighbors.

It follows that the foregoing non-adaptive POT has detection probability at least $\rho/2$. Applying this POT for $O(1/\rho(\epsilon))$ times, we obtain a non-adaptive tester of query complexity $O(N^{1-(1/q)}/\rho(\epsilon))$.   ∎

Conclusion. Recall that all subgraph-freeness properties do have a proximity-oblivious testers of constant-query complexity in the bounded-degree graph model. Our conclusion is that non-adaptive testers are not totally useless in that model.

A.3   Testing Strong Connectivity with Forward Queries Only

We show that, for any constant $\epsilon > 0$, strong connectivity in bounded-degree digraphs can be tested by using $N^{1-\Omega(1)}$ forward queries (and no backward queries). Needless to say, the same holds for using only backward queries, and in both cases the tester has two-sided error (which is unavoidable).[23]

Proposition A.3. *In the directed bounded-degree model where only forward queries are allowed, strong connectivity can be tested in query complexity* $\exp(1/\epsilon)\cdot N^{1-\frac{1}{t}}$, *where* $t = \lceil 4/\epsilon d\rceil \cdot d < d+(1/\epsilon)$ *and* d *is the in-degree and out-degree bound.*

Proof Sketch: Our starting point is the observation that if a graph is ϵ-far from being strongly connected, then it contains at least $\epsilon dN/4$ source and sink components each containing at most $\lceil 4/\epsilon d\rceil$ vertices (cf. [16, Cor. 9]).[24] The easy case is when the graph contains at least $\epsilon dN/8$ small sink component, since these are easy to detect by forward queries. The problematic case is the one in which the graph contains $\epsilon dN/8$ source components, and we start by considering the simple case in which each of these source components consists of a single vertex.

In the latter case we can estimate the number of vertices having in-degree zero, by estimating the number of vertices having in-degree d, $d-1$, all through 1. To estimate the number of vertices having in-degree $i > 1$, we estimate the number of i-way collisions at the head of randomly selected[25] directed edges, and use the information we already gathered regarding in-degree j for every $j > i$. The number of vertices having in-degree 1 is estimated by estimating the collisions between a uniformly selected vertex and the vertex at the head of a uniformly selected random edge. Note that, for every $i \geq 2$, the number of i-way collisions can be estimated by a sample of size $O(N^{1-\frac{1}{i}})$.

In the foregoing, we have relied on the fact that a vertex has zero in-degree if and only if it is a source vertex, and on the hypothesis that many source vertices exist. But, in general, we only know that there are many small source components. So the intuitive idea is to "contract" all small components, and

---

[23] The distributions used in [16, Sec. 5.2] can be used to prove an $\Omega(N)$ query bound for one-sided error. The point is that we can find no direct evidence to the fact that a vertex has in-degree zero.

[24] Throughout this proof, the word component means a strongly connected component, and source (resp., sink) components are components that have no in-coming (resp., out-going) edges.

[25] We may select a random directed edge by selecting a vertex uniformly, and selecting each of its out-going edges with probability $1/d$.

consider in-coming edges at the component level. One small difficulty is that we cannot determine the components of the input graph, and so the following modification is used.

For every vertex v, we let C_v denote the set of vertices u such that v and u reside on a directed cycle of size at most $s \stackrel{\text{def}}{=} \lceil 4/\epsilon d \rceil$. We say that v is good if for every $u \in C_v$ it holds that $C_u = C_v$. Note that, given a vertex v, we can determine C_v as well as whether v is good by using d^s queries. Also note that every vertex that resides in a small source component is good. We now emulate the foregoing procedure on the directed graph in which for every good v the set C_v is contracted to a new vertex, and note that a vertex has in-degree zero in the resulting graph if and only if it represents a small source of G. Noting that the maximum degree in this graph is $s \cdot s$, the claim follows.　∎

Conclusion. Our lesson is that some non-trivial testing can be carried out also in the model that allows forward queries only.

Property Testing of Massively Parametrized Problems – A Survey

Ilan Newman

Department of Computer Science, University of Haifa, Haifa, Israel
ilan@cs.haifa.ac.il

Abstract. We survey property testing results for the so called 'massively parametrized' model (or problems). The massively-parametrized framework studies problems such as: Given a graph $G = (V, E)$, consider the property $BI(G)$ containing all subgraphs of G that are bipartite.

Massively parametrized Properties, such as $BI(G)$, are defined by two ingredients: a 'general' property of inputs (e.g, 'being bipartite' for the example above), and an underlying given structure (the graph G for the property $BI(G)$). Then the property is the restriction of the general property to inputs that are associated with the given structure.

In such situation, keeping the general property fixed and varying the underlying structure, the property and the corresponding tests might vary significantly in their structure. The focus of the study in this framework is how, for a fixed general property, the testing problem changes while changing the underlying structure.

1 Introduction

A Massively Parametrized property is a restriction of a 'general' property to a subset of inputs, associated with a given structure. E.g., consider the general graph property of being bipartite. A structure in this case is a graph $G = (V, E)$. Then the corresponding property $BI(G)$ is a property of all subgraphs of G that are bipartite.

Property Testing of *massively parametrized* problems falls formally into the combinatorial property testing definitions of [15]. The focus of the study in this framework is how, for a fixed general property, the testing problem changes while changing the underlying structure.

To better understand the different focus between this 'model' and the standard property testing model, let us outline the main relevant features of a problem in the standard model[1]: To this end, each such testing 'problem' contains three main elements:

- A *fixed structure:* For each problem size n there is a fixed structure that determines all inputs.

---

[1] We use the standard definitions of property testing as in [15], and assume basic knowledge of it. For details see [15,8,21].

O. Goldreich (Ed.): Property Testing, LNCS 6390, pp. 142–157, 2010.

- *Inputs:* A set of vectors associated with the structure, e.g, coloring of its points, (or sometimes viewed as a function from the points of the structure to a certain range).
- The *Property:* A subset of all possible input vectors.

As an example consider the 'dense graph' model and the graph property of being bipartite. Given the problem size $m = n^2$, the *fixed structure* is K_n, the complete graph on n vertices. (alternatively, the fixed structure can be viewed also as an $n \times n$ array). The collection of *inputs* is the set of all subgraphs of K_n which are just all possible graphs on n vertices. Alternatively, the inputs are all $0/1$ coloring of the $n \times n$ array, viewed as the adjacency matrix of graphs on n vertices. The *property* is the set of all bipartite graphs on n vertices, or all labelings of the corresponding array.

Any other graph property induces a similar example. Thus all graph properties of graphs on n vertices share the same fixed structure (that is K_n) and the same set of inputs.

For digraph properties, the fixed structure is again an array as above, and the set of inputs contains all the Boolean labelings of it, with the corresponding interpretation of an input as a digraph. When properties of Boolean functions are being considered, the fixed structure is the Boolean cube (of a given dimension). The set of inputs are all $0/1$ coloring of vertices of the cube, where each coloring is viewed as a Boolean function. A property is then, any subset of functions, e.g., the linear functions, small-degree polynomials, monotone functions etc.

Lets us now examine a typical situation in the 'massively-parametrized' model. For a general property, e.g, the graph property of 'being bipartite', and a given input length m, the structure *is not* fixed in advance. It could be any graph $G = (V, E)$ with m edges. The *inputs* would be all $0/1$ coloring of the edges of G, each viewed as defining a subgraph of G. The property is then all those subgraphs $G' \subseteq G$ that are bipartite. Likewise, any graph property can be considered, e.g., Eulerianity, being connected, being H-free for some fixed graph H etc.

Properties of digraphs can be considered in a similar way. The structure is any given digraph, the inputs: all subgraphs defined by $0/1$ coloring of its edges, and the property - any collection of subgraphs. One can also consider properties of *induced* subgraphs, rather than just subgraphs. Namely, taking the same structure (either for the undirected or directed case), with the set of inputs being the set of all Boolean coloring of the vertices, each viewed as an encoding of an induced (di)subgraph.

Examples of a different flavor are these that are associated with the general property of satisfiability. Let the structure be a given Boolean circuit, formula, or any other computational mechanism. The set of inputs will be the set of all assignments to the Boolean variables (or appropriate input words to the computation - e.g., words over some alphabet for, say, a given grammar/ automata, etc.). The property that is considered in this setting is the set of all inputs accepted by the computation. We call this property the *satisfiability* (or sometimes 'membership) property of the corresponding structure.

There are several points to be emphasized.

1. The *distance* - In all the examples above, the set of inputs can be viewed as all possible Boolean vectors of a certain dimension that is defined by the structure. (E.g., the set of Boolean coloring of the edges for the example of subgraph properties, or the set of Boolean assignments to the variables for a given Boolean formula). Thus, the distance between inputs is just the hamming distance, the same as it is in the standard model of property-testing (as per the definitions of [15]). In this respect, all the examples considered above, as well as these that are discussed in what follows, fall directly into the standard framework of property-testing.

2. By fixing the property and varying the structure, the testability (or non-testability) of it becomes a property of the structure. In particular, for computational mechanisms as described above, the satisfiability property is the unique and fixed general property, while the structure varies significantly (both in terms of different types, e.g., circuits vs. formulae etc. and in terms of different instances within a type, e.g., the specific given 3CNF formula).

3. Note that a test may be viewed as being composed of two parts: a *preprocessing* part which depends on the structure, and the 'standard testing' part (usually some sampling of the input). As the structure varies, one would expect that there is much to be changed in the test, which, in turn, shifts somewhat the focus towards the input-independent algorithmic preprocessing of the structure. Indeed in all positive examples, the tests exhibit significantly complicated and interesting preprocessing.

In the rest of this survey we outline some of the results that may be viewed as falling into the framework of massively parametrized property testing. We start (Section 3) with some results that in retrospect fall into the framework of massively-parametrized problems although they were not presented as such. Following, in Section 4, are the results that were done in full awareness of this framework. This includes results on testing orientations properties in Section 4.1, and a 'general' lower bound in Section 4.2. Then in Section 4.3 we survey some results on the satisfiability property for several representations of Boolean formulae. Finally, in Section 5 we list some of the open problems in this area.

2   General Notations

In what follows we adopt the definitions of [15] with regards to property testing. A property, as discussed above, will always be formally represented by a set $P = \cup_{n \geq 1} P_n$ of 0/1 vectors where $P_n \subseteq \{0,1\}^n$ is a subset of vectors of length n. An ϵ-test will be described for a given input length n, that is for P_n. Our focus is on the asymptotic complexity with respect to n (and the distance parameter ϵ).

In what follows a graph $G = (V, E)$ is always undirected, unless explicitly said otherwise. Directed graphs will be denoted similarly but, we will explicitly say that $G = (V, E)$ is digraph (directed graph) when applies. For a graph $G = (V, E)$, we will usually denote $n = |V|$, and the complexity will be in terms of n although the input length will be different (e.g., n^2 for the dense-graph model).

A property P will be said to be *testable* if there is an ϵ-test for P making $q = q(\epsilon)$ queries, for every ϵ. Namely, its query complexity depends only on ϵ but not on the input length n.

For two vectors $u, v \in \{0,1\}^n$, their hamming distance is defined as $dist(u, v) = |\{i|\ u_i \neq v_i\}|$. For $0 < \epsilon < 1$, and $u, v \in \{0,1\}^n$, we say that u is ϵ-far from v if $dist(u, v) \geq \epsilon \cdot n$, and ϵ-close otherwise. For $P_n \subseteq \{0,1\}^n$ we say that u is ϵ-far from P_n if u is ϵ-far from every $v \in P_n$. Otherwise, we say that u is ϵ-close to P_n.

3 Results That Are in Retrospect Massively-Parametrized Testing

As made clear above, massively-parametrized testing falls strictly into the area of combinatorial property testing as defined formally in [15]. In particular, some older results may be viewed in retrospect as results on massively-parametrized problems, although they have not been considered or stated as such at the time. We review some of these results.

3.1 Testing Membership in Read-Once Branching Programs

In [20] the following problem is considered: Given a read-once deterministic oblivious branching program P on n Boolean variables[2], the set of inputs is the set of all Boolean assignments to the variables. It is proved in [20] that the satisfiability (or membership) problem can be tested (non-adaptively, 1-sided error test) using $O(exp(2^w/\epsilon))$ queries.

The preprocessing is rather heavy, but can be done in time $poly(n)$.

We note that this result is a generalization of a result of [2] stating that membership in regular languages is testable. The result of [2] can already be viewed as fits the massively-parametrized framework. There however, the preprocessing is 'light' as the same fixed (hence constant size) automaton serves as the defining mechanism for every input length word.

3.2 Testing Monotonicity in General Posets

Monotonicity testing in general posets generalizes monotonicity-testing of Boolean functions. Here a poset $P = (V, \leq)$ is the given structure and the set of inputs is the collection of functions C^V where C is typically a total ordered set (e.g., $C = [n]$ with the natural order, and in particular the Boolean case $n = 2$) but could be taken as any poset. The *fixed* property here is the set of all functions that are monotone, namely all such f for which $f(x) >_C f(y)$ implies that $x >_P y$.

---

[2] For readers that are not familiar with the concept of branching programs, this concept can be replaced with, say, a depth $d = O(1)$ Boolean formula where each variable appears only once.

The main results in [9] are lower bounds for testing monotonicity for Boolean labelings of general posets, and in particular, a construction of some posets for which monotonicity requires relatively large query complexity (previous lower bounds where only for large range [11]). In addition it is shown in [9] that monotonicity is testable when the poset is the transitive closure of an orientation of a tree (or a forest) and C is Boolean. It is also shown that there is a test with $O(\log |V|/\epsilon)$ queries when P is the transitive closure of a directed acyclic graph (DAG) that as an undirected graph is of bounded tree-width[3].

There are some recent interesting generalization of the positive results above, as well as some improvements (E.g., for posets induced by planar graphs) in [3], using 2TC-spanners. This is the main topic of another article in this volume and will not be further described here.

4   Contemporary Results

Here we survey some of the more contemporary results.

As already discussed in the previous sections, one model for graph related properties is the following. The given but variable underlying structure is an *undirected* graph $G = (V, E)$. The set of inputs contains all Boolean assignments to the *edges*, namely $\{0,1\}^E$. The interpretation of an input $\alpha \in \{0,1\}^E$ is either the subgraph of G that is defined by (V, E_1) where $E_1 = \{e \in E| \alpha(e) = 1\}$, or as an orientation of the edges relative to a fixed orientation. E.g., one can interpret α as the digraph $G_\alpha = (V, E_\alpha)$ where

$$E_\alpha = \{(i \to j)| i < j, \alpha(i,j) = 1\} \cup \{(i \to j)| i > j, \alpha(i,j) = 0\}.$$

Most work in the model was done with respect to the later interpretation, which was often referred to as the *orientation* model.

4.1   Main Results in the Orientation Model

For an undirected graph $G = (V, E)$ and a property P of digraphs, one can consider the property P_G containing these orientations of G that have P. E.g., Eulerianity, being strongly connected, not containing a forbidden subgraph H, etc. We refer to the set of all orientation of G as the G-orientations, and denote such an orientation by $\vec{G}$.

Testing H-Freeness. Let $H = (V_H, E_H)$ be a fixed directed graph. The digraph property of *H-freeness* contains all digraphs that do not have a subgraph isomorphic to H, and is denoted by $F_H(G)$. A sink in a digraph is a vertex v for which $d_{out}(v) = 0$, namely, has no outgoing edges. Similarly v is a source if $d_{in}(v) = 0$. A graph is source-free (analogously sink-free) if it contains no source.

---

[3] The test here has the same complexity even if the range is any total ordered set. There are other positive results in [9] for special types of posets.

The following theorem is proved in [16].

Theorem 4.1. *[16] Let H be a fixed digraph that is either source-free or sink-free. Then, for any $O(1)$-bounded degree graph G and $\epsilon > 0$, there is a 1-sided error ϵ-test for $F_H(G)$ making $poly(1/\epsilon)$ queries.*

Proof. **(Sketch.)** Assume w.l.o.g that H is source-free. To make a G-orientation $\vec{G}$, H-free, one can pick a vertex v from each copy of H in $\vec{G}$ and reorient the necessary edges so to make v a source. Obviously this should be done sequentially picking no two adjacent vertices. This implies that if $\vec{G}$ is ϵ-far from being H-free, every subset $C \subseteq V$ that hits all copies of H in $\vec{G}$ must have $|C| = \Omega(n)$. Hence, sampling a random vertex and checking its neighborhood of size $|H|$ to discover if it belongs to a H-copy (and reject if it does) is a test for H-freeness.   □

Note that the test relies on the fact that G is of bounded degree and that H is either source free or sink free. One may wonder whether these conditions are truly necessary. We first address the issue of being either source or sink free.

The simplest interesting digraph H that has both a source and a sink is P_2, the path of length 2 on 3 vertices (every digraph with non-empty edge set is not P_1-free). Testing P_2-freeness is easy since a G-orientation is P_2-free if and only if G is bipartite $G = (X, Y; E)$ and all its edges are oriented, say, from X to Y. Thus sampling a random edge is a good test. Going however one step further, it is shown in [16] that testing P_3-freeness requires a linear number of queries for some bounded degree graphs. The reduction is from testing whether a 3-coloring of a 3-colorable graph is indeed a proper coloring, which is shown to require a linear number of queries (implicit in [4]).

Regarding the degree restriction, we don't know if this restriction is absolutely required, however, [16] observed:

Let C_6 be the directed cycle on 6 vertices.

Observation 4.1. *[16] There exists an infinite family of graphs, each with $O(1)$-average degree, for which every 1-sided error ϵ-test for the G-orientations property of C_6-freeness makes $(1/\epsilon)^{\Omega(\log(1/\epsilon))}$ queries.*

The proof is immediate: testing triangle-freeness in the dense graph model is reducible to testing C_6-freeness in the orientation model where the underlying graph is a subdivided K_n. The result of [1] completes the proof.

Testing Strong Connectivity. Strong connectivity is a very natural and basic property of digraphs. We are not aware of any universal result for this property of G-orientations (namely, testability for every underlying graph). [16] presents some initial positive results which we describe below.

We start with a few observations. Obviously, one may assume that the underlying graph $G = (V, E)$ is 2 edge-connected as otherwise no orientation is strongly connected. We can also assume that $|E| = O(|V|)$ as otherwise every G-orientation is ϵ-close to be strongly connected (by simply orienting a minimal 2-edge connected subgraph of G to be strongly connected). Thus it is enough to consider 2-edge connected sparse graphs (that is, graphs for which the degree is $O(1)$ for every vertex).

Let D be a directed graph. We denote by $SC(D)$ the DAG that is defined on the strongly connected components of D in the standard way. A *source component* of D is a strongly connected component C of D that corresponds to a source vertex in $SC(D)$ (in other words, every edge between a vertex in C and a vertex in $V(D) \setminus C$ is directed away from C). A *sink component* of D is defined similarly. The following is an observation (used in the context of testing directed graphs properties) from [5].

Observation 4.2. *[5] Let $G = (V, E)$ be a graph on n vertices, and $\vec{G}$ a G-orientation. If $\vec{G}$ has at least $\Omega(n)$ sources or sinks components then a source or sink component of $\vec{G}$ can be found in $O(1)$ queries.* □

The only known test for strong connectivity of G-orientations is based on Observation 4.2. The following property identifies a class of underlying graphs on n vertices for which any orientation that is far from being strongly connected has $\Omega(n)$ sources or sinks.

Definition 4.1. *A family of undirected 2-edge connected graphs $\{G = (V, E)\}$ is called efficiently-Steiner-connected if for every $\delta < 1$ and graph G in the family, for every $S \subseteq V$ such that $|S| \leq \delta^2 |V|$, there is a connected subgraph $T = (V', E')$ with $S \subseteq V'$ and $|E'| \leq 10\delta |V|$.*

We note that the constant 10 and the function types of δ in the definition are somewhat arbitrary.

Theorem 4.2. *[16] If G is efficiently-Steiner-connected then the G-orientations property of being strongly connected is testable by a 1-sided error test.*

We note that any 'slightly expanding' graph is strongly-Steiner-connected, while for example, the cycle is not. A simple application of Theorem 4.2 is given by the following theorem.

Theorem 4.3. *[16] For any G that is a linear expander graph as well for the $\sqrt{n} \times \sqrt{n}$ two dimensional grid, the G-orientations property of being strongly connected is testable by a 1-sided error test.*

Testing $s - t$ Connectivity. Let $G = (V, E)$ be an underlying graph, and $s, t \in V(G)$. The G-orientations property of being $s - t$ connected is another natural property of (labeled) digraphs. It is shown in [6] that this property is testable for any underlying graph G. The test cannot easily be described. It is composed of a series of non trivial reduction steps that finally reduces the problem to that of testing membership in a bounded width branching program. The preprocessing, however, is in polynomial time. It seems that the dependence on ϵ might be unnecessarily high, although we don't know of a better algorithm or any non-trivial lower bound.

We state the main theorem of [6]. Let P_G^{st} be the G-orientations property of being $s - t$-connected.

Theorem 4.4. *[6] For any undirected graph G, two vertices $s, t \in V(G)$ and every $\epsilon > 0$, there is a 1-sided error ϵ-test for the property P_G^{st} with query complexity q such that,*

$$q = (2/\epsilon)^{2^{O((1/\epsilon)^{(1/\epsilon)})}}$$

Testing Eulerianity. A digraph $G = (V, E)$ is Euler if for every $v \in V$, $d_{out}(v) = d_{in}(v)$. It is shown in [12] that the orientation problem of being Eulerian is not testable in general, even by 2-sided error tests. However, the upper and lower bounds are still quite far apart. We state some of the results in [12].

Theorem 4.5. *[12] There exists an infinite family of graphs for which every 1-sided error test for being Eulerian must make $\Omega(|E|)$ queries.*

The proof is based on the following observation: every 1-sided error test on far inputs must discover a witness for non-Eulerianity. Clearly if a digraph is Eulerian then for every cut $C = (S, \bar{S})$, the number of edges oriented from S to $\bar{S}$ equals the number of edges that are oriented in the opposite direction. Such a cut will be called a *balanced cut*. Thus for a cut C, any orientation of at most half the edges of C does not exclude the possibility that C is balanced. Indeed, it is shown in [12] that any witness for non-Eulerianity must contain at least half the edges of some unbalanced cut.

The lower bound follows by constructing an underlying graph and set of orientations that are ϵ-far from being Eulerian, and such that each unbalanced cut is of size $\Omega(|E|)$.

The lower bound for 2-sided error test is much weaker, however, unlike Theorem 4.5, it holds for bounded degree graphs as well.

Theorem 4.6. *[12] There exists an infinite family of graphs of bounded degree $\{G_n\}_{n \geq 1}$, such that, for every $0 < \epsilon \leq 1/64$, every non-adaptive (2-sided error) ϵ-test for the property of G_n-orientations of being Eulerian, uses $\Omega\left(\sqrt{\frac{\log n}{\log \log n}}\right)$ queries. Consequently, every adaptive test requires $\Omega(\log \log n)$ queries.*

We note that [12] also shows a stronger lower bound for 1-sided error tests for bounded degree graphs, which will not be quoted here.

As for positive results, [12] include several (sublinear) upper bounds for testing Eulerianity. Their complexity depends on the maximum and average degree of the underlying graph. There is, however, a large gap between the upper bounds and lower bounds in this case.

4.2 Some Lower Bounds – An On-Going Work

Proving lower bounds for the orientation model has been so far with limited success. In what follows we present a fairly general lower bound for 1-sided error *non-adaptive* tests. We don't know the corresponding right bound for adaptive tests.

It will be easier now to switch to the interpretation of an input as a subgraph of the underlying graph, rather than an orientation of it. Namely, let $G = (V, E)$

be an underlying graph, an input $g \in \{0, 1\}^E$ is identified with a subgraph G_g of G containing the edges that are assigned 1 by g. A property will be interpreted as a collection of subgraphs of G.

Consider the property of being bipartite. Any 1-sided error test must find an odd cycle in every far input. However, the graph G could be taken to be a bounded degree Ramanujan expander, hence, with no short cycles. Thus an obvious lower bound for a 1-sided error test is the length of the shortest cycle which would be $\Omega(\log n)$ in that case. Can one do better?

[13] prove a fairly general lower bound for any graph property in which every 0-witness contains a cycle (e.g., bipartiteness). The lower bound holds only for *non-adaptive* tests and gives a result of the type $\Omega(n^\alpha)$ for some $0 < \alpha < 1$.

We present here the results of [13]. For this we need to define the following distribution, analog to the Erdös-Rényi random graph model $\mathcal{G}(n, p)$ for arbitrary underlying graph rather than for the complete graph.

Definition 4.2. *Let $0 < \delta < 1$ and $G = (V, E)$. We define the distribution $D(\delta)$ on subgraphs of G as follows. We select each edge of E independently, with probability δ, and set E' to include all the edges that were selected. $G' = (V, E')$ is the resulting subgraph.*

Theorem 4.7. *[13] Let $G = (V, E)$ be an underlying fixed graph of girth $\geq g$. Let $\mathcal{P}$ a property of subgraphs of G such that every 0-witness for $\mathcal{P}$ contains a cycle. If $\delta < 1$ is such that $Prob_{D(\delta)}[dist(G', P) > \epsilon|E|] \geq 0.8$ then every 1-sided error ϵ-test for P requires at least $\Omega(\delta^{-g/3})$ queries.*

We note that the constant 0.8 in the theorem is somewhat arbitrary.

Proof. **(Sketch.)** Since any witness for ϵ-far inputs contains a cycle, a 1-sided error test must find a cycle on any far input. Let D be any distribution on ϵ-far inputs, and $q > 0$. Suppose that for every fixed set of queries Q of size q, the probability (over D) that Q contains a cycle is bounded by, say $1/3$ then, using Yao's argument, every non-adaptive algorithm fails to find a cycle with probability $2/3$ on ϵ-far inputs, implying that any 1-sided error ϵ-test for $\mathcal{P}$ must make at least q queries.

Now let $D = D(\delta)$, with δ such that $D(\delta)$ is essentially concentrated on ϵ-far inputs, as stated by the assumption of the theorem. For any fixed set $Q \subseteq E(G)$, Q contains at most $2q$ end-points. Let v be an end-point of an edge in Q. The probability that v is in a cycle $C \subseteq Q$ is bounded by $2q^2 \cdot \delta^g$, since there are only $2q^2$ possible end-points for the two disjoint path of length $g/2$ that start at v and are part of a cycle of length at least g. Hence, by the union bound (on the $2q$ different v's), we conclude that the probability for finding a witness is bounded by $4q^3\delta^g$, implying that $q = \Omega(\delta^{-g/3})$ to guarantee constant probability of finding a witness. We note that although D is not strictly concentrated on far inputs but rather has most of its mass there, the proof does follow e.g, by formally using $D^* = D|Far$ where Far is the event that the graph chosen according to D is indeed ϵ-far from P. It is standard to show that D well approximates D^*. $\square$

Corollary 4.1. *Let $G = (V, E)$ be an underlying fixed graph of girth $\geq g$. Let P be a property of subgraphs of G for which every 0-witness for P contains a cycle. Assume further that $dist(G, P) > (1 - \delta)|E|$ for some fixed $0 < \delta < 1/2$. Then every 1-sided error δ-test for P requires at least $\Omega(\delta)^{-g/3})$ queries.*

Proof. **(Sketch.)** Let G as above and δ for which the assumption of the corollary holds. By Chernoff bound $D(3\delta)$ will be essentially concentrated on subgraphs of G that have at least $2\delta|E|$ edges. Hence by the assumption are δ-far from P. Thus the corollary is implied by Theorem 4.7. $\qquad\square$

Here are some applications of Corollary 4.1.

Let H be a fixed graph. A graph G contains H as a minor if after the deletion and contraction of some edges in G one arrives at a graph isomorphic to H. The graph property of not containing specific minors is extensively studied in graph theory, e.g., extending planarity.

Theorem 4.8. *Let $H = (V_H, E_H)$ be a fixed graph containing a cycle. Then for some fixed $\epsilon, \alpha > 0$, there is an infinite sequence of graphs $\{G_n\}$, where G_n is an n vertex graph, such that any non-adaptive 1-sided error ϵ-test for the G_n-subgraphs property of not having H as a minor must make $\Omega(n^\alpha)$ queries.*

We sketch the proof for $H = K_r$ the complete graph on $r \geq 3$ vertices. The same proof holds for any such H.

Proof. **(Sketch.)** In view of Corollary 4.1 it is enough to show a high girth underlying graph that is sufficiently far from being K_r-free. Lemma 4.1 asserts this for $\delta = 1/3$.

Lemma 4.1. *Let r be fixed, then there is a graph on n vertices G that has girth $\Omega(\log n)$, and is $(2/3)$-far from being K_r minor free.*

Proof. Let G be a random graph from the standard Erdös-Rényi model of random graphs, $\mathcal{G}(n, c/n)$ (namely, each edges is chosen with probability c/n and independent of any other edge), for $c = c(r)$ to be defined later.

It is standard to show that after deleting $o(n)$ edges such a random graph will have (with very high probability) girth $g = \Omega(\log n)$, and at least $cn/3$ edges (or average degree at least $2c/3$).

A theorem of Kostochka, Thomason [18,22], asserts that every graph of average degree d contains a K_ℓ for $\ell = \ell(d) = \frac{\beta d}{\sqrt{\log d}}$ as a minor, for some fixed universal constant β. This means that if we fix $c = \frac{a}{\beta} r \sqrt{\log r}$, for large enough constant a, we are guaranteed that $\ell(2c/9) > r$. Hence, not only that G has a K_r minor but also every subgraph of it containing $1/3$ of the edges has such minor. We conclude that G is $2/3$-far from being K_r minor free. $\qquad\square$

As every 0-witness for not being H-free has a cycle, corollary 4.1 ends the proof. $\qquad\square$

For an underlying graph $G = (V, E)$, let $\text{Bi}(G)$ be the property of G-subgraphs containing these subgraphs of G that are bipartite. Another application is for testing $\text{Bi}(G)$.

Theorem 4.9. *For some fixed $\epsilon, \alpha > 0$, there is an infinite sequence of graphs $\{G_n\}$, where G_n is an n vertex graph, such that any non-adaptive 1-sided error ϵ-test for $\mathrm{Bi}(G_n)$ must query $\Omega(n^\alpha)$ queries.*

Proof. (Sketch.)   As any witness for non-bipartiteness must contain an odd cycle, using Corollary 4.1 it is enough to show a high girth underlying graph that is sufficiently far from being bipartite. To construct such graph, again we choose one from the random graph model $\mathcal{G}(n, c/n)$. It is not hard to see that for large enough constant c, one can ensure that such a graph is 2/3-far from being bipartite, simply since each partition will contain many edges with both endpoints inside the parts. To ensure logarithmic girth, as before, one may delete $o(n)$ edges to get rid of all small cycles.   □

A similar application with a similar proof is the following: For $G = (V, E)$ and $\phi : E \to \{0, 1\}$, we now interpret ϕ as defining an orientation rather then a subgraph. The property $Ac(G)$ contains all $\vec{G}$-orientations that are acyclic.

Theorem 4.10. *For some fixed $\epsilon, \alpha > 0$, there is an infinite sequence of graphs $\{G_n\}$, where G_n is an n vertex graph, such that any non-adaptive 1-sided error test for $Ac(G_n)$ must query $\Omega(n^\alpha)$ queries.*

4.3   Testing Membership in Boolean Formulae

Testing membership in Boolean formula is a natural property and is extensively studied in TCS. We have seen some corresponding testability results for the more general situation of membership in different models of language representations in Section 1. Here we concentrate on CNF formulae, and formulae that are the conjunction of typically many small subformlae.

Constraint Graph Formulae. We start with a model for formulae that is quite related to the orientation model discussed in Section 4.1, and that was formulated in [17].

A constraint-graph is a labeled multi-graph (a graph where loops and parallel edges are allowed), where each edge is labeled by a distinct Boolean variable, and every vertex is associate with a Boolean function over the variables that label its adjacent edges[4]. A Boolean assignment to the variables satisfies the constraint graph if it satisfies every vertex function[5]. We associate with a constraint-graph G the property of all assignments satisfying G, denoted $SAT(G)$.

Thus, e.g., the property of subgraphs (or of G-orientations) of being Eulerian can be expressed in an obvious way. Similarly, the property of G-orientations of

---

[4] One should not confuse this definition of 'constraint-graphs' with the definition used by Dinur in [7]. There, each vertex is associated with a *distinct* Boolean variable and the edges are labeled by constraints.

[5] Thus formulae are conjunction of the individual vertex formulae. One may similarly consider other types of top gates to connect the vertex formulae, e.g, threshold gates, etc.

not having a source vertex can be expressed, as well as the related property of edge bi-coloring so that each vertex sees both colors.

Obviously, formulae represented by constraint graphs are general enough to represent any Boolean formula (as one can take a two vertex graph, with many parallel edges between the two vertices, and any complex formula on one of the vertices). Hence, the interesting case is when further restrictions are put on the complexity of the vertex formulae.

The main result of [17] is a test for the following class of Boolean formulae. For a vertex $v \in V(G)$ let $E_v = \{(v, w) \in E(G)\}$, namely all edges adjacent to v. For an assignment $\sigma \in \{0, 1\}^E$ and $F \subseteq E$ we denote by $\sigma(F)$ the restriction of the assignment σ on F.

Definition 4.3. *Let $\mathcal{LD}_i$ be the set of constraint graph formulae obtained from an underlying graph G such that for every two assignments σ_1, σ_2 and every vertex v with $\deg_G(v) \geq 3$, if σ_1, σ_2 do not satisfy f_v then either $\sigma_1(E_v) = \sigma_2(E_v)$ or they differ on at least i variables.*

Theorem 4.11. *[17] For every constraint-graph $G \in \mathcal{LD}_3$ there exists a 1-sided error, non-adaptive ϵ-test for $SAT(G)$, making $2^{\tilde{O}(1/\epsilon)}$ queries.*

The test is too complicated to be described here. We further mention that it is a generalization of a test for the property of G-orientations of not containing a source. The preprocessing involved in constructing the test, given G, is in polynomial time. We also note that this property, as well as the property of edge bi-coloring in which each vertex sees edges of both colors are in the family $\mathcal{LD}_3$ (while the property of being Eulerian is not).

An interesting application of Theorem 4.11 is to better understand the testability 'boundary' of the membership problem for restricted CNF formulae. This is the focus of the next subsection.

Testing Membership in CNF Formulae. General CNF formulae are hard to test. In [4] the authors show that there exists a family of 3-CNF formulae that are highly non-testable (every test requires a linear number of queries). Hence further restriction on the formulae has to be put in order to allow testability.

A Read-k-times formula is a formula in which every variable appears at most k times. By standard arguments the result of [4] can also be extended to read-three times formulae (and any constant $k \geq 3$ size clauses).

What about 2CNF ? It turns out that this does not yet allow testability either. A corollary from the lower bound of [9] for testing monotonicity over general posets is that even *monotone* 2CNF are hard to test (although the lower bound is far from being linear). This is so as in [9] a two leveled poset is constructed for which monotonicity is hard to test. Thus this hard to test property can be expressed as 2CNF. Moreover, by renaming the variables it is, in fact, a monotone 2CNF[6].

---

[6] Alternatively, the 2CNF can also be transformed into a read-3-times (but not monotone), hard to test 2CNF, using the standard reduction to bounded number of occurrences of a variable.

Theorem 4.11 implies that restricting the CNF to be Read-2-times (a.k.a 'read-twice') already guarantee testability, disregarding the clause size. In view of the discussion above, this seems best possible in this respect.

Corollary 4.2. *Every read-twice CNF formulae is testable by a 1-sided error test.*

Proof. A read-twice formula can obviously be represented by a constraint graph. Further, as per the definition of $\mathcal{LD}_3$, clauses of size smaller than 3 pose no restrictions. Since every clause as only one 0-assignment, every two non identical 0-assignments that falsify a clause of size 3 or more differ on at least 3 variables. Thus Theorem 4.11 applies, and implies the claim. $\square$

One may wonder whether $\mathcal{LD}_3$ is coincidental, namely are $\mathcal{LD}_2$ formulae testable. Indeed [10] (see more details in [17]) noted that this is not the case: For every Boolean formula θ there is a formulae $\phi \in \mathcal{LD}_2$ so that for any q, there is an ϵ-test for θ-membership with q queries if and only if there is an ϵ-test for ϕ-membership using q queries. To see this let θ be a Boolean formula over a set variables $X = \{x_1, \ldots, x_n\}$. Let G be the constraint-graph on two vertices $\{v, t\}$, that has $n+1$ parallel edges between v and t, one that is associated with the variable y and the rest n edges are labeled each with a distinct variable in X. Let $f_v = y \oplus (\bigoplus_{i=1}^{n} x_i)$ and $f_t = \theta(x_1, \ldots, x_n) \vee (y = \bigoplus_{i=1}^{n} x_i)$. Obviously the resulting property is in $\mathcal{LD}_2$. Given a test for one of the properties it is straightforward to build a test for the other property.

Testing General Read-Once Formulae. Here we discuss testing membership in other restricted formulae that are not the results of [20] or [17].

Read-Once formulae are Boolean formulae with $\wedge$ (AND) and $\vee$ (OR) gates in which every variable appears at most once. In addition, in what follows, we will assume that each variable appear unnegated, thus such formulae represent monotone Boolean functions.[7]

Read-Once formulae have been studied extensively in the past, but surprisingly, not with respect to whether their corresponding membership problem is testable. We note that bounded depth read-once formulae are represented by constant width branching program, thus their testability follows from [20]. We present here a result of Fischer, Lachish and Nimbhorkar that is much stronger.

Theorem 4.12. *[14] For every read-once formula F, the membership problem for F is 1-sided error testable.*

We present a relatively detailed proof as the idea here is somewhat different from the standard ideas one meets in property testing. The analysis we bring here imply an ϵ-test that has query complexity $(1/\epsilon)^{O(1/\epsilon)}$, and that the test can be constructed (that is, the preprocessing) in polynomial time. A variation of this idea and better analysis [14] implies a much better dependence on ϵ.

---

[7] This is w.l.o.g as the formula is known in advance, and by renaming we arrive to the desired situation.

Proof. (**Sketch**), Along the sequel we identify the formula F with the Boolean function that it computes $F : \{0,1\}^n \mapsto \{0,1\}$. We say that F is an $\vee$-formula if its top gate is $\vee$, similarly $\wedge$-formulae are defined. A single variable is considered both a $\vee$-formula and an $\wedge$-formula.

We first note that the formula F may be assumed to be layered so that the entries to each $\vee$-gate are variables or come from $\wedge$-gates, and the entries for $\wedge$-gates are variable or come from $\vee$-gates.

The test will be recursive w.r.t ϵ, that is, an ϵ-test for a $\vee$-formula will call a δ-test for a $\wedge$-subformula, with δ significantly larger than ϵ, while the test for $\wedge$-formula will call a test for $\vee$-formula with nearly the same ϵ. For this to end we first note that for any read-once $\vee$-formula F on at least two variables, any assignment is $1/2$-close to $F^{-1}(1)$.

Let $F = \circ_{i=1}^k F_k$ where $\circ$ is either $\vee$ or $\wedge$. For any assignment x to F, x is naturally being partitioned to $x_1, \ldots x_k$, where x_i, $i = 1, \ldots, k$ is an assignment to the variables in F_i. We keep this notation and for each assignment x to F, $x_1, \ldots, x_k$ refer to the corresponding assignments for $F_1, \ldots, F_k$.

We next describe a $\wedge$-test for $\wedge$-formulae and $\vee$-test for $\vee$-formula. In the following n is the number of variables of F.

ϵ-$\wedge$-**test:** Let $F = \wedge_{i=1}^k F_k$ with F_i on s_i variables. If $k = 1$ and F_1 is a single variable test deterministically F_1 by probing the variable.

Otherwise, independently for $9/\epsilon^2$ times pick an $i \in [k]$ with probability s_i/n and δ-$\vee$-test F_i independently twice, with $\delta = \epsilon(1 - \epsilon/3)$. If for the chosen i, both tests answer 1, set $T_i = 1$ and otherwise set $T_i = 0$. If for any of the chosen i's $T_i = 0$ stop and output 0 (that is reject x as not being in $F^{-1}(1)$), and otherwise (if for all i's $T_i = 1$) accept x (as being ϵ-close to $F^{-1}(1)$).

ϵ-$\vee$-**test:** Let $F = \vee_{i=1}^k F_k$ with F_i on s_i variables.
If $k = 1$ and F_1 is a single variable, deterministically test F_1 using one query. If for some i, $s_i < \epsilon n$, or if $\epsilon \geq 1/2$ stop and accept x (as being ϵ-close to $F^{-1}(1)$).

Otherwise, for every $i \in [k]$ set $\alpha_i = \epsilon \cdot n/s_i$ and perform $\ell = 1 + \log_3 k$ independent times α_i-$\wedge$-test for F_i. For such i set $T_i = 1$ if all the ℓ independent tests return 1 and 0 otherwise. If any of the T_i, $i = 1, \ldots, k$ has returned 1 accept x (as being ϵ-close to $F^{-1}(1)$). Otherwise (if for every i, $T_i = 0$) reject x as being ϵ-far from $F^{-1}(1)$.

To assert the correctness of the tests, assume inductively that both tests are 1-sided error and have error probability at most $1/3$ on smaller formulae for any ϵ (the reader may check this for the base case of one gate formulae). It is easy to verify that both tests have 1-sided error assuming inductively that they do on subformulae.

To assess the error probability of the $\wedge$-test, assume that x is ϵ-far from the $\wedge$-formula F. Then x_i is ϵ_i-far from being a member of F_i with $\sum \epsilon_i \cdot \frac{s_i}{n} = \epsilon$. From this it follows, by simple average-argument, that there are at most $1 - \epsilon^2/3$-fraction of the variables that may appear in F_i for which x_i is $(1 - \epsilon/3)\epsilon$-close to F_i. Thus an error will occur on x only if all $9/\epsilon^2$ i's that are sampled resulted in F_i for which x_i is $(1 - \epsilon/3)\epsilon$-close to F_i, or when an i for which x_i is $(1 - \epsilon/3)\epsilon$-far from F_i is sampled, and the $\vee$-test for F_i errs twice. The former event happens with probability smaller than e^{-3} and the later with probability at most $1/9$, resulting a total error bounded by $1/3$.

For the $\vee$-test, note that if x is ϵ-far from being a member of F, namely one needs to change at least ϵn bits of x in order to become in $F^{-1}(1)$, then in each x_i one should change at least ϵn bits to become a member of $F_i^{-1}(1)$. It follows that x_i is at least α_i-far from F_i. Hence, an error on x occurs only if at least one of the T_i's is wrong, meaning that all ℓ tests were wrong for F_i. By the union bound this will happen with probability at most $k \cdot 1/(3k) \leq 1/3$.

We end by estimating the query complexity. Let $t_\vee(\epsilon), t_\wedge(\epsilon)$ denote the corresponding complexity of the $\vee$ and $\wedge$ tests. Note first that for $\epsilon \geq 1/2$ $t_\vee(\epsilon) = 1$. Also, this is the case if F is $\vee$-formula with $k > 1/\epsilon$, or $s_i < \epsilon n$. Thus we get $t_\vee(\epsilon) \leq \frac{\log(1/\epsilon)}{\epsilon} \cdot t_\wedge(\epsilon/(1-\epsilon))$. (Since $s_i \geq (1-\epsilon)n$ for all i's, which implies that $\alpha_i \leq \frac{\epsilon}{1-\epsilon}$).

Similarly, $\wedge$-test takes 1-query for $k = 1$, and for general ϵ it takes $t_\wedge(\epsilon) \leq \frac{6}{\epsilon^2} \cdot t_\vee(\epsilon(1-\epsilon))$. Solving the recurrence implies the result.   □

5   Open Problems

There are many open problems that arise in view of the results above. I will mention here but some.

1. There are nearly no testability results for any interesting subgraph-property of a given undirected underlying graph. There are no interesting results at all on the *induced* analog (namely, where the subgraph is defined by vertex labeling).
2. Testing monotonicity is not a main theme here. However, testing monotonicity in general posets does fall into this context. In view of the large gap between the upper and lower bounds in [9], closing this gap remains an interesting open problem. One should mention that even for the Boolean cube, the complexity of testing monotonicity of Boolean functions (2-sided error) is far from being understood.
3. In [20] the dependence in the width is doubly exponential. There is no matching lower bound for this dependence (see also [19] for further related details). As this result is what partially determines the dependence on ϵ in the result for testing $s - t$-connectivity (Section 4.1), this further motivates resolving the question of the exact dependence in ϵ of both testing problems.
4. A general result on testing strong connectivity for G-orientations, or a non-testability result are still missing. It still might be the case that for every underlying graph this problem is testable. On the other hand, we don't have any non-trivial upper bound even of the form n^α for $\alpha < 1$. Similarly the variants of $s - t$ strong connectivity and s-connectivity (namely, that s can reach every other vertex) are open.

Acknowledgments

This work was partially supported by the Israel Science Foundation (grants No. 1011/06).

References

1. Alon, N.: Testing subgraphs in large graphs. Random Struct. Algorithms 21(3-4), 359–370 (2002)
2. Alon, N., Krivelevich, M., Newman, I., Szegedy, M.: Regular languages are testable with a constant number of queries. SIAM Journal on Computing 30, 1842–1862 (2001)
3. Bhattacharyya, A., Grigorescu, E., Jung, K., Raskhodnikova, S., Woodruff, D.P.: Transitive-closure spanners. In: SODA, pp. 932–941 (2009)
4. Ben-Sasson, E., Harsha, P., Raskhodnikova, S.: Some 3CNF properties are hard to test. SIAM J. Comput. 35(1), 1–21 (2005)
5. Bender, M.A., Ron, D.: Testing properties of directed graphs: acyclicity and connectivity. Random Struct. Algorithms 20(2), 184–205 (2002)
6. Chakraborty, S., Fischer, E., Lachish, O., Matsliah, A., Newman, I.: Testing s-t -connectivity. In: Charikar, M., Jansen, K., Reingold, O., Rolim, J.D.P. (eds.) RANDOM 2007 and APPROX 2007. LNCS, vol. 4627, pp. 380–394. Springer, Heidelberg (2007)
7. Dinur, I.: The PCP theorem by gap amplification. J. ACM 54(3), 12 (2007)
8. Fischer, E.: The art of uninformed decisions: A primer to property testing. BEATCS: Bulletin of the European Association for Theoretical Computer Science 75, 97–126 (2001)
9. Fischer, E., Lehman, E., Newman, I., Raskhodnikova, S., Rubinfeld, R., Samorodnitsky, A.: Monotonicity testing over general poset domains. In: Proceedings of the 34th ACM STOC, pp. 474–483 (2002)
10. Fischer, E.: Personal communication
11. Fischer, E.: On the strength of comparisons in property testing. Inf. Comput. 189(1), 107–116 (2004)
12. Fischer, E., Lachish, O., Matsliah, A., Newman, I., Yahalom, O.: On the query complexity of testing orientations for being Eulerian. In: Goel, A., Jansen, K., Rolim, J.D.P., Rubinfeld, R. (eds.) APPROX and RANDOM 2008. LNCS, vol. 5171, pp. 402–415. Springer, Heidelberg (2008) (to appear in Algorithmica)
13. Fischer, E., Lachish, O., Newman, I., Rosenberg, E.: Lower bound technique for properties of underlying graphs. In Preparations
14. Fischer, E., Lachish, O., Nimbhorkar, P.: In Preparations
15. Goldreich, O., Goldwasser, S., Ron, D.: Property testing and its connection to learning and approximation. J. ACM 45(4), 653–750 (1998)
16. Halevy, S., Lachish, O., Newman, I., Tsur, D.: Testing orientation properties. Electronic Colloquium on Computational Complexity (ECCC) (153) (2005)
17. Halevy, S., Lachish, O., Newman, I., Tsur, D.: Testing properties of constraint-graphs. In: IEEE Conference on Computational Complexity, pp. 264–277 (2007)
18. Kostochka, A.: The minimum Hadwiger number for graphs with a given mean degree of vertices. Metody Diskret. Analiz. 38, 37–58 (1982)
19. Lachish, O., Newman, I., Shapira, A.: Space Complexity vs. Query Complexity. Computational Complexity 17(1), 70–93 (2008)
20. Newman, I.: Testing membership in languages that have small width branching programs. SIAM J. Comput. 31(5), 1557–1570 (2002)
21. Ron, D.: Property testing (A Tutorial). In: Rajasekaran, S., Pardalos, P.M., Reif, J.H., Rolin, J.D.P. (eds.) Handbook of Randomized Computing. Kluwer Press, Dordrecht (2001)
22. Thomason, A.: An extremal function for contractions of graphs. Math. Proc. Cambridge Philos. Soc. 95, 261–265 (1984)

Sublinear Graph Approximation Algorithms

Krzysztof Onak*

Massachusetts Institute of Technology, Cambridge, MA, USA

Abstract. We survey the recent research on algorithms that approximate the optimal solution size for problems such as vertex cover, maximum matching, and dominating set. Techniques developed for these problems have found applications in property testing in the bounded-degree graph model.

Keywords: constant-time algorithms, maximum matching, vertex cover, minor-freeness, hereditary properties.

1 Introduction

Classical optimization problems on graphs include maximum matching, vertex cover, and dominating set. Solving them—even approximately—requires in most cases reading an amount of information that is at least linear in the number of vertices. A natural question that arises in this context is whether this amount of processing is necessary if one is only interested in the approximate size of the optimal solution rather than the solution itself.

An early example of an efficient approximation algorithm in this setting is given by Chazelle, Rubinfeld, and Trevisan [1]. Their algorithm computes a $(1 + \varepsilon)$-approximation to the minimum spanning tree cost in a connected graph of maximum degree bounded by d and weights in $\{1, \ldots, w\}$. The running time of the algorithm is $\tilde{O}(dw/\varepsilon^2)$, which does not depend on the number of vertices in the graph.

In this brief survey, we focus on the line of research that was initiated by Parnas and Ron [2] as a result of observing a connection between sublinear approximation algorithms and distributed algorithms.

1.1 Preliminaries

The Approximation Notion. We say that an algorithm is an $(\alpha, \varepsilon n)$-*approximation* algorithm for a problem $\mathcal{X}$ if with probability $2/3$, it outputs a value A such that $\mathrm{OPT} \leq A \leq \alpha \cdot \mathrm{OPT} + \varepsilon n$, where OPT is the optimal solution size for the input instance of $\mathcal{X}$, and n is the number of vertices in the graph.

---

* Supported in part by NSF grants 0514771, 0732334, and 0728645, an Akamai fellowship, and a Symantec fellowship.

O. Goldreich (Ed.): Property Testing, LNCS 6390, pp. 158–166, 2010.

Constant Time Algorithms. If the running time of an algorithm is a function of only the average vertex degree, and the approximation parameter ε, then such an algorithm is traditionally described as a *constant-time* algorithm, since for graphs with constant average degree, it runs in time independent of the graph size.

Throughout the survey, we focus on the case when the *maximum* degree is bounded by d. In all considered problems, whenever only a bound $\tilde{d}$ on the *average* degree is known, one can slightly modify the problem and remove from the graph all vertices of degree greater than $C \cdot \tilde{d}/\varepsilon$, where C is a sufficiently large constant. The number of such vertices is sufficiently small not to impact the solution size significantly.

The Model. A sublinear-time algorithm does not have time to go over the entire input to preprocess it in an suitable way. Therefore, we have to assume something about how the algorithm accesses the input. Two kinds of queries are allowed: degree queries and neighbor queries; both answered in constant time. In a *degree query*, the algorithm specifies a vertex v and it obtains the degree of v. In a *neighbor query*, the algorithm specifies both a vertex v and a positive integer k. In reply, the algorithm obtains the label of the k-th neighbor of v. We do not assume anything about the ordering of any vertex's neighbors.

Furthermore, we assume that the algorithm can select a vertex in the graph uniformly at random in constant time. Also, we assume that it is possible to maintain a efficient dictionary with vertices as keys.

2   General Bounded-Degree Graphs

We start by describing constant-time approximation algorithms for general graphs.

2.1   Vertex Cover

The framework. In this section, we present constant-time $(2, \varepsilon n)$-approximation algorithms for vertex cover. Constant-time approximation algorithms for this and other problems are often created by constructing an oracle that provides query access to an approximately optimal solution to the problem. For vertex cover, an oracle provides query access to a vertex cover S by answering queries of the form "Is v in S?" The cover S can be a function of the graph and random bits, but it is important that it not be a function of queries. Given such an oracle, one can use sampling to approximate the size of S, and indirectly, the optimal solution size if S is approximately optimal. If the oracle can answer each query very quickly, reading just a tiny part of the graph, then this approach yields a sublinear-time approximation algorithm.

A connection to local distributed algorithms. We say that a distributed algorithm is *local* if it runs in a number of communication rounds that is independent of the number of vertices in the graph. Parnas and Ron [2] observe that an oracle as the above can be constructed from a local distributed algorithm for vertex cover. Let

$\mathcal{A}$ be a local distributed algorithm that computes a vertex cover of size at most α times the optimum in at most t communication rounds. To find out whether $\mathcal{A}$ includes a given vertex v, it suffices to simulate $\mathcal{A}$ on the neighborhood of v of radius $t + 1$. Since the distributed algorithm runs for t rounds, running the algorithm on the limited neighborhood does not affect $\mathcal{A}$'s decision on whether $v \in S$.

Then an $(\alpha, \varepsilon n)$-approximation to the vertex cover size can be computed by sampling $O(1/\varepsilon^2)$ vertices and estimating the fraction of those that the distributed algorithm includes in the vertex cover. If d is the maximum degree of the graph, an algorithm only needs to make $d^{O(t)}/\varepsilon^2$ queries to the graph.

By applying a local distributed algorithm of Kuhn, Moscibroda, and Wattenhofer [3], Parnas and Ron obtain a $(2, \varepsilon n)$-approximation algorithm for vertex cover that runs in $d^{O(\frac{\log d}{\varepsilon^3})}$ time.

Marko and Ron [4] improve on the above result by using Luby's distributed maximal independent set algorithm [5] to locally find a *maximal* matching (which is a maximal independent set in the line graph). The set of vertices matched in any maximal matching is well known to form a vertex cover of size at most twice the optimum [6]. The running time of the $(2, \varepsilon n)$-approximation algorithm of Marko and Ron is $d^{O(\log(d/\varepsilon))}$.

A Local Greedy Approach. In [7], we propose a new method for locally computing a maximal independent set. Compared to Luby's algorithm, our method turns out to be more efficient in terms of the number of queries, while Luby's algorithm needs fewer communication rounds. Our starting point is the simplest classical algorithm for finding a maximal independent set. The algorithm considers vertices one by one in arbitrary order, and greedily adds them to the independent set being constructed whenever possible. Checking if a vertex v is in the maximal independent set can be achieved by recursively checking if any of the neighbors of v considered before v is in the maximal independent set. If none of them are, v is in the maximal independent set. Otherwise, it is not. Unfortunately, if the vertices are considered in arbitrary order, this can create a long chain of dependencies, and may require exploring almost the entire graph to answer a single query. We avoid this by considering vertices in random order. The randomization makes the probability of visiting vertices distant from a query point very small. Instead of explicitly maintaining an ordering of vertices, we independently assign a uniform random number in $[0, 1]$ (after a proper discretization) to every vertex. The numbers generate a random ordering of the vertices.

It can be shown that for a fixed query vertex, the above method requires exploring $2^{O(d)}$ vertices in expectation, and therefore, the query complexity is $2^{O(d)}$ in expectation as well. This implies that our algorithm runs in $2^{O(d)}/\varepsilon^2$ time. This is better than the algorithm of Marko and Ron [4] in terms of the dependence on ε, but worse in terms of the dependence on d.

In [7], we also propose the following simple pruning heuristic. For a given vertex v, the algorithm recursively considers neighbors of v that were assigned a number lower than that of v. In the heuristic, the algorithm recursively considers the neighbors in order of their random numbers, starting from the lowest. Once it finds a neighbor that belongs to the maximal independent set, it terminates the recursive search, because this already implies that v is not in the maximal independent set. Yoshida, Yamamoto, and Ito [8] prove that the expected number of vertices visited for a *random* query point is $O(d)$. Using this fact, they show that the algorithm with the pruning heuristic runs in $\tilde{O}(d^4/\varepsilon^2)$ expected time, which gives the first algorithm of running time polynomial in both d and $1/\varepsilon$.

Lower Bounds. It is also worth noting that the multiplicative constant 2 is the best possible. Trevisan (the result appeared in [2]) proves that there is no constant-time $(2 - \delta, \varepsilon n)$-approximation algorithm for this problem for any constant $\delta > 0$. Any such algorithm has to make at least $\Omega(\sqrt{n})$ queries for a sufficiently high constant d. Furthermore, Parnas and Ron [2] prove that any $(O(1), \varepsilon n)$-approximation algorithm must make at least $\Omega(d)$ queries to the graph.

2.2   Maximum Matching

Our paper [7] gives the first $(1, \varepsilon n)$-approximation algorithm for maximum matching. In the algorithm, we construct a series of oracles that provide query access to better and better matchings.

Recall than an *augmenting path* for a given matching M is a path that starts at an unmatched vertex and alternates between edges not in M and those in M until it reaches another unmatched node. Such a path can be used to increase the matching size by 1. A crucial role in our approach is played by a lemma proved by Hopcroft and Karp [9]. They showed that if M is a matching that has no augmenting paths of length less than t, and one applies to M a *maximal* set of vertex-disjoint augmenting paths of length t, then the resulting matching can only have augmenting paths of length greater than t.

It is also easy to show that if a matching does not have any augmenting path of length less than $2k+1$, then its size is at least $\frac{k}{k+1}$ times the optimum. One can construct a large matching, of size at least the optimum minus εn, by starting with an empty matching and applying a maximal set of disjoint augmenting paths of length first 1, then 3, 5, 7, and so on until there is no augmenting path of length less than $2\lceil 1/\varepsilon \rceil + 1$.

In our constant-time algorithm, we create a separate oracle for each of the matchings obtained in the above process. The k-th oracle provides query access to a matching that has no augmenting paths of length less than $2k + 1$. The oracle creates the matching locally by querying the previous oracle, and applying a maximal vertex-disjoint set of augmenting paths of length $2k - 1$. To find such a set locally, the oracle uses any method for locally finding a maximal independent set.

We show that the algorithm runs in $2^{d^{O(1/\varepsilon)}}$ time if it uses our local greedy approach. Yoshida, Yamamoto, and Ito [8] show that the application of the pruning heuristic improves the algorithm's running time to $d^{O(1/\varepsilon^2)}$.

The Property of Having a Perfect Matching. For graphs on an even number of vertices, a natural property of interest is the property of having a perfect matching. Note that the distance to this property can be decreased only by adding edges that increase the maximum matching size. If an $O(1)$-degree graph is ε-*far* from having this property (e.g., at least $\Omega(\varepsilon n)$ edges of the graph must be modified so that it has the property), this means that the matching size is at most $n/2 - \Omega(\varepsilon n)$. If the graph has this property, then the matching size is $n/2$. It is clear that a constant-time tester for this property can be obtained by using any of the matching algorithms to distinguish the above two cases. Moreover, Yoshida *et al.* [8] show that testing for this property with *one-sided error* (e.g., never rejecting graphs that have a perfect matching) requires $\Omega(n)$ queries.

2.3   Other Problems

The method we use to construct an algorithm for matchings can be used to transform other greedy approximation algorithms into constant-time approximation algorithms. One example is the greedy $O(\log n)$-approximation algorithm for set cover [10,11], which can then be used to obtain a $(O(\log n), \varepsilon n)$-approximation algorithm for dominating set. See [7] for more details.

Alon [12] shows that there is no constant-time $(o(\log d), \varepsilon n)$-approximation algorithm for dominating set. For maximum independent set, he shows that there is a constant-time $(O(\frac{d \log \log d}{\log d}), \varepsilon n)$-approximation algorithm, but there is no constant-time $(o(\frac{d}{\log d}), \varepsilon n)$-approximation algorithm.

3   Algorithms for Hyperfinite Graphs

The results of Alon [12] and Trevisan [2] show that there are no constant-time $(1, \varepsilon n)$-approximation algorithms for vertex cover, dominating set, and maximum matching. We now look at an important class of graphs for which such algorithms do exist.

3.1   Hyperfinite Graphs

We say that a family of constant-degree graphs is *hyperfinite* if there is a function $\delta : (0,1) \rightarrow [1, +\infty)$ such that for every graph in the family and every $\varepsilon > 0$, one can remove at most εn edges from the graph, and obtain connected components of size at most $\delta(\varepsilon)$. Hyperfinite families of graphs include graphs of subexponential growth[1] [13], constant-degree graphs with an excluded minor (this follows from the separator theorem [14,15]), and the non-expanding graphs considered by Czumaj, Shapira, and Sohler [16].

3.2   Approximation Algorithms and Partitioning Oracles

In [17], we show that for any $\varepsilon > 0$, and any hyperfinite family of graphs, there are constant-time $(1, \varepsilon n)$-approximation algorithms for vertex cover, dominating

---

[1] For a family of graphs, the *growth* is a function $g(r)$ equal to the maximum number of vertices at distance at most r from any vertex in any of the graphs. If $g(r) = o((1 + \delta)^r)$ for every $\delta > 0$, then we say that the growth is *subexponential*.

set, and maximum independent set. In particular, for any family of graphs with an excluded minor, we show algorithms that run in $2^{\text{poly}(1/\varepsilon)}$ time. Note that since these problems are NP-hard even in the planar setting, it is unlikely that algorithms that run in $2^{(1/\varepsilon)^{o(1)}}$ time exist, since by setting $\varepsilon = 1/(100n)$, they could be used to solve the problems exactly in subexponential time with constant probability.

Independently, Czygrinow, Hańćkowiak, and Wawrzyniak show how to construct local distributed $(1 + \varepsilon)$-approximation algorithms for these problems in planar graphs, which implies constant-time $(1, \varepsilon n)$-approximation algorithms for planar graphs via the connection discovered by Parnas and Ron [2]. Elek [18] shows constant-time approximation algorithms for graphs with subexponential growth.

The main tool of [17] is *partitioning oracles*. For every hyperfinite family $\mathcal{F}$ of graphs and every $\varepsilon > 0$, we design a partitioning oracle that provides query access to a partition of the input graph. It has the following properties:

- The oracle answers queries of the form: "What vertices belong to the component in the partition that contains v?"
- The partition has components of constant size, where the bound on the component size depends on the family of graphs and ε.
- The partition can only be a function of the graph and random bits of the oracle, but does not depend on the order of queries.
- The oracle works even if the input does not belong to $\mathcal{F}$ (this is useful for the application to minor-freeness testing, which we describe below), but if it does belong, then with probability $99/100$, the number of edges cut in the partition is εdn.
- To answer every query, the oracle makes only a constant number of queries to the input graph. The total amount of computation that the oracle uses is only a function of the number of queries to the oracle.

The simplest construction of a partitioning oracle employs the technique from our previous paper [7] to simulate a global greedy partitioning method locally. We also show more efficient oracle constructions for graphs with an excluded minor, among other families of graphs.

Given a partitioning oracle for a family that the input graph belongs to, an approximation algorithm is constructed as follows. We know that the oracle is likely to cut only a small fraction of edges of the input graph. One can show that this changes the solution size for all considered problems by $O(\varepsilon n)$. Moreover, an optimal solution for the entire graph is the union of optimal solutions for all components in the partition. Therefore, the algorithm samples $O(1/\varepsilon^2)$ vertices, computes optimal solutions for the components they belong to, and returns the fraction of vertices that are included in the solution. This up to an additive $O(\varepsilon)$ is the fraction of vertices in the optimal solution for the input graph.

3.3 Other Applications of Partitioning Oracles

Our partitioning oracles find applications to other problems. Here, we describe two of them.

Testing Minor-Freeness. Recall that a graph H is a *minor* of a graph G if H can be obtained from G via a sequence of edge deletions, vertex deletions, and edge contractions. Moreover, G is H-*minor-free* if H is not a minor of G.

We now consider the following problem in the bounded-degree model of Goldreich and Ron [19]. For every fixed non-empty graph H, we want to construct an algorithm that

- accepts with probability at least $2/3$ if the input is H-minor-free,
- rejects with probability at least $2/3$ if the input graph needs to have at least εdn edges removed to achieve H-minor-freeness.

Goldreich and Ron [19] show an $\tilde{O}(1/\varepsilon^3)$-time tester for cycle-freeness, which is equivalent to K_3-minor-freeness. The breakthrough result of Benjamini, Schramm, and Shapira [20] shows that H-minor-freeness is testable for any H in $2^{2^{2^{\mathrm{poly}(1/\varepsilon)}}}$ time. Our techniques give a much simpler proof of this fact, and also give a tester that runs in $2^{\mathrm{poly}(1/\varepsilon)}$ time.

Note that every *minor-closed* property can be expressed via a finite set S of forbidden minors due to the celebrated result of Robertson and Seymour [21]. One can show that this implies that each minor-closed property can be tested by running testers for each of the forbidden minors. Well known minor-closed properties include planarity, outerplanarity, and constant treewidth.

As an example, we now sketch how the tester for K_5-minor-freeness works. The tester is given a partitioning oracle for the family of K_5-minor-free graphs with an appropriately chosen parameter. First, the tester estimates the fraction of the edges cut by the partition that is provided by the oracle. If this fraction is large, the tester rejects. Otherwise, it samples a small number of vertices in the graph, and checks whether their components in the partition are K_5-free. If all of them are, the tester accepts, and otherwise, it rejects.

Why does this work? If the input is K_5-minor-free, then all components in the partition are K_5-minor-free as well, so the tester can only reject if the number of edges cut by the partition is large, but this is unlikely because the input is K_5-minor-free. On the other hand, if the input must be significantly modified to obtain K_5-minor-freeness, then either the tester rejects it with high probability because the number of edges cut is large or many components must have K_5 as a minor, and at least one of them can easily be spotted by sampling.

Approximating Distance to Hereditary Properties for Hyperfinite Graphs. We say that a graph property is *hereditary* if it is closed under vertex removal. For example, all minor-closed properties, perfectness, and k-colorability, for any constant k, are hereditary. All hereditary properties are testable in constant time in the dense graph model with one-sided error [22]. This turns out not to be the case in the sparse graph model. Bipartiteness requires $\Omega(\sqrt{n})$ queries [19], and 3-colorability requires $\Omega(n)$ queries [23]. The paper of Czumaj, Shapira, and Sohler [16] shows that in hyperfinite families of graphs, hereditary properties are testable in constant time.

In [17], we show that the result of Czumaj *et al.* can be extended for hereditary properties that hold for all empty graphs. Using partitioning oracles, we prove that for each such property, it is possible to approximate—up to an additive εn—the number of edges that must be modified in the input graph from a given hyperfinite family of graphs to obtain the property. As a sample application of our result, we obtain a constant-time $(1, \varepsilon n)$-approximation algorithm for the number of edges that must be removed from a given constant-degree planar graph to make it 3-colorable.

Independently, Elek [18] proves that it is possible to approximate the distance to union-closed monotone properties in graphs of subexponential growth. Not all union-closed monotone properties are hereditary, but almost all natural ones are. In particular, all union-closed monotone properties that Elek lists in his paper are hereditary. On the other hand, perfectness is hereditary, but not monotone.

4   Open Problems

The following questions remain open in this line of research:

- What graph problems and to what extent can be approximated in sublinear time?
- For what other popular classes of graphs can one design approximation algorithms better than those in the general case?
- What is the exact complexity of approximating the minimum vertex cover size and the maximum matching size?
- Can one design partitioning oracles for minor-free families of graphs with complexity polynomial in $1/\varepsilon$?

References

1. Chazelle, B., Rubinfeld, R., Trevisan, L.: Approximating the minimum spanning tree weight in sublinear time. SIAM J. Comput. 34(6), 1370–1379 (2005)
2. Parnas, M., Ron, D.: Approximating the minimum vertex cover in sublinear time and a connection to distributed algorithms. Theor. Comput. Sci. 381(1-3), 183–196 (2007)
3. Kuhn, F., Moscibroda, T., Wattenhofer, R.: The price of being near-sighted. In: SODA, pp. 980–989 (2006)
4. Marko, S., Ron, D.: Approximating the distance to properties in bounded-degree and general sparse graphs. ACM Transactions on Algorithms 5(2) (2009)
5. Luby, M.: A simple parallel algorithm for the maximal independent set problem. SIAM J. Comput. 15(4), 1036–1053 (1986)
6. Garey, M.R., Johnson, D.S.: Computers and Intractability: A Guide to the Theory of NP-Completeness. W. H. Freeman, New York (1979)
7. Nguyen, H.N., Onak, K.: Constant-time approximation algorithms via local improvements. In: FOCS, pp. 327–336 (2008)
8. Yoshida, Y., Yamamoto, M., Ito, H.: An improved constant-time approximation algorithm for maximum matchings. In: STOC, pp. 225–234 (2009)
9. Hopcroft, J.E., Karp, R.M.: An $n^{5/2}$ algorithm for maximum matchings in bipartite graphs. SIAM J. Comput. 2(4), 225–231 (1973)

10. Johnson, D.S.: Approximation algorithms for combinatorial problems. J. Comput. Syst. Sci. 9(3), 256–278 (1974)
11. Lovász, L.: On the ratio of optimal integral and fractional covers. Discrete Mathematics 13, 383–390 (1975)
12. Alon, N.: On constant time approximation of parameters of bounded degree graphs
13. Elek, G.: L^2-spectral invariants and convergent sequences of finite graphs. Journal of Functional Analysis 254(10), 2667–2689 (2008)
14. Lipton, R.J., Tarjan, R.E.: A separator theorem for planar graphs. SIAM Journal on Applied Mathematics 36, 177–189 (1979)
15. Alon, N., Seymour, P.D., Thomas, R.: A separator theorem for graphs with an excluded minor and its applications. In: STOC, pp. 293–299 (1990)
16. Czumaj, A., Shapira, A., Sohler, C.: Testing hereditary properties of nonexpanding bounded-degree graphs. SIAM J. Comput. 38(6), 2499–2510 (2009)
17. Hassidim, A., Kelner, J.A., Nguyen, H.N., Onak, K.: Local graph partitions for approximation and testing. In: FOCS (2009)
18. Elek, G.: Parameter testing with bounded degree graphs of subexponential growth. arXiv:0711.2800v3 (2009)
19. Goldreich, O., Ron, D.: Property testing in bounded degree graphs. Algorithmica 32(2), 302–343 (2002)
20. Benjamini, I., Schramm, O., Shapira, A.: Every minor-closed property of sparse graphs is testable. In: STOC, pp. 393–402 (2008)
21. Robertson, N., Seymour, P.D.: Graph minors. XX. Wagner's conjecture. Journal of Combinatorial Theory, Series B 92(2), 325–357 (2004); Special Issue Dedicated to Professor W.T. Tutte
22. Alon, N., Shapira, A.: A characterization of the (natural) graph properties testable with one-sided error. SIAM J. Comput. 37(6), 1703–1727 (2008)
23. Bogdanov, A., Obata, K., Trevisan, L.: A lower bound for testing 3-colorability in bounded-degree graphs. In: FOCS, pp. 93–102 (2002)

Transitive-Closure Spanners: A Survey*

Sofya Raskhodnikova**

Pennsylvania State University, USA
sofya@cse.psu.edu

Abstract. We survey results on transitive-closure spanners and their applications. Given a directed graph $G = (V, E)$ and an integer $k \geq 1$, a *k-transitive-closure-spanner (k-TC-spanner)* of G is a directed graph $H = (V, E_H)$ that has (1) the same transitive-closure as G and (2) diameter at most k. These spanners were studied implicitly in different areas of computer science, and properties of these spanners have been rediscovered over the span of 20 years. The common task implicitly tackled in these diverse applications can be abstracted as the problem of constructing sparse TC-spanners.

In this article, we survey combinatorial bounds on the size of sparsest TC-spanners, and algorithms and inapproximability results for the problem of computing the sparsest TC-spanner of a given directed graph. We also describe multiple applications of TC-spanners, including property testing, property reconstruction, key management in access control hierarchies and data structures.

1 Introduction

A *spanner* is a sparse backbone of a graph that approximately preserves distances between every pair of vertices. More precisely, a subgraph $H = (V, E_H)$ is a *k-spanner* of $G = (V, E)$ if for every pair of vertices $u, v \in V$, the shortest path distance $d_H(u, v)$ from u to v in H is at most $k \cdot d_G(u, v)$. Since they were introduced by Awerbuch [10] and Peleg and Schäffer [43] in the context of distributed computing, spanners for undirected graphs have found numerous applications, including efficient routing [22,23,45,46,55], simulating synchronized protocols in unsynchronized networks [44], parallel and distributed algorithms for approximating shortest paths [20,21,27], and algorithms for distance oracles [11,56].

In the setting of directed graphs, three notions of spanners have been proposed: the direct generalization of the above definition [43], *roundtrip spanners* [23,46] and *transitive-closure spanners* [15]. In this survey, we focus on the latter definition. It captures the notion that a spanner should have a small diameter but preserve

---

* Parts of this survey are adapted from [15,16,17,14].

** Supported by National Science Foundation (NSF/CCF award 0729171 and NSF/CCF CAREER award 0845701).

O. Goldreich (Ed.): Property Testing, LNCS 6390, pp. 167–196, 2010.

the connectivity of the original graph. By diameter[1], we mean the largest distance between a pair (u, v) of nodes in a directed graph such that v is reachable from u.

Recall that the *transitive closure* of a graph $G = (V, E)$ is a graph (V, E_{TC}) where $(u, v) \in E_{TC}$ if and only if there is a directed path from u to v in G.

Definition 1.1 (TC-spanner [15]). *Given a directed graph $G = (V, E)$ and an integer $k \geq 1$, a k-**transitive-closure-spanner** (k-**TC-spanner**) is a directed graph $H = (V, E_H)$ with the following properties:*

1. *E_H is a subset of the edges in the transitive closure of G.*
2. *For all vertices $u, v \in V$, if $d_G(u, v) < \infty$, then $d_H(u, v) \leq k$.*

*The edges from the transitive closure of G that are added to G to obtain a TC-spanner are called **shortcuts**, and the parameter k is called the **stretch**.*

Notice that a k-TC-spanner of G is a directed k-spanner of the transitive-closure of G. Nevertheless, TC-spanners are interesting in their own right due to the multiple TC-spanner-specific applications.

Before TC-spanners were introduced in [15], they were studied implicitly in access control, property testing, and data structures, and properties of these combinatorial objects have been discovered and rediscovered over the span of 20 years. In this work, we survey results on TC-spanners and their applications. We start by discussing a simple example of a TC-spanner of a directed line, which has been studied in different areas under different guises.

1.1   A Simple Example: TC-spanners of the Directed Line

Directed acyclic graphs (DAGs) represent the most interesting case in applications of TC-spanners. There is also a reduction from constructing TC-spanners of graphs with cycles to constructing TC-spanners of DAGs, with a small loss in stretch, which we present in Section 3.2. In this section, we illustrate the definition of TC-spanners by constructing sparse TC-spanners of the simplest DAG—the directed line. The directed line L_n consists of nodes[2] $[n]$ and edges $\{(i, i+1) : 1 \leq i \leq n-1\}$. As discussed in Section 3.1, TC-spanners of the line were implicitly studied in different contexts by multiple authors, many of whom discovered the optimal constructions we describe here.

An additional motivation for considering optimal TC-spanners of the directed line in a separate section is that it gives one of the simplest settings where inverse Ackermann's functions (see Definitions 1.2) arise naturally—simple enough to explain in an undergraduate algorithms class.

It is easy to see that the transitive closure of L_n has $\binom{n}{2} = \Theta(n^2)$ edges. A TC-spanner, even of the smallest possible stretch—stretch 2, can be much sparser then the transitive closure.

---

[1] The definition of diameter used in this survey and other papers on transitive-closure spanners is nonstandard. The diameter is usually defined as the largest distance between a pair of nodes in a graph, and is set to infinity if a graph contains a pair of nodes with no path from one to the other.

[2] We use $[n]$ to denote $\{1, 2, \ldots, n\}$.

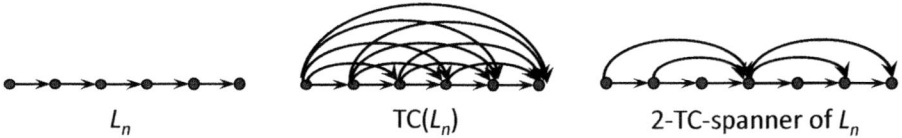

L_n $TC(L_n)$ 2-TC-spanner of L_n

Lemma 1.1 (2-TC-spanner of the line). *For all $n \geq 3$, the directed line L_n has a 2-TC-spanner with at most $n \log n$ edges[3].*

Proof. Our 2-TC-spanner, H, is a graph with vertex set $[n]$. We construct the edge set of H recursively. First, define the middle node $v_{mid} = \lceil \frac{n}{2} \rceil$. Use this node as a *hub*: namely, add edges (v, v_{mid}) for all nodes $v < v_{mid}$ and edges (v_{mid}, v) for all nodes $v > v_{mid}$. Then recurse on the two line segments resulting from removing v_{mid} from the current line. Proceed until each line segment contains exactly one node.

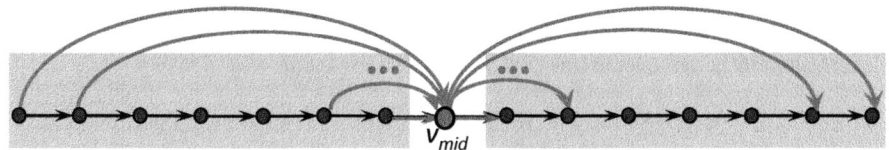

H is a 2-TC-spanner of the line L_n, since every pair of nodes $u, v \in [n]$ is connected by a path of length at most 2 via a hub. This happens in the stage of the recursion during which u and v are separated into different line segments, or one of these two nodes is removed.

To get the bound on the size of the 2-TC-spanner, observe that there are $\lfloor \log n \rfloor$ stages of the recursion. In each stage, every non-hub node connects to the hub in its current line segment, adding a total of at most n edges. Therefore, the constructed spanner has at most $n \log n$ edges. □

The same idea can be extended to construct a 3-TC-spanner of the line graph:

Lemma 1.2 (3-TC-spanner of the line). *The directed line L_n has a 3-TC-spanner with $O(n \log \log n)$ edges.*

Proof. Again we construct the edge set of our 3-TC-spanner H recursively. For simplicity, assume that $\sqrt{n}$ is an integer. Designate nodes which are the multiples of $\sqrt{n}$ as hubs. Connect each non-hub node to the nearest hub before it and the nearest hub after it. More precisely, for each non-hub node v, let v_ℓ be v rounded down to the nearest multiple of $\sqrt{n}$ and let $v_r = v_\ell + \sqrt{n}$. Add edge (v_ℓ, v) if $v_\ell \in [n]$ and edge (v, v_r) if $v_r \in [n]$.

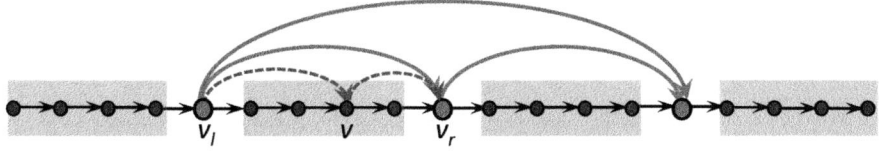

---

[3] Logarithms in this article are base 2 unless indicated otherwise.

Also, add edges between all hubs, orienting them from the smaller to the larger. Finally, remove the hubs from the current line and recurse on the $\sqrt{n}$ resulting line segments. Proceed until each line segment contains exactly one node.

H is a 3-TC-spanner of the line L_n, since every pair of nodes $u, v \in [n]$ is connected by a path of length at most 3 via a pair of hubs. This happens in the stage of the recursion where u and v are separated into different line segments, or one of these two nodes is removed. For example, we add a path (u, u_r, v_ℓ, v) if u and v are not hubs.

Denote the number of edges in the spanner by $T(n)$. At the first stage of recursion, we add $\binom{\sqrt{n}}{2} \leq n$ edges to connect the hubs and at most 2 edges per non-hub node to connect non-hubs to hubs. Therefore, $T(n)$ satisfies the following recurrence:

$$T(n) \leq \begin{cases} 0 & \text{if } n \leq 1; \\ 3n + \sqrt{n} \cdot T(\sqrt{n}) & \text{if } n > 1. \end{cases}$$

The solution to this recurrence is $T(n) \leq 3n \log \log n$. □

This construction generalizes to TC-spanners of arbitrary constant stretch k, giving k-TC-spanners of size $O(n \cdot \lambda_k(n))$, where $\lambda_k(n)$ are very slowly growing functions of n, called the kth-row inverse Ackermann functions.

Definition 1.2 (Inverse Ackermann functions[4]). *Let $\mathbb{R}^{\geq 0}$ be the set of non-negative real numbers. For every function $f : \mathbb{R}^{\geq 0} \to \mathbb{R}^{\geq 0}$ satisfying $f(n) < n$ for all $n > 1$, define the function $f^*(n) : \mathbb{R}^{\geq 0} \to \mathbb{R}^{\geq 0}$ as:*

$$f^*(n) = \min\{k \in \mathbb{Z}^{\geq 0} : f^{(k)}(n) < 2\},$$

where $f^{(k)}$ denotes f composed with itself k times.

Define the kth-row inverse Ackermann function $\lambda_k(n)$ as follows:

$$\lambda_0(n) = n/2, \lambda_1(n) = \sqrt{n}, \text{ and } \lambda_k(n) = \lambda_{k-2}^*(n) \text{ for } k \geq 2.$$

Intuitively, $f^*(n)$ represents the number of times f can be applied before the answer drops below 2. If $f(n) = n/2$ this number is $\Theta(\log n)$, if $f(n) = \sqrt{n}$ it is $\Theta(\log \log n)$, and if $f(n) = \log n$ it is $\log^* n$. Therefore, $\lambda_2(n) = \Theta(\log n)$, $\lambda_3(n) = \Theta(\log \log n)$ and $\lambda_4(n) = \Theta(\log^* n)$. Also note that $\lambda_k(n)$ is a non-decreasing function of n for all $k \geq 0$.

Lemma 1.3 (k-TC-spanner of the line). *The directed line L_n has a k-TC-spanner with at most $k \cdot n \cdot \lambda_k(n)$ edges.*

Proof. Lemmas 1.1 and 1.2 imply Lemma 1.3 for $k = 2$ and 3. (Recall that in the proof of Lemma 1.3, we gave an upper bound of $3n \log\log n$ on the size of the

---

[4] We define these functions, following the spirit of the presentation by Seidel [50]. The prevalent definition (see, e.g., [5]) is complicated and yields asymptotically equivalent functions.

sparsest 3-TC-spanner of the line, even though we chose to hide the constant in the statement of the lemma.) We prove Lemma 1.3 by induction on k, using the constructions of 2- and 3-TC-spanners as base cases. Our induction hypothesis is that one can construct a $(k-2)$-TC-spanner of the line L_n with at most $(k-2) \cdot n \cdot \lambda_{k-2}(n)$ edges. To construct a k-TC-spanner for $k > 3$, we proceed as for $k = 3$, but select even more nodes as hubs, connect them using an optimal $(k-2)$-TC-spanner, add edges from each node to the nearest hub before and the nearest hub after it, and recurse. Then each pair of nodes (u, v) will be connected by a path that jumps from u to the smallest hub greater than u, then follows a path of length at most $k-2$ in the $(k-2)$-TC-spanner on the hubs to reach the largest hub smaller than v, and uses one more edge to jump to v.

It remains to specify the number of hubs, which we will denote by h, and to analyze the size of the spanner. Let $f(n) = \lambda_{k-2}(n)$. We set $h = \frac{n}{f(n)} - 1$ and recurse on the segments of size $f(n)$. By the induction hypothesis, the size of the optimal $(k-2)$-TC-spanner on the hubs is at most

$$(k-2) \cdot h \cdot \lambda_{k-2}(h) \leq (k-2) \cdot \frac{n}{f(n)} \cdot f\left(\frac{n}{f(n)}\right) \leq (k-2) \cdot \frac{n}{f(n)} \cdot f(n) \leq (k-2)n.$$

As before, each non-hub connects to at most 2 hubs. Therefore, the number of edges in constructed spanner, $T(n)$, satisfies the following recursion:

$$T(n) \leq \begin{cases} 0 & \text{if } n \leq 1; \\ kn + \frac{n}{f(n)} \cdot T(f(n)) & \text{if } n > 1. \end{cases}$$

The solution is $T(n) = k \cdot n \cdot f^*(n)$. This follows from the fact that $f^*(f(n)) = f^*(n) - 1$ for $n > 1$. Thus, $T(n) = k \cdot n \cdot \lambda_{k-2}^*(n) = k \cdot n \cdot \lambda_k(n)$.   □

As discussed in Section 3.1, the bound in Lemma 1.3 is tight when k is a constant.

1.2   A Brief Overview

TC-spanners were defined by Bhattacharyya *et al.* [15] as a common abstraction for several applications. Prior to that, Thorup [52] considered a special case of TC-spanners of graphs G that have at most twice as many edges as G, and conjectured that for all directed graphs G on n nodes there are such k-TC-spanners with k polylogarithmic in n. He proved his conjecture for planar graphs [53], but later Hesse [37] gave a counterexample to Thorup's conjecture for general graphs. TC-spanners were also studied for directed trees: implicitly in [5,9,19,25,60] and explicitly in [18,54]. The implicit results were interpreted as TC-spanner constructions in [15].

[15] presented several applications of TC-spanners: testing monotonicity of functions, key management in an access hierarchy and data structures for computing partial products in a semigroup. They also studied the computational problem of finding a sparsest k-TC-spanner for a given directed graph. They presented algorithms and inapproximability results for this problem. Finally, they gave sparse TC-spanner constructions for new graph families.

Later, [16,14] studied TC-spanners for the directed hypercube and hypergrid and presented an application of TC-spanners to property reconstruction. Steiner TC-spanners (see Definition 3.1) were formally introduced in [17], but studied before that in the context of access control hierarchies by [7] and [49]. Berman, Raskhodnikova and Ruan [13] improved algorithms presented in [15]. Finally, Jha and Raskhodnikova [39] pointed out the application of TC-spanners to testing if a function is Lipschitz.

1.3   Organization of This Survey

We start by introducing notation and basic graph-theoretic background in Section 2. In Section 3, we describe structural results on TC-spanners, that is, combinatorial bounds on their size. Structural results for specific graph families are surveyed in Section 3.1, while general structural results, applicable to all graphs, appear in Section 3.2. In Section 4, we survey results on the computational problem of finding a sparsest TC-spanner of a given directed graph and the more general problem of finding directed spanners. We describe approximation algorithms and hardness results for these problems. Finally, Section 5 presents multiple applications of TC-spanners, including property testing, property reconstruction, key management in access control hierarchies and data structures for computing partial product in a semigroup.

2   Preliminaries and Notation

We write $u \preceq_G v$ to denote that vertex v is reachable from vertex u in graph G. When the graph is clear from the context, we omit G. The *transitive closure* of a directed graph $G = (V, E)$, denoted $TC(G)$, is the directed graph (V, E'), where $E' = \{(u, v) : u \preceq_G v\}$. Vertices u and v are *comparable* if either $(u, v) \in TC(G)$ (that is, u is *below* v or, equivalently, *smaller than* v) or $(v, u) \in TC(G)$ (that is, u is *above* v or, equivalently, *larger than* v). This terminology and notation is usually used for partially-ordered sets (posets), which are equivalent to directed acyclic graphs, but can be also applied to general directed graphs.

A digraph G is *weakly connected* if replacing each directed edge in G with an undirected edge results in a connected undirected graph. A digraph is *strongly connected* if each vertex in the graph is reachable from every other vertex via a directed path. The graph of strongly connected components of a digraph G is the digraph obtained by contracting each strongly connected component into one vertex, while maintaining all the edges between these components.

A *transitive reduction* of G is a digraph G' with the fewest edges for which $TC(G') = TC(G)$. As shown by Aho *et al.* [2], a transitive reduction of a given graph can be computed efficiently via a greedy algorithm. The algorithm contracts each strongly connected component C to a vertex $v(C)$ to get a supergraph H, obtains a supergraph H' by greedily removing edges in H that do not change its transitive closure, and finally uncontracts $v(C)$ to an arbitrary directed cycle on vertices in C, choosing a representative vertex of C to be incident to the edges incident to $v(C)$ in H'. Directed acyclic graphs have a unique transitive reduction. We say G is *transitively reduced* if G is equal to its own transitive reduction.

3   Overview of Structural Results on TC-spanners

For a directed graph G, we denote the number of edges in G by $|G|$ and the size of the sparsest k-TC-spanner of G by $S_k(G)$. (The size refers to the number of edges.) To put the following results in proper context, observe that if G has n vertices, $S_k(G) = O(n^2)$. Unlike in the undirected setting, where for every $k \geq 1$, all graphs on n vertices have $(2k - 1)$-spanners with $O(n^{1+1/k})$ edges [6,42,56], sparsest TC-spanners (and hence, sparsest directed spanners) can have $\Omega(n^2)$ edges. An example of a graph with a sparsest 2-TC-spanner of size $\Omega(n^2)$ is the complete bipartite graph $K_{\frac{n}{2},\frac{n}{2}}$ with $n/2$ vertices in each part and all edges directed from the first part to the second. Therefore, most constructions surveyed below are for TC-spanners of specific graph families. Nevertheless, there are several general results, described in Section 3.2.

3.1   TC-spanners of Specific Graph Families

TC-spanners of lines and trees. TC-spanners of the directed lines and directed trees were discovered under many different guises. They were studied implicitly in [5,9,19,25,60] and explicitly in [18,54]. Alon and Schieber [5] implicitly gave tight bounds on $S_k(L_n)$. They showed that, for constant k, the size of the sparsest k-TC-spanner of the directed line is $\Theta(n \cdot \lambda_k(n))$, where $\lambda_k(n)$ is the kth-row inverse Ackermann function (see Definition 1.2 and Lemma 1.3). [5] also showed that the smallest k for which $S_k(L_n) = O(n)$ is $O(\alpha(n))$, where $\alpha(n)$ is the inverse Ackermann function. (The inverse Ackermann function is defined by $\alpha(n) = \min\{k \in \mathbb{Z}^{\geq 0} : \lambda_{2k}(n) \leq 3\}$.) Note that the size of any TC-spanner of L_n is at least $n - 1$, since all edges of the form $(i, i + 1)$ must be present in a TC-spanner to ensure the same connectivity as in L_n. [5,19,54] proved that sparsest k-TC-spanners of rooted directed trees asymptotically have the same number of edges as k-TC-spanners of the line.

TC-spanners of planar graphs. Thorup [52] considered a special case of TC-spanners of graphs G that have at most twice as many edges as G. In [53], he proved that all directed planar graphs G on n nodes have such TC-spanners with stretch polylogarithmic in n.

TC-spanners of graphs with small separators (H-minor-free graphs). A graph H is a *minor* of G if H is a (not necessarily induced) subgraph of a graph obtained from G by a sequence of edge contractions. A graph family $\mathcal{F}$ is *minor-closed* if it contains every minor of every graph in $\mathcal{F}$. For a fixed graph H (e.g., K_5), a minor-closed family $\mathcal{F}$ is *H-minor-free* if $H \notin \mathcal{F}$. Examples of such families include planar graphs, bounded treewidth graphs, and bounded genus graphs, explicitly studied in applications in Section 5. Bhattacharyya *et al.* [15] gave an efficient construction of k-TC-spanners of H-minor-free graphs. For constant k, the size of the spanners is $O(n \cdot \log n \cdot \lambda_k(n))$, where $\lambda_k(\cdot)$ is the kth-row inverse Ackermann function. This result allowed [15] to drastically improve monotonicity testers of Fischer et al. [33]. The application to monotonicity testing is described in Section 5.

The construction in [15] uses divide-and-conquer approach. A natural first attempt would be to use separators of Lipton and Tarjan [41]. Recall that an s-separator for a graph G on n nodes is a set of s nodes whose removal disconnects G into connected components of size at most $2n/3$. Observe that the proofs of Lemmas 1.1–1.3 implicitly use this approach for the special case of the line graph. There, at every stage graph separators play a role of hubs. To come up with efficient constructions for a wider family of graphs, Bhattacharyya et al. use the *path separators* for undirected H-minor free graphs due to Abraham and Gavoille [1]. An s-path-separator[5] for a graph G on n nodes is a set of s paths whose removal disconnects G into connected components of size at most $2n/3$. For some graph families, path separators can be much smaller than ordinary separators. For example, planar graphs require ordinary separators of size $\Theta(\sqrt{n})$, but are 3-path separable [1]. For a simple case of a 2-dimensional $m \times m$ grid, a Lipton-Tarjan separator has size of $\Omega(m)$, but it is enough to remove m nodes on one path, say, a horizontal line that cuts through the middle, to separate it.

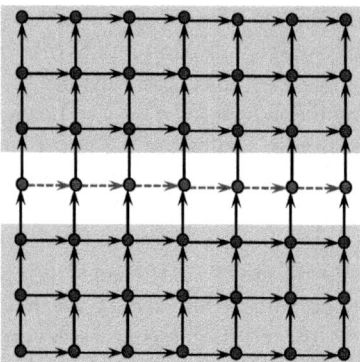

The path separators of Abraham and Gavoile were constructed for undirected graphs. Bhattacharyya et al. employ this construction on the undirected graph, resulting from ignoring the directions of the edges of the input directed graph. The resulting path separator for the original directed graph may be the union of many directed paths. Here we only explain the construction for the simple case when a separator for the graph (and a separator for every subgraph obtained by removing a separator) consists of a small number of directed paths, as is the case, for instance, for a 2-dimensional grid $[m] \times [m]$ where all edges are directed

---

[5] For a graph G to be s-path-separable one needs to be able to disconnect the graph by removing nodes on at most s paths from any minimum spanning tree of G. To keep our high-level overview simple, we do not get into details.

towards vertices with larger coordinates. More precisely, we focus on the case when there exists an integer s, such that every graph obtained at any recursion stage has a separator with at most s directed paths, and moreover, this separator can be found efficiently. (See [15] for the general treatment.)

Even though we use a 2-dimensional grid as an example in this proof, TC-spanners of d-dimensional hypergrids are treated separately in the current section, after TC-spanners of H-minor-free graphs.

If a separator consists of a constant number of (say, at most s) directed paths, we can construct a k-TC-spanner of each path P in the separator as in the proof of Lemma 1.3. We also need to make sure that our TC-spanner contains short paths between all pairs of nodes that were using P to connect. To accomplish this, for each node u with a path to some node in a separator path P, let u' be the first node in P reachable from u. As we are constructing a k-TC-spanner of P, at each stage of the recursion, we add an edge from u to a hub h whenever we add an edge (u', h). We deal symmetrically with each node v with a path from some node in P.

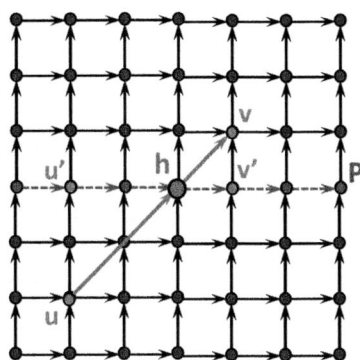

Now, if there is a path from u to v via some vertex in P, there is a path of length at most k in the spanner we are constructing. This is because u and v are connected to the same hubs as u' and v' and, as demonstrated in the proof of Lemma 1.3, u' and v' are connected by a path of length at most k via the hubs. Now, we can safely remove the paths in the separator and recurse on the resulting components. To distinguish this recursion from the recursion in the construction of the TC-spanners of the paths, we call it an *outer* recursion. Observe that at each stage of outer recursion we are adding no more edges per node than in the construction for the line— namely, $O(\lambda_k(n))$ edges. This results in $O(n \cdot \lambda_k(n))$ edges per stage. Since there are $O(\log n)$ stages of outer recursion, the constructed k-TC-spanner has size $O(n \cdot \log n \cdot \lambda_k(n))$.

TC-spanners of hypergrids. The *directed hypergrid*, denoted $\mathcal{H}_{m,d}$, has vertex set $[m]^d$ and edge set $\{(x, y) : \exists i \in [d] \text{ such that } y_i - x_i = 1 \text{ and for } j \neq i, y_j = x_j\}$. For the special case $m = 2$, $\mathcal{H}_{2,d}$ is called a *hypercube* and is also denoted by $\mathcal{H}_d$. 2-TC-spanners of hypergrids are especially relevant for applications in property testing and property reconstruction. TC-spanners of hypergrids of general stretch

k are used in the application to key management in an access hierarchy. The following results on TC-spanners of hypergrids are from [16,14].

As a comparison point for bounds below, note that the obvious bounds on $S_2(\mathcal{H}_d)$ are the number of edges in the d-dimensional hypercube, $2^{d-1}d$, and the number of edges in the transitive closure of $\mathcal{H}_d$, which is $3^d - 2^d$. (An edge in the transitive closure of $\mathcal{H}_d$ has 3 possibilities for each coordinate: both endpoints are 0, both endpoints are 1, or the first endpoint is 0 and the second is 1. This includes self-loops, so we subtract the number of vertices in $\mathcal{H}_d$ to get the desired quantity.) Thus, $2^{d-1}d \leq S_2(\mathcal{H}_d) \leq 3^d - 2^d$. Similarly, the straightforward bounds on the number of edges in a 2-TC-spanner of $\mathcal{H}_{m,d}$ in terms of the number of edges in the directed grid and in its transitive closure are $dm^{d-1}(m-1)$ and $\left(\frac{m^2+m}{2}\right)^d - m^d$, respectively.

The following theorem gives upper and lower bounds on $S_2(\mathcal{H}_{m,d})$:

Theorem 3.1 (Hypergrid [16,14]). *Let $S_2(\mathcal{H}_{m,d})$ denote the number of edges in the sparsest 2-TC-spanner of $\mathcal{H}_{m,d}$. Then for $m \geq 3$,*

$$S_2(\mathcal{H}_{m,d}) = \Omega\left(\frac{m^d \log^d m}{(2d \log\log m)^{d-1}}\right) \quad \text{and} \quad \leq m^d \log^d m.$$

The upper bound in Theorem 3.1 follows from a general construction of k-TC-spanners for graph products for arbitrary $k \geq 2$. The lower bound is proved by a reducing the 2-TC-spanner construction for $[m]^d$ to that for the $[2] \times [m]^{d-1}$ grid and then directly analyzing the number of edges required for a 2-TC-spanner of $[2] \times [m]^{d-1}$. The authors show a tradeoff between the number of edges in the 2-TC-spanner of the $[2] \times [m]^{d-1}$ grid that stay within the hyperplanes $\{1\} \times [m]^{d-1}$ and $\{2\} \times [m]^{d-1}$ versus the number of edges that cross from one hyperplane to the other. The proof proceeds in multiple stages. Assuming an upper bound on the number of edges staying within the hyperplanes, each stage is shown to separately contribute a substantial number of edges crossing between the hyperplanes.

Theorem 3.1 is most useful when m is large. When m is small, it is superseded by another set of bounds on $S_2(\mathcal{H}_{m,d})$, given in [16,14], which are optimal up to a factor of d^{2m}. These bounds are formulated in terms of a complicated combinatorial expression, but value of this expression can be estimated numerically. Specifically, $S_2(\mathcal{H}_{m,d}) = 2^{c_m d} \text{poly}(d)$, where $c_2 \approx 1.1620$, $c_3 \approx 2.03$, $c_4 \approx 2.82$ and $c_5 \approx 3.24$, each significantly smaller than the exponents corresponding to the transitive closure sizes for the different m. More precisely, for the hypercube, $S_2(\mathcal{H}_d) = O(d^3 2^{c_2 d})$ and $\Omega(2^{c_2 d})$. The upper bound on $S_2(\mathcal{H}_d)$ is proved via a randomized construction of a 2-TC-spanner of the directed hypercube. Curiously, even though the upper and lower bounds above differ by a factor of $O(d^3)$, it is known that the randomized construction yields a 2-TC-spanner of $\mathcal{H}_d$ of size within $O(d^2)$ of the optimal.

Steiner TC-spanners of d-dimensional posets. In some applications (in particular, to access control hierarchies [8,9,49,7]), the shortcuts can use *Steiner* vertices,

that is, vertices not in the original graph G. The resulting spanner is called a
Steiner TC-spanner.

Definition 3.1 (Steiner TC-spanner [17]). *Given a directed graph $G = (V, E)$
and an integer $k \geq 1$, a **Steiner k-transitive-closure-spanner (Steiner k-
TC-spanner)** of G is a directed graph $H = (V_H, E_H)$ satisfying:*

1. *$V \subseteq V_H$;*
2. *for all vertices $u, v \in V$, if $d_G(u, v) < \infty$ then $d_H(u, v) \leq k$ and if $d_G(u, v) = \infty$ then $d_H(u, v) = \infty$.*

*Vertices in $V_H \backslash V$ are called **Steiner vertices**.*

For some graphs, Steiner TC-spanners can be significantly sparser than ordinary
TC-spanners. Before, our example of a graph with a 2-TC-spanner of size $\Omega(n^2)$
was a complete bipartite graph $K_{\frac{n}{2}, \frac{n}{2}}$ with $n/2$ vertices in each part and all edges
directed from the first part to the second. This graph has a Steiner 2-TC-spanner
of size n: it is enough to add one Steiner vertex v, edges to v from all nodes in
the left part, and edges from v to all nodes in the right part. Thus, for $K_{\frac{n}{2}, \frac{n}{2}}$
there is a linear gap between the size of the sparsest Steiner 2-TC-spanner and
the size of an ordinary 2-TC-spanner.

However, Bhattacharyya *et al.* [17] show that for directed hypergrids, Steiner
vertices do not help: sparsest Steiner TC-spanners have the same size as TC-
spanners with no Steiner vertices.

Lemma 3.1 ([17]). *If $\mathcal{H}_{m,d}$ has a Steiner k-TC spanner H, it also has a k-TC
spanner of size $|H|$.*

Proof. We show how to replace one Steiner vertex in H with a grid vertex while
keeping the same number of edges in the Steiner k-TC-spanner. This step can
be repeated to remove all Steiner vertices.

Since $\mathcal{H}_{m,d}$ is acyclic, a cycle in H can contain at most one non-Steiner vertex,
and therefore H will still remain a Steiner k-TC-spanner of $\mathcal{H}_{m,d}$ if this cycle is
contracted to one vertex. Thus, we can assume without loss of generality that
H is acyclic.

Let s be a Steiner vertex in H which does not have any other Steiner vertices
below it. Let s' be the smallest vertex in $\mathcal{H}_{m,d}$ which is above all vertices v in
$\mathcal{H}_{m,d}$ satisfying $v \preceq s$. (If there are no such v then s' is the grid vertex with all
coordinates equal to 1.) Observe that s' always exists and is unique. Moreover,
every vertex in $\mathcal{H}_{m,d}$ that is above all such v is also above s'. By definition, s'
is above all vertices in $\mathcal{H}_{m,d}$ which have a path to s. It is also below all vertices

in $\mathcal{H}_{m,d}$ which are reachable from s. We replace all edges in H that have s as an endpoint with the corresponding edges with s' as an endpoint, and remove s from H. Every pair of vertices that was connected via a path of length at most k is still connected via the same path, with s replaced by s' if necessary. No new pair (u, v) of vertices in $\mathcal{H}_{m,d}$ got connected via s' since if u was below s and v was above s then $u \preceq s \preceq v$. The number of edges in H has not increased.   □

Atallah *et al.* [7], De Santis *et al.* [49] and Bhattacharyya *et al.* [17] study Steiner TC-spanners of directed *acyclic* graphs or, equivalently, partially ordered sets. Motivated by the application to access control hierarchies (described in Section 5), they focus on the relationship between the dimension of a poset and the size of its sparsest Steiner TC-spanner.

Definition 3.2 (Poset dimension). *The* dimension *of a poset G is the smallest d such that G can be embedded into a d-dimensional hypergrid $\mathcal{H}_{m,d}$ via an* order-preserving embedding. *A mapping from a poset G to a poset G' is called an* order-preserving embedding *if it respects the partial order, that is, all $x, y \in G$ are mapped to $x', y' \in G'$ such that $x \preceq_G y$ iff $x' \preceq_{G'} y'$.*

Each poset has a dimension. In particular, each poset with n elements can be embedded into a hypergrid $\mathcal{H}_{n,d}$, so that for all $i \in [d]$, the ith coordinates of images of all points are distinct.

Poset dimension is a fundamental and well-studied parameter in poset theory. For instance, Dilworth's famous chain partitioning theorem was originally intended as a lemma for proving a theorem about the dimension of distributive lattices [24]. A survey on poset dimension can be found in Trotter's monograph [57]. One important result for the discussion below, proved by Dushnik and Miller [26], is that that for all m, the hypergrid $\mathcal{H}_{m,d}$ has dimension exactly d. Atallah *et al.* argue that many access control hierarchies are low-dimensional posets that come equipped with an embedding demonstrating low dimensionality.

Observe that the only poset of dimension 1 is the directed line. Tight bounds for the size of (Steiner) TC-spanners of directed lines were discussed in the beginning of Section 3. Table 1 summaries the best bounds for $d \geq 2$. The upper

Table 1. The size of the sparsest Steiner k-TC-spanners of d-dimensional posets on n vertices for $d \geq 2$

Stretch k	Upper Bounds on $S_k(G)$	Lower Bounds on $S_k(G)$	Reference
$k = 2$	$O(n \log^d n)$	$\Omega\left(n \left(\frac{\log n}{\log \log n}\right)^d\right)$ for constant d	[17]
constant $k \geq 3$		$n \log^{\Omega(d)} n$ for constant d	[17]
$k = 2t + 1$ for $t \in [d]$	$O(3^{d-t} t \cdot n \log^{d-1} n \log \log n)$		[49]

bounds hold for all posets of dimension d. The TC-spanners in the upper bounds can be constructed efficiently, given an explicit embedding of the poset into a d-dimensional grid. (Finding such an embedding is NP-hard [59].) Furthermore, paths of length at most k between all pairs of vertices in the resulting k-TC-spanners can be found efficiently. This is important for the application to access control hierarchies.

The lower bounds mean that there exists a poset of dimension d for which every Steiner k-TC-spanner has the specified number of edges. The lower bound for Steiner 2-TC-spanners holds for the hypergrid $\mathcal{H}_{m,d}$ and follows from the lower bound on $S_2(\mathcal{H}_{m,d})$ in Theorem 3.1 and the fact that Steiner vertices do not help for directed hypergrids (Lemma 3.1). The lower bound on the size of a Steiner k-TC-spanner for $k \geq 3$ holds for a poset obtained by a randomized construction.

Note that the Steiner vertices used in the constructions for d-dimensional posets are necessary to obtain sparse TC-spanners. Recall our example of a bipartite graph $K_{\frac{n}{2},\frac{n}{2}}$ for which every 2-TC-spanners required $\Omega(n^2)$ edges. $K_{\frac{n}{2},\frac{n}{2}}$ is a poset of dimension 2, and thus, by the upper bound in [17], has a Steiner 2-TC spanner of size $O(n \log^2 n)$. (As we mentioned before, for this graph there is an even better Steiner 2-TC spanner with $O(n)$ edges.) To see that $K_{\frac{n}{2},\frac{n}{2}}$ is embeddable into a $[n] \times [n]$ grid, map each of the $n/2$ left vertices of $K_{\frac{n}{2},\frac{n}{2}}$ to a distinct grid vertex in the set of incomparable vertices $\{(i, n/2 + 1 - i) : i \in [n/2]\}$, and similarly map each right vertex to a distinct vertex in the set $\{(n + 1 - i, i + n/2) : i \in [n/2]\}$. It is easy to see that this is a proper embedding.

3.2 General TC-spanner Constructions

Graphs that require a large number of shortcuts. We have seen in the beginning of Section 3 that, in general, TC-spanners can be large. However, in the example of $K_{\frac{n}{2},\frac{n}{2}}$ we looked at, the graph itself was large and, in fact, we did not have to add any shortcuts to construct a 2-TC-spanner of that graph. Can one always construct a TC-spanner by adding a small number of edges to the original graph? Thorup [52] conjectured that all directed graphs G on n nodes have TC-spanners with stretch polylogarithmic in n and size at most $2|G|$. As mentioned before, he proved his conjecture for planar graphs [53], but later Hesse [37] gave a counterexample to Thorup's conjecture for general graphs. He constructed a family of graphs with $n^{1+\epsilon}$ edges for which all n^ϵ-TC-spanners require $\Omega(n^{2-\epsilon})$ edges, for small $\epsilon > 0$.

2-TC-spanners from $2k$-TC-spanners. Berman, Raskhodnikova and Ruan [13] show how to (efficiently) obtain a 2-TC-spanner of a graph G with diameter at most $2k$ by adding $O(n^{1-1/k} \cdot |G|)$ shortcuts. They prove that this relationship is nearly tight in the following sense: for every sufficiently small positive ϵ, there are graphs with $2k$-TC-spanners of size $n^{1+(1-\epsilon)/k}$ and no 2-TC-spanners of size less than $n^{2-\epsilon}$. These graphs are obtained by adjusting the parameters in the construction by Hesse mentioned above.

Moreover, as shown in [13], their upper bound is completely tight for the transformation from 3-TC-spanners to 2-TC-spanners: the number of added edges is asymptotically optimal, as evidenced by the 4-layered graph with m^2 nodes in layers 1 and 4 and m nodes in layers 2 and 3, where the edges are directed from smaller to larger layers and are formed as follows. There is a complete bipartite graph between layers 2 and 3. Each node in layer 2 is connected to m nodes in layer 1, and each node in layer 1 has outdegree 1. The edges between layers 3 and 4 are constructed in the same manner. The resulting graph has $3m^2$ edges and is a 3-TC-spanner. A 2-TC-spanner of this graph must connect m^4 pairs of vertices in layers 1 and 4 via paths of length at most 2. Each shortcut edge can be used by at most m such pairs. Therefore, at least m^3 shortcuts are required. Setting $n = 2m^2 + 2m$, we obtain a graph with a 3-TC-spanner of size $O(n)$, for which every 2-TC-spanner requires $\Omega(n^{3/2})$ edges.

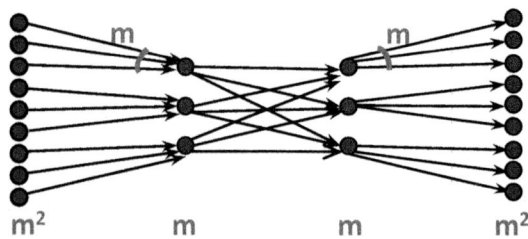

TC-spanners with large stretch. Improving on the first result in this vein in [15], Berman, Raskhodnikova and Ruan show that one can obtain a k-TC-spanner of any graph by adding $O(n^2/k^2)$ shortcut edges. This construction is efficient.

TC-spanners of Graphs with Cycles. Here we give a reduction from constructing TC-spanners of general directed graphs to constructing TC-spanners of directed acyclic graphs (DAGs).

Lemma 3.2. *Let G be a directed graph on n vertices, and G' be the graph of strongly connected components of G. Then $S_{k+2}(G) \leq S_k(G') + 2n$. Moreover, given a k-TC-spanner H' of G', one can efficiently construct a $(k+2)$-TC-spanner H of G with at most $|H'| + 2n$ edges.*

Proof. For each strongly connected component C of G, pick an arbitrary vertex v_C and call it a representative of C. To construct a $(k+2)$-TC-spanner H of G from a k-TC-spanner H' of G', first connect representatives of connected components to mimic the structure of H': namely, add an edge (v_{C_1}, v_{C_2}) to H for every edge (C_1, C_2) in H'. Second, for every vertex u in the component C, where $u \neq v_C$, add edges (u, v_C) and (v_C, u) to H.

The resulting H has the same number of edges as H' plus at most 2 edges per vertex, added to connect each vertex to the representative of its strongly connected component. That is, $|H| \leq |H'| + 2n$. To see that H is a $(k+2)$-TC-spanner of G, consider vertices u_1, u_2 in G, where u_2 is reachable from u_1. Let v_1 and v_2 be the representatives of the components of u_1 and u_2, respectively.

Since H' is a k-TC-spanner of G', there is a path of length at most k from the component of u_1 to the component of u_2 in H'. Therefore, H contains a path of length at most k from v_1 to v_2. Since H also contains edges (u_1, v_1) and (v_2, u_2), it contains a path of length at most $k + 2$ from u_1 to u_2. $\square$

4 Overview of Computational Results on Directed Spanners

The computational problem of finding the size of the sparsest k-TC-spanner of a given graph, called k-TC-SPANNER, was first considered in [15]. k-TC-SPANNER is a special case of a well-studied problem, called DIRECTED k-SPANNER, of finding the size of the sparsest k-spanner of a given (not necessarily transitively closed) directed graph [29,28,15,13]. In this section, we survey approximation algorithms and inapproximability results for these two problems. All known algorithms on DIRECTED k-SPANNER also apply to two other variants, CLIENT/SERVER DIRECTED k-SPANNER and k-DIAMETER SPANNING SUBGRAPH, defined by Elkin and Peleg [29].

Algorithms for DIRECTED k-SPANNER *and* k-TC-SPANNER. All algorithms for DIRECTED k-SPANNER immediately yield algorithms for k-TC-SPANNER with the same approximation ratio because k-TC-SPANNER on input graph G is equivalent to DIRECTED k-SPANNER on input $TC(G)$. Table 2 summarizes the best known approximation algorithms for these problems for different stretch k. Elkin and Peleg [29] gave an $O(\log n)$-approximation algorithm for DIRECTED 2-SPANNER. For $k = 3$, approximation algorithms were proposed in [28,15,13] with the best ratio, $O(\sqrt{n} \cdot \log n)$, due to [13]. In general, for $k > 3$, [13] prove an approximation ratio $O(kn^{1-1/\lceil k/2 \rceil} \cdot \log n)$, improving the first non-trivial polynomial time algorithm for this problem, given in [15]. For the special case of k-TC-SPANNER, [13] give a slightly better ratio of $O(n^{1-1/\lceil k/2 \rceil} \cdot \log n)$. For large k, the best approximation ratio is $O(n/k^2)$, due to [13], again an improvement over the first non-trivial algorithm for this range of parameters, proposed in [15].

We briefly describe the two TC-spanner-specific approximation algorithms from [13]. They are based on the structural results mentioned in Section 3.2. The first algorithm runs the $O(\log n)$-approximation algorithm from [29] for

Table 2. Summary: Algorithmic Results on DIRECTED k-SPANNER and k-TC-SPANNER

Problem	Stretch k	Approximability		Previous Work
DIRECTED k-SPANNER (and k-TC-SPANNER)	$k = 2$	$O(\log n)$	[29]	
	$k = 3$	$O(\sqrt{n} \cdot \log n)$	[13]	[28,15]
	$k \geq 3$	$O(kn^{1-1/\lceil k/2 \rceil} \cdot \log n)$	[13]	[15]
k-TC-SPANNER only	$k \geq 3$	$O(n^{1-1/\lceil k/2 \rceil} \cdot \log n)$	[13]	[15]
	$k = \Omega\left(\frac{\log n}{\log \log n}\right)$	$O(n/k^2)$	[13]	[15]

DIRECTED 2-SPANNER on the transitive closure of the input graph G. The analysis relies on the construction of 2-TC-spanners from k-TC-spanners, mentioned in Section 3.2. This construction proves that $S_2(G) \leq S_k(G) + O(n^{1-1/\lceil k/2 \rceil} \cdot TR(G))$, where $TR(G)$ denotes the size of a transitive reduction of G. (A transitive reduction was defined and discussed in Section 2.) Since the algorithm is guaranteed to output a 2-TC-spanner of size $O(\log n \cdot S_2(G)) = O(\log n \cdot S_k(G) + n^{1-1/\lceil k/2 \rceil} \log n \cdot TR(G))$, the result is an $O(n^{1-1/\lceil k/2 \rceil} \log n)$-approximation. (Recall that $TR(G)$ is a lower bound on the size of a TC-spanner.)

The algorithm for large k is based on an efficient procedure that obtains a k-TC-spanner by adding $O(n^2/k^2)$ shortcut edges. It can be run on each weakly connected component separately. For a weakly connected component with n nodes, the size of a k-TC-spanner is at least $n - 1$, so the resulting graph is a $O(n/k^2)$-approximation.

It is important to note that the algorithm for large k has a better approximation ratio than the corresponding hardness result for DIRECTED k-SPANNER, demonstrating that k-TC-SPANNER is a strictly easier problem for this range of parameters.

Inapproximability of DIRECTED k-SPANNER. For completeness, we state the inapproximability results for DIRECTED k-SPANNER, even though they do not imply anything for k-TC-SPANNER. Kortsarz [40] showed that the $O(\log n)$ approximation ratio for DIRECTED 2-SPANNER cannot be improved unless P=NP. For all $\delta, \epsilon \in (0, 1)$ and $3 \leq k \leq n^{1-\delta}$, it is impossible to approximate DIRECTED k-SPANNER within a factor of $2^{\log^{1-\epsilon} n}$ in polynomial time, assuming NP$\not\subseteq$DTIME$(n^{\text{polylog} n})$ [28,30]. ($DTIME(f(n))$ denotes the class of languages decidable deterministically in time $f(n)$.) Thus, according to Arora and Lund's classification [38] of NP-hard problems, DIRECTED k-SPANNER is in class III, for $k \in [3, n^{1-\delta}]$. Moreover, [30] showed that proving that DIRECTED k-SPANNER is in class IV, that is, inapproximable within n^{δ} for some $\delta \in (0, 1)$, would resolve a long standing open question in complexity theory: namely, cause classes III and IV to collapse into a single class.

Inapproximability of k-TC-SPANNER. Table 3 summarizes inapproximability results for k-TC-SPANNER for different values of k. For constant k, the hardness results are the same as for DIRECTED k-SPANNER, even though the reductions are much more technically involved. Observe that a stronger inapproximability result

Table 3. Summary of Hardness Results on k-TC-SPANNER; all results are from [15]

Stretch k	Inapproximability	Assumption	Notes
$k = 2$	$\Omega(\log n)$	P$\neq$NP	Matches the upper bound
constant $k \geq 3$	$\Omega(2^{\log^{1-\epsilon} n})$ $\forall \epsilon \in (0, 1)$	NP$\not\subseteq$DTIME$(n^{\text{polylog} n})$	Improvement implies breakthrough
$k \leq n^{1-\delta}$ $\forall \delta \in (0, 1)$	$\Omega(1 + \delta)$	P$\neq$NP	

for $k > 2$ would imply the same inaproximability for DIRECTED-k-SPANNER and, as shown in [30], collapse classes III and IV in Arora and Lund's classification. For nonconstant k for which there exists a sufficiently small $\gamma > 0$ such that $k \leq n^{1-\gamma}$, we know that the problem is NP-hard, but not much beyond that. This contrasts sharply with the known hardness of DIRECTED k-SPANNER, but, as mentioned previously, k-TC-SPANNER is known to be strictly easier for some (but not all) k in that range.

The $2^{\log^{1-\epsilon} n}$-inapproximability of k-TC-SPANNER for constant $k \geq 3$ in [15] matches the inapproximability of DIRECTED k-SPANNER for the same stretch in [30]. As is the case for DIRECTED k-SPANNER, the reduction is from a problem called MIN-REP, whose inapproximability DIRECTED k-SPANNER inherits. However, as illustrated in [15], all known hard instances for DIRECTED k-SPANNER cannot imply anything better than $\Omega(1)$-hardness for k-TC-SPANNER. Intuitively, inapproximability of k-TC-SPANNER is harder to prove than inapproximability of DIRECTED k-SPANNER because an instance of k-TC-SPANNER must be transitively-closed, and thus, have more "shortcut" routes between pairs of vertices. The construction of hard instances of k-TC-SPANNER in [15] uses so-called *generalized butterfly* and *broom* graphs. The paths in these graphs are well-structured, making it possible to analyze many different routes in the transitive closure of a hard instance.

The reduction from MIN-REP to k-TC-SPANNER in [15] is quite involved. We briefly describe some of the ideas behind the reduction. An instance of MIN-REP is a bipartite graph G, where each part consists of n nodes partitioned into r *clusters* of size n/r. The clusters in the left part are called $\mathcal{A}_1, \ldots, \mathcal{A}_r$ and the clusters in the right part are $\mathcal{B}_1, \ldots, \mathcal{B}_r$.

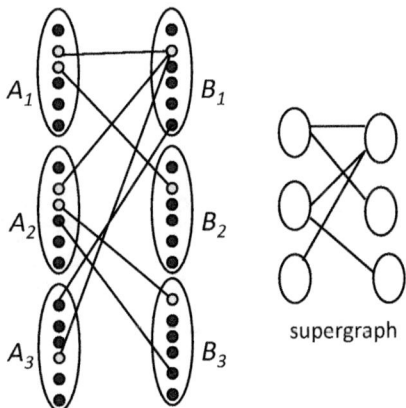

Define the *supergraph* to have nodes $\mathcal{A}_1, \ldots, \mathcal{A}_r, \mathcal{B}_1, \ldots, \mathcal{B}_r$, with a *superedge* $(\mathcal{A}_i, \mathcal{B}_j)$ iff there is a node in $\mathcal{A}_i$ adjacent to a node in $\mathcal{B}_j$. A *rep-cover* is a vertex set S in the graph such that whenever $(\mathcal{A}_i, \mathcal{B}_j)$ is an edge in the supergraph, there is an edge between some $u, v \in S$ with $u \in \mathcal{A}_i$ and $v \in \mathcal{B}_j$. A solution to MIN-REP is a smallest rep-cover. Elkin and Peleg [28] showed that MIN-REP is $2^{\log^{1-\epsilon} n}$-inapproximable.

We now describe *generalized butterfly* and *broom* graphs used in the reduction. *Generalized butterflies* were defined by Woodruff [58]. Each node in a generalized butterfly has k coordinates: $(a_1, \ldots, a_{k-1}, i)$, where $a_1, \ldots, a_{k-1} \in [d]$ and $i \in [k]$. There is an edge from node $(a_1, \ldots, a_{k-1}, i)$ to node $(b_1, \ldots, b_{k-1}, i+1)$ iff for all $j \neq i$, $a_j = b_j$.

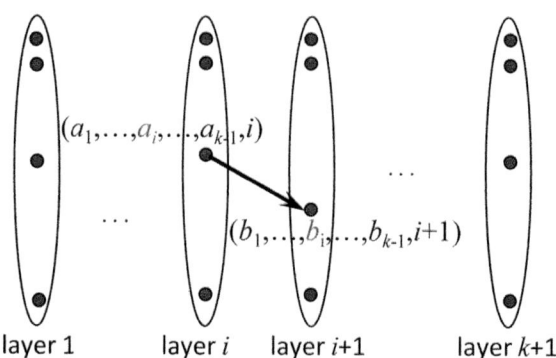

Since there are d possibilities for b_i, each node has outdegree d. Similarly, each node has indegree d. It is easy to see that there is a unique shortest path of length $k - 1$ from any node in layer 1 to any node in layer k. Moreover, any shortcut is on at most d^{k-3} such paths because if it connects layer i to layer $i + \ell$ (where $\ell \geq 2$) it fixes all but $i - 1$ coordinates of the first node and all but $k - (i + \ell)$ coordinates of the second. Thus, at least d^{k+1} shortcuts are needed to reduce the diameter from $k - 1$ to $k - 2$.

A *broom* is a 3-layer graph, where the two leftmost layers form a bipartite clique, and the right layer consists of degree-1 nodes, attached to nodes in the middle layer. Each node in the first and second layer has outdegree d. All edges are directed from left to right.

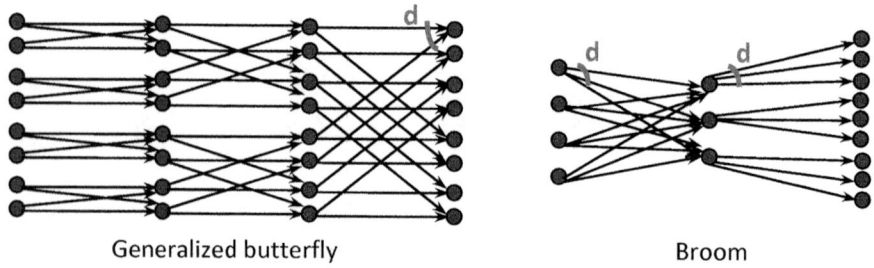

Generalized butterfly   Broom

Given an instance of MIN-REP, we construct an instance G of k-TC-SPANNER as follows. We attach a disjoint copy of a generalized butterfly of diameter $k - 1$ to each $\mathcal{A}_i$ in the MIN-REP instance graph; that is, we identify the vertices in $\mathcal{A}_i$ with the vertices in layer k of the butterfly. The parameter d is determined by the size of $\mathcal{A}_i$ and k. (We can add isolated vertices to each cluster of the given

MIN-REP instance to ensure that $|\mathcal{A}_i|$ is a $(k-1)$st power.) Next, each $\mathcal{B}_j$ is identified with the leftmost layer of a disjoint broom graph. All edges of G are directed towards the rightmost nodes of the brooms. The resulting graph has diameter $k+2$.

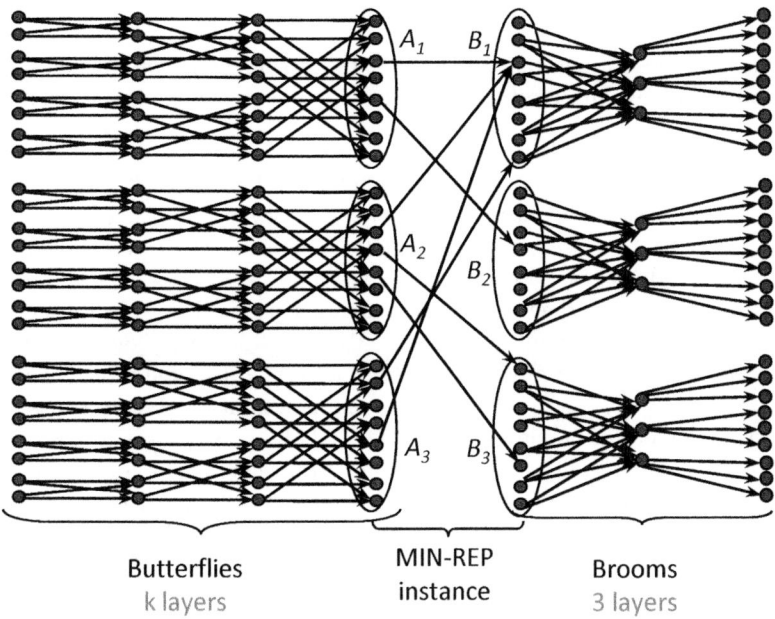

Butterflies	MIN-REP	Brooms
k layers	instance	3 layers

A k-TC-spanner H of G is formed as follows. Let O be a minimum rep-cover of the underlying MIN-REP instance. For each butterfly, include all shortcuts from layer $k-2$ to comparable vertices in layer k which are also in O. In addition, include all shortcuts from vertices in layer $k+1$ which are also in O to all comparable nodes in the last layer. Since O is a rep-cover, H is a k-TC-spanner. The size of H is $|G| + d^2|O|$ because for each vertex in O we add shortcuts to d^2 vertices (in layer $k-2$ for vertices in the left clusters of MIN-REP, and in layer $k+3$ for vertices in the right clusters).

If H were optimal, then approximating its size would approximate a minimum rep-cover of the original MIN-REP instance within the same factor. To ensure that H is optimal, [15] carefully modify the original MIN-REP instance and only then apply the reduction we described.

5   Applications of TC-spanners

We describe four types of applications that use sparse TC-spanners: property testing, property reconstruction, key management in an access hierarchy and data structures for computing partial products in a semigroup. For property testing, we give two applications: to testing monotonicity of functions and to testing if a function is Lipschitz. All these applications, with the exception of testing Lipschitz functions and property reconstruction, were pointed out and described in [15]. The application to Lipschitz functions is from [39]. The application to property reconstruction is from [14].

5.1 Applications to Property Testing

We start by describing the application to testing monotonicity of functions. We also point out the limitations of TC-spanner techniques and related open questions in the area.

Monotonicity testing. Monotonicity of functions [31,35,25,12,32,33,36,4,15,16,51] is one of the most widely studied properties in property testing [34,47]. Fischer *et al.* [33] prove that testing monotonicity is equivalent to several other testing problems. Let V_n be a poset of n elements and $G_n = (V_n, E)$ be the relation graph, i.e., the Hasse diagram, for V_n. A function $f : V_n \to \mathbb{R}$ is called *monotone* if $f(x) \le f(y)$ for all $(x, y) \in E$. We say f is ϵ-far from monotone if f has to be changed on $\ge \epsilon$ fraction of the domain to become monotone, that is, $\min_{\text{monotone } g} |\{x : f(x) \ne g(x)\}| \ge \epsilon n$. A monotonicity tester on G_n is an algorithm that, given an oracle for a function $f : V_n \to \mathbb{R}$, passes if f is monotone but fails with probability $\ge \frac{2}{3}$ if f is ϵ-*far* from monotone. For instance, if G_n is a directed line L_n, the tester needs to determine whether the input sequence, specified by f, is sorted or ϵ-far from sorted. If G_n is a 2-dimensional grid $\mathcal{H}_{m,2}$, the goal is to determine whether the input matrix has non-decreasing rows and columns. The optimal monotonicity tester for the directed line L_n, proposed by Dodis *et al.* [25], is based on the sparsest 2-TC-spanner for that graph. Implicit in the proof of Proposition 9 in [25] is a lemma relating the complexity of a monotonicity tester for L_n to the size of a 2-TC-spanner of L_n. Bhattacharyya *et al.* [15] generalized this lemma by observing that a sparse 2-TC-spanner for any partial order graph G_n implies an efficient monotonicity tester on G_n.

Lemma 5.1 ([15]). *If a directed acyclic graph G_n has a 2-TC-spanner with $s(n)$ edges, then there exists a monotonicity tester on G_n that runs in time $O\left(\frac{s(n)}{\epsilon n}\right)$.*

Proof. The tester selects $\frac{8s(n)}{\epsilon n}$ edges of the 2-TC-spanner H uniformly at random. It queries function f on the endpoints of all the selected edges and rejects if some selected edge (x, y) is *violated* by f, that is, $f(x) > f(y)$.

   If the function f is monotone on G_n, the algorithm always accepts. The crux of the proof is to show that functions that are ϵ-far from monotone are rejected with probability at least $\frac{2}{3}$. Let $f : V_n \to \mathbb{R}$ be a function that is ϵ-far from monotone. It is enough to demonstrate that f violates at least $\frac{\epsilon n}{4}$ edges in H. Then each selected edge is violated with probability $\frac{\epsilon n}{4s(n)}$, and the lemma follows by elementary probability theory.

   Denote the transitive closure of G_n by $TC(G_n)$. We say a vertex $x \in V_n$ is assigned a *bad* label by f if x has an incident violated edge in $TC(G_n)$; otherwise, x has a *good* label. Let V' be a set of vertices with good labels. Observe that f is monotone on the induced subgraph $G' = (V', E')$ of $TC(G_n)$. This implies

([33], Lemma 1) that f can be changed into a monotone function by modifying it on at most $|V_n - V'|$ vertices. Since f is ϵ-far from monotone, it shows that there are at least ϵn vertices with bad labels.

Every function that is ϵ-far from monotone has a matching M of at least $\frac{\epsilon n}{2}$ violated edges in $TC(G_n)$ [25]. We will establish a map from the set of edges in M to the set of violated edges in H, so that each violated edge in H is the image of at most 2 edges in M. For each edge (x, y) in the matching, consider the corresponding path from x to y of length at most 2 in the 2-TC-spanner H. If the path is of length 1, (x, y) is the violated edge in H corresponding to the matching edge (x, y). Otherwise, let (x, z, y) be a path of length 2 in H. At least one of the edges (x, z) and (z, y) is violated, and we map (x, y) to that edge. Since M is a matching, at most 2 edges in M can be mapped to one violated edge in H. Thus, the 2-TC-spanner H has $\geq \frac{\epsilon n}{4}$ violated edges, as required.   □

The fact that H is a 2-TC-spanner is crucial for the proof. If it was a k-TC-spanner for $k > 2$, the path of length k from x to y might not have any violated edges incident to x or y, even if $f(x) > f(y)$. Consider $G_{2n} = (V_{2n}, E)$ where $V_{2n} = \{x_1, \ldots, x_{2n}\}$, $E = \{(x_i, x_n) \mid i < n\} \cup (x_n, x_{n+1}) \cup \{(x_{n+1}, x_j) \mid j > n+1\}$. G_{2n} is a 3-TC-spanner of itself. Now set $f(x_i) = 1$ for $i \leq n$ and $f(x_i) = 0$ otherwise. Clearly, this function is $\frac{1}{2}$-far from monotone, but only one edge, (x_n, x_{n+1}) is violated in the 3-TC-spanner.

As demonstrated by Lemma 5.1, all the 2-TC-spanner constructions yield monotonicity testers for functions defined on the corresponding posets. This lemma led to significant improvements in monotonicity testers for several graph families, including planar graphs and, in general, H-minor-free graphs [15]. Indeed, [15] achieve testers with $O(\log^2 n/\epsilon)$ queries for H-minor-free graphs using their construction of sparse 2-TC-spanners for this graph family, whereas the previous tester, due to Fischer et al. [33], worked only for planar graphs and required $\Theta(\sqrt{n}/\epsilon)$ queries.

We briefly discuss the limitations of the TC-spanner method for constructing monotonicity testers. The lower bounds in [16,14] on the size of the sparsest 2-TC-spanners for the hypercube and the hypergrid (described in Theorem 3.1) rule out the TC-spanner approach for improving monotonicity testers on the hypercube and hypergrid. Currently, the running time of the best tester for monotonicity of functions of the form $f : \{0, 1\}^d \to R$ and, more generally, $f : [m]^d \to R$, where R is an arbitrary range, is $O\left(\frac{d}{\epsilon} \log m \cdot \log |R|\right)$ [25]. The best known lower bound (for the hypercube with range $R = \{0, 1\}$) is $\Omega(\log \log d)$ [33]. (There are better bounds for restricted classes of tests in [33] and [51].) Even though for a fixed d, it is known that the optimal monotonicity tester for the grid runs in time $\Theta(\frac{\log m}{\epsilon})$ [36,32], bridging the gap between the lower and upper bounds for arbitrary d has remained elusive. Lemma 5.1 showed that if a 2-TC-spanner of size $o(2^d d^2)$ for the hypercube or, more generally, a 2-TC-spanner of size $o(m^d d^2 \log^2 m)$ for the hypergrid were found, the monotonicity tester of [25] would be improved. In the light of the lower bounds for the hypercube and the hypergrid, a fundamentally new approach is required.

Testing if a function is Lipschitz. In the important special case when G_n is the directed line, Lemma 5.1 yields an optimal tester for whether a function of the form $f : [n] \to R$ is monotone or, equivalently, of whether a list of n elements is sorted, that runs in time $O(\log n/\epsilon)$. (There is another optimal tester for this problem that was discovered first [31].) Jha and Raskhodnikova [39] observe that the test and analysis in Lemma 5.1 apply to any property of a list of numbers if (a) it can be expressed in terms of pairs of list elements and (b) it is transitive: namely, for all $x \prec y \prec z$, whenever (x,y) and (y,z) are not *violated*, (x,z) is also not violated. In particular, it applies to testing whether a function of the form $f : [n] \to \mathbb{R}$ is Lipschitz. A function $f : \mathcal{D} \to \mathcal{R}$ is called *Lipschitz* if $dist_{\mathcal{R}}(f(x), f(y)) \leq dist_{\mathcal{D}}(x,y)$ for all x, y in $\mathcal{D}$, where $dist_{\mathcal{R}}$ and $dist_{\mathcal{D}}$ denote the distance functions on the range and domain of f, respectively. Testing the Lipschitz property has applications to programs with noisy inputs and to data privacy.

Note that the Lipschitz property was defined in terms of pairs of domain elements. Consider a function $f : [n] \to \mathbb{R}$, where the domain and range are equipped with distance functions $dist_{\mathcal{D}}(x,y) = |x - y|$ and $dist_{\mathcal{R}}(f(x), f(y)) = |f(y) - f(x)|$. We say a pair (x,y) is *violated* if $|f(y) - f(x)| > |y - x|$. Then if (x,y) and (y,z) are not violated, it implies that neither is (x,z). Thus, the requirements (a) and (b) above hold and, using their observation, Jha and Raskhodnikova get a $O(n/\epsilon)$ *Lipschitz* test for functions of the form $f : [n] \to \mathbb{R}$ via the optimal 2-TC-spanner construction of the line.

5.2 Application to Property Reconstruction

Property-preserving data reconstruction was introduced by Ailon, Chazelle, Comandur and Liu [3]. In this model, a reconstruction algorithm, called a *filter*, sits between a *client* and a *dataset*. A dataset is viewed as a function $f : \mathcal{D} \to \mathcal{R}$. Client accesses the dataset using *queries* of the form $x \in \mathcal{D}$ to the filter. The filter *looks up* a small number of values in the dataset and outputs $g(x)$, where g must satisfy some fixed *structural* property $\mathcal{P}$. Extending this notion, Saks and Seshadhri [48] defined *local* reconstruction. A filter is *local* if it allows for a local (or distributed) implementation: namely, if the output function g does not depend on the order of the queries.

Definition 5.1 (Local filter). *A* local filter *for reconstructing property $\mathcal{P}$ is an algorithm A that has oracle access to a function $f : \mathcal{D} \to \mathcal{R}$, and to an auxiliary random string ρ (the "random seed"), and takes as input $x \in \mathcal{D}$. For fixed f and ρ, A runs deterministically on input x to produce an output $A_{f,\rho}(x) \in \mathcal{R}$. As x varies over the domain $\mathcal{D}$, this defines a function $g : \mathcal{D} \to \mathcal{R}$, where $g(x) = A_{f,\rho}(x)$. (Note that a local filter has no internal state to store previously made queries.) The filter must satisfy the following conditions:*

– *For each f and ρ, the function g output by the filter satisfies $\mathcal{P}$.*
– *If f satisfies $\mathcal{P}$, then g is identical to f with probability at least $1 - \delta$, for some $\delta \leq 1/3$. The parameter δ is called* error probability.

In answering query $x \in \mathcal{D}$, the filter A may ask for values of f at domain points of its choice using its oracle access to f. Each such access made to the oracle is called a *lookup* to distinguish it from the client query x. A local filter is *non-adaptive* if the set of domain points that the filter looks up to answer an input query x does not depend on answers given by the oracle.

Saks and Seshadhri also required that g must be sufficiently close to f: *With high probability (over the choice of ρ), $Dist(g, f) \leq B(n) \cdot Dist(f, \mathcal{P})$, where $B(n)$ is called the* error blow-up. ($Dist(g, f)$ is the number of points in the domain on which f and g differ. $Dist(f, \mathcal{P})$ is $\min_{g \in \mathcal{P}} Dist(g, f)$.) If a local filter satisfies this condition along with Definition 5.1, we call it *distance-respecting*.

Local Monotonicity Reconstructors. The first property considered in the reconstruction [3] and local reconstruction [48] models was monotonicity of functions. (See Section 5.1 for a definition.) A (distance-respecting) filter for monotonicity can be used, for example, when a program will run correctly only if its input is sorted. Then, instead of accessing the input directly, the program can access it via a filter, which will ensure that the program always sees a sorted input, making small corrections when necessary. A local filter can be implemented in a distributed manner with an additional guarantee that every program run on the same not-quite-sorted input will see the same corrected version. This can be done by supplying the same random string to each copy of the filter.

To motivate monotonicity reconstructors for hypergrids, consider the scenario of rolling admissions: An admissions office assigns d scores to each application, such as the applicant's GPA, SAT results, essay quality, etc. Based on these scores, some complicated (third-party) algorithm outputs the probability that a given applicant should be accepted. The admissions office wants to make sure "on the fly" that strictly better applicants are given higher probability, that is, probabilities are *monotone* in scores. A hypergrid monotonicity filter may be used here. And, as before, if the filter is local, it can be implemented in a distributed manner, guaranteeing the same results for all filters running in parallel.

Saks and Seshadhri [48] give a *distance-respecting* local monotonicity filter for the directed hypergrid, $\mathcal{H}_{m,d}$, that makes $(\log m)^{O(d)}$ lookups per query. No non-trivial monotonicity filter for the hypercube $\mathcal{H}_d$ (performing $o(2^d)$ lookups per query) is known. One of the monotonicity filters in [3] is a local filter for the directed line $\mathcal{H}_{m,1}$ with $O(\log m)$ lookups per query (but a worse error blow up than in [48]). As observed in [48], this upper bound is tight. Notably, all known local filters for monotonicity property are *non-adaptive*. A lower bound of $2^{\alpha d}$, on the number of lookups per query for a *distance-respecting* local monotonicity filter on $\mathcal{H}_d$ with *error blow-up* $2^{\beta d}$, where α, β are sufficiently small constants, appeared in [48].

[14] show how to construct sparse 2-TC-spanners from local monotonicity reconstructors with low lookup complexity. These constructions, in conjunction with lower bounds on the size of 2-TC-spanners of the hypergrid and hypercube, described in Section 3.1, imply lower bounds on lookup complexity of local monotonicity reconstructors with arbitrary blow-up. The transformations

from non-adaptive and adaptive reconstructors are stated in Theorems 5.1 and 5.2, respectively.

Theorem 5.1 (Transformation from non-adaptive Local Monotonicity Reconstructors to 2-TC-spanners, [14]). *Let $G_n = (V_n, E)$ be a poset on n nodes. Suppose there is a* non-adaptive *local monotonicity reconstructor A for G_n that looks up at most $\ell(n)$ values to answer any query $x \in V_n$ and has error* probability at most δ. *Then there is a 2-TC-Spanner of G_n with at most $O(n\ell(n) \cdot \lceil \log n / \log(1/\delta) \rceil)$ edges.*

Proof. Let A be a local reconstructor given by the statement of the theorem. Let $\mathcal{F}$ be the set of pairs (x, y) with x, y in V_n such that $x \prec y$. Then, $\mathcal{F}$ is of size at most $\binom{n}{2}$. Given $(x, y) \in \mathcal{F}$, let $\mathsf{cube}(x, y)$ be the set $\{z \in V_n : x \preceq z \preceq y\}$. Define function $f^{(x,y)}(v)$ to be 1 on all $v \succeq x$ and all $v \succeq y$, and 0 everywhere else. Also, define function $f^{(\overline{x},\overline{y})}(v)$, which is identical to $f^{(x,y)}(v)$ for all $v \notin \mathsf{cube}(x, y)$ and 0 for $v \in \mathsf{cube}(x, y)$. Both, $f^{(x,y)}$ and $f^{(\overline{x},\overline{y})}$, are monotone functions for all $(x, y) \in \mathcal{F}$. Let A_ρ be the deterministic algorithm which runs A with the random seed fixed to ρ. We say a string ρ is *good* for $(x, y) \in \mathcal{F}$ if filter A_ρ on input $f^{(x,y)}$ returns $g = f^{(x,y)}$ *and* on input $f^{(\overline{x},\overline{y})}$ returns $g = f^{(\overline{x},\overline{y})}$.

Now we show that there exists a set S of size $s \leq \lceil 2 \log n / \log(1/2\delta) \rceil$, consisting of strings used as random seeds by A, such that for every $(x, y) \in \mathcal{F}$ some string $\rho \in S$ is good for (x, y). We choose S by picking strings used as random seeds uniformly and independently at random. Since A has error probability at most δ, we know that for every monotone f, with probability at least $1 - \delta$ (with respect to the choice of ρ), the function $A_{f,\rho}$ is identical to f. Then, for fixed $(x, y) \in \mathcal{F}$ and uniformly random ρ,

$$\Pr[\rho \text{ is not } good \text{ for } (x, y)] \leq \Pr[A_\rho \text{ on input } f^{(x,y)} \text{ fails to output } f^{(x,y)}]$$
$$+ \Pr[A_\rho \text{ on input } f^{(\overline{x},\overline{y})} \text{ fails to output } f^{(\overline{x},\overline{y})}] \leq 2\delta.$$

Since strings in S are chosen independently, $\Pr[\text{no } \rho \in S \text{ is good for } (x, y)] \leq (2 \cdot \delta)^s$, which, for $s = \lceil 2 \log n / \log(1/2\delta) \rceil$, is at most $1/n^2 < 1/|\mathcal{F}|$. By a union bound over $\mathcal{F}$,

$$\Pr[\text{for some } (x, y) \in \mathcal{F}, \text{ no } \rho \in S \text{ is good for } (x, y)] < 1.$$

Thus, there exists a set S with required properties.

We construct our 2-TC-spanner $H = (V_n, E_H)$ of G_n using set S described above. Let $\mathcal{N}_\rho(x)$ be the set consisting of x and all vertices looked up by A_ρ on query x. For each string $\rho \in S$ and each vertex $x \in V_n$, connect x to all comparable vertices in $\mathcal{N}_\rho(x)$ (other than itself) and orient these edges according to their direction in G_n.

We prove H is a 2-TC-Spanner as follows. Suppose not, *i.e.*, there exists $(x, y) \in \mathcal{F}$ with no path of length at most 2 in H from x to y. Consider $\rho \in S$ which is *good* for (x, y). Define function h by setting $h(v) = f^{(x,y)}(v)$ for all $v \notin \mathsf{cube}(x, y)$. Then $h(v) = f^{(\overline{x},\overline{y})}(v)$ for all $v \notin \mathsf{cube}(x, y)$, by definition of $f^{(\overline{x},\overline{y})}$. For a $v \in \mathsf{cube}(x, y)$, set $h(v)$ to 1 for $v \in \mathcal{N}_\rho(x)$ and to 0 for $v \in \mathcal{N}_\rho(y)$.

All unassigned points are set to 0. By the assumption above, $\mathcal{N}_\rho(x) \cap \mathcal{N}_\rho(y)$ does not contain any points in $\mathsf{cube}(x,y)$. Therefore, h is well-defined. Since, ρ is *good* for (x,y) and h is identical to $f^{(x,y)}$ for all look ups made on query x, $A_\rho(x) = h(x) = 1$. Similarly, $A_\rho(y) = h(y) = 0$. But $x \prec y$, so $A_{h,\rho}(v)$ is not monotone. Contradiction.

The number of edges in H is at most

$$\sum_{x \in V_n, \rho \in S} |\mathcal{N}_\rho(x)| \leq n \cdot \ell \cdot s \leq n\ell \cdot \lceil 2\log n / \log(1/2\delta) \rceil. \qquad \square$$

The next theorem applies even to *adaptive* local monotonicity reconstructors. It takes into account how many lookups on query x are points incomparable to x. In particular, if there are no such lookups, then the constructed 2-TC-spanner is of the same size as in Lemma 5.1.

Theorem 5.2 (Transformation from adaptive Local Monotonicity Reconstructors to 2-TC-spanners, [14]). *Let $G_n = (V_n, E)$ be a poset on n nodes. Suppose there is a (possibly adaptive) local monotonicity reconstructor A for G_n that, for any query $x \in V_n$, looks up at most $\ell_1(n)$ vertices comparable to x and at most $\ell_2(n)$ vertices incomparable to x, and has error probability at most δ. Then there is a 2-TC-Spanner of G_n with at most $O(n\ell_1(n) \cdot 2^{\ell_2(n)} \lceil \log n / \log(1/\delta) \rceil)$ edges.*

Proof. Define $\mathcal{F}$, $f^{(x,y)}$, $f^{(\overline{x},\overline{y})}$, A_ρ and S as in the proof of Theorem 5.1. As before, for each $x \in V_n$, we define sets $\mathcal{N}_\rho(x)$, and construct the 2-TC-Spanner H by connecting each x to comparable points in $\mathcal{N}(x)$ for all $\rho \in S$ and orienting the edges according to G_n. However, now $\mathcal{N}_\rho(x)$ is a union of several sets $\mathcal{N}_\rho^{b,w}(x)$, indexed by $b \in \{0,1\}$ and $w \in \{0,1\}^{\ell_2(n)}$. (In addition, $\mathcal{N}_\rho(x)$ contains x.) For each $x \in V_n$, $b \in \{0,1\}$ and $w \in \{0,1\}^{\ell_2(n)}$, let $\mathcal{N}_\rho^{b,w}(x) \subseteq V_n$ be the set of lookups performed by A_ρ on query x, assuming that the oracle answers all lookups as follows. When a lookup y is comparable to x, answer 0 if $y \prec x$, b if $y = x$, 1 if $x \prec y$. Otherwise, if y is the i'th lookup made to an incomparable point for some $i \in [\ell_2]$, answer $w[i]$. Recall that we set $N_\rho(x)$ to be the union of $N_\rho^{b,w}$ for all $b \in \{0,1\}$ and all $w \in \{0,1\}^{\ell_2(n)}$. This completes the description of $\mathcal{N}_\rho(x)$ and construction of H.

The argument that H is a 2-TC-spanner proceeds similarly to that in the proof of Theorem 5.1. The caveat is that an adaptive local filter might choose lookups based on the answers to previous lookups. The constructed function h sets all points comparable to x to 0 if they are below x and 1 if they are above x. However, points incomparable to x might be comparable to y and might be set to 0 or 1, depending on whether they are above or below y. Since we included sets of points queried under all these possibilities in $\mathcal{N}_\rho(x)$, we can now conclude that $A_\rho(x) = h(x) = 1$. The same applies for y. So, $A_{h,\rho}$ outputs a non-monotone function, witnessed the pair (x,y). Contradiction.

We proceed to bound the number of edges E_H in H. For each $\rho \in S$, $x \in V_n$, $b \in \{0,1\}$, and $w \in \{0,1\}^{\ell_2(n)}$, the number of vertices in $N^{\rho}_{b,w}(x)$ comparable to x is at most $\ell_1(n)$. Therefore,

$$|E_H| \le \ell_1(n) \cdot 2 \cdot 2^{\ell_2(n)} \cdot |S| \le O\left(n \cdot \ell_1(n) \cdot 2^{\ell_2(n)} \lceil \log n / \log(1/\delta) \rceil\right). \quad \square$$

In Theorems 5.1 and 5.2 when δ is sufficiently small the bounds on the 2-TC-Spanner size become $O(n\ell(n))$ and $O(n\ell_1(n) \cdot 2^{\ell_2(n)})$, respectively. As pointed out earlier, all known monotonicity reconstructors are non-adaptive. It is an open question whether it is possible to give a transformation from adaptive local monotonicity reconstructors to 2-TC-spanners without incurring an exponential dependence on the number of lookups made to points incomparable to the query point. It is not known if this dependence is an artifact of the proof or an indication that lookups to incomparable points might be helpful for adaptive local monotonicity reconstructors.

Theorems 5.1 and Theorem 5.2 imply the following lower bounds on the lookup complexity of local monotonicity reconstructors. These lower bounds hold for any error-blow up.

Corollary 5.1 ([14]). *Consider a nonadaptive local monotonicity filter with constant error probability δ. If the filter is for functions $f : \mathcal{H}_{m,d} \to \mathbb{R}$, it must perform $\Omega\left(\frac{\log^{d-1} m}{d^d (2 \log \log m)^{d-1}}\right)$ lookups per query. If the filter is for functions $f : \mathcal{H}_d \to \mathbb{R}$, it must perform $\Omega\left(2^{\alpha d}/d\right)$ lookups per query, where $\alpha \ge 0.1620$.*

Corollary 5.2 ([14]). *Consider an (adaptive) local monotonicity filter with constant error probability δ, that for every query $x \in V_n$, looks up at most ℓ_2 vertices incomparable to x. If the filter is for functions $f : \mathcal{H}_{m,d} \to \mathbb{R}$, it must perform $\Omega\left(\frac{\log^{d-1} m}{2^{\ell_2} d^d (2 \log \log m)^{d-1}}\right)$ lookups to vertices comparable to x per query x. If the filter is for functions $f : \mathcal{H}_d \to \mathbb{R}$, it must perform $\Omega\left(2^{\alpha d - \ell_2}/d\right)$ comparable lookups, where $\alpha \ge 0.1620$.*

Prior to [14], no lower bounds for monotonicity reconstructors on $\mathcal{H}_{m,d}$ with dependence on both m and d were known. Unlike the bound in [48], the TC-spanner-based lower bounds hold for any error blow-up. These bounds are tight for reconstructors that are either non-adaptive or perform the number of incomparable lookups that is polylogarithmic in the number of points in the domain. Specifically, for the hypergrid $\mathcal{H}_{m,d}$ of constant dimension d, the number of lookups is $(\log m)^{\Theta(d)}$, and for the hypercube $\mathcal{H}_d$, it is $2^{\Theta(d)}$ for any error blow-up.

5.3   Application to Key Management in Access Control Hierarchies

Atallah *et al.* [9] used sparse Steiner TC-spanners to construct efficient key management schemes for access control hierarchies. An *access hierarchy* is a partially ordered set G of access classes. Each user is entitled to access a certain class and all classes reachable from the corresponding node in G. One approach for devising a cryptographic protocol that enforces the access hierarchy is to have the users follow a *key management scheme* [8,9,49,7,17]. Here, each edge (i, j) has an associated public key $P(i, j)$, and each node i, an associated secret key k_i. Only users with the secret key for a node have the required permissions for the associated access class. The public and secret keys are designed so that there is an efficient algorithm A which takes k_i and $P(i, j)$ and generates k_j, but for each (i, j) in G, it is computationally hard to generate k_j without knowledge of k_i. Thus, a user can efficiently generate the required keys to access a descendant class, but not other classes. The number of runs of algorithm A needed to generate a secret key k_v from a secret key k_u is equal to the number of edges on the shortest path from u to v in G. To speed this up, Atallah *et al.* [7] suggest adding edges and nodes to G to increase connectivity. To preserve the access hierarchy represented by G, the new graph H must be a Steiner TC-spanner of G. The number of edges in H corresponds to the space complexity of the scheme, while the stretch k of the spanner corresponds to the time complexity.

We note that the time to find the path from u to v is also important in this application. In the upper bounds from [17] listed in Table 1, this time is $O(d)$, which for, say, constant d is likely to be much less than $2g(n)$ or $3g(n)$, where $g(n)$ is the time to run algorithm A. This is because algorithm A involves the evaluation of a cryptographic hash function, which is very expensive: any hash function secure against $\mathrm{poly}(n)$-time adversaries requires $g(n) \geq \mathrm{polylog}\, n$ evaluation time under existing number-theoretic assumptions.

5.4   Application to Computing Partial Products in a Semigroup

Yao [60] and Alon and Schieber [5] study space-efficient data structures for the following problem: Preprocess elements $\{s_1, \ldots, s_n\}$ of a semigroup $(S, \circ)$ to be able to compute partial products $s_i \circ s_{i+1} \circ \cdots \circ s_j$ for all $i, j \in [n]$ with at most k queries to a small database of pre-computed partial products. Examples of a semigroup $(S, \circ)$ include $(\mathbb{R}, \min)$, the space of real d-dimensional vectors with operation $(x_1, \ldots, x_d) \circ (y_1, \ldots, y_d) = (\min(x_1, y_1), \ldots, \min(x_d, y_d))$, and the space of real $d \times d$ matrices equipped with the multiplication operation.

Bhattacharyya *et al.* [15] point out that the problem of computing partial products in a semigroup reduces to finding a sparsest k-TC-spanner for a directed line L_{n+1}. If the database stores a product $s_u \circ \cdots \circ s_v$ for each k-TC-spanner edge $(u, v + 1)$, every product $s_i \circ \cdots \circ s_j$ can be computed by multiplying the products corresponding to the edges on a path of length at most k from i to $j + 1$ in the k-TC-spanner for L_{n+1}.

Chazelle [19] and Alon and Schieber [5] also consider a generalization of the above problem, where the input is an (undirected) tree T with an element s_i of a semigroup associated with each vertex i. The goal is to create a space-efficient data structure that allows us to compute the product of elements associated with all vertices on the path from i to j, for all vertices i, j in T. As before, only k queries to the data structure are allowed for each product computation. The generalized problem reduces to finding a sparsest k-TC-spanner for a directed

tree T' obtained from T by appending a new vertex to each leaf, and then selecting an arbitrary root and directing all edges away from it. A k-TC-spanner for
T' with $s(n)$ edges yields a preprocessing scheme with space complexity $s(n)$ for
computing products on T with at most $2k$ queries as follows. The database stores
a product $s_{v_1} \circ \cdots \circ s_{v_t}$ for each k-TC-spanner edge (v_1, v_{t+1}) if the endpoints
of that edge are connected by the path $v_1, \cdots, v_t, v_{t+1}$ in T'. Let $LCA(u,v)$
denote the lowest common ancestor of u and v in T. To compute the product
corresponding to a path from u to v in T, we consider 2 cases: (1) if u is an
ancestor of v (or vice versa) in T, query the products corresponding to the k-
TC-spanner edges on the shortest path from u to a child of v (from v to a child of
u, respectively); (2) otherwise, make queries corresponding to the k-TC-spanner
edges on the shortest path from $LCA(u,v)$ to a child of u and on the shortest
path from a child of $LCA(u,v)$ nearest to u to a child of u. This gives a total of
at most $2k$ queries.

Acknowledgment

The author would like to thank Oded Goldreich for persuading her to write this
survey and Adam Smith, Ramesh T.K. and Piotr Berman for useful comments.

References

1. Abraham, I., Gavoille, C.: Object location using path separators. In: PODC, pp.
   188–197 (2006)
2. Aho, A.V., Garey, M.R., Ullman, J.D.: The transitive reduction of a directed graph.
   SIAM J. Comput. 1(2), 131–137 (1972)
3. Ailon, N., Chazelle, B., Comandur, S., Liu, D.: Property-preserving data reconstruction. Algorithmica 51(2), 160–182 (2008)
4. Ailon, N., Chazelle, B.: Information theory in property testing and monotonicity
   testing in higher dimension. Inf. Comput. 204(11), 1704–1717 (2006)
5. Alon, N., Schieber, B.: Optimal preprocessing for answering on-line product
   queries. Tech. Rep. 71/87, Tel-Aviv University (1987)
6. Althöfer, I., Das, G., Dobkin, D., Joseph, D., Soares, J.: On sparse spanners of
   weighted graphs. Discrete & Computational Geometry 9(1), 81–100 (1993)
7. Atallah, M.J., Blanton, M., Fazio, N., Frikken, K.B.: Dynamic and efficient key
   management for access hierarchies. ACM Trans. Inf. Syst. Secur. 12(3), 1–43 (2009)
8. Atallah, M.J., Blanton, M., Frikken, K.B.: Key management for non-tree access
   hierarchies. In: SACMAT, pp. 11–18 (2006)
9. Atallah, M.J., Frikken, K.B., Blanton, M.: Dynamic and efficient key management
   for access hierarchies. In: ACM Conference on Computer and Communications
   Security, pp. 190–202 (2005)
10. Awerbuch, B.: Communication-time trade-offs in network synchronization. In:
    PODC, pp. 272–276 (1985)
11. Baswana, S., Sen, S.: Approximate distance oracles for unweighted graphs in ex-
    pected $\tilde{O}(n^2)$ time. ACM Transactions on Algorithms 2(4), 557–577 (2006)
12. Batu, T., Rubinfeld, R., White, P.: Fast approximate PCPs for multidimensional
    bin-packing problems. Inf. Comput. 196(1), 42–56 (2005)
13. Berman, P., Raskhodnikova, S., Ruan, G.: Finding sparser directed spanners (2010)
    (manuscript)

14. Bhattacharyya, A., Grigorescu, E., Jha, M., Jung, K., Raskhodnikova, S., Woodruff, D.: Lower bounds for local monotonicity reconstruction from transitive-closure spanners. In: RANDOM (2010)
15. Bhattacharyya, A., Grigorescu, E., Jung, K., Raskhodnikova, S., Woodruff, D.: Transitive-closure spanners. In: SODA, pp. 932–941 (2009)
16. Bhattacharyya, A., Grigorescu, E., Jung, K., Raskhodnikova, S., Woodruff, D.: Transitive-closure spanners of the hypercube and the hypergrid (2009), eCCC Report TR09-046
17. Bhattacharyya, A., Grigorescu, E., Raskhodnikova, S., Woodruff, D.: Steiner transitive-closure spanners of d-dimensional posets (2010) (manuscript)
18. Bodlaender, H.L., Tel, G., Santoro, N.: Tradeoffs in non-reversing diameter. Nordic Journal of Computing 1(1), 111–134 (1994)
19. Chazelle, B.: Computing on a free tree via complexity-preserving mappings. Algorithmica 2, 337–361 (1987)
20. Cohen, E.: Fast algorithms for constructing t-spanners and paths with stretch t. SIAM J. Comput. 28(1), 210–236 (1998)
21. Cohen, E.: Polylog-time and near-linear work approximation scheme for undirected shortest paths. JACM 47(1), 132–166 (2000)
22. Cowen, L.: Compact routing with minimum stretch. J. Algorithms 38(1), 170–183 (2001)
23. Cowen, L., Wagner, C.G.: Compact roundtrip routing in directed networks. J. Algorithms 50(1), 79–95 (2004)
24. Dilworth, R.P.: A decomposition theorem for partially ordered sets. The Annals of Mathematics, Second Series 51(1), 161–166 (1950)
25. Dodis, Y., Goldreich, O., Lehman, E., Raskhodnikova, S., Ron, D., Samorodnitsky, A.: Improved testing algorithms for monotonicity. In: Hochbaum, D.S., Jansen, K., Rolim, J.D.P., Sinclair, A. (eds.) RANDOM 1999 and APPROX 1999. LNCS, vol. 1671, pp. 97–108. Springer, Heidelberg (1999)
26. Dushnik, B., Miller, E.: Concerning similarity transformations of linearly ordered sets. Bulletin Amer. Math. Soc. 46, 322–326 (1940)
27. Elkin, M.: Computing almost shortest paths. In: PODC, pp. 53–62 (2001)
28. Elkin, M., Peleg, D.: Strong inapproximability of the basic k-spanner problem. In: Welzl, E., Montanari, U., Rolim, J.D.P. (eds.) ICALP 2000. LNCS, vol. 1853, pp. 636–647. Springer, Heidelberg (2000)
29. Elkin, M., Peleg, D.: The client-server 2-spanner problem with applications to network design. In: SIROCCO, pp. 117–132 (2001)
30. Elkin, M., Peleg, D.: The hardness of approximating spanner problems. Theory Comput. Syst. 41(4), 691–729 (2007)
31. Ergun, F., Kannan, S., Kumar, S.R., Rubinfeld, R., Viswanathan, M.: Spot-checkers. JCSS 60(3), 717–751 (2000)
32. Fischer, E.: On the strength of comparisons in property testing. Inf. Comput. 189(1), 107–116 (2004)
33. Fischer, E., Lehman, E., Newman, I., Raskhodnikova, S., Rubinfeld, R., Samorodnitsky, A.: Monotonicity testing over general poset domains. In: STOC, pp. 474–483 (2002)
34. Goldreich, O., Goldwasser, S., Ron, D.: Property testing and its connection to learning and approximation. JACM 45(4), 653–750 (1998)
35. Goldreich, O., Goldwasser, S., Lehman, E., Ron, D., Samorodnitsky, A.: Testing monotonicity. Combinatorica 20(3), 301–337 (2000)
36. Halevy, S., Kushilevitz, E.: Testing monotonicity over graph products. In: Díaz, J., Karhumäki, J., Lepistö, A., Sannella, D. (eds.) ICALP 2004. LNCS, vol. 3142, pp. 721–732. Springer, Heidelberg (2004)

37. Hesse, W.: Directed graphs requiring large numbers of shortcuts. In: SODA, pp. 665–669 (2003)
38. Hochbaum, D. (ed.): Approximation Algorithms for NP-hard Problems. PWS Publishing Company, Boston (1997)
39. Jha, M., Raskhodnikova, S.: Testing and reconstruction of lipschitz functions with applications to data privacy (2010) (manuscript)
40. Kortsarz, G.: On the hardness of approximating spanners. Algorithmica 30(3), 432–450 (2001)
41. Lipton, R.J., Tarjan, R.E.: A separator theorem for planar graphs. SIAM Journal on Applied Mathematics 36(2), 177–189 (1979), http://www.jstor.org/stable/2100927
42. Peleg, D.: Distributed computing: a locality-sensitive approach. Society for Industrial and Applied Mathematics, Philadelphia (2000)
43. Peleg, D., Schäffer, A.A.: Graph spanners. Journal of Graph Theory 13(1), 99–116 (1989)
44. Peleg, D., Ullman, J.D.: An optimal synchronizer for the hypercube. SIAM J. Comput. 18(4), 740–747 (1989)
45. Peleg, D., Upfal, E.: A trade-off between space and efficiency for routing tables. JACM 36(3), 510–530 (1989)
46. Roditty, L., Thorup, M., Zwick, U.: Roundtrip spanners and roundtrip routing in directed graphs. In: SODA, pp. 844–851 (2002)
47. Rubinfeld, R., Sudan, M.: Robust characterization of polynomials with applications to program testing. SIAM Journal on Computing 25(2), 252–271 (1996)
48. Saks, M.E., Seshadhri, C.: Parallel monotonicity reconstruction. In: Proceedings of the 19th Annual Symposium on Discrete Algorithms (SODA), pp. 962–971 (2008)
49. Santis, A.D., Ferrara, A.L., Masucci, B.: Efficient provably-secure hierarchical key assignment schemes. In: Kučera, L., Kučera, A. (eds.) MFCS 2007. LNCS, vol. 4708, pp. 371–382. Springer, Heidelberg (2007)
50. Seidel, R.: Understanding the inverse Ackermann function (2006), http://cgi.di.uoa.gr/~ewcg06/invited/Seidel.pdf
51. Soriano, D.G., Matsliah, A., Chakraborty, S., Briet, J.: Monotonicity testing and shortest-path routing on the cube (2010), eCCC Report TR10-048
52. Thorup, M.: On shortcutting digraphs. In: Mayr, E.W. (ed.) WG 1992. LNCS, vol. 657, pp. 205–211. Springer, Heidelberg (1993)
53. Thorup, M.: Shortcutting planar digraphs. Combinatorics, Probability & Computing 4, 287–315 (1995)
54. Thorup, M.: Parallel shortcutting of rooted trees. J. Algorithms 23(1), 139–159 (1997)
55. Thorup, M., Zwick, U.: Compact routing schemes. In: ACM Symposium on Parallel Algorithms and Architectures, pp. 1–10 (2001), http://citeseer.ist.psu.edu/thorup01compact.html
56. Thorup, M., Zwick, U.: Approximate distance oracles. JACM 52(1), 1–24 (2005)
57. Trotter, W. (ed.): Combinatorics and Partially Ordered Sets: Dimension Theory. Johns Hopkins University Press, Baltimore (1992)
58. Woodruff, D.P.: Lower bounds for additive spanners, emulators, and more. In: FOCS, pp. 389–398 (2006)
59. Yannakakis, M.: The complexity of the partial order dimension problem. JMAA 3(3), 351–358 (1982)
60. Yao, A.C.C.: Space-time tradeoff for answering range queries (extended abstract). In: STOC, pp. 128–136 (1982)

Testing by Implicit Learning: A Brief Survey

Rocco A. Servedio

Department of Computer Science
Columbia University
New York, NY, U.S.A
rocco@cs.columbia.edu

Abstract. We give a high-level survey of the "testing by implicit learning" paradigm, and explain some of the property testing results for various Boolean function classes that have been obtained using this approach.

Keywords: Boolean functions, computational learning theory, Occam's Razor.

1 Introduction

This brief survey is about an approach by which proper learning algorithms from computational learning theory can sometimes be leveraged to obtain query-efficient property testing algorithms. It does not contain full proofs but gives a high-level overview of the main ideas along with complete statements of the key results. Readers who are interested in more details are referred to the full proofs in [DLM+07, DLM+08, GOS+09].

After presenting some relevant background from earlier work in property testing, in Section 2 we first give an explanation of the basic approach, called "testing by implicit learning," and describe a general condition under which a class C of functions can be efficiently tested using the basic approach. (Roughly speaking, the condition is that functions in the class must have some sort of concise representation, and must be well-approximated by juntas in the class.) The results in Section 2 are from [DLM+07].

We then present two extensions of the basic "testing by implicit learning" approach. First, in Section 3 we describe how in one particular case (testing the class of s-sparse polynomials over $\mathbb{F}_2$) it is possible to augment the basic approach to obtain a testing algorithm which is *computationally* efficient as well as query-efficient. These results were first presented in [DLM+08]. For the second extension, we explain (at a high level) how the "testing by implicit learning" approach can be carried out in the Fourier domain; very roughly speaking, in this extension the 2^n parity functions over $x_1, \ldots, x_n$ play the role that the n Boolean variables $x_1, \ldots, x_n$ play in the original approach. This extension, and the testing results obtained using it, were first established in [GOS+09] and are presented here in Section 4. We close in Section 5 with suggested directions for future work.

O. Goldreich (Ed.): Property Testing, LNCS 6390, pp. 197–210, 2010.
© Springer-Verlag Berlin Heidelberg 2010

Notation and Conventions. In [DLM+07] the basic "testing by implicit learning" approach is presented in a rather general setting in which the functions being tested are mappings from Ω^n to X, where Ω and X may be arbitrary finite sets. For simplicity and ease of exposition, in this survey we will only consider testing Boolean functions which map $\{0,1\}^n \to \{0,1\}$. Thus for us a "property of Boolean functions" is a class $\mathcal{C}$ of Boolean functions over $\{0,1\}^n$ as described above.

We work in the usual property testing model for Boolean functions. To recap, we view $\{0,1\}^n$ as endowed with the uniform distribution. Two functions $f_1, f_2 : \{0,1\}^n \to \{0,1\}$ are said to be ϵ-*close* if $\Pr[f_1(x) \neq f_2(x)] \leq \epsilon$, and are ϵ-*far* if $\Pr[f_1(x) \neq f_2(x)] > \epsilon$. A *testing algorithm for class* $\mathcal{C}$ is an algorithm A which takes black-box oracle access to an unknown and arbitrary function $f : \{0,1\}^n \to \{0,1\}$. If f belongs to $\mathcal{C}$ then A must output "yes" with probability at least $2/3$ (over its internal coin tosses), and if f is ϵ-far from every $g \in \mathcal{C}$ then A must output "no" with probability at least $2/3$ (thus we consider testers with two-sided error).

A function $f : \{0,1\}^n \to \{0,1\}$ is said to be a *J-junta* if there exists a set $\mathcal{J} \subseteq [n]$ of size at most J such that $f(x) = f(y)$ for every two assignments x, y that agree on all the coordinates in $\mathcal{J}$ (in other words, f is a J-junta if it depends on at most J out of the n input variables).

1.1   Some Previous Work on Testing Classes of Boolean Functions

The "testing by implicit learning" approach is useful for classes of Boolean functions that have some sort of "concise representation." There has been considerable previous research on testing functions for properties corresponding to different notions of having a concise representation. An important precursor of the "testing by implicit learning" work is the paper of Parnas *et al.* [PRS02], which gave algorithms for testing whether Boolean functions $f : \{0,1\}^n \to \{0,1\}$ have certain very simple representations as Boolean formulae. They gave an $O(1/\epsilon)$-query algorithm for testing whether f is a single Boolean literal or a Boolean conjunction, and an $\tilde{O}(s^2/\epsilon)$-query algorithm for testing whether f is an s-term monotone DNF. Parnas *et al.* posed as an open question whether a similar testing result can be obtained for the broader class of general (non-monotone) s-term DNF formulas; as we will see, the "testing by implicit learning" method gives an affirmative answer to this question.

Another closely related work is that of Fischer *et al.* [FKR+04], who gave an algorithm to test whether a Boolean function $f : \Omega^n \to \{0,1\}$ is a J-junta (i.e. depends only on at most J of its n arguments) with query complexity polynomial in J and $1/\epsilon$. As described below, the "testing by implicit learning" approach makes crucial use of techniques from [FKR+04], in combination with ideas from computational learning theory.

1.2   Relevant Earlier Work Relating Property Testing and Learning

The basic idea of using a learning algorithm to do property testing goes back to Goldreich *et al.* [GGR98]. They observed that any proper learning algorithm for

Table 1. Selected results on testing various classes of Boolean functions over $\{0,1\}^n$. The acronyms DNFs, DTs, BPs stand for Disjunctive Normal Form Boolean formulas, Decision Trees, and Branching Programs respectively. The "testing by implicit learning" method is how the upper bounds in lines marked by "[DLM$^+$07]" and "[GOS$^+$09]" are achieved. The upper bounds for those results are for adaptive algorithms (though it is shown in [DLM$^+$07] that very similar bounds can be achieved by non-adaptive algorithms), and the lower bounds are for non-adaptive algorithms unless otherwise indicated by (adaptive).

Class of functions	Number of Queries	Reference
Boolean literals (dictators), conjunctions	$O(1/\epsilon)$	[PRS02]
s-term monotone DNFs	$\tilde{O}(s^2/\epsilon)$	[PRS02]
J-juntas	$\tilde{O}(J^2/\epsilon)$, $\Omega(J)$ (adaptive)	[FKR$^+$04], [CG04]
	$\tilde{O}(J/\epsilon)$	[Bla09]
decision lists	$\tilde{O}(1/\epsilon^2)$	[DLM$^+$07]
size-s DTs, size-s BPs, s-term DNFs, size-s Boolean formulas	$\tilde{O}(s^4/\epsilon^2)$, $\tilde{\Omega}(\log s)$ (adaptive)	[DLM$^+$07]
s-sparse polynomials over $\mathbb{F}_2$	$\tilde{O}(s^4/\epsilon^2)$, $\tilde{\Omega}(\sqrt{s})$	[DLM$^+$07]
size-s Boolean circuits	$\tilde{O}(s^6/\epsilon^2)$	[DLM$^+$07]
functions with Fourier degree $\leq d$	$\tilde{O}(2^{6d}/\epsilon^2)$, $\tilde{\Omega}(\sqrt{d})$	[DLM$^+$07]
induced subclasses of functions with k-dimensional Fourier spectra	$2^{O(k)} \cdot \mathrm{poly}(1/\epsilon)$, $\Omega(2^{k/2})$ (adaptive)	[GOS$^+$09]

a class $\mathcal{C}$ can immediately be used as a testing algorithm for $\mathcal{C}$. (Recall that a proper learning algorithm for $\mathcal{C}$ is one which outputs a hypothesis h that itself belongs to $\mathcal{C}$.) The idea behind this observation is that if the function f being tested belongs to $\mathcal{C}$ then a proper learning algorithm will succeed in constructing a hypothesis that is close to f, while if f is ϵ-far from every $g \in \mathcal{C}$ then any hypothesis $h \in \mathcal{C}$ that the learning algorithm outputs must necessarily be far from f.

This observation shows that any class $\mathcal{C}$ can be tested to accuracy ϵ using essentially the same number of queries that are required to properly learn the class to accuracy $\Theta(\epsilon)$. However, it is well known that proper learning algorithms for virtually every interesting class of n-variable Boolean functions (such as all the classes listed in Table 1, including such simple classes as Boolean literals) must make at least $\Omega(\log n)$ queries. In many cases the "testing by implicit learning" approach can be used to test a class $\mathcal{C}$ with fewer queries than are required for learning – in particular, with query complexity independent of n (again see Table 1).

2   The Basic "Testing by Implicit Learning" Approach

2.1   Overview of the Approach

An observation which is at the heart of the approach is that many interesting classes $\mathcal{C}$ of functions are "well-approximated" by juntas in the following sense:

every function in $\mathcal{C}$ is close to some function in $\mathcal{C}_J$, where $\mathcal{C}_J \subseteq \mathcal{C}$ and every function in $\mathcal{C}_J$ is a J-junta. For example, every s-term DNF over $\{0,1\}^n$ is τ-close to an s-term DNF that depends on only $s \log(s/\tau)$ variables, since each term with more than $\log(s/\tau)$ variables can be removed from the DNF at the cost of at most τ/s error.

Roughly speaking, the "testing by implicit learning" approach to testing whether f belongs to $\mathcal{C}$ works by attempting to learn the "structure" of the junta in $\mathcal{C}_J$ that f is close to *without actually identifying the relevant variables on which the junta depends*. If the algorithm finds such a junta function, it accepts, and if it does not, it rejects. The approach is described as "implicit learning" (as opposed to the explicit proper learning of Goldreich *et al.* [GGR98]), since it learns the structure of the junta to which f is close without explicitly identifying its relevant variables. Indeed, avoiding identifying the relevant variables is what makes it possible to have query complexity independent of n.

The basic algorithm finds the structure of the junta f' in $\mathcal{C}_J$ that f is close to by using the techniques of [FKR+04]. As in [FKR+04], it begins by randomly partitioning the variables of f into subsets and identifying which subsets contain an influential variable (the random partitioning ensures that with high probability, each subset contains at most one such variable if f is indeed in $\mathcal{C}$). Next, the algorithm creates a sample of random labeled examples $(x^1, y^1), (x^2, y^2), ..., (x^m, y^m)$, where each x^i is a string of length J (not length n; this is crucial to the query complexity of the algorithm) whose bits correspond to the influential variables of f, and where y^i corresponds with high probability to the value of junta f' on x^i. We note that the number m of examples created is the number of examples required to learn the class $\mathcal{C}_J$ of J-variable functions using "Occam's Razor" [BEHW87]; it is independent of n. Finally, the algorithm exhaustively checks whether any function in $\mathcal{C}_J$ – over J input variables – is consistent with this labeled sample. This step takes at least $|\mathcal{C}_J|$ time steps, which is exponential in s for the classes in Table 1; but since $|\mathcal{C}_J|$ is independent of n the overall approach has query complexity that is independent of n. (The overall time complexity is linear as a function of n; note that such a runtime dependence on n is inevitable since it takes n time steps simply to prepare a length-n query string to the black-box function.)

In the rest of this section we give some more details on the algorithm and its performance.

2.2   Subclass Approximators

Let $\mathcal{C}$ denote a class of functions from $\{0,1\}^n$ to $\{0,1\}$. We will be interested in classes of functions that can be closely approximated by juntas in the class. We have the following:

Definition 1. *For $\tau > 0$, we say that a subclass $\mathcal{C}(\tau) \subseteq \mathcal{C}$ is a $(\tau, J(\tau))$-approximator for $\mathcal{C}$ if*

- *$\mathcal{C}(\tau)$ is closed under permutation of variables, i.e. if $f(x_1, \ldots, x_n) \in \mathcal{C}(\tau)$ then $f(x_{\sigma_1}, \ldots, x_{\sigma_n})$ is also in $\mathcal{C}(\tau)$ for every permutation σ of $[n]$; and*

> – *for every function $f \in \mathcal{C}$, there is a function $f' \in \mathcal{C}(\tau)$ such that f' is τ-close to f and f' is a $J(\tau)$-junta.*

Typically for us $\mathcal{C}$ will be a class of functions with size bound s in some particular representation, and $J(\tau)$ will depend on s and τ. (A good running example to keep in mind is that $\mathcal{C}$ is the class of all functions that have s-term DNF representations. In this case we may take $\mathcal{C}(\tau)$ to be the class of all s-term $\log(s/\tau)$-DNFs, and we have $J(\tau) = s \log(s/\tau)$.) The approach succeeds on function classes $\mathcal{C}$ for which $J(\tau)$ is a slowly growing function of $1/\tau$ such as $\log(1/\tau)$.

We write $\mathcal{C}(\tau)_k$ to denote the subclass of $\mathcal{C}(\tau)$ consisting of those functions that depend only on variables in $\{x_1, \ldots, x_k\}$. We may (and will) view functions in $\mathcal{C}(\tau)_k$ as taking k arguments rather than n.

2.3   More Detailed Explanation of the Basic Algorithm

Given $\epsilon > 0$ and black-box access to f, the algorithm performs three main steps:

1. **Identify critical subsets.** In Step 1, the algorithm first randomly partitions the variables $x_1, \ldots, x_n$ into r disjoint subsets $I_1, \ldots, I_r$. It then attempts to identify a set of $j \leq J(\tau^\star)$ of these r subsets, which we refer to as *critical subsets* because they each contain a "highly relevant" variable. (For now the value $\tau^\star$ should be thought of as a small quantity; we discuss how this value is selected below.) This step is essentially the same as the 2-sided test for J-juntas from Section 4.2 of Fischer *et al.* [FKR+04]. The analysis shows that if f is close to a $J(\tau^\star)$-junta then this step will succeed w.h.p., and if f is far from every $J(\tau^\star)$-junta then this step will fail w.h.p.

2. **Construct a sample.** Let $I_{i_1}, \ldots, I_{i_j}$ be the critical subsets identified in the previous step. In Step 2 the algorithm constructs a set S of m labeled examples $\{(x^1, y^1), \ldots, (x^m, y^m)\}$, where each x^i is independent and uniformly distributed over $\{0, 1\}^{J(\tau^\star)}$. The analysis shows that if f belongs to $\mathcal{C}$, then with high probability there is a fixed $f'' \in \mathcal{C}(\tau^\star)_{J(\tau^\star)}$ such that each y^i is equal to $f''(x^i)$. On the other hand, if f is far from $\mathcal{C}$, then the analysis shows that w.h.p. no such $f'' \in \mathcal{C}(\tau^\star)_{J(\tau^\star)}$ exists.

   Each labeled example is constructed by again borrowing a technique outlined in [FKR+04]. The algorithm starts with a uniformly random $z \in \{0, 1\}^n$. It then attempts to determine how the j highly relevant coordinates of z are set. Although the algorithm does not know which of the coordinates of z are highly relevant, it does know that, assuming the previous step was successful, there should be one highly relevant coordinate in each of the critical subsets. It uses the independence test of [FKR+04] repeatedly to determine the setting of the highly relevant coordinate in each critical subset. For example, suppose that I_1 is a critical subset. To determine the setting of the highly relevant coordinate of z in critical subset I_1, the algorithm subdivides I_1 into two sets: the subset $\Omega_0 \subseteq I_1$ of indices where z is set to 0, and the subset $\Omega_1 = I_1 \backslash \Omega_0$ of indices where z is set to 1. It then uses the independence test separately on both Ω_0 and Ω_1 to find out which one contains the highly relevant variable. This reveals whether the highly relevant coordinate of z in subset I_1 is set to 0 or 1. The algorithm repeats this process for each critical subset in order to find the settings of the j highly relevant coordinates of z; these form the string x. (The other $J(\tau^\star) - j$ coordinates of x are set to random values; intuitively, this is okay since they are essentially irrelevant.) The algorithm ultimately uses $(x, f(z))$ as the labeled example.

3. **Check consistency.** Finally, in Step 3 the algorithm searches through $\mathcal{C}(\tau^\star)_{J(\tau^\star)}$ looking for a function f'' over $\{0,1\}^{J(\tau^\star)}$ that is consistent with all m examples in S. (Note that this step takes $\Omega(|\mathcal{C}(\tau^\star)_{J(\tau^\star)}|)$ time but uses no queries.) If it finds such a function it accepts f, otherwise it rejects.

2.4   Sketch of the Analysis

We now give an intuitive explanation of the analysis of the test.

Completeness. Suppose f is in $\mathcal{C}$. Then there is some $f' \in \mathcal{C}(\tau^\star)$ that is $\tau^\star$-close to f. Intuitively, $\tau^\star$-close is so close that for the entire execution of the testing algorithm, the black-box function f might as well actually be f' (the algorithm only performs $\ll 1/\tau^\star$ many queries in total, each on a uniform random string, so w.h.p. the view of the algorithm will be the same whether the target is f or f'). Thus, for the rest of this intuitive explanation of completeness, we pretend that the black-box function is f'.

Recall that the function f' is a $J(\tau^\star)$-junta. Since f' is a junta, in Step 1 the test will be able to identify a collection of $j \leq J(\tau^\star)$ "critical subsets" with high probability. Intuitively, these subsets have the property that:

- each "highly relevant" variable occurs in one of the critical subsets, and each critical subset contains at most one highly relevant variable (in fact at most one relevant variable for f');
- the variables outside the critical subsets are so "irrelevant" that w.h.p. in all the queries the algorithm makes, it doesn't matter how those variables are set (randomly flipping the values of these variables would not change the value of f' w.h.p.).

Given critical subsets from Step 1 that satisfy the above properties, in Step 2 the test constructs a sample of labeled examples $S = \{(x^1, y^1), \ldots, (x^m, y^m)\}$ where each x^i is independent and uniform over $\{0,1\}^{J(\tau^\star)}$. In [DLM+07] it is shown that w.h.p. there is a $J(\tau^\star)$-junta $f'' \in C(\tau^\star)_{J(\tau^\star)}$ with the following properties:

- there is a permutation $\sigma : [n] \to [n]$ for which $f''(x_{\sigma(1)}, \ldots, x_{\sigma(J(\tau))})$ is close to $f'(x_1, \ldots, x_n)$;
- the sample S is labeled according to f''.

Finally, in Step 3 the test does a brute-force search over all of $\mathcal{C}(\tau^\star)_{J(\tau^\star)}$ to see if there is a function consistent with S. Since f'' is such a function, the search will succeed and the test outputs "yes" with high probability overall.

Soundness. Suppose now that f is ϵ-far from $\mathcal{C}$.

One possibility is that f is ϵ-far from every $J(\tau^\star)$-junta; if this is the case then w.h.p. the test will output "no" in Step 1.

The other possibility is that f is ϵ-close to a $J(\tau^\star)$-junta f' (or is itself such a junta). Suppose that this is the case and that the testing algorithm reaches Step 2. In Step 2, the algorithm tries to construct a set of labeled examples that is consistent with f'. The algorithm may fail to construct a sample at all; if this happens then it outputs "no." If the algorithm succeeds in constructing a sample S, then w.h.p. this sample is indeed consistent with f'; but in this case, w.h.p. in Step 3 the algorithm will not find any function $g \in \mathcal{C}(\tau^\star)_{J(\tau^\star)}$ that is consistent with all the examples. (If there were such a function g, then standard arguments in learning theory show that w.h.p. any such function $g \in \mathcal{C}(\tau^\star)_{J(\tau^\star)}$ that is consistent with S is in fact close to f'. Since f' is in turn close to f, this would mean that g is close to f. But g belongs to $\mathcal{C}(\tau^\star)_{J(\tau^\star)}$ and hence to $\mathcal{C}$, so this violates the assumption that f is ϵ-far from $\mathcal{C}$.)

2.5   The Main Theorem and Its Consequences

The main theorem about the "testing by implicit learning" algorithm sketched above is stated below. A full proof can be found in [DLM+07].

Theorem 1. *There is an algorithm $\mathcal{A}$ with the following properties:*
Let $\mathcal{C}$ be a class of functions from $\{0,1\}^n$ to $\{0,1\}$. Suppose that for every $\tau > 0$, $\mathcal{C}(\tau) \subseteq \mathcal{C}$ is a $(\tau, J(\tau))$-approximator for $\mathcal{C}$. Suppose moreover that for every $\epsilon > 0$, there is a τ satisfying

$$\tau \le \kappa \cdot \frac{\epsilon^2}{J(\tau)^2 \cdot \ln^2(J(\tau)) \cdot \ln\ln(J(\tau)) \cdot \ln^2(|\mathcal{C}(\tau)_{J(\tau)}|) \cdot \ln(\frac{1}{\epsilon}\ln|\mathcal{C}(\tau)_{J(\tau)}|)}, \quad (1)$$

where $\kappa > 0$ is a fixed absolute constant. Let $\tau^\star$ be the largest value τ satisfying (1) above. Then algorithm $\mathcal{A}$ makes:

$$\tilde{O}\left(\frac{1}{\epsilon^2}J(\tau^\star)^2 \ln^2(|\mathcal{C}(\tau^\star)_{J(\tau^\star)}|)\right)$$

many black-box queries to f, and satisfies the following:

- *If $f \in \mathcal{C}$ then $\mathcal{A}$ outputs "yes" with probability at least $2/3$;*
- *If f is ϵ-far from $\mathcal{C}$ then $\mathcal{A}$ outputs "no" with probability at least $2/3$.*

Here are some observations to help interpret the bound (1). Note that if $J(\tau)$ grows too rapidly as a function of $1/\tau$, e.g. $J(\tau) = \Omega(1/\sqrt{\tau})$, then there will be no $\tau > 0$ satisfying inequality (1). On the other hand, if $J(\tau)$ grows slowly as a function of $1/\tau$, e.g. $\log(1/\tau)$, then it is may be possible to satisfy (1).

In many cases, including all of the applications stated below, $J(\tau)$ will grow as $O(\log(1/\tau))$, and $\ln|\mathcal{C}(\tau)_{J(\tau)}|$ will always be at most $\mathrm{poly}(J(\tau))$, so (1) will always be satisfiable. The most typical case is that $J(\tau) \le \mathrm{poly}(s)\log(1/\tau)$ (where s is a size parameter for the class of functions in question) and $\ln|\mathcal{C}(\tau)_{J(\tau)}| \le \mathrm{poly}(s) \cdot \mathrm{poly}\log(1/\tau)$, which yields $\tau^\star = \tilde{O}(\epsilon^2)/\mathrm{poly}(s)$ and an overall query bound of $\mathrm{poly}(s)/\tilde{O}(\epsilon^2)$.

Theorem 1 can be used to establish that many well-studied classes of Boolean functions have query-efficient testing algorithms. The following results, which are proved in [DLM$^+$07], are also summarized in Table 1.

Theorem 2. *For any s and any $\epsilon > 0$, Algorithm $\mathcal{A}$ yields a testing algorithm for*

(i)   *decision lists using $\tilde{O}(1/\epsilon^2)$ queries;*

(ii) *size-s decision trees using $\tilde{O}(s^4/\epsilon^2)$ queries;*

(iii) *size-s branching programs using $\tilde{O}(s^4/\epsilon^2)$ queries;*

(iv) *s-term DNF using $\tilde{O}(s^4/\epsilon^2)$ queries;*

(v)   *size-s Boolean formulas (with unbounded-fanin AND, OR and NOT gates) using $\tilde{O}(s^4/\epsilon^2)$ queries;*

(vi) *size-s Boolean circuits (with unbounded-fanin AND, OR and NOT gates) using $\tilde{O}(s^6/\epsilon^2)$ queries;*

(vii) *functions with Fourier degree at most d using $\tilde{O}(2^{6d}/\epsilon^2)$ queries;*

(viii) *s-sparse $\mathbb{F}_2$ polynomials using $\tilde{O}(s^4/\epsilon^2)$ queries.*

3   Efficiently Testing Sparse $\mathbb{F}_2$ Polynomials

The main drawback of the basic "testing by implicit learning" approach described in Section 2 is its time complexity. The original [DLM$^+$07] algorithm has running time $2^{\omega(s)}$ as a function of s and $\omega(\text{poly}(1/\epsilon))$ as a function of ϵ for each of the "size-s" function classes (ii) through (viii) in Theorem 2.[1]

[DLM$^+$07] asked whether any of these classes can be tested with both time complexity and query complexity poly$(s, 1/\epsilon)$. This question was answered in [DLM$^+$08], where it was shown that the class of *s-sparse $\mathbb{F}_2$ polynomials* can be so tested. Recall that a $\mathbb{F}_2$ polynomial $p : \{0,1\}^n \to \{0,1\}$ is a multilinear polynomial with coefficients from $\mathbb{F}_2$, i.e. all nonzero coefficients are 1. Such a polynomial may be viewed as a parity of monotone conjunctions (monomials). It is *s-sparse* if it contains at most s monomials (including the constant-1 monomial if it is present). The main result of [DLM$^+$08] is a time-efficient and query-efficient tester for $\mathbb{F}_2$ polynomials:

Theorem 3. *There is a poly$(s, 1/\epsilon)$-query algorithm which has the following performance guarantee: given parameters s, ϵ and black-box access to any $f : \{0,1\}^n \to \{-1,1\}$, it runs in time poly$(s, 1/\epsilon)$ and tests whether f is an s-sparse $\mathbb{F}_2$ polynomial versus ϵ-far from every s-sparse polynomial.*

At a high level, the algorithm of [DLM$^+$08] augments the basic "testing by implicit learning" approach by using a sophisticated proper learning algorithm due to Schapire and Sellie [SS96] in place of the naive brute-force search which is

---

[1] As discussed in the previous section, an $\Omega(n)$ running time is necessary since the testing algorithm must prepare n-bit strings for the black-box oracle for f. The algorithms of this section and the next one both have running times that are linear in n for this reason; henceforth we discuss the running times of our testers only as a function of the other parameters.

used as the learning step in the basic approach. However, significant complications arise in the attempt to "implicitly" run the [SS96] algorithm that do not arise with the brute-force search of [DLM+07]. In the rest of this subsection we briefly describe those complications and how they are addressed in [DLM+08].

We first note that if f is an s-sparse $\mathbb{F}_2$ polynomial, an easy argument shows that there is a function f' - obtained by discarding from f all monomials of degree more than $\log(s/\tau)$ - that is τ-close to f and depends on at most $r = s \log(s/\tau)$ variables. As described in the previous section, the basic "testing by implicit learning" approach uses ideas of [FKR+04] for testing juntas to construct a sample of uniform random examples over $\{0,1\}^r$ which with high probability are all labeled according to f'. At this point, the [DLM+07] algorithm uses a naive brute-force search to check all s-sparse $\mathbb{F}_2$ polynomials over r (as opposed to n) variables, to see if any one of them is consistent with the labeled sample. This leads to a running time of roughly $2^{\omega(s)} \cdot (1/\epsilon)^{\log\log(1/\epsilon)}$.

The proper learning algorithm of Schapire and Sellie [SS96] runs in time polynomial in r and s to exactly learn any unknown s-sparse $\mathbb{F}_2$ polynomial over r variables; thus a natural idea is to use it instead of the computationally inefficient brute-force search of [DLM+07]. However, the [SS96] learning algorithm requires access to a *membership query* oracle, i.e. a black-box oracle for the function being learned. Thus, in order to run the Schapire/Sellie algorithm in the "testing by implicit learning" framework, it is necessary to simulate membership queries to an approximating function f' which is close to f but depends on only r variables. This is significantly more challenging than generating uniform random examples labeled according to f', which is all that is required in the original [DLM+07] approach.

To see why membership queries to f' are more difficult to simulate than uniform random examples, recall that f and the f' described above (obtained from f by discarding high-degree monomials) are τ-close. Intuitively this is extremely close, disagreeing only on a $1/m$ fraction of inputs for an m that is much larger than the number of random examples required for learning f' via brute-force search (this number is "small" – independent of n – because f' depends on only r variables). Thus in the [DLM+07] approach it suffices to use f, the function to which we actually have black-box access, rather than f' to label the random examples used for learning f'; since f and f' are so close, and the examples are uniformly random, with high probability all the labels will also be correct for f'. However, in the membership query scenario of [DLM+08], things are no longer that simple. For any given f' which is close to f, one can no longer assume that the learning algorithm's queries to f' are uniformly distributed and hence unlikely to hit the error region – indeed, it is possible that the learning algorithm's membership queries to f' are clustered on the few inputs where f and f' disagree.

Thus, in order to successfully simulate membership queries, the algorithm must consistently answer queries according to a particular approximator f', even though it only has oracle access to f. This must be done implicitly in a query-efficient way, since explicitly identifying even a single variable relevant to f' requires at least

$\Omega(\log n)$ queries. [DLM+08] does this by showing that for any s-sparse polynomial f, an approximating f' can be obtained as a restriction of f by setting certain carefully chosen subsets of variables to zero. Roughly speaking, this restriction is obtained by randomly partitioning all of the input variables into r subsets and zeroing out all subsets whose variables have small "collective influence."

3.1   The [DLM+08] Algorithm and Its Analysis

In the rest of this section we give a high-level description of the actual [DLM+08] testing algorithm, called **Test-Sparse-Poly**, and its analysis.

Test-Sparse-Poly is based on the idea that if f is a sparse polynomial then it only has a small number of "high-influence" variables, and it is close to another sparse polynomial f' (obtained from f by fixing some input variables to zero) that depends only on those high-influence variables. Roughly speaking, the algorithm works by first isolating the high-influence variables into distinct subsets, and then attempting to implicitly learn f' using the [SS96] algorithm.

We now describe the testing algorithm in tandem with a sketch of why the test is complete, i.e. why it accepts s-sparse polynomials (later we will give a sketch of the soundness argument). The first thing **Test-Sparse-Poly** does (Step 1) is to randomly partition the variables into $r = \text{poly}(s/\tau)$ subsets. If f is an s-sparse polynomial, then it indeed has few high-influence variables, so with high probability at most one such variable will be present in each subset.

Next (Step 2) the algorithm attempts to distinguish subsets that contain a high-influence variable from subsets that do not; this is done by using the independence test of [FKR+04] as described in Section 2.

Once the subsets that contain high-influence variables have been identified, next (Step 3) the algorithm defines a function f' which "zeroes out" all of the variables in all low-influence subsets. Note that if the original function f is an s-sparse polynomial, then f' will be one too. Step 4 of **Test-Sparse-Poly** checks that f is close to f'; it is shown in [DLM+08] that this is indeed the case with high probability if f is an s-sparse polynomial.

The final step (Step 5) of **Test-Sparse-Poly** is to implicitly run the [SS96] algorithm to learn a sparse polynomial, which we call f'', which is isomorphic to f' but is defined only over the high-influence variables of f (recall that there is at most one from each high-variation subset). The overall **Test-Sparse-Poly** algorithm accepts f if and only if the learning algorithm successfully returns a final hypothesis (i.e. does not halt and output "fail"). It is shown in [DLM+08] that for f an s-sparse polynomial, with high probability the subsets that are not restricted in Step 3 have a certain "nice structure:" essentially, they each have one variable with very high influence and all other variables with very low influence. [DLM+08] show that this makes it possible to simulate the membership queries that the [SS96] algorithm requires, so it is possible to implicitly run the [SS96] learning algorithm. Thus, for f an s-sparse polynomial the [SS96] algorithm can run successfully, and the test will accept.

Sketch of soundness. We close this section with a brief sketch of the soundness argument, that if **Test-Sparse-Poly** accepts f with high probability then f must be close to some s-sparse polynomial.

If f passes Step 2 with high probability, then [DLM+08] shows that **Test-Sparse-Poly** must have obtained a partition of variables into subsets that contain a high-influence variable, and subsets that have low "collective influence." If f passes Step 4, then it must moreover be the case that f is close to the function f' obtained by zeroing out the low-influence subsets.

In the last step, **Test-Sparse-Poly** attempts to run the [SS96] algorithm to learn f'' using the high-influence subsets; in the course of doing this, it attempts to simulate membership queries to f''. Since f could be an arbitrary function, we do not know whether each high-influence subset has at most one variable relevant to f'. However, [DLM+08] show that, if the routine to simulate membership queries with high probability never returns "fail," then f' must be close to a junta g whose relevant variables are the individual "highest-influence" variables in each of the high-influence subsets. Now, given that the [SS96] algorithm halts successfully, it must be the case that it constructs a final hypothesis h that is itself an s-sparse polynomial and that agrees with a large sample of many random examples. From this it is possible to argue that h must be close to g (using standard arguments from learning theory), hence close to f', and hence close to f. So indeed if **Test-Sparse-Poly** accepts f with high probability then f must be close to some s-sparse polynomial.

4   Testing Induced Subclasses of Functions with k-Dimensional Fourier Spectra

In [DLM+07] and [DLM+08] the learning is "implicit" in the sense that a learning algorithm is executed to learn some function depending on r of the n input variables, without the identity of those r variables ever being explicitly determined by the algorithm. [GOS+09] extended this methodology to learn some function depending on k of the 2^n *parity functions* $\{\chi_\alpha\}$ over input variables without ever explicitly identifying those parity functions, and obtained testing results using this extended notion of "implicit learning."

We establish some terminology and recall some background in order to state the results of [GOS+09]. We view the domain $\{0,1\}^n$ as $\mathbb{F}_2^n$, so a real-valued function over the Boolean cube is a mapping $\mathbb{F}_2^n \to \mathbb{R}$. Every such function f has a unique representation as

$$f(x) = \sum_{\alpha \in \mathbb{F}_2^n} \hat{f}(\alpha)\chi_\alpha(x) \qquad \text{where} \qquad \chi_\alpha(x) \stackrel{\text{def}}{=} (-1)^{\langle \alpha, x \rangle} = (-1)^{\sum_{i=1}^n \alpha_i x_i}.$$

The coefficients $\hat{f}(\alpha)$ are the *Fourier coefficients* of f, and the functions $\chi_\alpha(\cdot)$ are sometimes referred to as *linear functions* or *characters*; they are simply parity functions over all possible subsets of the n input variables. In addition to treating input strings x as lying in $\mathbb{F}_2^n$, we also index the characters by vectors

$\alpha \in \mathbb{F}_2^n$; this is to emphasize the fact that we are concerned with the linear-algebraic structure. We write $\mathrm{Spec}(f)$ for the Fourier spectrum of f, i.e. the set $\{\alpha \in \mathbb{F}_2^n : \hat{f}(\alpha) \neq 0\}$.

A Boolean function $f : \mathbb{F}_2^n \to \{0, 1\}$ is said to be *k-dimensional* if $\mathrm{Spec}(f)$ lies in a k-dimensional subspace of $\mathbb{F}_2^n$. An equivalent definition is that f is k-dimensional if it is a function of k characters $\chi_{\alpha_1}, \ldots, \chi_{\alpha_k}$, i.e. f is a junta over k parity functions (this is easily seen by picking $\{\alpha_i\}$ to be a basis for $\mathrm{Spec}(f)$). Thus the class of all k-dimensional Boolean functions consists of all Boolean functions of the form $g(\chi_{\alpha_1}, \ldots, \chi_{\alpha_k})$ where g is any k-junta and $\chi_{\alpha_1}, \ldots, \chi_{\alpha_k}$ are any parity functions from $\mathbb{F}_2^n$ to $\mathbb{F}_2$.

Let $\mathcal{C}$ be a class of n-variable Boolean functions. We say that $\mathcal{C}$ is an *induced subclass of k-dimensional functions* if there is some collection $\mathcal{C}'$ of k-variable Boolean functions such that $\mathcal{C}$ is the class of all functions $f = g(\chi_{\alpha_1}, \ldots, \chi_{\alpha_k})$ where g is any function in $\mathcal{C}'$ and $\chi_{\alpha_1}, \ldots, \chi_{\alpha_k}$ are any parity functions from $\mathbb{F}_2^n$ to $\mathbb{F}_2$ as before. For example, let $\mathcal{C}$ be the class of all k-sparse polynomial threshold functions over $\{-1, 1\}^n$; i.e., each function in $\mathcal{C}$ is the sign of a *real* polynomial with at most k nonzero terms. This is an induced subclass of k-dimensional functions, corresponding to the collection $\mathcal{C}' = \{$all linear threshold functions over k Boolean variables$\}$.

[GOS$^+$09] shows that any induced subclass of k-dimensional functions has a query-efficient testing algorithm:

Theorem 4. *Let $\mathcal{C}$ be any induced subclass of k-dimensional functions. There is a nonadaptive* $\mathrm{poly}(2^k, 1/\epsilon)$-*query algorithm for ϵ-testing $\mathcal{C}$.*

The algorithm combines the "testing by implicit learning" approach with a technique from [GOS$^+$09] (building on [FGKP06]) of pairwise independently hashing the Fourier coefficients of f. Very roughly speaking, this pairwise independent hashing is used to "isolate" each nonzero Fourier coefficient of a low-dimensional function f, similar to how a random partition of the variables of a junta into disjoint subsets is used in [FKR$^+$04, DLM$^+$07] to "isolate" each of the relevant variables of a junta. With the Fourier coefficients isolated in this way, given an n-bit example $(z, f(z))$ it is possible to determine the value $\chi_\alpha(z)$ that each character function corresponding to a nonzero Fourier coefficient $\hat{f}(\alpha)$ takes on z, without explicitly identifying the string α. This makes it possible to build a "data set" for implicitly learning the function $f = g(\chi_{\alpha_1}, \ldots, \chi_{\alpha_k})$; each example in the data set consists of a vector of values for all of the parity functions corresponding to nonzero Fourier coefficients, and the corresponding label is the value $f = g(\chi_{\alpha_1}, \ldots, \chi_{\alpha_k})$. This is actually done in an "exhaustive" way, using $2^{\Theta(k)}$ examples, so the "data set" really is more akin to a truth table – it contains an entry for each of the 2^k possible vectors of values that $(\chi_{\alpha_1}, \ldots, \chi_{\alpha_k})$ can take. (It is shown in [GOS$^+$09] that if f is far from any k-dimensional function, then with high probability the construction of the "data set" will reveal this.) Thus one obtains a complete truth table for the k-variable function g, and from this it is trivial (without making any additional queries) to determine whether g belongs to $\mathcal{C}'$.

5 Open Problems and Directions for Future Work

There are natural goals for future work related to each of the three main results described above.

As described in Section 2, the basic "testing by implicit learning" approach gives poly(s/ϵ)-query upper bounds for testing many natural classes such as size-s decision trees, s-term DNF, size-s Boolean formulas, and more. What can be said about lower bounds for these classes? By adapting arguments of Chockler and Gutfreund [CG04], Diakonikolas et al. [DLM+07] establish $\tilde{\Omega}(\log s)$ lower bounds (for adaptive algorithms), but it is quite possible that a poly(s) lower bound in fact holds. (We note that [DLM+07] does establish a $\tilde{\Omega}(\sqrt{s})$ lower bound for nonadaptive algorithms that test the class of s-sparse $\mathbb{F}_2$ polynomials.)

An obvious goal related to [DLM+08] is to obtain poly$(s, 1/\epsilon)$-time testing algorithms for other classes beyond just s-sparse $\mathbb{F}_2$ polynomials. Polynomial-time proper learning algorithms are not known for classes such as s-term DNF or size-s decision trees (even if membership queries are allowed), and thus it seems that new ideas may be needed for these classes. A (potentially more modest) goal is to extend the [DLM+08] results to testing s-sparse polynomials over other finite fields; see the conclusion of [DLM+08] for more discussion of this.

Finally, an interesting direction related to the use of "testing by implicit learning" over the Fourier domain, as in [GOS+09], is whether sharper query complexity bounds can be obtained for specific classes of interest. [GOS+09] give a $2^{\Omega(k)}$ lower bound for testing the entire class of all k-dimensional functions, and thus the $2^{O(k)} \cdot \text{poly}(1/\epsilon)$-query upper bound for induced subclasses of k-dimensional functions cannot be improved much in the worst case. But what about specific classes of k-dimensional functions, such as the class of all k-sparse polynomial threshold functions over $\{-1, 1\}^n$? It would be interesting to determine whether this class can be tested using only poly$(k, 1/\epsilon)$ queries.

References

[BEHW87] Blumer, A., Ehrenfeucht, A., Haussler, D., Warmuth, M.: Occam's razor. Information Processing Letters 24, 377–380 (1987)

[Bla09]   Blais, E.: Testing juntas nearly optimally. In: Proc. 41st Annual ACM Symposium on Theory of Computing (STOC), pp. 151–158 (2009)

[CG04]   Chockler, H., Gutfreund, D.: A lower bound for testing juntas. Information Processing Letters 90(6), 301–305 (2004)

[DLM+07] Diakonikolas, I., Lee, H., Matulef, K., Onak, K., Rubinfeld, R., Servedio, R., Wan, A.: Testing for concise representations. In: Proc. 48th Ann. Symposium on Computer Science (FOCS), pp. 549–558 (2007)

[DLM+08] Diakonikolas, I., Lee, H., Matulef, K., Servedio, R., Wan, A.: Efficiently testing sparse GF(2) polynomials. In: Aceto, L., Damgård, I., Goldberg, L.A., Halldórsson, M.M., Ingólfsdóttir, A., Walukiewicz, I. (eds.) ICALP 2008, Part I. LNCS, vol. 5125, pp. 502–514. Springer, Heidelberg (2008)

[FGKP06] Feldman, V., Gopalan, P., Khot, S., Ponnuswami, A.: New results for learning noisy parities and halfspaces. In: Proc. 47th IEEE Symposium on Foundations of Computer Science (FOCS), pp. 563–576 (2006)

[FKR+04]   Fischer, E., Kindler, G., Ron, D., Safra, S., Samorodnitsky, A.: Testing juntas. Journal of Computer & System Sciences 68, 753–787 (2004)

[GGR98]   Goldreich, O., Goldwasser, S., Ron, D.: Property testing and its connection to learning and approximation. Journal of the ACM 45, 653–750 (1998)

[GOS+09]   Gopalan, P., O'Donnell, R., Servedio, R., Shpilka, A., Wimmer, K.: Testing Fourier dimensionality and sparsity. In: Albers, S., Marchetti-Spaccamela, A., Matias, Y., Nikoletseas, S., Thomas, W. (eds.) ICALP 2009. LNCS, vol. 5555, pp. 500–512. Springer, Heidelberg (2009)

[PRS02]   Parnas, M., Ron, D., Samorodnitsky, A.: Testing basic boolean formulae. SIAM J. Disc. Math. 16, 20–46 (2002)

[SS96]   Schapire, R., Sellie, L.: Learning sparse multivariate polynomials over a field with queries and counterexamples. J. Comput. & Syst. Sci. 52(2), 201–213 (1996)

Invariance in Property Testing

Madhu Sudan

Microsoft Research New England, One Memorial Drive, Cambridge, MA 02142, USA
madhu@mit.edu

Abstract. Property testing considers the task of testing rapidly (in particular, with very few samples into the data), if some massive data satisfies some given property, or is far from satisfying the property. For "global properties", i.e., properties that really depend somewhat on every piece of the data, one could ask how it can be tested by so few samples? We suggest that for "natural" properties, this should happen because the property is invariant under "nice" set of "relabellings" of the data. We refer to this set of relabellings as the "invariance class" of the property and advocate explicit identification of the invariance class of locally testable properties. Our hope is the explicit knowledge of the invariance class may lead to more general, broader, results.

After pointing out the invariance classes associated with some the basic classes of testable properties, we focus on "algebraic properties" which seem to be characterized by the fact that the properties are themselves vector spaces, while their domains are also vector spaces and the properties are invariant under affine transformations of the domain. We survey recent results (obtained with Tali Kaufman, Elena Grigorescu and Eli Ben-Sasson) that give broad conditions that are sufficient for local testability among this class of properties, and some structural theorems that attempt to describe which properties exhibit the sufficient conditions.

1 Introduction: Property Testing and Invariance

We assume the reader of this article has some passing familiarity with some of the basic motivations and nature of questions in Property Testing, and jump directly to establishing our notations.

In this article, we will consider testing properties of *functions* mapping some finite domain D to a finite range R. We let $\{D \to R\}$ denote the set of all such functions. A property will be specified by the set of functions $\mathcal{P} \subseteq \{D \to R\}$. (More generally, we may consider a parameterized family of domains D_n one for each positive integer n, and the property will be given by $\mathcal{P} = \{\mathcal{P}_n\}_n$, where $\mathcal{P}_n \subseteq \{D_n \to R\}$.)

We will measure distance between functions via the normalized Hamming distance (as is standard in Property Testing). Specifically, for $f, g : D \to R$, the *distance* between f and g, denoted $\delta(f, g)$, is given by $\delta(f, g) = \Pr_{x \leftarrow_U D}[f(x) \neq g(x)]$, where the notation $x \leftarrow_U D$ denotes a random variable x drawn uniformly from the domain D. The distance from f to a family $\mathcal{F} \subseteq \{D \to R\}$, denoted $\delta(f, \mathcal{F})$, is the quantity $\min_{g \in \mathcal{F}}\{\delta(f, g)\}$. We say f is δ-*close* to $\mathcal{F}$ if $\delta(f, \mathcal{F}) \leq \delta$ and δ-*far* otherwise.

O. Goldreich (Ed.): Property Testing, LNCS 6390, pp. 211–227, 2010.

Definition 1.1. *A $(k, \epsilon_1, \epsilon_2, \delta)$ tester for a property $\mathcal{P}$ is a probabilistic algorithm T with oracle access to a function $f : D \to R$ that makes at most k queries to the oracle for f, and accepts $f \in \mathcal{P}$ with probability at least $1 - \epsilon_2$, while rejecting f that is δ-far from $\mathcal{P}$ with probability at least ϵ_1.*

A principal focus in property testing is on properties that are defined for infinitely many n where the tests are parameterized by δ. and for every $\delta > 0$, $k = O(1)$ while $\epsilon_1 - \epsilon_2 = \Omega(1) > 0$. (In particular, k and $\epsilon_1 - \epsilon_2$ do not depend on n.) We will also focus mostly on one-sided error tests, i.e., tests where $\epsilon_2 = 0$. In such a case, we simply refer to the tester as a (k, ϵ, δ)-tester.

1.1   Invariances

We now move to the definition of central interest to this article, namely the invariances of a property.

We say that $\mathcal{P}$ is *invariant* under a function $\pi : D \to D$ if for every $f \in \mathcal{P}$ it is the case that the function $f \circ \pi$, defined as $f \circ \pi(x) = f(\pi(x))$, is also in $\mathcal{P}$. We say that $\mathcal{P}$ is invariant under a set $G \subseteq \{D \to D\}$ if for every $\pi \in G$, $\mathcal{P}$ is invaraint under π. The set of all functions π under which $\mathcal{P}$ is invariant is termed the invariance class of $\mathcal{P}$. (The invariance class is a semi-group under composition.) The set of all *permutations* (bijections) π under which $\mathcal{P}$ is invariant is the *automorphism group* of $\mathcal{P}$.

The notion of examining testability of properties with explicit attention on their invariance is a slowly emerging theme. An early result of Babai, Shpilka and Stefankovic [7] gave lower bounds on rates of locally testable codes for cyclic codes is perhaps the first to explicitly relate testability to invariances, albeit to give negative results. The work by Goldreich and Sheffet [33] also uses symmetries to give lower bounds on query complexity. Alon et al. [3] were possibly the first to suggest this might lead to positive results. The work by Kaufman and Sudan [40] seems to be the first to to explicitly focus on invariances to derive positive results.

The goal of this article is two-fold: The first is a collection of observations pointing out that several earlier results in property testing describe natural properties that have nice invariance classes (and in some cases, the invariance classes characterize the properties completely). The second, more technical aspect, is to describe the invariances of algebraic properties. In this part we survey several recent works [40,36,37,15], joint with Ben-Sasson, Grigorescu, and Kaufman, that study the relationship between testability and the invariance classes of the property.

2   Invariances of Some Well-Studied Properties

2.1   Statistical Properties

One of the oldest examples of a "property test" may be "polling", which tests for "approximate majority". This test can be formalized by considering functions mapping some finite universe D to the range $R = \{0,1\}$ and the property $\mathcal{P}$ includes all functions f that take the value 1 on at least $|D|/2$ inputs. (Thus the set D may be thought of as the names of a set of people, f denotes their preference among the two choices in the set R. $\mathcal{P}$ then consists of all possible preference functions in which the majority prefers the choice $1 \in R$.) The standard test (sample f on k random inputs and accept if the majority is 1) and analysis shows that if $k = \Omega(1/\delta^2)$, then we can get $\epsilon_1 - \epsilon_2 = \Omega(1)$.

The invariance class of this property equals its automorphism group and is the full group of permutations from $D \to D$. We now assert that this group of permutations is what leads to the testability of this property.

Indeed any property $\mathcal{P}$ of functions mapping D to R that is invariant under the full group of permutations from $D \to D$ depends on only $|R|$ frequency counts $\{\eta_y\}_{y \in R}$ where $\eta_y = \Pr_{x \in_U D}[f(x) = y]$. To separate $f \in \mathcal{P}$ from f that is δ-far from $\mathcal{P}$ it suffices to get an approximation $\{\nu_y\}_{y \in R}$ to the vector $\{\eta_y\}_{y \in R}$ of ℓ_1 error at most 2δ (i.e., $\sum_{y \in R} |\eta_y - \nu_y| \le 2\delta$). It is straightforward to get such an approximation with $O(|R| \log |R|)$ queries into f (by getting a pointwise approximation of $O(\delta/|R|)$. A better approximation in time $O(|R|)$ is also not too hard (see [48]). We note that the recent results in testing properties of distributions [30,10,8,9,11,1,52,50,55] have revealed many properties that can be tested in $o(|R|)$ samples, and in several cases given nearly-tight bounds (to within $|R|^{o(1)}$ factors) on the query complexity of such tests.

2.2   Graph Property Testing

One of the most actively investigated themes in property testing is the testing of "graph properties", initiated in [28], with recent progress [2,20] characterizing the properties of "dense graphs" that can be tested with constant queries.

The basic model (the "dense graph" model) considers functions from $D = \binom{V}{2}$ to the range $R = \{0,1\}$ (where V is some finite set and $\binom{V}{2}$ denotes the collection of all subsets of size of V).

We note that a property $\mathcal{P}$ is considered a graph property if and only if it is invariant under permutations on $\binom{V}{2}$ "induced" by permutations of V. Specifically, say that a permutation $\pi : \binom{V}{2} \to \binom{V}{2}$ is a graph-permutation if there exists a permutation $\sigma : V \to V$ such that for all $u, v \in V$, $\pi(u,v) = (\sigma(u), \sigma(v))$. A property $\mathcal{P}$ is a graph property if and only if its automorphism group contains all graph-permutations. It follows (trivially) that the success of graph-property testing (i.e., the understanding of testability to the extent of getting a necessary and sufficient condition for testing with $O(1)$ queries) is attributable to the underlying automorphism group.

We note that the above explanation of graph-properties in terms of symmetries seems to apply only to the "dense-graph" model, but not the "bounded-degree" graph model [31], where it is natural to think of the inputs as functions from $V \times [d] \to V$. To include graph properties on such representations of graphs, one could expand the notion of symmetries to consider permutations from $D \times R$ to $D \times R$, but we won't attempt to do so here.

We note that an upcoming work of Goldreich and Kaufman [29] investigates various aspects of property testing and in particular graph-property testing, even in the bounded-degree case, in terms of invariances.

2.3 Properties of Boolean Functions

Another broad class of properties that have been explored recently are properties of "Boolean functions". Sample properties here include monotonicity testing, dictator-testing, junta-testing, testing if a function is given by a real half-space, testing various forms of concise representations etc. [24,27,23,12,49,25,18,46,21]. Boolean properties have nice symmetries too.

Here the domain is the set $D = \{0,1\}^n$ and the invariant group includes all permutations π that are induced by permutations on the coordinates, i.e., permutations $\pi : \{0,1\}^n \to \{0,1\}^n$ for which there exists a corresponding permutation $\sigma : [n] \to [n]$ such that $\pi(\langle b_1, \ldots, b_n \rangle) = \langle b_{\sigma(1)}, \ldots, b_{\sigma(n)} \rangle$.

Unlike in the previous settings, where the invariant group leads to a complete characterization of testable properties, such a characterization is notably missing in this setting.

3 Algebraic Properties

We now move to the topic of focus of this article, namely a large class of "algebraic properties". This class of property tests indeed form the origins of property testing with the seminal work of Blum, Luby and Rubinfeld [19] proposing the now famous "linearity-test" (a test for homomorphisms between groups). Somewhat independently, and with significantly different motivation, Babai, Fortnow, and Lund [6] proposed and analyzed a test to check if a multivariate function over a finite subset of the integers was a multilinear function (linear in each variable). This result was one of the key technical ingredients behind the remarkable result "MIP=NEXP" which formed the predecessor for the modern PCP theory and its connection to inapproximability. Subsequently, Babai, Fortnow, Levin and Szegedy [5] analyzed a property test for when a multivariate function over a vector space over a finite field was a polynomial of a specified (low) degree in each variable. While both of these tests were quite efficient they could work with "constant" queries only when both the degree and the number of variables were constant. Partly to remedy this, Rubinfeld and Sudan [53], proposed and analyzed a low-degree test generalizing the test of [19]. This test would test if a multivariate function over a finite field was a polynomial of low "total" degree (with degree being somewhat smaller than field size). If the degree bound specified was a constant, then the query complexity of this test was a constant independent of the number of variables. Both the linearity test and the low-degree test played a crucial role in the work of Arora et al. [4] leading to the PCP theorem. Indeed a significant component of PCP theory focusses on new/improved analyses of various linearity/low-degree tests.

The ability to test algebraic functions in constant time for constant degree, is not restricted only to the case where the degree is smaller than the field size.

This was first shown by Alon et al. [3] for the case of multivariate functions over the binary field, and then independently by Kaufman and Ron [39] and Jutla et al. [38] for functions over arbitrary fields as well. The work reported below is an attempt to unify the properties, tests and analyses reported in the many works above, in particular those of [19,53,3,39,38].

3.1 A Generalization of Algebraic Properties

From this point onwards, throughout this section we will be consider functions from an n-dimensional vector space over a field $\mathbb{K}$ of size Q to a subfield $\mathbb{F}$ of size q and characteristic p.[1] Let $q = p^s$ and $Q = q^t$. Throughout we will think of q as a constant. The two extreme cases of interest to us will be (1) Q is also a constant and $n \to \infty$; and (2) $n = 1$ and Q (or t) is the parameter going to infinity.

We will consider properties $\mathcal{P}$ of functions mapping $\mathbb{K}^n \to \mathbb{F}$ that are "linear" and "affine-invariant", where we define the terms below.

Linear Properties. A property $\mathcal{P} \subseteq \{\mathbb{K}^n \to \mathbb{F}\}$ is said to be ($\mathbb{F}$-)linear if for every $f, g \in \mathcal{P}$ and $\alpha, \beta \in \mathbb{F}$ the function $\alpha \cdot f + \beta \cdot g$ is also in $\mathcal{P}$, where $\alpha \cdot f + \beta \cdot g$ is the function given by $(\alpha \cdot f + \beta \cdot g)(x) = \alpha f(x) + \beta g(x)$

Affine-invariant Properties. A function $A : \mathbb{K}^n \to \mathbb{K}^n$ is said to be affine if there exists a matrix $M \in \mathbb{K}^{n \times n}$ and a vector $b \in \mathbb{K}^n$ such that $A(x) = Mx + b$ for $x \in \mathbb{K}^n$. A property $\mathcal{P} \subseteq \{\mathbb{K}^n \to \mathbb{F}\}$ is said to be affine-invariant (over $\mathbb{K}^n$) if for every $f \in \mathcal{P}$ and affine function $A : \mathbb{K}^n \to \mathbb{K}^n$ it is the case that $f \circ A \in \mathcal{P}$, where $f \circ A(x) = f(A(x))$.

Since both linearity and affine-invariance seem to impose some sort of "vector-space" restrictions, we stress the different role of the two restrictions. Note that while linearity depends on the range of the functions, the invariance only depends on the domain of the function. And while the latter property (invariance) is more close to the focus of this article, the former assumption (linearity) will be crucial to the rest of this section. Indeed it is possible to consider properties that are linear without focussing on invariances (as was done by Ben-Sasson et al. [14]), or on properties that are affine-invariant, while not being linear (as done in Bhattacharyya et al. [16] and Shapira [54]). We will discuss the latter setting in a later section, and use some of the results in the former setting in this section to motivate our analysis.

In the future, we refer to the set of affine transformations from $\mathbb{K}^n \to \mathbb{K}^n$ as the *affine semi-group*. (They form a semi-group under multiplication.)

3.2 Constraints and Characterizations

One of the basic and very useful observations from the work of Ben-Sasson et al. [14] for linear properties is that tests for such properties might as well

---

[1] We note that it is possible to consider a broader class of properties allowing $\mathbb{F}$ to be an arbitrary field, and not just a subfield of $\mathbb{K}$. However we are not aware of any results that work in this more general setting.

be non-adaptive, and make one-sided error. In other words, a k-query tester would pick (based on its internal randomness) some k points $\alpha_1, \ldots, \alpha_k \in \mathbb{K}^n$, and a predicate $P : \mathbb{F}^k \to \{0, 1\}$ and accept a function f if and only if $P(f(\alpha_1), \ldots, f(\alpha_k)) = 1$. Non-adaptivity refers to the fact that $\alpha_1, \ldots, \alpha_k$ are chosen without knowledge of f on any of the other points. One-sided error implies that if $P(f(\alpha_1), \ldots, f(\alpha_k)) = 0$ then $f \notin \mathcal{P}$. Finally [14] also show that the acceptance predicate can also be chosen to be a linear system, i.e., the set $V = P^{-1}(1)$ is a vector subspace of $\mathbb{F}^k$. This motivates our notion of a (k-local) constraint.

Definition 3.1. *A k-local constraint C is given by $C = (\langle \alpha_1, \ldots, \alpha_k \rangle; V)$ where $\alpha_i \in \mathbb{K}^n$ and $V \subsetneq \mathbb{F}^k$ is a vector subspace of $\mathbb{F}^k$. A function f is said to satisfy the constraint C if $\langle f(\alpha_1), \ldots, f(\alpha_k) \rangle \in V$. A property $\mathcal{P}$ satisfies a constraint C if every function $f \in \mathcal{P}$ satisfies C.*

In the language of constraints, the above-mentioned result of [14] could be interpreted as asserting that a k-query tester for a property $\mathcal{P}$ is simply a distribution on k-local constraints. Given oracle access to a function f, the tester simply picks a k-local constraint (according to the distribution) and accepts if f satisfies the chosen constraint. Thus in order for a test to exist, a property $\mathcal{P}$ must satisfy many k-local constraints; and while it is not (a prioiri) necessary that the constraints completely determine the property $\mathcal{P}$, for many properties considered above, local constraints do seem to determine the property. The notion of a characterization below formalizes this concept.

Definition 3.2. *A collection of k-local constraints $C_1, \ldots, C_m$ form a k-local characterization of a a property $\mathcal{P}$ if $f \in \mathcal{P}$ if and only if f satisfies C_j for every $j \in [m]$.*

In general, the fact that a property satisfies even one local constraint may seem to be a rare event. To find a whole collection of local constraints satisfied by a property, to the extent that they even characterize it, may seem even more so. But for properties that exhibit some (invariances), this is not as suprising (and indeed this is what motivates some of the study of invariances of properties in the context of local testing). If a property $\mathcal{P}$, invariant under a function π satisfies the constraint $C = (\langle \alpha_1, \ldots, \alpha_k \rangle; V)$ then it also satisfies the constraint $C \circ \pi = (\langle \pi(\alpha_1), \ldots, \pi(\alpha_k) \rangle; V)$. This motivates our definition below of the *orbit* of a constraint C under a set of invariances G.

Definition 3.3. *Given a property $\mathcal{P}$ invariant under a set G that satisfies a constraint C, we say that the orbit of C under G is the set of constraints $\{C \circ \pi | \pi \in G\}$.*

Of course the definition above makes sense even when no "property" is mentioned, but the definition makes the most sense when applied to some property $\mathcal{P}$ satisfying C and invariant under G. If we are (seemingly incredibly) lucky, then the orbit of a constraint may provide enough constraints to actually characterize a property $\mathcal{P}$. This concept, while seemingly too restrictive turns out to be central to our analysis of property testing.

Definition 3.4 (Single-orbit characterization). *A property $\mathcal{P} \subseteq \{D \to R\}$ is said to have a k-single orbit characterization under a set G of invariances if there exists a k-ary constraint C such that $f \in \mathcal{P}$ if and only if f satisfies $C \circ \pi$ for every $\pi \in G$.*

For this section, properties of interest will be those with a k-single orbit characterization over the affine semi-group. As we explain below a k-single-orbit characterization over the affine semi-group immediately leads to local testability; and this explains the results of [53,3,39,38]. We then describe structural results about affine-invariant properties and give examples of new properties that end up being testable as a consequence.

3.3   Testability of Linear Affine-Invariant Properties

In joint work with Tali Kaufman [40] we show that a property that has a k-single orbit characterization under the affine semi-group has a k-query tester. The following theorem gives the precise soundness condition of the test.

Theorem 3.1 ([40, Theorem 2.9]). *Let $\mathcal{P} \subseteq \{\mathbb{K}^n \to \mathbb{F}\}$ have a k-single orbit characterization under the affine semi-group. Then there exists a k-query test T that accepts $f \in \mathcal{P}$ with probability one, while rejecting f that is δ-far from $\mathcal{P}$ with probability $\min\left\{\delta/2, \frac{1}{(2k+1)(k-1)}\right\}$.*

The test T above is the "natural" one. Recall that the k-single orbit characterization implies that there exists a constraint $C = (\langle \alpha_1, \ldots, \alpha_k \rangle; V)$ such that for every $g \in \mathcal{P}$ and every affine map $A : \mathbb{K}^n \to \mathbb{K}^n$ it is the case that $\langle g(A(\alpha_1)), \ldots, g(A(\alpha_k)) \rangle \in V$. Given oracle access to a function $f : \mathbb{K}^n \to \mathbb{F}$, the test T simply picks a random affine map $A : \mathbb{K}^n \to \mathbb{K}^n$ and accepts if and only if $\langle f(A(\alpha_1)), \ldots, f(A(\alpha_k)) \rangle \in V$. The completeness analysis follows from the definition, while the soundness analysis (though not so small that we can summarize it here) essentially abstracts the common elements of the proofs of [19,53,3,39,38] while unifying the seemingly different parts by using the concept of "tensor products of linear spaces".[2]

To apply the theorem above to recover the results of [19,53,3,39,38] one needs appropriate single orbit characterizations for the appropriate families. Below we list some of the known ones.

Example 3.1 (Affine functions from $\mathbb{K}^n \to \mathbb{K}$, for $n \geq 2$). Let $\mathcal{P} \subseteq \{\mathbb{K}^n \to \mathbb{K}\}$ be the set of affine functions, i.e., $\mathcal{P} = \{f(x) | \exists a_1, \ldots, a_n, b \in \mathbb{K} \text{ s.t. } f(x_1, \ldots, x_n) = \sum_{i=1}^{n} a_i x_i + b\}$.

---

[2] Given two vector spaces $U \subseteq \mathbb{K}^n$ and $V \subseteq \mathbb{K}^m$, their tensor product $U \otimes V \subseteq \mathbb{K}^{n \times m}$ can be thought of as the collection of $n \times m$ matrices each of whose rows is an element of V and columns is an element of U. The "key" (though simple) fact about this tensor product space is that its dimension is the product of dimensions of U and V. This fact turns out be the heart of the "creative steps" in the analyses of of [19,53,3,39,38].

For $n \geq 2$, let $\alpha, \beta, \in \mathbb{K}^n$ be (any) two linearly independent vectors in $\mathbb{K}^n$. Let $V \subseteq \mathbb{K}^4$ be the set $\{\langle a, b, c, a + b + c \rangle | a, b, c \in \mathbb{K}\}$. Let C be the constraint $(\langle 0, \alpha, \beta, \alpha + \beta \rangle; V)$.

Then $\mathcal{P}$ has a 4-single orbit characterization under the affine semi-group, given by the constraint C.

The characterization above, combined with Theorem 3.1 above, effectively captures the essential elements of the linearity test of [19], though in fact it is only a variation (affineness, instead of linearity) of a special case (linearity of maps over finite fields, as opposed to homomorphisms between abelian groups) of their main result.

Example 3.2 (Degree d polynomials from $\mathbb{K}^n \to \mathbb{K}$, for $|\mathbb{K}| = p^r \geq d + 1 + p^{r-1}$.). Let $\mathcal{P} \subseteq \{\mathbb{K}^n \to \mathbb{K}\}$ be the set of n-variate polynomials of degree at most d.

Let $\alpha \in \mathbb{K}^n$ be any non-zero vector, and let $\omega \in \mathbb{K}$ be a primitive element (i.e., $\omega^i \neq 1$ for $i < |\mathbb{K}| - 1$).

Let $V = \{\langle p(1), p(\omega), p(\omega^2), \ldots, p(\omega^{d+1}) \rangle | p : \mathbb{K} \to \mathbb{K}$ is a univariate polynomial of degree at most $d\}$. Then the constraint $C = (\langle \alpha, \omega \cdot \alpha, \omega^2 \cdot \alpha, \ldots, \omega^{d+1} \cdot \alpha \rangle, V)$ is a $d + 2$-single orbit characterization of $\mathcal{P}$.

The above characterization follows essentially from [53,26] and implies that the property of being a degree d polynomial is testable with $O(d)$ queries over any large enough field (of size greater than d). What about the case when the field size $Q < d$? In such a case also one can get a single orbit characterization, where the locality of the queries is however exponential in d.

Example 3.3 (Degree d polynomials from $\mathbb{K}^n \to \mathbb{K}$ [39]). Let $\mathcal{P} \subseteq \{\mathbb{K}^n \to \mathbb{K}\}$ be the set of n-variate polynomials of degree at most d. Let $\ell = (d + 1 + Q/p)/(Q-1)$ (recall $Q = |\mathbb{K}|$ and p is its characteristic). Let U be an arbitrary ℓ dimensional subspace of $\mathbb{K}^n$, and let $\alpha \in (\mathbb{K}^n)^{Q^\ell}$ be an (arbitrary) enumeration of the points of U. Let $V \subseteq \mathbb{K}^{Q^\ell}$ be the set of all evaluations of degree d, n-variate polynomials on the sequence α. Then the constraint $C = (\alpha, V)$ is a Q^ℓ-single orbit characterization of $\mathcal{P}$.

Applying Theorem 3.1 to the characterization above one can get the main result of [39] (which in turn subsumes the results of [3] and [38]).

3.4   Structure of Linear Affine-Invariant Properties

While the examples of the previous section describe how many previously known results can be unified under the perspective of invariance, to get new families that can be testable, one needs to understand more about affine-invariant properties. Here we describe some of the basic results that characterize affine invariant properties in terms of the supporting monomials.

Recall first that every function from $\mathbb{K}^n \to \mathbb{K}$, and hence every function from $\mathbb{K}^n \to \mathbb{F}$, is a polynomial in n-variables of degree at most $Q - 1$ in each variable. Thus the "monomials", i.e., functions of the form $m(x_1, \ldots, x_n) = \prod_{i=1}^n x_i^{d_i}$

for some $d_1, \ldots, d_n \in \{0, \ldots, Q-1\}$ form a linear basis of all functions from $\{\mathbb{K}^n \to \mathbb{F}\}$. We let $\mathcal{M}$ denote the set of all monomials. Recalling that every function $f : \mathbb{K}^n \to \mathbb{K}$ can be written uniquely as $f(x) = \sum_{m \in \mathcal{M}} c_m m(x)$. We say that the *support* of f is the set of all monomials m whose coefficient c_m is non-zero.

Most polynomials would not map the the domain $\mathbb{K}^n$ to elements of $\mathbb{F}$ (and would often take on values from $\mathbb{K} - \mathbb{F}$). To get a basis for functions from $\mathbb{K}^n \to \mathbb{F}$, one needs to look at "traces" of monomials. The Trace function mapping $\mathbb{K} \to \mathbb{F}$ is defined as $\mathrm{Tr}(x) = x + x^q + x^{q^2} + \cdots + x^{q^{t-1}}$. (Recall that $\mathbb{F} = \mathbb{F}_q$ and $\mathbb{K} = \mathbb{F}_Q = \mathbb{F}_{q^t}$.) The reader may verify that the Trace function indeed maps all $\mathbb{K}$ to $\mathbb{F}$ (by verifying that $\mathrm{Tr}(x)^q = \mathrm{Tr}(x)$ for every $x \in \mathbb{K}$), and that it is linear, i.e., $\mathrm{Tr}(x + y) = \mathrm{Tr}(x) + \mathrm{Tr}(y)$ and $\mathrm{Tr}(\alpha x) = \alpha \mathrm{Tr}(x)$ for every $x, y \in \mathbb{K}$ and $\alpha \in \mathbb{F}$. It follows that $\mathrm{Tr}(g(x))$ is a function mapping $\mathbb{K}^n$ to $\mathbb{F}$ for every n-variate polynomial g. The following proposition establishes the converse.

Proposition 3.1. *Every function f from $\mathbb{K}^n \to \mathbb{F}$ is the trace of some polynomial g from $\mathbb{K}^n \to \mathbb{K}$. Furthermore, there always exists such a polynomial g whose support is contained in the support of f.*

Of course, given that the number of polynomials from $\mathbb{K}^n \to \mathbb{K}$ is much more than the number of functions from $\mathbb{K}^n \to \mathbb{F}$, it must be the case that different polynomials have the same trace. Some explicit examples include $\mathrm{Tr}(x) = \mathrm{Tr}(x^q)$, and $\mathrm{Tr}((\alpha + \alpha^q).x^{1+q+q^2+\cdots+q^{t-1}}) = \mathrm{Tr}(0)$. (This is why the proposition only claims that some polynomial g has its support contained in the support of f.) Nevertheless the traces give a very useful understanding of affine invariant families thanks to the following lemma (essentially from [40]) which shows that the affine-invariant properties are captured by the monomials in their support.

Lemma 3.1 (Monomial Extraction [40]). *For every affine-invariant property $\mathcal{P} \subseteq \{\mathbb{K}^n \to \mathbb{F}\}$ there exists a set $\mathcal{D} \subseteq \mathcal{M}$ such that a function $f \in \mathcal{P}$ if and only if there exists a polynomial $g : \mathbb{K}^n \to \mathbb{K}$ supported on $\mathcal{D}$ such that $f = \mathrm{Tr}(g)$. Furthermore there is a unique maximal such set $\mathcal{D}$ for any affine-invariant property $\mathcal{P}$.*

We refer to the unique maximal set as the *degree set* of $\mathcal{P}$.

To see some examples, first lets consider the simpler case of $\mathbb{K} = \mathbb{F}$. In this case the Trace function is simply the identity function; and the set $\mathcal{D}$ is simply the union of the support of all functions in $\mathcal{P}$. Thus in this case if the function, say, $3x^5 + 2x^2 + 1$ is in $\mathcal{P}$, it follows that $\{1, x^2, x^5\} \subseteq \mathcal{D}$ and thus the functions x^5, x^2 and $2x^5 + x^2 + 4$ are also in $\mathcal{P}$.

One of the uses of the lemma above, is that it allows us to focus on the degree set of an affine invariant property to understand its local-testability (and in particular in understanding when it may have a single orbit characterization).

But before investigating the locality of tests we first note that not every set $\mathcal{D}$ is a degree set of some affine-invariant property $\mathcal{P}$. While a compact description of exactly which degree sets are valid sets for affine-invariant properties is not easy to describe (and depends on n, p, q, Q etc.) this is well-understood. When $n = 1$ this is somewhat easier to describe, and we do so next.

Definition 3.5. $\mathcal{D} \subseteq \mathbb{K}[x]$ *is* (q, Q)-*modular if* $x^d \in \mathcal{D} \Rightarrow x^{q \cdot d (\mathrm{mod} Q - 1)} \in \mathcal{D}$.

Definition 3.6. *For non-negative integers e and d and prime p, let $e_0, \ldots, e_i, \ldots$, and $d_0, \ldots, d_i, \ldots$ denote the p-ary representation of e and d (i.e., $e_0, \ldots, e_i, \ldots \in \{0, \ldots, p-1\}$ and $e = \sum_{i=0}^{\infty} e_i p^i$). We say that e is in the p-shadow of d if $e_i \leq d_i$ for every i. We say that a set $\mathcal{D} \subseteq \mathbb{Z}^{\geq 0}$ is p-shadow-closed if for every d and e in the p-shadow of d, we have $x^d \in \mathcal{D} \Rightarrow x^e \in \mathcal{D}$.*

We are now ready to describe degree sets of univariate affine-invariant properties.

Lemma 3.2. $\mathcal{D}$ *is the degree set of an affine-invariant property $\mathcal{P} \subseteq \{\mathbb{F}_Q \to \mathbb{F}_q\}$ where $Q = q^t$ and $q = p^s$ for prime p if and only if $\mathcal{D}$ is (q, Q)-modular and $\mathcal{D}$ is p-shadow-closed.*

We note in passing that the case $n = 1$ is really the most general case, since every affine-invariant property from $\{K^n \to \mathbb{F}\}$ can also be viewed as an affine-invariant property from $\{\mathbb{L} \to \mathbb{F}\}$ where $\mathbb{L}$ is the nth degree extension of $\mathbb{K}$ (i.e. $\mathbb{L} = \mathbb{F}_{Q^n}$).

While the lemmas above describe some basic features of affine-invariant properties, they don't explain when they may be locally testable. In particular when can they have local constraints, local characterizations, and even single-orbit characterizations? These questions are more novel, and less well-understood.

If one considers the case where Q, q are just constants, and only n is going to infinity, then, as shown in [40], the degree of the highest degree monomial in $\mathcal{D}$ roughly determines the best possible locality of the constraints and single orbit characterizations. Specifically they show:

Lemma 3.3. *Let $\mathcal{P} \subseteq \{\mathbb{K}^n \to \mathbb{F}\}$ be an affine-invariant property with degree set $\mathcal{D}$ and let d be the largest degree of a monomial in $\mathcal{D}$. Then every constraint on $\mathcal{P}$ has locality at least $Q^{(d/Q^2 - 1)}$. Conversely it has a $Q^{2(d+Q)/p}$-local single orbit characerization.*

To understand the above lemma, note that it implies that if an affine-invariant propery has a *single* k-local constraint then it has k'-local characterization, and in fact, a k'-single orbit characterization, and is hence k'-locally testable for $k' \approx Q^2 \cdot k^{2Q^2}$. This appears to be far from tight and indeed the analysis in [40] is quite sloppy allowing for tighter characterizations. Indeed, we believe it should be possible to get a k'-local characterization for $k' = \mathrm{poly}(Q, k)$.

A somewhat more optimistic conjecture might be that one can get $k' = \mathrm{poly}(k)$ (or some other function of k which is independent of Q). Indeed such a relationship was effectively conjectured by [3] (for a broader class of properties than affine-invariant ones). However this turned out be false as shown by [36].

Theorem 3.2. *There exists an affine invariant family mapping $\{\mathbb{F}_{2^t} \to \mathbb{F}_2\}$ with an 8-local constraint, but no $(t/2 - 2)$-local characterization.*

The family given by [36] is easy to describe in the language developed so far: Their family $\mathcal{P}$ has, as its degree set, the set $\mathcal{D} = \{x^{2^i + 2^{i+j}} | i \in \{0, \ldots, t-1\}, j \in$

$\{0, \ldots, t/2 - 2\}\} \cup \{x^{2^i} | i \in \{0, \ldots, t - 1\} \cup \{x^0\}$. It is easy to check that $\mathcal{D}$ is 2-shadow closed, and $(2, 2^t)$-modular, and so $\mathcal{P}$ is indeed an affine invariant family. It is also easy to show that every function $f \in \mathcal{P}$ satisfies the constraint $f(x + y + z) = f(x + y) + f(x + z) + f(y + z) + f(x) + f(y) + f(z) + f(0)$ and thus $\mathcal{P}$ has an 8-local constraint. The main contribution of [36] is to show that $\mathcal{P}$ has no $t/2 - 2$ local constraints that are not constraints also on the larger family $\mathcal{P}'$ given by its degree set $\mathcal{D}' = \mathcal{D} \cup \{x^{2^i + 2^{i + (t/2 - 1)}} | i\}$. Since $\mathcal{P}'$ is strictly larger than $\mathcal{P}$ it follows that $\mathcal{P}$ can not have a $t/2 - 2$-local characterization. Furthermore, since $\mathcal{P}'$ contains functions that are quite far from functions in $\mathcal{P}$, it follows that $\mathcal{P}$ also does not have any tests of locality $t/2 - 2$.

We remark that most examples of single orbit characterizations have been natural ones; i.e., the properties have a natural characterization that happens to be a single orbit one. Indeed all the examples given above (affine functions, low-degree polynomials etc.) had this property. One significant class of exceptions is given in [37] who show that every "sparse" affine invariant property from $\mathcal{P} \subseteq \{\mathbb{F}_{2^t} \to \mathbb{F}_2\}$, with $|\mathcal{P}| \leq 2^{t\ell}$, has a $k = k(\ell)$-single orbit characterization, if t is prime. (Here "sparse" refers to the fact that the size of $\mathcal{P}$ is a polynomial in the domain size. Note that the locality of the characterization is independent of the domain size and depends only on the exponent relating the size of $\mathcal{P}$ with the size of the domain.) The result of [37] is also obtained by analyzing the degree sets of sparse affine-invariant properties and noticing that functions satisfying such properties can be expressed as traces of sparse polynomials, and then combining recent results from additive number theory with classical results from coding theory to conclude that these properties have a local single orbit characterization. These results are interesting in that they yield single orbit characterizations for a very rich class of properties - so rich that it would take $\Omega(\log t)$ bits to describe a typical such property and so a "totally uniform" characterization (with $O(1)$-bits) would be out of question. Natural "local" characterizations of such properties involve describing roughly 2^t constraints each requiring $O(t)$ bits to specify. The single-orbit characterization, in contrast, only requires $O(t)$ bits to describe giving a somewhat more uniform, and yet local, description of the property and the tester for the property.

Moving on, affine-invariant properties offer a clean generalization of "low-degree" polynomials, while being significantly richer, rich enough to counter some natural conjectures about reasons for local testability in codes/algebraic properties. Furthermore, the class still offers the possibility of some locally testable codes of constant distance that may outperform Reed-Muller codes (codes derived from low-degree polynomials) in terms of their rate. To investigate this possibility one needs a significantly better understanding of the relationship between the locality of characterizations and the degree sets of affine-invariant properties. In the case of univariate functions, no non-trivial upper bounds are known for general degree sets (the trivial one being the size of the degree set), and till recently no general lower bounds were known either. A recent result with Ben-Sasson gives the first general lower bound on the locality of constraints for a general degree set, in terms of the notion of the p-weight of elements in the degree set. We define this notion next, and give their main theorem afterwards.

Definition 3.7. *For integer d and prime p, the p-weight of d, denoted $wt_p(d)$, is defined to be the sum of the non-zero elements in the p-ary expansion of d, i.e., $wt_p(d) = \sum_i d_i$ where $d_0, \ldots, d_i, \ldots, \in \{0, \ldots, p-1\}$ s.t. $d = \sum_i d_i p^i$.*

Theorem 3.3 ([15]). *Let $\mathcal{P} \subseteq \{\mathbb{F}_{p^t} \to \mathbb{F}_p\}$ be an affine-invariant property with degree set $\mathcal{D}$. Let k be the maximum p-weight of elements of $\mathcal{D}$. Then every constraint on $\mathcal{P}$ has locality at least $k+1$. Conversely, $\mathcal{P}$ does have a constraint of locality p^{k+1}.*

When the range p is a constant, the above theorem thus pins down a necessary and sufficient condition for an affine-invariant family to have a $O(1)$-local constraint. Of course, it does not say anything about characterizations and this remains an open question. Indeed the following question remains open.

Question 3.1. Let $\mathcal{P} \subseteq \{\mathbb{K} \to \mathbb{F}\}$ be an affine-invariant property with a k-local characterization. Let k' be the smallest integer such that $\mathcal{P}$ has a k'-single orbit characterization. Give the best possible upper bound on k' as a function of k, q and Q. Can we get a bound independent of Q? Can it be independent of q?

To understand such questions a significantly better understanding of the relationship between local characterizations and degree sets is needed. The following question is an example of some very basic questions about this relationship which is still not understood.

Question 3.2. Let $\mathcal{P} \subseteq \{\mathbb{F}_{2^t} \to \mathbb{F}_2\}$ be an affine-invariant property with degree set $\mathcal{D} \subsetneq \{x^{2^i + 2^j} | i, j \in \{0, \ldots, t-1\}\} \cup \{1, x, x^2, \ldots, x^{2^{t-1}}\}$. Further, suppose t is prime. Then does the locality of the characterization of $\mathcal{P}$ grow with $|\mathcal{D}|$? (I.e., is the following statement true? For every k, there exists a prime t and a $(2, 2^t)$-modular, 2-shadow closed set $\mathcal{D} \subsetneq \{x^{2^i + 2^j} | i, j \in \{0, \ldots, t-1\}\} \cup \{1, x, x^2, \ldots, x^{2^{t-1}}\}$ such that the affine invariant property $\mathcal{P}$ with degree set $\mathcal{D}$ has no k-local characterization.)

We remark that if t is not a prime, then the question above does have negative answers; and understanding the exact reson for such negative answers also appears important to understanding affine-invariant properties.

4   Non-linear Affine-Invariant Properties

Finally, we remark briefly on the setting where the properties of interest are invariant under affine/linear transformations, but are not necessarily linear. Examples of such testable non-linear properties that are affine-invariant can be generated easily by taking the union of two (or more) affine-invariant (linear) properties. (Thanks to Noga Alon for this class of examples). More interesting cases, motivated by learning theory and additive number theory, have also been explored in the literature and we mention these results briefly below.

Locally characterized properties: A broad class of affine-invariant properties that seem potentially testable can be obtained by generalizing the notion of constraints and characterizations to the non-linear setting as follows: Let $\mathbb{K}$ be a finite field and Σ be an arbitrary finite set. A k-local constraint C is given by $C = (\alpha_1, \ldots, \alpha_k; S)$ where $\alpha_1, \ldots, \alpha_k \in \mathbb{K}^n$ and $S \subsetneq \Sigma^k$. We say a function $f : \mathbb{K}^n \to \Sigma$ satisfies the constraint C if $\langle f(\alpha_1), \ldots, f(\alpha_k) \rangle \in S$. A collection of constraints $C_1, \ldots, C_m$ characterizes a property $\mathcal{P} \subseteq \{\mathbb{K}^n \to \Sigma\}$ if $f \in \mathcal{P}$ if and only if f satisfies C_j for every $j \in [m]$. Of course, our interest here is in affine-invariant property that are k-locally characterized.

A very simple example of an affine invariant property considered in Green [35] (we note that this is merely an example of property considered there, not the broadest class considered there) is the following: Let $\mathbb{K} = \mathbb{F}_2$ and $\Sigma = \{0, 1\}$. Let $C = \{\alpha, \beta, \alpha + \beta; \{0, 1\}^3 - \{111\}\}$, where $\alpha, \beta \in \mathbb{K}^n$ are an arbitrary pair of linearly independent elements. Now consider the property $\mathcal{P}$ characterized by $\{C \circ \pi | \pi : \mathbb{K}^n \to \mathbb{K}^n$ is an affine map$\}$. This is a 3-locally characterized affine-invariant property from $\mathbb{K}^n \to \Sigma$ consisting of all functions f that are "triangle-free", i.e., $f^{-1}(1)$ does not contain a triple of the form $x, y, x + y$. Green showed that the natural test (pick a random affine map $\pi : \mathbb{K}^n \to \mathbb{K}^n$ and verify that f satisfies $C \circ \pi$) does reject functions that are ϵ-far with positive probability. Somewhat intriguingly the analysis of this test, is quite different from the analyses in the linear cases and more reminiscent of the analyses in graph property testing.

This result was then generalized in various works [16,44,43,54]) with perhaps the strongest result being due to [43,54] who considers any constant number of "freeness" constraints $C_1, \ldots, C_\ell$, and their affine shifts and shows that any such property is locally testable. (A constraint $C = (\alpha_1, \ldots, \alpha_k; S)$ is said to be a freeness constraint if $\Sigma = \{0, 1\}$ and $S = \Sigma^k - \{1^k\}$.) In the process, these works also tighten the connection to graph property testing, by deriving their main results as a corollary of a new hypergraph removal lemma (a typical ingredient in hypergraph property testing).

Of course, despite all this progress, this area abounds with questions with some basic ones being: Which subclass of locally characterizaed affine-invariant properties are locally testable? When can the rejection probability of the test be lower bounded by a polynomial in the distance to the property? Some progress in this direction is reported in Bhattacharyya and Xing [17].

Sparse linear functions: As part of their investigations of properties of Boolean functions Gopalan et al. [34] investigate functions that are represented as sparse functions (e.g., k-juntas) of linear functions of their input. (Formally, their properties are given by some collection of functions $\mathcal{G} \subseteq \{\mathbb{F}_2^\ell \to \mathbb{F}_2\}$ and the property of functions $\mathcal{P} \subseteq \{\mathbb{F}_2^n \to \mathbb{F}_2\}$ is given by $\mathcal{P} = \{g \circ L | g \in \mathcal{G}$ and $L : \mathbb{F}_2^n \to \mathbb{F}_2^\ell$ is linear $\}$. For a broad collection of "sparse" functions (i.e., classes of sets $\mathcal{G}$), they show that the associated property $\mathcal{P}$ is testable. Since these classes of functions are naturally closed under linear transforms, it follows that this is yet another broad class of properties that is linear-invariant. Typically, these properties (e.g., being representable as a k-junta, or as a k-sparse polynomial) are not closed under addition, and so these properties are non-linear. Also interestingly, these properties are testable, at least to within current knowledge, only with two-sided error.

5   Conclusions

We summarize with the main message: Testing of natural properties is often intimately related to the invariances shown by the property. When the class of invariances is the full symmetric group of permutations, testing ends up representing the classical problems of statistics (though even here some improvements are feasible). But modern property testing highlights the ability to do much better when the underlying class of invariances is no the full set, and often exponentially smaller than the full set. Among the rich variety of such symmetries that can be explored, we emphasize the role of the affine-invariance (or linear-invariance) as a natural way to unify many known results. Understanding affine-invariance further might be one way of making progress on the design of locally testable codes.

We stress that we don't believe that invariance is necessary for local testability, hence the appeal to the weakening clause of "natural properties". We are aware of a wide class of properties that are known to be testable, where we are not aware of a nice invariance class. For example, a typical PCP verifier tends to accept encodings of any satisfying assignment of a SAT formula (this is actually a requirement for PCPP verifier [13] or assignment testers [22]) with local tests. Depending on the SAT formula being checked for satisfiability, the property "tested" by such a verifier is unlikely to have rich invariances. Even in the algebraic setting, we have the example of functional equations that are known to be testable [51], but where the invariance class is not known or has not been determined explicitly. In contrast to the above, we highlight the work of Goldreich and Kaufman [29] who give examples of properties that are testable, but provably have no non-trivial invariances, and also exhibit testable properties which do not have a "local single-orbit characterization".

Our point is not that such exceptions may not exist, but rather that *when invariances exist* clean rules may be found explaining when a property is testable. For graph properties properties the characterization of testability in terms of regularity instances [2] gives such a rule. For affine-invariant linear properties, the existence of local characterizations may be necessary as well as sufficient (both directions being open). The results in [40] show this to be the case, when the field size in the domain is of constant size. More importantly, the invariance class cleanly separates the many different contexts in which property testing results have been found; and gives a general approach to extracting general techniques.

We hope that in future work, the invariance classes may help further the understanding of property testing, while also help in the design of novel classes of testable codes. (A promising result here is that of Kaufman and Wigderson [41] that have given some novel codes that exhibit symmetries.) At the moment, we lack a broad understanding of group theoretic properties that help analyze testability of properties; and indeed the collection of groups for which we are able to derive testing results still remains quite limited. We hope this is remedied in future work.

Acknowledgments

Thanks to Tali Kaufman, Elena Grigorescu, and Ben-Sasson for their collaboration and comments. Thanks to Oded Goldreich and Dana Ron for advice and discussions, and to Arnab Bhattacharyya for valuable suggestions.

References

1. Alon, N., Andoni, A., Kaufman, T., Matulef, K., Rubinfeld, R., Xie, N.: Testing k-wise and almost k-wise independence. In: Johnson, D.S., Feige, U. (eds.) STOC, pp. 496–505. ACM, New York (2007)
2. N. Alon, E. Fischer, I. Newman, A. Shapira. A combinatorial characterization of the testable graph properties: it's all about regularity. In: Kleinberg [42], pp. 251–260
3. Alon, N., Kaufman, T., Krivelevich, M., Litsyn, S., Ron, D.: Testing reed-muller codes. IEEE Transactions on Information Theory 51(11), 4032–4039 (2005)
4. Arora, S., Lund, C., Motwani, R., Sudan, M., Szegedy, M.: Proof verification and the hardness of approximation problems. Journal of the ACM 45(3), 501–555 (1998)
5. Babai, L., Fortnow, L., Levin, L.A., Szegedy, M.: Checking computations in polylogarithmic time. In: Proceedings of the 23rd ACM Symposium on the Theory of Computing, pp. 21–32. ACM Press, New York (1991)
6. Babai, L., Fortnow, L., Lund, C.: Non-deterministic exponential time has two-prover interactive protocols. Computational Complexity 1(1), 3–40 (1991)
7. Babai, L., Shpilka, A., Stefankovic, D.: Locally testable cyclic codes. IEEE Transactions on Information Theory 51(8), 2849–2858 (2005)
8. Batu, T., Dasgupta, S., Kumar, R., Rubinfeld, R.: The complexity of approximating the entropy. SIAM J. Comput. 35(1), 132–150 (2005)
9. Batu, T., Fortnow, L., Fischer, E., Kumar, R., Rubinfeld, R., White, P.: Testing random variables for independence and identity. In: FOCS, pp. 442–451 (2001)
10. Batu, T., Fortnow, L., Rubinfeld, R., Smith, W.D., White, P.: Testing that distributions are close. In: FOCS, pp. 259–269 (2000)
11. Batu, T., Kumar, R., Rubinfeld, R.: Sublinear algorithms for testing monotone and unimodal distributions. In: Babai, L. (ed.) STOC, pp. 381–390. ACM, New York (2004)
12. Bellare, M., Goldreich, O., Sudan, M.: Free bits, PCP's and non-approximability — towards tight results. SIAM Journal on Computing 27(3), 804–915 (1998)
13. Ben-Sasson, E., Goldreich, O., Harsha, P., Sudan, M., Vadhan, S.P.: Robust pcps of proximity, shorter pcps, and applications to coding. SIAM Journal on Computing 36(4), 889–974 (2006)
14. Ben-Sasson, E., Harsha, P., Raskhodnikova, S.: Some 3CNF properties are hard to test. SIAM Journal on Computing 35, 1–21 (2005); Preliminary version in Proc. STOC 2003 (2003)
15. Ben-Sasson, E., Sudan, M.: Limits on the rate of locally testable affine-invariant codes (November 2009) (manuscript)

16. Bhattacharyya, A., Chen, V., Sudan, M., Xie, N.: Testing linear-invariant non-linear properties. In: Albers, S., Marion, J.-Y. (eds.) STACS. LIPIcs, vol. 3, pp. 135–146. Schloss Dagstuhl - Leibniz-Zentrum fuer Informatik, Germany (2009)

17. Bhattacharyya, A., Xie, N.: Lower bounds for testing triangle-freeness in boolean functions. In: SODA 2010: Proceedings of the twenty-first Annual ACM-SIAM Symposium on Discrete Algorithms, pp. 87–98. Society for Industrial and Applied Mathematics, Philadelphia (2010)

18. Blais, E.: Testing juntas nearly optimally. In: Mitzenmacher [47], pp. 151–158

19. Blum, M., Luby, M., Rubinfeld, R.: Self-testing/correcting with applications to numerical problems. Journal of Computer and System Sciences 47(3), 549–595 (1993)

20. Borgs, C., Chayes, J.T., Lovász, L., Sós, V.T., Szegedy, B., Vesztergombi, K.: Graph limits and parameter testing. In: Kleinberg [42], pp. 261–270

21. Diakonikolas, I., Lee, H.K., Matulef, K., Onak, K., Rubinfeld, R., Servedio, R.A., Wan, A.: Testing for concise representations. In: FOCS, pp. 549–558. IEEE Computer Society, Los Alamitos (2007)

22. Dinur, I., Reingold, O.: Assignment testers: Towards a combinatorial proof of the PCP-theorem. In: Proceedings of the 45th Annual IEEE Symposium on Foundations of Computer Science, pp. 155–164. IEEE Press, Los Alamitos (2004)

23. Dodis, Y., Goldreich, O., Lehman, E., Raskhodnikova, S., Ron, D., Samorodnitsky, A.: Improved testing algorithms for monotonicity. In: Hochbaum, D.S., Jansen, K., Rolim, J.D.P., Sinclair, A. (eds.) RANDOM 1999 and APPROX 1999. LNCS, vol. 1671, pp. 97–108. Springer, Heidelberg (1999)

24. Ergün, F., Kannan, S., Kumar, R., Rubinfeld, R., Viswanathan, M.: Spot-checkers. J. Comput. Syst. Sci. 60(3), 717–751 (2000)

25. Fischer, E., Kindler, G., Ron, D., Safra, S., Samorodnitsky, A.: Testing juntas. J. Comput. Syst. Sci. 68(4), 753–787 (2004)

26. Friedl, K., Sudan, M.: Some improvements to total degree tests. In: Proceedings of the 3rd Annual Israel Symposium on Theory of Computing and Systems, Washington, DC, USA, January 4-6, pp. 190–198. IEEE Computer Society, Los Alamitos (1995), http://people.csail.mit.edu/madhu/papers/friedl.ps

27. Goldreich, O., Goldwasser, S., Lehman, E., Ron, D., Samorodnitsky, A.: Testing monotonicity. Combinatorica 20(3), 301–337 (2000)

28. Goldreich, O., Goldwasser, S., Ron, D.: Property testing and its connection to learning and approximation. JACM 45(4), 653–750 (1998)

29. Goldreich, O., Kaufman, T.: Proximity oblivious testing and the role of invariances (March 2010) (manuscript)

30. Goldreich, O., Ron, D.: On testing expansion in bounded-degree graphs. Electronic Colloquium on Computational Complexity (ECCC) 7(20) (2000)

31. Goldreich, O., Ron, D.: Property testing in bounded degree graphs. Algorithmica 32(2), 302–343 (2002)

32. Goldreich, O., Ron, D.: On proximity oblivious testing. In: Mitzenmacher [47], pp. 141–150

33. Goldreich, O., Sheffet, O.: On the randomness complexity of property testing. In: Charikar, M., Jansen, K., Reingold, O., Rolim, J.D.P. (eds.) RANDOM 2007 and APPROX 2007. LNCS, vol. 4627, pp. 509–524. Springer, Heidelberg (2007)

34. Gopalan, P., O'Donnell, R., Servedio, R.A., Shpilka, A., Wimmer, K.: Testing fourier dimensionality and sparsity. In: Albers, S.E., Marchetti-Spaccamela, A., Matias, Y., Nikoletseas, S., Thomas, W. (eds.) ICALP 2009, Part I. LNCS, vol. 5555, pp. 500–512. Springer, Heidelberg (2009)

35. Green, B.: A Szemerédi-type regularity lemma in abelian groups, with applications. Geometric and Functional Analysis 15(2), 340–376 (2005)
36. Grigorescu, E., Kaufman, T., Sudan, M.: 2-transitivity is insufficient for local testability. In: CCC 2008: Proceedings of the 23rd IEEE Conference on Computational Complexity, June 23-26. IEEE Computer Society, Los Alamitos (2008) (to appear)
37. Grigorescu, E., Kaufman, T., Sudan, M.: Succinct representation of codes with applications to testing. In: Dinur, I., Jansen, K., Naor, J., Rolim, J.D.P. (eds.) APPROX-RANDOM 2009. LNCS, vol. 5687, pp. 534–547. Springer, Heidelberg (2009)
38. Jutla, C.S., Patthak, A.C., Rudra, A., Zuckerman, D.: Testing low-degree polynomials over prime fields. In: FOCS 2004: Proceedings of the Forty-Fifth Annual IEEE Symposium on Foundations of Computer Science, pp. 423–432. IEEE Computer Society, Los Alamitos (2004)
39. Kaufman, T., Ron, D.: Testing polynomials over general fields. SIAM J. Comput. 36(3), 779–802 (2006)
40. Kaufman, T., Sudan, M.: Algebraic property testing: The role of invariance. Technical Report TR07-111, Electronic Colloquium on Computational Complexity, November 2 (2007); Extended abstract in Proc. 40th STOC (2008)
41. Kaufman, T., Wigderson, A.: Symmetric LDPC codes and local testing. In: Proceedings of ICS 2010 (January 2010)
42. Kleinberg, J.M. (ed.): Proceedings of the 38th Annual ACM Symposium on Theory of Computing, Seattle, WA, USA, May 21-23. ACM, New York (2006)
43. Král', D., Serra, O., Vena, L.: A removal lemma for systems of linear equations over finite fields. arxiv.org:0809.1846v1 [math.CO]
44. Král', D., Serra, O., Vena, L.: A combinatorial proof of the removal lemma for groups. Journal of Combinatorial Theory, Series A 116(4), 971–978 (2009)
45. Ladner, R.E., Dwork, C. (eds.): Proceedings of the 40th Annual ACM Symposium on Theory of Computing, Victoria, British Columbia, Canada, May 17-20. ACM, New York (2008)
46. Matulef, K., O'Donnell, R., Rubinfeld, R., Servedio, R.A.: Testing halfspaces. In: Mathieu, C. (ed.) SODA, pp. 256–264. SIAM, Philadelphia (2009)
47. Mitzenmacher, M. (ed.): Proceedings of the 41st Annual ACM Symposium on Theory of Computing, STOC 2009, Bethesda, MD, USA, May 31-June 2. ACM, New York (2009)
48. Onak, K., Sudan, M.: Learnability of general discrete distributions (March 2010) (manuscript)
49. Parnas, M., Ron, D., Samorodnitsky, A.: Testing basic boolean formulae. SIAM J. Discrete Math. 16(1), 20–46 (2002)
50. Raskhodnikova, S., Ron, D., Shpilka, A., Smith, A.: Strong lower bounds for approximating distribution support size and the distinct elements problem. SIAM J. Comput. 39(3), 813–842 (2009)
51. Rubinfeld, R.: Robust functional equations and their applications to program testing. SIAM Journal on Computing 28(6), 1972–1997 (1999)
52. Rubinfeld, R., Servedio, R.A.: Testing monotone high-dimensional distributions. Random Struct. Algorithms 34(1), 24–44 (2009)
53. Rubinfeld, R., Sudan, M.: Robust characterizations of polynomials with applications to program testing. SIAM Journal on Computing 25(2), 252–271 (1996)
54. Shapira, A.: Green's conjecture and testing linear-invariant properties. In: Mitzenmacher [47], pp. 159–166
55. Valiant, P.: Testing symmetric properties of distributions. In: Ladner and Dwork [45], pp. 383–392

Testing Monotone Continuous Distributions on High-Dimensional Real Cubes[*]

Michał Adamaszek[1], Artur Czumaj[2], and Christian Sohler[3]

[1] Centre for Discrete Mathematics and its Applications (DIMAP) and
Warwick Mathematics Institute, University of Warwick
`M.J.Adamaszek@warwick.ac.uk`
[2] Department of Computer Science and Centre for Discrete Mathematics and
its Applications (DIMAP), University of Warwick
`A.Czumaj@warwick.ac.uk`
[3] Department of Computer Science, Technische Universität Dortmund
`sohler@informatik.uni-bonn.de`

Abstract. We study the task of testing properties of probability distributions and our focus is on understanding the role of continuous distributions in this setting. We consider a scenario in which we have access to independent samples of an unknown distribution $\mathfrak{D}$ with infinite (perhaps even uncountable) support. Our goal is to test whether $\mathfrak{D}$ has a given property or it is ε-far from it (in the statistical distance, with the L_1-distance measure).

It is not difficult to see that for many natural distributions on infinite or uncountable domains, no algorithm can exist and the central objective of our study is to understand if there are any nontrivial distributions that can be efficiently tested. For example, it is easy to see that there is no algorithm that tests if a given probability distribution on $[0, 1]$ is uniform. We show however, that if some additional information about the input distribution is known, testing uniform distribution is possible. We extend the recent result about testing uniformity for monotone distributions on Boolean n-dimensional cubes by Rubinfeld and Servedio (STOC'2005) to the case of *continuous* $[0, 1]^n$ cubes. We show that if a distribution $\mathfrak{D}$ on $[0, 1]^n$ is monotone, then one can test if $\mathfrak{D}$ is uniform with the sample complexity $\mathcal{O}(n/\varepsilon^2)$. This result is optimal up to a polylogarithmic factor. We also extend the result of Rubinfeld and Servedio (STOC'2005) to test if a distribution $\mathfrak{D}$ on $\{0, 1, \ldots, k\}^n$ is monotone with the sample complexity $\mathcal{O}(n/\varepsilon^2)$.

Keywords: testing distributions.

1 Introduction

We study the task of testing properties of probability distributions. We consider a scenario in which we have access to independent samples of an unknown

---

[*] A preliminary, conference version of this work appeared in [1]. Research supported by EPSRC award EP/G064679/1, DFG grant So 514/3-1, and by the Centre for Discrete Mathematics and its Applications (DIMAP), EPSRC award EP/D063191/1.

O. Goldreich (Ed.): Property Testing, LNCS 6390, pp. 228–233, 2010.

distribution $\mathfrak{D}$ with infinite (perhaps even uncountable) support. Our goal is to test whether $\mathfrak{D}$ has a given property or it is ε-far from it (in the statistical distance, with the L_1-distance measure).

The topic of testing basic properties of the underlying probability distributions has been extensively studied for many decades. While the standard approach in statistics (and also more modern approaches, e.g., in data mining) have led to the development of many high quality techniques and algorithms, until very recently little attention has been paid to the computational complexity of testing in the situations when the underlying distributions are over very large domains. Motivated by these considerations, a number of new studies have emerged that aim at developing efficient testers for various properties of distributions with the focus on the small number of samples used for testing. In particular, it has been shown that for a number of fundamental properties, such as independence, entropy estimation, and the closeness between distributions, it is possible to test the underlying distribution with the number of samples sublinear in the domain size.

While these studies lead to very efficient testers for various properties for distributions on finite support, they seem to be useless when the underlying distribution is on a continuous, or infinite, or even uncountable domain. In this paper, our goal is to study the phenomenon of testability of continuous distributions. We assume that there is an underlying probability distribution $\mathfrak{D}$ from which we can draw *independent identically distributed samples* (see, e.g., [5]). We assume that each sample is of infinite precision and we will not consider the issue of representation of the real numbers. The *complexity of the tester* is measured in terms of the *number of samples* required in order to obtain a desired information about the distribution. We study probability distributions over a domain Ω which will be either finite or infinite; our main focus is on the domain $\Omega = [0,1]^n$, $n \in \mathbb{N}$, that is, (continuous) n-dimensional unit cube.

We study the similarity and dissimilarity between various distributions. Following the mainstream research of testing properties of distributions in theoretical computer science, we use the *total variation distance to measure the similarity between distributions* (L_1-distance). For any two *discrete* distributions $\mathcal{X}$ and $\mathcal{Y}$ over Ω, defined by the probability functions $\mathbf{Pr}_{\mathcal{X}}$ and $\mathbf{Pr}_{\mathcal{Y}}$, respectively, we say $\mathcal{Y}$ *is ε-far from* $\mathcal{X}$ if $\frac{1}{2} \cdot \sum_{\omega \in \Omega} |\mathbf{Pr}_{\mathcal{X}}[\omega] - \mathbf{Pr}_{\mathcal{Y}}[\omega]| \geq \varepsilon$. For general distributions, the definition is analogous: for any two distributions $\mathcal{X}$ and $\mathcal{Y}$ over Ω, with density functions $\mathfrak{f}_{\mathcal{X}}$ and $\mathfrak{f}_{\mathcal{Y}}$, respectively, we say $\mathcal{Y}$ *is ε-far from* $\mathcal{X}$ if

$$\frac{1}{2} \cdot \int_{\mathbf{x} \in \Omega} |\mathfrak{f}_{\mathcal{X}}(\mathbf{x}) - \mathfrak{f}_{\mathcal{Y}}(\mathbf{x})| \, d\mathbf{x} \geq \varepsilon \ . \tag{1}$$

Note that inequality (1) is equivalent to $\int_{\mathbf{x} \in \Omega : \mathfrak{f}_{\mathcal{X}}(\mathbf{x}) \geq \mathfrak{f}_{\mathcal{Y}}(\mathbf{x})} (\mathfrak{f}_{\mathcal{X}}(\mathbf{x}) - \mathfrak{f}_{\mathcal{Y}}(\mathbf{x})) \, d\mathbf{x} \geq \varepsilon$.

We say $\mathcal{Y}$ *is ε-close to* $\mathcal{X}$ if $\mathcal{Y}$ is not ε-far from $\mathcal{X}$.

Let us remind that a distribution $\mathfrak{D}$ over $[0,1]^n$ with density function $\mathfrak{f}$ is *uniform* if $\mathfrak{f}$ is identically 1. Therefore, for the uniform distribution, (1) can be rephrased as follows: A distribution $\mathfrak{D}$ over $[0,1]^n$ with density function $\mathfrak{f}$ is *ε-far from uniform* if

$$\frac{1}{2} \cdot \int_{\mathbf{x} \in [0,1]^n} |f(\mathbf{x}) - 1| \, d\mathbf{x} \geq \varepsilon \ .$$

Our goal is to design an algorithm that for a given ε and a given underlying distribution $\mathfrak{Q}$ and a distribution $\mathfrak{D}$ available through random sampling, is able to distinguish between the case when $\mathfrak{Q} = \mathfrak{D}$ and when $\mathfrak{D}$ is ε-far from it $\mathfrak{Q}$. The algorithm is allowed to be randomized and can err with probability at most $\frac{1}{4}$.

2 Continuous Distributions Are Typically Not Testable

In general, when using the total variation distance to measure the similarity between distributions, it is infeasible to investigate interesting properties of distributions on infinite domains without any assumptions on the density function. For example, one can show that for every integer t there is no tester A that distinguishes with at most t samples between uniform distribution $\mathfrak{D}_U$ on $[0,1]$ and any distribution that is ε-far from uniform (for example, take a uniform distribution on t^3 randomly chosen points from the interval $[0,1]$; such distribution is discrete and hence it is $\frac{1}{2}$-far from uniform). This observation can be easily generalized to testing a number of natural properties for distributions on infinite domains.

One can also derive similar impossibility results from the existing lower bounds for testing properties of discrete distributions. For example, Batu et al. [5] (see also [7]) show that testing if a given distribution on the support of size n is uniform requires $\Omega(\sqrt{n}/\varepsilon^2)$ samples. With that, by taking $n \to \infty$, the lower bound in [5] immediately implies that no algorithm can test if a given distribution on $[0,1]$ is uniform. This approach implies also similar impossibility results for testing if a given distribution is monotone, unimodal, or if two distributions are identical, are independent, and so on (see [2,3,4,5,6,8,9,10] for more examples).

Once we see these negative result, the natural question is: what properties of distributions on infinite domains can be tested?

3 Testing If a Distribution Is Discrete on N Points

In order to understand the problem of testing distributions on infinite domains, the very first question should be to test if a given distribution has infinite support. We first briefly consider a dual question: to verify if a given distribution has support of up to a given size.

Recall that by the Radon-Nikodym theorem, every distribution on Ω has a Lebesgue decomposition into a sum of two parts:

- continuous (with respect to the standard Lebesgue measure), that is, given by a measurable density function f, and
- singular (concentrated on a set of Lebesgue measure 0).

A point $\mathbf{x}$ is called an *atom* of $\mathfrak{D}$ if $\mathbf{Pr}_{\mathfrak{D}}[\mathbf{x}] > 0$. Detection of a single atom is not possible, since its probability may be arbitrarily small, beyond the resolution of any given algorithm. Instead we may try to determine whether a large part of the probability mass is concentrated on the atoms: for a given parameter N, distinguish between distributions that have the entire support on at most N points (*discrete on N points*) and those that are ε-far from discrete on N points.

A related question has been studied recently by Raskhodnikova et al. [8], who were interested in estimating the size of the support of a given distribution under the assumption that every element in the support is an atom (distribution is singular) with the probability at least $\frac{1}{M}$. For such problem, Raskhodnikova et al. [8] (see also [11]) show that one needs at least $\Omega(M^{1-o(1)})$ samples to estimate the size of the support. On the other hand, it is easy to compute (exactly) the size of the support with $\mathcal{O}(M \log M)$ samples (e.g., by using the approach from the coupon collector problem). Our goal is different than that in [8], because on one hand, we do not have any lower bound on the probability of the points in the support (which makes the task of even estimating the size of the support impossible), and on the other hand, we want to test if a given distribution has at most N points in the support (rather than estimate the size of the support). Still, one can rather easily prove that the lower bound result from Raskhodnikova et al. [8] carries over for our problem and gives a lower bound for the sample size of $\Omega(N^{1-o(1)})$. One can also show that the following algorithm sampling $\mathcal{O}(N/\varepsilon)$ elements is a testing algorithm that distinguishes between a discrete distribution on N points and any distribution that is ε-far from discrete on N points with $\mathcal{O}(N/\varepsilon)$ samples.

Testing discreteness (N):

- Draw a sample (according to the distribution $\mathfrak{D}$) $S = \langle s_1, \ldots, s_\ell \rangle$ from Ω with $\ell = \lceil 2N/\varepsilon \rceil$
- If S has more than N distinct elements then **Reject**
- else **Accept**

Observe that this result immediately implies that we can estimate the smallest number $\mathcal{N}$ of points in the domain of $\mathfrak{D}$ such that $\mathfrak{D}$ has $\mathcal{N}$ points that have the total probability at least $1 - \varepsilon$ using $\mathcal{O}(\mathcal{N}/\varepsilon)$ samples.

An interesting **open question** is whether the upper bound is tight, that is, if every algorithm testing if a distribution is discrete on N points requires $\Omega(N/\varepsilon)$ samples.

4 Testing If a Monotone High-Dimensional Distribution on a Real Hypercube Is Uniform

The main goal of this work is to investigate if there are any interesting distributions on infinite domains that are testable. One of a very few properties of discrete distributions considered in the Computer Science literature that has only a light dependency on the size of the support (the condition that by our discussion above seems to be necessary to hope for a fast tester) is that of *testing if a monotone distribution*[1] *on the Boolean cube is uniform*. Rubinfeld and Servedio [10] consider the following problem: *given a monotone distribution $\mathfrak{D}$ on a Boolean n-dimensional cube $\{0, 1\}^n$, test if $\mathfrak{D}$ is uniform.*

Rubinfeld and Servedio [10] show that without any assumption about the monotonicity of $\mathfrak{D}$, every testing algorithm requires $2^{\Omega(n)}$ samples (because the domain's size is 2^n), however, if $\mathfrak{D}$ is monotone, then one distinguishes between the case when $\mathfrak{D}$ is uniform and when $\mathfrak{D}$ is ε-far from uniform using $\mathcal{O}(n \log(1/\varepsilon)/\varepsilon^2)$ samples. Furthermore, this result is almost optimal in the sense that $\Omega(n/\log^2 n)$ samples are necessary [10].

Our main contribution is the analysis of this problem in the setting when $\mathfrak{D}$ is a monotone distribution on an n-dimensional (*real*) cube $[0, 1]^n$. A distribution $\mathfrak{D}$ on $[0, 1]^n$ with density function $\mathfrak{f}$ is *monotone* if for any $\mathbf{x} = (x_1, \ldots, x_n)$, $\mathbf{y} = (y_1, \ldots, y_n)$, if $x_i \leq y_i$ for every i then $\mathfrak{f}(\mathbf{x}) \leq \mathfrak{f}(\mathbf{y})$. On high-level our approach is similar to that used by Rubinfeld and Servedio [10] in the case of Boolean n-cubes. However, the fact that we have to deal with continuous domain makes our proof of the main result, Lemma 1, more complicated.

We characterize monotone distributions that are ε-far from uniform:

Lemma 1. *Let $\mathfrak{D}$ be a monotone distribution on $[0, 1]^n$ with density function $\mathfrak{f}$. If $\mathfrak{D}$ is ε-far from uniform then*

$$\mathbb{E}_{\mathfrak{f}}[\|\mathbf{x}\|_1] = \int_{\mathbf{x}} \|\mathbf{x}\|_1 \cdot \mathfrak{f}(\mathbf{x}) \, d\mathbf{x} \geq \frac{n}{2} + \frac{\varepsilon}{2} \ .$$

The proof of this lemma can be deduced from the following result (which is the main technical contribution of the paper; for a proof, see [1]) by substituting $\mathfrak{g}(\mathbf{x}) = \mathfrak{f}(\mathbf{x}) - 1$.

Lemma 2. *Let $\mathfrak{g}: [0, 1]^n \to \mathbb{R}$ be a monotone function with $\int_{\mathbf{x}} \mathfrak{g}(\mathbf{x}) \, d\mathbf{x} = 0$. Then*

$$\int_{\mathbf{x}} \|\mathbf{x}\|_1 \cdot \mathfrak{g}(\mathbf{x}) \, d\mathbf{x} \geq \frac{1}{4} \int_{\mathbf{x}} |\mathfrak{g}(\mathbf{x})| \, d\mathbf{x} \ .$$

By combining Lemma 1 with the fact that for uniform distribution $\mathfrak{Q}$ on $[0, 1]^n$ we have $\mathbb{E}_{\mathfrak{Q}}[\|\mathbf{x}\|_1] = \frac{n}{2}$, we can show that the following simple algorithm tests if a distribution is uniform or it is ε-far from uniform:

Testing uniformity:

- **Repeat** $r = 20$ times:
  Draw a sample (according to the distribution $\mathfrak{D}$) $S = \langle \mathbf{x}_1, \ldots, \mathbf{x}_s \rangle$ from $[0, 1]^n$ with $s = \lceil \frac{40n}{\varepsilon^2} \rceil$
  If $\sum_{i=1}^s \|\mathbf{x}_i\|_1 \geq s(\frac{n}{2} + \frac{\varepsilon}{4})$ then **Reject** and exit
- **Accept**

---

[1] Distribution $\mathfrak{D}$ *is* monotone if for any $\mathbf{x} = (x_1, \ldots, x_n)$, $\mathbf{y} = (y_1, \ldots, y_n)$, if $x_i \leq y_i$ for every i then $\mathbf{Pr}_{\mathfrak{D}}[\mathbf{x}] \leq \mathbf{Pr}_{\mathfrak{D}}[\mathbf{y}]$.

Theorem 1. Testing uniformity *distinguishes between uniform distribution on* $[0,1]^n$ *and any monotone distribution over* $[0,1]^n$ *that is ε-far from uniform. Its sample complexity is $\mathcal{O}(n/\varepsilon^2)$ and it errs with probability at most $\frac{1}{4}$.*

Our analysis does not only extend the result from [10] to real cubes, but also leads to an algorithm slightly faster than that from [10] (we shave off an $\mathcal{O}(\log(n/\varepsilon))$ factor) for both the Boolean and real hypercube. We observe that since the lower bound from [10] can be directly carried over to the case of real $[0,1]^n$ cubes, our upper bound is almost optimal.

Let us also notice that using a very similar analysis, our tester will work with the same complexity if the input is a monotone distribution on a discrete cube $\{0,1,\ldots,k\}^n$. The obtained sample size is independent of k:

Theorem 2. *Let k be any positive integer and consider any n-dimensional finite grid $\{0,1,2,\ldots,k\}^n$. One can test if a given monotone distribution $\mathfrak{D}$ over $\{0,1,2,\ldots,k\}^n$ is uniform with $\mathcal{O}(n/\varepsilon^2)$ samples.*

References

1. Adamaszek, M., Czumaj, A., Sohler, C.: Testing monotone continuous distributions on high-dimensional real cubes. In: Proc. 21st Annual ACM-SIAM Symposium on Discrete Algorithms, pp. 56–65 (2010)
2. Alon, N., Andoni, A., Kaufman, T., Matulef, K., Rubinfeld, R., Xie, N.: Testing k-wise and almost k-wise Independence. In: Proc. 39th Annual ACM Symposium on Theory of Computing, pp. 496–505 (2007)
3. Batu, T., Dasgupta, S., Kumar, R., Rubinfeld, R.: The complexity of approximating the entropy. In: Proc. 34th Annual ACM Symposium on Theory of Computing, pp. 678–687 (2002)
4. Batu, T., Fischer, E., Fortnow, L., Kumar, R., Rubinfeld, R., White, P.: Testing random variables for independence and identity. In: Proc. 42nd IEEE Symposium on Foundations of Computer Science, pp. 442–415 (2001)
5. Batu, T., Fortnow, L., Rubinfeld, R., Smith, W.D., White, P.: Testing that distributions are close. In: Proc. 41st IEEE Symposium on Foundations of Computer Science, pp. 259–269 (2000)
6. Batu, T., Kumar, R., Rubinfeld, R.: Sublinear algorithms for testing monotone and unimodal distributions. In: Proc. 36th Annual ACM Symposium on Theory of Computing, pp. 381–390 (2004)
7. Goldreich, O., Ron, D.: On testing expansion in bounded-degree graphs. Electronic Colloquium on Computational Complexity, Report No. 7 (2000)
8. Raskhodnikova, S., Ron, D., Shpilka, A., Smith, A.: Strong lower bounds for approximating distribution support size and the distinct elements problem. SIAM Journal on Computing 39(3), 813–842 (2009)
9. Rubinfeld, R.: Sublinear time algorithms. In: Proc. International Congress of Mathematicians, Madrid, Spain, August 22-30 (2006)
10. Rubinfeld, R., Servedio, R.A.: Testing monotone high-dimensional distributions. In: Proc. 37th Annual ACM Symposium on Theory of Computing, pp. 147–156 (2005)
11. Valiant, P.: Testing symmetric properties of distributions. In: Proc. 40th Annual ACM Symposium on Theory of Computing, pp. 383–392 (2008)

On Constant Time Approximation of Parameters of Bounded Degree Graphs

Noga Alon

Schools of Mathematics and Computer Science, Raymond and Beverly Sackler
Faculty of Exact Sciences, Tel Aviv University, Tel Aviv 69978, Israel
nogaa@tau.ac.il

Abstract. How well can the maximum size of an independent set, or
the minimum size of a dominating set of a graph in which all degrees are
at most d be approximated by a randomized constant time algorithm ?
Motivated by results and questions of Nguyen and Onak, and of Parnas, Ron and Trevisan, we show that the best approximation ratio that
can be achieved for the first question (independence number) is between
$\Omega(d/\log d)$ and $O(d \log \log d/\log d)$, whereas the answer to the second
(domination number) is $(1 + o(1)) \ln d$.

Keywords: independence number of a graph, dominating set in a graph,
constant time approximation.

1 Introduction

The question of identifying the properties of bounded degree graphs in the model
of [7] that can be tested efficiently, is that of recognizing the properties that are
local in nature. These are properties for which the local structure of the graph
supplies meaningful information about the global property. A related problem
deals with efficient approximation algorithms for graph parameters, like the independence number, or the domination number of a given bounded degree graph.
The question here is to decide how well we can approximate these quantities by
observing the local structure of the graph. In this short paper we discuss several
problems of this type, continuing the work in several earlier papers including
[13] and [12] on related questions.

1.1 Notation and Definitions

Following [13] and [12] we call a real number $\bar{y}$ an (α, β)-approximation for a number y if $y \le \bar{y} \le \alpha y + \beta$. A randomized algorithm A is an (α, β)-approximation
algorithm for a graph parameter $P(G)$ if given an input graph G the algorithm
computes a value y which is an (α, β)-approximation for $P(G)$ with probability
at least $2/3$ for any proper input graph G. Let $G(n, d)$ denote the family of all
graphs on n vertices with maximum degree at most d, represented by their adjacency lists. In this note we are interested in randomized $(\alpha, \epsilon n)$-approximation
algorithms for some graph parameters, that work on input graphs $G \in G(n, d)$ in

O. Goldreich (Ed.): Property Testing, LNCS 6390, pp. 234–239, 2010.

constant expected time, that is, in time bounded by a function $f = f(d, \epsilon)$ of d and ϵ only, which is independent of n. In order to make the discussion cleaner, we will not be interested in the precise function f as long as it is independent of n. Our objective is to try to determine or estimate, for a given graph parameter P, the best possible α so that for any positive $\epsilon > 0$ there is a constant time, randomized $(\alpha, \epsilon n)$-approximation algorithm for $P(G)$ for inputs $G \in G(n, d)$. In some cases the same α actually works for $\epsilon = 0$ as well.

We note that it is possible to replace the success probability $2/3$ by any larger number smaller than 1 (by running the algorithm several times, taking the median of all values computed) and it is also possible to replace the expected running time by worst case running time by stopping the algorithm if its running time exceeds its expectation by a large constant factor.

1.2   Examples

- Simple sampling of vertices, checking their degrees, shows that for every $\epsilon > 0$ there is a constant time randomized $(1, \epsilon n)$ approximation algorithm for estimating the sum of degrees of a graph $G \in G(n, d)$. The analysis is essentially trivial, and the example is listed here mainly to practice the definitions. A similar approach provides a constant time randomized $(1, \epsilon n)$ approximation algorithm for the number of triangles (or copies of any fixed connected graph) in a given $G \in G(n, d)$.
- As shown in [13] and [11] for every $\epsilon > 0$ there is a constant time randomized $(2, \epsilon n)$ approximation algorithm for the vertex cover of a graph $G \in G(n, d)$. Trevisan (c.f. [13]) observed that there is no such $(2 - \delta, \epsilon n)$-approximation algorithm, for any fixed $\epsilon < \delta$.
- In [12] it is proved that for every $\epsilon > 0$ there is a constant time randomized $(H(d + 1), \epsilon n)$ approximation algorithm for the domination number of a graph $G \in G(n, d)$, where $H(d + 1) = 1 + \frac{1}{2} + \ldots + \frac{1}{d+1} = \ln d + \Theta(1)$ and the domination number of $G = (V, E)$ is the minimum cardinality of a set of vertices $U \subset V$ so that each vertex $v \in V - U$ has a neighbor in U.
- Another result proved in [12] is that for every $\epsilon > 0$ there is a constant time randomized $(1, \epsilon n)$ approximation algorithm for the maximum size of a matching in a graph $G \in G(n, d)$.

1.3   A Useful Tool

The basic tool applied in the algorithms of [13] and [12] is the result that for every $\epsilon > 0$ there is a constant time randomized algorithm that computes, for a given $G \in G(n, d)$, a $(1, \epsilon n)$-approximation of the size of some maximal (with respect to containment) independent set in G. The same proof provides such an approximation even in the weighted case, which we'll need here, where the weight of each vertex is, say, an integer between 1 and d.

1.4   The New Results

Our first result deals with approximation of domination numbers, showing that the $(H(d + 1), \epsilon n)$ approximation proved in [12] is essentially tight.

Theorem 1.1. *The smallest α so that for every $\epsilon > 0$ there is a constant time randomized $(\alpha, \epsilon n)$ approximation algorithm for the domination number of a graph $G \in G(n, d)$ is $(1 + o(1)) \ln d$, where the $o(1)$-term tends to zero as d tends to infinity.*

The second result is about approximation of independence numbers.

Theorem 1.2. *There are two positive constants c_1, c_2 so that the following holds. The smallest α so that for every $\epsilon > 0$ there is a constant time randomized $(\alpha, \epsilon n)$ approximation algorithm for the independence number of a graph $G \in G(n, d)$ is at least $c_1 \frac{d}{\log d}$, and at most $c_2 \frac{d \log \log d}{\log d}$.*

2 Proofs

In this section we present the proofs of the two theorems stated above. Recall that the girth of a graph G is the length of a shortest cycle in it. We start with the proof of Theorem 1.1.

2.1 Dominating Set

Lemma 2.1. *For every fixed d and large n divisible by $2d + 2$, there are two d-regular graphs $G = (V, E)$ and $G' = (V', E')$ in $G(n, d)$ so that the following holds.*
(i) The girth of G is at least, say, $0.5 \log n / \log d$ and its domination number is precisely $\frac{n}{d+1}$.
(ii) The girth of G' is at least $0.5 \log n / \log d$ and its domination number is $(1 + o(1)) \frac{n \ln d}{d}$.

Proof: (i) Start with an arbitrary set of $n/(d+1)$ pairwise vertex disjoint stars, each containing d edges. Let U denote the set of centers of these stars, and let W denote the set of all their end vertices. Note that $|W| = \frac{nd}{d+1}$ is even, and thus there is a $d - 1$-regular graph H on the set W. Let G be the d-regular graph on the n vertices $U \cup W$ consisting of all edges of the initial stars as well as all edges of H. The domination number of G is clearly $n/(d+1)$, as U is a dominating set in it. The girth, however, can be small, and our objective is to show that it can be increased without changing the domination number. This will be done by modifying the graph G, without touching the edges of the stars.

The method is similar to that of Erdős and Sachs in [6]. Call a cycle short if its length is at most $0.5 \log n / \log d$. As long as there is a short cycle, take arbitrarily a shortest cycle C, and an arbitrary non-star edge in it uv. Since the graph is d-regular, there is another non-star edge xy in it whose distance from uv is at least $\log n / \log d$. Omit the two edges uv and xy from G and replace them by the two new edges xu and yv. It is easy to check that this replacement does not create any new cycles of length at most that of C, and destroys the cycle C itself. Continuing in this manner until there are no short cycles left we obtain the desired graph G.

(ii) Let G' be a random d-regular graph on n vertices. It is known that with high probability the domination number of G' is $(1 + o(1))\frac{n \ln d}{d}$ (see, e.g., [4]). It is also not difficult to check that the expected number of cycles of length at most $0.5 \log n / \log d$ in G' is at most $\sqrt{n}$, and hence, by Markov's Inequality, with probability at least 0.5 there are at most $2\sqrt{n}$ such cycles. Fix a graph with domination number $(1 + o(1))\frac{n \ln d}{d}$ and at most $2\sqrt{n}$ short cycles, and modify it according to the process described in the proof of part (i), destroying all short cycles. Since each modification step touches at most 2 edges, and cannot create too many short cycles, the domination number changes during this process by at most $o(n)$, providing the desired result. □

Proof of Theorem 1.1: The existence of the required approximation algorithm is proved in [12], as described in Section 1. It remains to show that no better approximation is possible.

Let G and G' be as in Lemma 2.1 and consider two distributions on graphs in $G(n, d)$: the first is a permuted copy of G, and the second is a permuted copy of G'. By the girth and regularity conditions the statistics of subgraphs of less than $0.5 \log n / \log d$ vertices and edges in both distributions is identical. Hence no constant time algorithm can distinguish between them. This completes the proof. Note that the two distributions here are not only computationally indistinguishable for randomized constant time algorithms, but are completely identical. It is possible to use computational indistinguishability here and show that in fact in order to obtain an $(\alpha, \epsilon n)$-approximation for the domination of $G \in G(n, d)$, where $\alpha < (1 - \delta) \ln d$ and $\epsilon < \epsilon(\delta)$, one needs to inspect at least $\Omega(\sqrt{n})$ vertices and edges, but as we care here only about algorithms whose running time is independent of n, we do not include the detailed analysis of this stronger claim. □

2.2 Independence Number

Proof of Theorem 1.2: The proof of the lower bound is simple: there is a d-regular bipartite graph G on n vertices with girth $\Omega(\log n)$ (and independence number $n/2$), and it is well known that a random d-regular graph on n vertices has, with high probability, independence number at most $O(\frac{n \log d}{d})$ and only a small number of cycles of length shorter than $0.5 \log n / \log d$. We can thus modify the graph as in the previous subsection and get a d-regular graph G' on n vertices with girth at least $\Omega(\log n / \log d)$ and independence number at most $O(\frac{n \log d}{d})$. As in the previous proof, no constant time algorithm will be able to distinguish between a permuted copy of G and a permuted copy of G', providing the lower bound.

The proof of the upper bound requires a bit more work. It is, in fact, non trivial to get any constant time $(\alpha, \epsilon n)$- approximation algorithm to the independence number, with $\alpha = o(d)$. We first describe a sequential deterministic algorithm and then observe that it can be converted into a randomized, constant time procedure.

Let $G \in G(n, d)$ be a given input graph. As long as there is a nonempty set X of vertices of G with a common neighbor in the graph, satisfying $|X| \leq \log^3 d$, so that the induced subgraph on X contains no independent set of size at least $|X|/\log d$, omit it. (When there are many choices for such a set X, pick one arbitrarily). Suppose that when there is no such set X left, there are t vertices in the remaining graph. By the result in [2] (see the remark following the proof of Theorem 1.1 in [2]), the induced graph left on the t remaining vertices contains an independent set of size at least $\Omega(\frac{t \log d}{d \log \log d})$. It is also obvious that G has an independent set of size at least $\frac{n}{d+1}$. Thus, the independence number of G is at least the average between these two, that is, at least $\Omega(\frac{n}{d+1} + \frac{t \log d}{d \log \log d})$. On the other hand we know that there is no independent set of size bigger than $\frac{n-t}{\log d} + t$, providing the required approximation if the value of t is known with sufficient accuracy.

It thus remains to show how to approximate t in randomized constant time. Define an auxiliary weighted graph F whose set of vertices is the set of all nonempty subsets X of at most $\log^3 d$ vertices of G with a common neighbor, so that the induced subgraph of G on X contains no independent set of size at least $|X|/\log d$. Two such vertices X and X' of F are adjacent iff the two sets X and X' have a nonempty intersection. The weight of each vertex X is the cardinality $|X|$ of the subset corresponding to it. Note that one can easily check the adjacency relations in the graph F by observing the original graph G locally. We can therefore find, in randomized constant time, a good approximation to the weight of a maximal (with respect to containment) independent set of vertices of F, which will enable us to approximate the value of t defined in the sequential procedure described above. This completes the proof.   □

3   Concluding Remarks and Open Problems

We have investigated the best possible approximation ratios that can be obtained for two graph parameters by randomized, constant time algorithms on bounded degree graphs represented by their adjacency lists. The problem for domination number is quite well understood, whereas in the case of independence number there is still a $\Theta(\log \log d)$ gap between the upper and lower bounds. It will be interesting to close this gap. We suspect that the $\log \log d$ term can be omitted, but this will require an additional argument. It is worth noting that the best known polynomial time algorithm for approximating the independence number of graphs $G \in G(n, d)$ provides an approximation ratio of $\Theta(d \log \log d / \log d)$, matching the approximation ratio obtained here. This has been found by Vishwanathan (first recorded in [8], see also [9]), and by Halperin [10], and is based on the method of [3] that applies semidefinite programming. It seems unlikely that these algorithms can be converted into constant time randomized ones. Austrin, Khot and Safra [5] have recently proved a hardness result of $\Omega(d/\log^2 d)$ for the problem, under the unique games conjecture.

A purely combinatorial problem, that seems related to the question above (although we do not know any direct relation) is the conjecture raised in [1] that for any fixed graph H, any graph $G \in G(n, d)$ that contains no copy of H has an independent set of size at least $c_H \frac{n \log d}{d}$. Here, too, the best known result, due to Shearer [14], is off by a factor of $\log \log d$, and it is only known that the independence number of any such graph is at least $c_H \frac{n \log d}{d \log \log d}$.

Acknowledgments. I would like to thank Krzysztof Onak for helpful discussions, comments and suggestions, and Per Austrin for pointing out several relevant references. Research supported in part by an ERC Advanced grant and by a USA-Israeli BSF grant.

References

1. Ajtai, M., Erdös, P., Komlós, J., Szemerédi, E.: On Turan's theorem for sparse graphs. Combinatorica 1, 313–317 (1981)
2. Alon, N.: Independence numbers of locally sparse graphs and a Ramsey type problem. Random Structures and Algorithms 9, 271–278 (1996)
3. Alon, N., Kahale, N.: Approximating the independence number via the θ-function. Math. Programming 80, 253–264 (1998)
4. Alon, N., Wormald, N.: High degree graphs contain large-star factors. In: Katona, G., Schrijver, A., Szönyi, T. (eds.) Fete of Combinatorics, Bolyai Soc. Math. Studies 20, pp. 9–21. Springer, Heidelberg (2010)
5. Austrin, P., Khot, S., Safra, M.: Inapproximability of Vertex Cover and Independent Set in Bounded Degree Graphs. In: IEEE Conference on Computational Complexity 2009, pp. 74–80 (2009)
6. Erdös, P., Sachs, H.: Reguläre Graphen gegebener Taillenweite mit minimaler Knotenzahl (German). Wiss. Z. Martin-Luther-Univ. Halle-Wittenberg Math.-Natur. Reihe 12, 251–257 (1963)
7. Goldreich, O., Ron, D.: Property testing in bounded degree graphs. In: Proc. 29th STOC, pp. 406–415 (1997)
8. Halldórsson, M.M.: Approximations of Independent Sets in Graphs. In: Jansen, K., Rolim, J.D.P. (eds.) APPROX 1998. LNCS, vol. 1444, pp. 1–13. Springer, Heidelberg (1998)
9. Halldórsson, M.M.: Approximations of Weighted Independent Set and Hereditary Subset Problems. Journal of Graphs Algorithms and Applications 4, 1–16 (2000)
10. Halperin, E.: Improved approximation algorithms for the vertex cover problem in graphs and hypergraphs. In: Proc. Eleventh ACM-SIAM Symp. on Discrete Algorithms, pp. 329–337 (2000)
11. Marko, S., Ron, D.: Distance approximation in bounded-degree and general sparse graphs. In: Díaz, J., Jansen, K., Rolim, J.D.P., Zwick, U. (eds.) APPROX 2006 and RANDOM 2006. LNCS, vol. 4110, pp. 475–486. Springer, Heidelberg (2006)
12. Nguyen, H.N., Onak, K.: Constant-Time Approximation Algorithms via Local Improvements. In: Proc. 49th Annual Symposium on Foundations of Computer Science (FOCS 2008), pp. 327–336 (2008)
13. Parnas, M., Ron, D.: Approximating the minimum vertex cover in sublinear time and a connection to distributed algorithms. Theoret. Comput. Sci. 381(1-3), 183–196 (2007)
14. Shearer, J.B.: On the independence number of sparse graphs. Random Structures and Algorithms 7, 269–271 (1995)

Sublinear Algorithms in the External Memory Model*

Alexandr Andoni[1], Piotr Indyk[2], Krzysztof Onak[2], and Ronitt Rubinfeld[2,3]

[1] Princeton University and Center for Computational Intractability,
Princeton, NJ, USA
[2] Massachusetts Institute of Technology, Cambridge, MA, USA
[3] Tel-Aviv University, Tel Aviv, Israel

Abstract. We initiate the study of sublinear-time algorithms in the external memory model. In this model, the data is stored in blocks of a certain size B, and the algorithm is charged a unit cost for each block access. This model is well-studied, since it reflects the computational issues occurring when the (massive) input is stored on a disk. Since each block access operates on B data elements in parallel, many problems have external memory algorithms whose number of block accesses is only a small fraction (e.g. $1/B$) of their main memory complexity.

However, to the best of our knowledge, no such reduction in complexity is known for *any* sublinear-time algorithm. One plausible explanation is that the vast majority of sublinear-time algorithms use random sampling and thus exhibit no locality of reference. This state of affairs is quite unfortunate, since both sublinear-time algorithms and the external memory model are important approaches to dealing with massive data sets, and ideally they should be combined to achieve best performance.

We show that such combination is indeed possible. In particular, we consider three well-studied problems: testing of *distinctness*, *uniformity* and *identity* of an empirical distribution induced by data. For these problems we show random-sampling-based algorithms whose number of block accesses is up to a factor of $1/\sqrt{B}$ smaller than the main memory complexity of those problems. We also show that this improvement is optimal for those problems.

Since these problems are natural primitives for a number of sampling-based algorithms for other problems, our tools improve the external memory complexity of other problems as well.

Keywords: external memory, sampling, distribution testing.

1 Introduction

Random sampling is one of the most fundamental methods for reducing task complexity. For a wide variety of problems, it is possible to infer an approximate

* The research was supported in part by David and Lucille Packard Fellowship, by MADALGO (Center for Massive Data Algorithmics, funded by the Danish National Research Association), by Marie Curie IRG Grant 231077, by NSF grants 0514771, 0728645, and 0732334, and by a Symantec Research Fellowship.

O. Goldreich (Ed.): Property Testing, LNCS 6390, pp. 240–243, 2010.

solution from a random sample containing only a small fraction of the data, yielding algorithms with sublinear running times. As a result, sampling is often the method of choice for processing massive data sets. Inferring properties of data from random sample has been a major subject of study in several areas, including statistics, databases [1,2], theoretical computer science [3,4,5,6], ...

However, using random sampling for massive data sets encounters the following problem: typically, massive data sets are not stored in main memory, where each element can be accessed at a unit cost. Instead, the data is stored on external storage devices, such as a hard disk. There, the data is stored in blocks of certain size (say, B), and each disk access returns a block of data, as opposed to an individual element. In such models [7], it is often possible to solve problems using roughly T/B disk accesses, where T is the time needed to solve the problem in main memory. The $1/B$ factor is often crucial to the efficiency of the algorithms, given that (a) the block size B tends to be large, on the order of thousands and (b) each block access is many orders of magnitude slower than a main memory lookup. Unfortunately, implementations of sampling algorithms typically need to perform[1] one block access per each sampled element [1]. Effectively, this means that out of B data elements retrieved by each block access, $B - 1$ elements are discarded by the algorithm. This makes sampling algorithms a much less attractive option for processing massive data sets.

Is it possible to improve the sampling algorithms by utilizing the *entire* information stored in each accessed block? At the first sight, it might not seem so. For example, consider the following basic sampling problem: the input data is a binary sequence such that the fraction of ones is either at most f or at least $2f$, and the goal is to detect which of these two cases occurs. A simple argument shows that any sampling algorithm for this problem requires $\Omega(1/f)$ samples to succeed with constant probability, since it may take that many trials to even retrieve one 1. It is also easy to observe that the same lower bound holds even if all elements within each block are equal (as long as the total number of blocks is $\Omega(1/f)$), in which case sampling blocks is equivalent to sampling elements. Thus, even for this simple problem, sampling blocks does not yield any reduction in the number of accesses.

2   Our Results

Contrary to the above impression, we show that there are natural problems for which it is possible to reduce the number of sampled blocks. Specifically, we consider the problem of testing properties of empirical distributions induced by the data sets. Consider a data set of size m with support size (i.e., the number of distinct elements) equal to n. Let p_i be the fraction of times an element i occurs in the data set. The vector p then defines a probability distribution over a set of distinct elements in the data set. We address the following three well-studied problems:

---

[1] It is possible to retrieve more samples per block if the data happens to be stored in a random order. Unfortunately, this is typically not guaranteed.

- Distinctness: are all data elements distinct (i.e., $n = m$), or are there at least ϵm duplicates?
- Uniformity: is p uniform over its support, or is it ϵ-far[2] from the uniform distribution?
- Identity: is p identical to an explicitly given distribution q, or is it ϵ-far from q?

Note that testing identity generalizes the first two problems. However, the algorithms for distinctness and uniformity are simpler and easier to describe.

It is known [8,9,10] that, if the elements are stored in main memory, then $\tilde{\Theta}(\sqrt{n})$ memory accesses are sufficient and necessary to solve both uniformity and identity testing. We give an external memory algorithm which uses only $\tilde{O}(\sqrt{m/B})$ block accesses. Thus, for m comparable to n, the number of accesses is reduced by a factor of $\sqrt{B}$. It also can be seen that this bound cannot be improved in general: if $B = m/n$, then each block could consist of equal elements, and thus the $\tilde{\Theta}(\sqrt{n}) = \tilde{\Theta}(\sqrt{m/B})$ main memory lower bound would apply.

From the technical perspective, our algorithms mimic the sampling algorithms of [10,11,9]. The key technical contribution is a careful analysis of those algorithms. In particular, we show that the additional information obtained from sampling blocks of data (as opposed to the individual elements) yields a substantial reduction of the variance of the estimators used by those algorithms.

3   Applications to Other Problems

The three problems from above are natural primitives for a number of other sampling-based problems. Thus, our algorithms improve the external memory complexity of other problems as well. Below we describe two examples of problems where our algorithms and techniques apply immediately to give improved guarantees in the external memory model.

The first such problem is testing graph isomorphism. In this problem, the tester is to decide, given two graphs G and H on n vertices, whether G are H are isomorphic or at least ϵn^2 edges of the graphs must be modified to achieve a pair of isomorphic graphs. Suppose one graph, G, is known to the tester (for instance, it is a fixed graph with an easily computable adjacency relation), and the other graph, H, is described by the adjacency matrix written in the row-major order on the disk. Then, our algorithm for identity testing improves the sample complexity of the Fischer and Matsliah algorithm [12] by essentially a factor of $\sqrt{B}$. Formally, in the main memory, the Fischer and Matsliah algorithm uses $O(\sqrt{n} \cdot \mathrm{poly}(\log n, 1/\epsilon))$ queries to H. Combined with our external memory identity tester, algorithm will use only $O((\sqrt{n/B}+1) \cdot \mathrm{poly}(\log n, 1/\epsilon))$ samples.

---

[2] We measure the distance between distribution using the standard variational distance, which is the maximum probability with which a statistical test can distinguish the two distributions. Formally, a distribution p is ϵ-far from a distribution q, if $\|p - q\|_1 \geq \epsilon$, where p and q are interpreted as vectors.

The second application is a set of questions on testing various properties of metric spaces, such as testing whether a metric is a tree-metric or ultrametric. In [13], Onak considers several such properties, for which he gives algorithms whose sampling complexity in main memory is of the form $O(\alpha/\epsilon + n^{(\beta-1)/\beta}/\epsilon^{1/\beta})$, where $\alpha \geq 1$ and $\beta \geq 2$ are constant integers. The additive term $n^{(\beta-1)/\beta}/\epsilon^{1/\beta}$ corresponds to sampling for a specific β-tuple. Using our techniques for distinctness testing, it can easily be shown that whenever an algorithm from [13] requires $O(\alpha/\epsilon + n^{(\beta-1)/\beta}/\epsilon^{1/\beta})$ samples, the sample complexity in external memory can be improved to $O(\alpha/\epsilon + (n/B)^{(\beta-1)/\beta}/\epsilon^{1/\beta})$, provided a single disk block contains B points.

References

1. Olken, F., Rotem, D.: Simple random sampling from relational databases. In: VLDB, pp. 160–169 (1986)
2. Olken, F.: Random Sampling from Databases. PhD thesis, U.C. Berkeley (1993)
3. Fischer, E.: The art of uninformed decisions: A primer to property testing. Bulletin of the European Association for Theoretical Computer Science 75, 97–126 (2001)
4. Ron, D.: Property testing (a tutorial). In: Rajasekaran, S., Pardalos, P.M., Reif, J.H., Rolim, J.D.P. (eds.) Handbook on Randomization, vol. II, pp. 597–649. Kluwer Academic Press, Dordrecht (2001)
5. Goldreich, O.: Combinatorial property testing—a survey. In: Randomization Methods in Algorithm Design, pp. 45–60 (1998)
6. Bar-Yossef, Z., Kumar, R., Sivakumar, D.: Sampling algorithms: lower bounds and applications. In: STOC, pp. 266–275 (2001)
7. Vitter, J.S.: External memory algorithms and data structures. ACM Comput. Surv. 33(2), 209–271 (2001)
8. Goldreich, O., Ron, D.: On testing expansion in bounded-degree graphs. Electronic Colloqium on Computational Complexity 7(20) (2000)
9. Batu, T.: Testing Properties of Distributions. PhD thesis, Cornell University (August 2001)
10. Batu, T., Fortnow, L., Rubinfeld, R., Smith, W.D., White, P.: Testing that distributions are close. In: FOCS, pp. 259–269 (2000)
11. Batu, T., Fortnow, L., Fischer, E., Kumar, R., Rubinfeld, R., White, P.: Testing random variables for independence and identity. In: FOCS, pp. 442–451 (2001)
12. Fischer, E., Matsliah, A.: Testing graph isomorphism. SIAM J. Comput. 38(1), 207–225 (2008)
13. Onak, K.: Testing properties of sets of points in metric spaces. In: Aceto, L., Damgård, I., Goldberg, L.A., Halldórsson, M.M., Ingólfsdóttir, A., Walukiewicz, I. (eds.) ICALP 2008, Part I. LNCS, vol. 5125, pp. 515–526. Springer, Heidelberg (2008)

Polylogarithmic Approximation for Edit Distance and the Asymmetric Query Complexity*

Alexandr Andoni[1], Robert Krauthgamer[2], and Krzysztof Onak[3]

[1] Princeton University and Center for Computational Intractability,
Princeton, NJ, USA
[2] The Weizmann Institute of Science, Rehovot, Israel
[3] Massachusetts Institute of Technology, Cambridge, MA, USA

Abstract. We present a near-linear time algorithm that approximates the edit distance between two strings within a polylogarithmic factor. More precisely, for strings of length n and every fixed $\varepsilon > 0$, it can compute a $(\log n)^{O(1/\varepsilon)}$-approximation in $n^{1+\varepsilon}$ time.

This result arises naturally in the study of a new *asymmetric query* model. In this model, the input consists of two strings x and y, and an algorithm can access y in an unrestricted manner, while being charged for querying every symbol of x. Our query lower bound for this model provides the first rigorous separation between edit distance and Ulam distance, which is edit distance on non-repetitive strings, i.e., permutations.

Keywords: edit distance, sampling, query complexity.

1 Introduction

Manipulation of strings has long been central to computer science, arising from the high demand to process texts and other sequences efficiently. For example, for the simple task of *comparing* two strings (sequences), one of the first methods emerged to be the *edit distance* (aka the Levenshtein distance) [1], defined as the minimum number of character insertions, deletions, and substitutions needed to transform one string into the other. This basic distance measure, together with its more elaborate versions, is widely used in a variety of areas such as computational biology, speech recognition, and information retrieval. Consequently, improvements in edit distance algorithms have the potential of major impact. As a result, computational problems involving edit distance have been studied extensively (see [2,3] and references therein).

The most basic problem is that of computing the edit distance between two strings of length n over some alphabet. It can be solved in $O(n^2)$ time by a classical algorithm [4]; in fact this is a prototypical dynamic programming algorithm,

---

* Alexandr Andoni was supported in part by NSF CCF 0832797. Robert Krauthgamer was supported in part by the Israel Science Foundation (grant #452/08), and by a Minerva grant. Krzysztof Onak was supported in part by NSF grants 0732334 and 0728645.

O. Goldreich (Ed.): Property Testing, LNCS 6390, pp. 244–252, 2010.

see, e.g., the textbook [5] and references therein. Despite significant research over more than three decades, this running time has so far been improved only slightly to $O(n^2 / \log^2 n)$ [6], which remains the fastest algorithm known to date.[1]

Still, a near-quadratic runtime is often unacceptable in modern applications that must deal with massive datasets, such as the genomic data. Hence practitioners tend to rely on faster heuristics [3,2]. This has motivated the quest for faster algorithms at the expense of approximation, see, e.g., [8, Section 6] and [9, Section 8.3.2]. Indeed, the past decade has seen a serious effort in this direction.[2] One general approach is to design linear time algorithms that approximate the edit distance. A linear-time $\sqrt{n}$-approximation algorithm immediately follows from the exact algorithm of [11], which runs in time $O(n+d^2)$, where d is the edit distance between the input strings. Subsequent research improved the approximation factor, first to $n^{3/7}$ [12], then to $n^{1/3+o(1)}$ [13], and finally to $2^{\tilde{O}(\sqrt{\log n})}$ [14] (building on [15]). Predating some of this work was the *sublinear-time* algorithm of [16] achieving n^ε approximation, but only when the edit distance d is rather large.

Better progress has been obtained on *variants* of edit distance, where one either restricts the input strings, or allows additional edit operations. An example from the first category is the edit distance on non-repetitive strings (e.g., permutations of $[n]$), termed *the Ulam distance* in the literature. The classical Patience Sorting algorithm computes the exact Ulam distance between two strings in $O(n \log n)$ time. An example in the second category is the case of two variants of the edit distance where certain block operations are allowed. Both of these variants admit an $\tilde{O}(\log n)$ approximation in near-linear time [17,18,19,20].

Despite the efforts, achieving a polylogarithmic approximation factor for the classical edit distance has eluded researchers for a long time. In fact, this is has been the case not only in the context of linear-time algorithms, but also in the related tasks, such as nearest neighbor search, ℓ_1-embedding, or sketching. From a lower bounds perspective, only a *sublogarithmic* approximation has been ruled out for the latter two tasks [21,22,23], thus giving evidence that a sublogarithmic approximation for the distance computation might be much harder or even impossible to attain.

2   Results

Our first and main result is an algorithm that runs in near-linear time and approximates edit distance within a *polylogarithmic factor*. Note that this is *exponentially better* than the previously known factor $2^{\tilde{O}(\sqrt{\log n})}$ (in comparable running time), due to [15,14].

---

[1] The result of [6] applies to constant-size alphabets. It was recently extended to arbitrarily large alphabets, albeit with an $O(\log \log n)^2$ factor loss in runtime [7].

[2] We shall not attempt to present a complete list of results for restricted settings (e.g., average-case/smoothed analysis, weakly-repetitive strings, and bounded distance-regime), for variants of the distance function (e.g., allowing more edit operations), or for related computational problems (such as pattern matching, nearest neighbor search, and sketching). See also the surveys of [2] and [10].

Theorem 1 (Main). *For every fixed $\varepsilon > 0$, there is an algorithm that approximates the edit distance between two input strings $x, y \in \Sigma^n$ within a factor of $(\log n)^{O(1/\varepsilon)}$, and runs in $n^{1+\varepsilon}$ time.*

This development stems from a principled study of edit distance in a computational model that we call the *asymmetric query* model, and which we shall define shortly. Specifically, we design a query-efficient procedure in the said model, and then show how this procedure yields a near-linear time algorithm. We also provide a query complexity lower bound for this model, which matches or nearly-matches the performance of our procedure.

A conceptual contribution of our query complexity lower bound is that it is the first one to expose hardness stemming from "repetitive substrings", which means that many small substrings of a string may be approximately equal. Empirically, it is well-recognized that such repetitiveness is a major obstacle for designing efficient algorithms. All previous lower bounds (in any computational model) failed to exploit it, while in our proof the strings' repetitive structure is readily apparent. More formally, our lower bound provides the first rigorous separation of edit distance from Ulam distance (edit distance on non-repetitive strings). Such a separation was not previously known in any studied model of computation, and in fact all the lower bounds known for the edit distance hold to (almost) the same degree for the Ulam distance. These models include: non-embeddability into normed spaces [21,22,23], lower bounds on sketching complexity [23,24], and (symmetric) query complexity [16,25].

Asymmetric Query Complexity. Before stating the results formally, we define the problem and the model precisely. Consider two strings $x, y \in \Sigma^n$ for some alphabet Σ, and let $\mathrm{ed}(x, y)$ denote the edit distance between these two strings. The computational problem is the promise problem known as the Distance Threshold Estimation Problem (DTEP) [26]: distinguish whether $\mathrm{ed}(x, y) > R$ or $\mathrm{ed}(x, y) \leq R/\alpha$, where $R > 0$ is a parameter (known to the algorithm) and $\alpha \geq 1$ is the *approximation factor*. We use DTEP_β to denote the case of $R = n/\beta$, where $\beta \geq 1$ may be a function of n.

In the *asymmetric query model*, the algorithm knows in advance (has unrestricted access to) one of the strings, say y, and has only *query access* to the other string, x. The *asymmetric query complexity* of an algorithm is the number of coordinates in x that the algorithm has to probe in order to solve DTEP with success probability at least $2/3$.

We now give complete statements of our upper and lower bound results. Both exhibit a smooth *tradeoff* between approximation factor and query complexity. For simplicity, we state the bounds in two extreme regimes of approximation ($\alpha = \mathrm{polylog}(n)$ and $\alpha = \mathrm{poly}(n)$). Full statements are available in the full paper.

Theorem 2 (Query complexity upper bound). *For every $\beta = \beta(n) \geq 2$ and fixed $0 < \varepsilon < 1$ there is an algorithm that solves DTEP_β with approximation $\alpha = (\log n)^{O(1/\varepsilon)}$, and makes βn^ε asymmetric queries. This algorithm runs in time $O(n^{1+\varepsilon})$.*

For every $\beta = O(1)$ and fixed integer $t \geq 2$ there is an algorithm for DTEP_β achieving approximation $\alpha = O(n^{1/t})$, with $O(\log^{t-1} n)$ queries into x.

It is an easy observation that our general edit distance algorithm in Theorem 1 follows immediately from the above query complexity upper bound theorem, by running the latter for all β that are a power of 2.

Theorem 3 (Query complexity lower bound). *For a sufficiently large constant $\beta > 1$, every algorithm that solves DTEP_β with approximation $\alpha = \alpha(n) > 2$ has asymmetric query complexity $2^{\Omega\left(\frac{\log n}{\log \alpha + \log \log n}\right)}$. Moreover, for every fixed non-integer $t > 1$, every algorithm that solves DTEP_β with approximation $\alpha = n^{1/t}$ has asymmetric query complexity $\Omega(\log^{\lfloor t \rfloor} n)$.*

We summarize in Table 1 our results and previous bounds for DTEP_β under edit distance and Ulam distance. For completeness, we also present known results for a common query model where the algorithm has query access to both strings (henceforth referred to as the *symmetric query* model). We point out two implications of our bounds on the asymmetric query complexity:

- There is a strong separation between edit distance and Ulam distances. In the Ulam metric, a *constant* approximation is achievable with only $O(\log n)$ asymmetric queries (see [27], which builds on [28]). In contrast, for edit distance, we show an exponentially higher complexity lower bound, of $2^{\Omega\left(\frac{\log n}{\log \log n}\right)}$, even for a larger (polylogarithmic) approximation.
- Our query complexity upper and lower bounds are nearly-matching, at least for a range of parameters. At one extreme, approximation $O(n^{1/2})$ can be achieved with $O(\log n)$ queries, whereas approximation $n^{1/2-\varepsilon}$ already requires $\Omega(\log^2 n)$ queries. At the other extreme, approximation $\alpha = (\log n)^{1/\varepsilon}$ can be achieved using $n^{O(\varepsilon)}$ queries, and requires $n^{\Omega(\varepsilon/\log \log n)}$ queries.

Table 1. Known results for DTEP_β and arbitrary $0 < \varepsilon < 1$

Model	Metric	Approx.	Complexity	Remarks
Near-linear time	Edit	$(\log n)^{O(1/\varepsilon)}$	$n^{1+\varepsilon}$	Theorem 1
	Edit	$2^{\tilde{O}(\sqrt{\log n})}$	$n^{1+o(1)}$	[14]
Symmetric query complexity	Edit	n^ε	$\tilde{O}(n^{\max\{1-2\varepsilon,(1-\varepsilon)/2\}})$	[16] (fixed $\beta > 1$)
	Ulam	$O(1)$	$\tilde{O}(\beta + \sqrt{n})$	[25]
	Ulam+edit	$O(1)$	$\tilde{\Omega}(\beta + \sqrt{n})$	[25]
Asymmetric query complexity	Edit	$n^{1/t}$	$O(\log^{t-1} n)$	Theorem 2 (fixed $t \in \mathbb{N}, \beta > 1$)
	Edit	$n^{1/t}$	$\Omega(\log^{\lfloor t \rfloor} n)$	Theorem 3 (fixed $t \notin \mathbb{N}, \beta > 1$)
	Edit	$(\log n)^{1/\varepsilon}$	$\beta n^{O(\varepsilon)}$	Theorem 2
	Edit	$(\log n)^{1/\varepsilon}$	$n^{\Omega(\varepsilon/\log \log n)}$	Theorem 3 (fixed $\beta > 1$)
	Ulam	$2 + \varepsilon$	$O_\varepsilon(\beta \log \log \beta \cdot \log n)$	[27]

3   Connections of Asymmetric Query Model to Other Models

The asymmetric query model is connected and has implications for two previously studied models, namely the communication complexity model and the symmetric query model (where the algorithm has query access to both strings). Specifically, the former is less restrictive than our model (i.e., easier for algorithms) while the latter is more restrictive (i.e., harder for algorithms). Our upper bound gives an $O(\beta n^\varepsilon)$ one-way communication complexity protocol for DTEP_β for polylogarithmic approximation.

Communication Complexity. In this setting, Alice and Bob each have a string, and they need to solve the DTEP_β problem by way of exchanging messages. The measure of complexity is the number of bits exchanged in order to solve DTEP_β with probability at least $2/3$.

The best non-trivial upper bound known is $2^{\tilde{O}(\sqrt{\log n})}$ approximation with constant communication via [15,29]. The only known lower bound says that approximation α requires $\Omega(\frac{\log n \; / \; \log\log n}{\alpha})$ communication [23,24].

The asymmetric model is "harder", in the sense that the query complexity is at least the communication complexity, up to a factor of $\log|\Sigma|$ in the complexity, since Alice and Bob can simulate the asymmetric query algorithm. In fact, our upper bound implies a communication protocol for the same DTEP_β problem with the same complexity, and it is a one-way communication protocol. Specifically, Alice can just send the $O(\beta n^\varepsilon)$ characters queried by the query algorithm in the asymmetric query model. This is the first communication protocol achieving polylogarithmic approximation for DTEP_β under edit distance with $o(n)$ communication.

Symmetric Query Complexity. In another related model, the measure of complexity is the number of characters the algorithm has to query in *both* strings (rather than only in one of the strings). Naturally, the query complexity in this model is at least as high as the query complexity in the asymmetric model. This model has been introduced (for the edit distance) in [16], and its main advantage is that it leads to *sublinear-time* algorithms for DTEP_β. The algorithm of [16] makes $\tilde{O}(n^{1-2\varepsilon} + n^{(1-\varepsilon)/2})$ queries (and runs in the same time), and achieves n^ε approximation. However, it only works for $\beta = O(1)$.

In the symmetric query model, the best query lower bound is of $\Omega(\sqrt{n/\alpha})$ for any approximation factor $\alpha > 1$ for both edit and Ulam distance [16,25]. The lower bound essentially arises from the birthday paradox. Hence, in terms of separating edit distance from the Ulam metric, this symmetric model can give at most a quadratic separation in the query complexity (since there exists a trivial algorithm with $2n$ queries). In contrast, in our asymmetric model, there is no lower bound based on the birthday paradox, and, in fact, the Ulam metric admits a constant approximation with $O(\log n)$ queries [28,27]. Our lower bound for edit distance is exponentially bigger.

4   Techniques

This section briefly highlights the main techniques and tools used in the course of proving our results.

Algorithm and Query Complexity Upper Bound. A high-level intuition for the near-linear time algorithm is as follows. The classical dynamic programming for edit distance runs in time that is the product of the lengths of the two strings. It seems plausible that, if we manage to "compress" one string to size n^ε, we may be able to compute the edit distance in time only $n^\varepsilon \cdot n$. Indeed, this is exactly what we accomplish. Specifically, our "compression" is achieved via a sampling procedure, which subsamples $\approx n^\varepsilon$ positions of x, and then computes $\mathrm{ed}(x,y)$ in time $n^{1+\varepsilon}$. Of course, the main challenge is, by far, subsampling x so that the above is possible.

Our asymmetric query upper bound has two major components. The first component is a *characterization* of the edit distance by a different "distance", denoted $\mathcal{E}$, which approximates $\mathrm{ed}(x,y)$ well. The characterization is parametrized by an integer parameter $b \geq 2$ governing the following tradeoff: a small b leads to a better approximation, whereas a large b leads to a faster algorithm. The second component is a *sampling algorithm* that approximates $\mathcal{E}$ for some settings of the parameter b, up to a constant factor, by querying a small number of positions in x.

Our characterization is based on a hierarchical decomposition of the edit distance computation, which is obtained by recursively partitioning the string x, each time into b blocks. We shall view this decomposition as a b-ary tree. Then, intuitively, the $\mathcal{E}$-distance at a node is the sum, over all b children, of the minima of the $\mathcal{E}$-distances at these children over a certain range of displacements (possible "shifts" with respect to the other strings). At the leaves (corresponding to single characters of x), the $\mathcal{E}$-distance is simply the Hamming distance to corresponding positions in y.

We show that our characterization is an $O(\frac{b}{\log b} \log n)$ approximation to $\mathrm{ed}(x,y)$. Intuitively, the characterization manages to break-up the edit distance computation into *independent* distance computations on smaller substrings. The independence is crucial here as it removes the need to find a *global* alignment between the two strings, which is one of the main reasons why computing edit distance is hard. We note that while the high-level approach of recursively partitioning the strings is somewhat similar to the previous approaches from [16,15,14], the technical development here is quite different. The previous hierarchical approaches all relied on the following recurrence relation for the approximation factor α:

$$\alpha(n) = c \cdot \alpha(n/b) + O(b),$$

for some $c \geq 2$. It is easy to see that one obtains $\alpha(n) \geq 2^{\Omega(\sqrt{\log n})}$ for any choice of $b \geq 2$. In contrast, our characterization is much more refined and has *no multiplicative factor loss*, i.e., $c = 1$ and hence $\alpha(n) = O(b \log_b n)$. We note that our characterization achieves a *logarithmic* approximation for $b = O(1)$ (although, we do not know efficient algorithms for this setting of b).

The second component of our query algorithm is a careful sampling procedure that approximates $\mathcal{E}$-distance up to a constant factor. The basic idea is to prune the above tree by subsampling at each node a subset of its children. In particular, for a tree with arity $b = (\log n)^{1/\varepsilon}$, the hope is to subsample $(\log n)^{O(1)}$ children and use Chernoff-type bounds to argue that the subsample approximates well the $\mathcal{E}$-distance at that node. We note that $\Omega(\log n)$ samples of children seem necessary due to the minimum operation taken at each node. The estimate at each node has to hold with high probability so that we can apply the union bound. After such a pruning of the tree, we would be left with only $(\log n)^{O(\log_b n)} = n^{O(\varepsilon)}$ leaves, i.e., $n^{O(\varepsilon)}$ positions of x to query.

However, this natural approach of subsampling $(\log n)^{O(1)}$ children at each node does not work when $\beta \gg 1$. Instead, we develop a *non-uniform subsampling technique*: for different nodes we subsample children at different, carefully-chosen rates. From a high-level, our deployed technique is somewhat reminiscent of the hierarchical decomposition and subsampling technique introduced by Indyk and Woodruff [30] in the context of sketching and streaming algorithms.

Query Complexity Lower Bound. The gist of our lower bound is designing two "hard distributions" $\mathcal{D}_0$ and $\mathcal{D}_1$, on strings in Σ^n, for which it is hard to distinguish with only a few queries to x whether $x \in \mathcal{D}_0$ or $x \in \mathcal{D}_1$. At the same time, every two strings x, y in the support of the same $\mathcal{D}_i$ are at a small edit distance: $\mathrm{ed}(x, y) \le n/(\alpha\beta)$; but for a mixed pair $x \in \mathcal{D}_0$ and $y \in \mathcal{D}_1$, the distance is large: $\mathrm{ed}(x, y) > n/\beta$.

We start by making the following core observation. Take two random strings $z_0, z_1 \in \{0, 1\}^n$. Each $\mathcal{D}_i$, $i \in \{0, 1\}$, is generated by applying a cyclic shift by a random displacement $r \in [1, n/100]$ to the corresponding z_i. We show that in order to discover, for an input string, from which $\mathcal{D}_i$ it came from, one has to make at least $\Omega(\log n)$ queries. Intuitively, this follows from the fact that if the number q of queries is small ($q = o(\log n)$) then the algorithm's view is close to the uniform distribution on $\{0, 1\}^q$, no matter which positions are queried. Nevertheless, the edit distance between the two random strings is likely to be large, and a small shift will not change this significantly.

We then amplify the above query lower bound by applying the same idea recursively. In a string generated according to $\mathcal{D}_i$'s, we replace every symbol $a \in \{0, 1\}$ by a random string selected independently from $\mathcal{D}_a$. This way we obtain two distributions on strings of length $n' = n^2$, that require $\Omega(\log^2 n) = \Omega(\log^2 n')$ queries to be told apart. We call the above operation of replacing symbols by strings that come from other distributions a *substitution product*. Strings created this way consist of n blocks of length n each. Intuitively, to distinguish from which of the new distributions an input string comes from, one has to discover for at least $\Omega(\log n)$ blocks which distribution $\mathcal{D}_a$ the respective block comes from. By applying the recursive step multiple times, we obtain a $2^{\Omega(\frac{\log n}{\log \log n})}$ lower bound for a polylogarithmic approximation factor.

To formally prove our result, we develop several tools. First, we need tools for analyzing the behavior of edit distance under the product substitution. It turns out that to control edit distance under the substitution product, we need to work with a large alphabet Σ. In the final step of the construction, we map the large alphabet to sufficiently long random binary strings, thereby extending the lower bound to the binary alphabet as well.

Second, we need tools for analyzing indistinguishability of our distributions under a small number of queries. For this, we introduce a notion of *similarity* of distributions. This notion smoothly composes with the substitution product operation, which amplifies the similarity. We also show that random acyclic shifts of random strings are likely to produce strings with high similarity. Finally, we show that if an algorithm is able to distinguish distributions meeting our similarity notion, then it must make many queries. We believe that these tools and ideas behind them may find applications in showing query lower bounds for other problems.

References

1. Levenshtein, V.I.: Binary codes capable of correcting deletions, insertions, and reversals (in russian). Doklady Akademii Nauk SSSR 4(163), 845–848 (1965); Levenshtein, V.I.: Binary codes capable of correcting deletions, insertions, and reversals. Soviet Physics Doklady 10(8), 707–710 (1966) (Appeared in English)
2. Navarro, G.: A guided tour to approximate string matching. ACM Comput. Surv. 33(1), 31–88 (2001)
3. Gusfield, D.: Algorithms on strings, trees, and sequences. Cambridge University Press, Cambridge (1997)
4. Wagner, R.A., Fischer, M.J.: The string-to-string correction problem. Journal of the ACM 21(1), 168–173 (1974)
5. Cormen, T.H., Leiserson, C.E., Rivest, R.L., Stein, C.: Introduction to Algorithms, 2nd edn. MIT Press, Cambridge (2001)
6. Masek, W.J., Paterson, M.: A faster algorithm computing string edit distances. J. Comput. Syst. Sci. 20(1), 18–31 (1980)
7. Bille, P., Farach-Colton, M.: Fast and compact regular expression matching. Theoretical Computer Science 409(28), 486–496 (2008)
8. Indyk, P.: Algorithmic aspects of geometric embeddings (tutorial). In: Proceedings of the Symposium on Foundations of Computer Science (FOCS), pp. 10–33 (2001)
9. Indyk, P., Matoušek, J.: Low distortion embeddings of finite metric spaces. In: CRC Handbook of Discrete and Computational Geometry (2003)
10. Sahinalp, S.C.: Edit distance under block operations. In: Kao, M.Y. (ed.) Encyclopedia of Algorithms. Springer, Heidelberg (2008)
11. Landau, G.M., Myers, E.W., Schmidt, J.P.: Incremental string comparison. SIAM J. Comput. 27(2), 557–582 (1998)
12. Bar-Yossef, Z., Jayram, T.S., Krauthgamer, R., Kumar, R.: Approximating edit distance efficiently. In: Proceedings of the Symposium on Foundations of Computer Science (FOCS), pp. 550–559 (2004)
13. Batu, T., Ergün, F., Sahinalp, C.: Oblivious string embeddings and edit distance approximations. In: Proceedings of the ACM-SIAM Symposium on Discrete Algorithms (SODA), pp. 792–801 (2006)
14. Andoni, A., Onak, K.: Approximating edit distance in near-linear time. In: Proceedings of the Symposium on Theory of Computing (STOC), pp. 199–204 (2009)
15. Ostrovsky, R., Rabani, Y.: Low distortion embedding for edit distance. J. ACM 54(5) (2007); Preliminary version appeared in STOC 2005

16. Batu, T., Ergün, F., Kilian, J., Magen, A., Raskhodnikova, S., Rubinfeld, R., Sami, R.: A sublinear algorithm for weakly approximating edit distance. In: Proceedings of the Symposium on Theory of Computing (STOC), pp. 316–324 (2003)

17. Cormode, G., Paterson, M., Sahinalp, S.C., Vishkin, U.: Communication complexity of document exchange. In: Proceedings of the ACM-SIAM Symposium on Discrete Algorithms (SODA), pp. 197–206 (2000)

18. Muthukrishnan, S., Sahinalp, C.: Approximate nearest neighbors and sequence comparison with block operations. In: Proceedings of the Symposium on Theory of Computing (STOC), pp. 416–424 (2000)

19. Cormode, G., Muthukrishnan, S.: The string edit distance matching problem with moves. ACM Trans. Algorithms 3(1) (2007); Special issue on SODA 2002

20. Cormode, G.: Sequence Distance Embeddings. Ph.D. Thesis, University of Warwick (2003)

21. Khot, S., Naor, A.: Nonembeddability theorems via Fourier analysis. Math. Ann. 334(4), 821–852 (2006); Preliminary version appeared in FOCS 2005

22. Krauthgamer, R., Rabani, Y.: Improved lower bounds for embeddings into L_1. In: Proceedings of the ACM-SIAM Symposium on Discrete Algorithms (SODA), pp. 1010–1017 (2006)

23. Andoni, A., Krauthgamer, R.: The computational hardness of estimating edit distance. SIAM Journal on Computing 39(6), 2398–2429 (2010); Previously appeared in FOCS 2007

24. Andoni, A., Jayram, T., Pǎtraşcu, M.: Lower bounds for edit distance and product metrics via Poincaré-type inequalities. Accepted to ACM-SIAM Symposium on Discrete Algorithms (SODA 2010) (2010)

25. Andoni, A., Nguyen, H.L.: Near-tight bounds for testing Ulam distance. Accepted to ACM-SIAM Symposium on Discrete Algorithms (SODA 2010) (2010)

26. Saks, M., Sun, X.: Space lower bounds for distance approximation in the data stream model. In: Proceedings of the Symposium on Theory of Computing (STOC), pp. 360–369 (2002)

27. Ailon, N., Chazelle, B., Comandur, S., Liu, D.: Estimating the distance to a monotone function. Random Structures and Algorithms 31, 371–383 (2007); Previously appeared in RANDOM 2004

28. Ergün, F., Kannan, S., Kumar, R., Rubinfeld, R., Viswanathan, M.: Spot-checkers. J. Comput. Syst. Sci. 60(3), 717–751 (2000)

29. Kushilevitz, E., Ostrovsky, R., Rabani, Y.: Efficient search for approximate nearest neighbor in high dimensional spaces. SIAM J. Comput. 30(2), 457–474 (2000); Preliminary version appeared in STOC 1998

30. Indyk, P., Woodruff, D.: Optimal approximations of the frequency moments of data streams. In: Proceedings of the Symposium on Theory of Computing (STOC) (2005)

Comparing the Strength of Query Types in Property Testing: The Case of Testing k-Colorability

Ido Ben-Eliezer[1], Tali Kaufman[2], Michael Krivelevich[3], and Dana Ron[4]

[1] School of Computer Science, Tel Aviv University, Tel Aviv 69978, Israel
idobene@post.tau.ac.il
[2] Institute for Advanced Study, Princeton, New Jersey, USA
kaufmant@ias.edu
[3] School of Mathematical Sciences, Tel Aviv University, Tel Aviv 69978, Israel
krivelev@post.tau.ac.il
[4] School of Electrical Engineering, Tel Aviv University, Tel Aviv 69978, Israel
danar@eng.tau.ac.il

Abstract. We study the power of four query models in the context of property testing in general graphs (i.e., with arbitrary edge densities), where our main case study is the problem of testing k-colorability. Two query types, which have been studied extensively in the past, are *pair queries* and *neighbor queries*. The former corresponds to asking whether there is an edge between any particular pair of vertices, and the latter to asking for the i'th neighbor of a particular vertex. We show that while for pair queries, testing k-colorability requires a number of queries that is a monotone decreasing function in the average degree d, the query complexity in the case of neighbor queries remains roughly the same for every density and for large values of k. We also consider a combined model that allows both types of queries, and we propose a new, stronger, query model, related to the field of Group Testing. We give one-sided error upper and lower bounds for all the models, where the bounds are nearly tight for three of the models. In some of the cases our lower bounds extend to two-sided error algorithms.

The problem of testing k-colorability was previously studied in the contexts of dense and sparse graphs, and in our proofs we unify approaches from those cases, and also provide some new tools and techniques that may be of independent interest.

Keywords: pair queries, neighbor queries, group queries, k-colorability.

This is an abridged version of [4] in the Proceedings of the 19th ACM-SIAM Symposium on Discrete Algorithms (SODA'2008). The reader is advised to consult [4] for technical details missing in this contribution.

1 Introduction

Property testing [7,13] deals with the problem of deciding whether a certain object has a prespecified property P or it is far (i.e., differs significantly) from

O. Goldreich (Ed.): Property Testing, LNCS 6390, pp. 253–259, 2010.

any object that has P. Namely, the algorithm should *accept* objects that have the property, and should *reject* objects that are far from having the property with respect to some predetermined distance measure, where the algorithm is allowed a small probability of failure. The algorithm is given query access to the object, and it should make the decision after observing only a small part of the object. Thus, the main complexity measure studied in the context of property testing is the *query complexity* of the algorithm, which is normally expected to be *sublinear* in the size of the object, and the question is what exact form does this complexity take.

In this work we compare the power of different query types in the context of testing graph properties of general graphs (i.e., with arbitrary edge densities). To this end we focus on the problem of testing k-colorability for $k \geq 3$, and study the query complexity of this problem for different query types as a function of the number of vertices and the average degree of the graph.

2   The Distance Measure and Query Types Studied in This Work

When defining models for property testing of graphs there are two issues to consider: the distance measure between graphs (which determines what graphs should be rejected by the testing algorithm) and the types of queries that the algorithm is allowed to make. Since we study graphs of varying edge densities and vertex degrees, we follow [11,10] and define our distance measure with respect to the total number of edges in the graph. Namely, if n denotes the number of graph vertices and d denotes the average degree, then we say that a graph is ϵ-*far* from being k-colorable for a given $0 \leq \epsilon \leq 1$, if it is necessary to remove more than ϵdn edges so as to obtain a k-colorable graph.[1]

We consider the following types of queries where the first two have been considered in the past and the third is a new query type we introduce.

- Pair queries. These are queries of the form "Is there an edge between the pair of vertices u and v?".
- Neighbor queries. These are queries of the form: "Who is the i'th neighbor of vertex v?". If v has less than i neighbors then a special symbol is returned, and no assumption is made about the order of the neighbors of a vertex.
- Group queries. We propose a new query type that extends pair queries. These queries are of the form "Is there at least one edge between a vertex u and a set of vertices S?".

The study of group queries is partially motivated by the field of Group Testing (see, e.g., [6]), where similar queries are allowed. Problems of group testing can

---

[1] Another well studied distance measure is the fraction of edge modification as a function of n^2. This measure is appropriate for dense graphs (i.e., that satisfy $d = \Theta(n)$). In what can be viewed as the other extreme, where all vertices have bounded degree $d_{\max}$ (in particular, $d_{\max} = O(1)$), distance is measured with respect to $d_{\max}n$).

be found in various fields such as Statistics and Biology. Another motivation for studying group-queries is that lower bounds on the query complexity of algorithms that use group queries also apply (up to poly-logarithmic factors in n) to algorithms that use pair queries and/or neighbor queries, and hence they can be viewed as a tool for obtaining query complexity lower bounds. This follows from the (easily proven) fact that pair queries are a special case of group queries and neighbor queries can be emulated using group queries. To be precise, the emulation is for random neighbor queries (that is, the query is of the form: "Give me a uniformly selected neighbor of vertex v").

For the sake of simplicity of the presentation, we allow algorithms that perform neighbor queries and algorithms that perform group queries to perform degree queries as well. That is, they may ask for the degree of any vertex of their choice. Clearly, a degree query can be emulated by $O(\log n)$ neighbor queries (using a simple binary search). The degree of any vertex can be approximated with high probability within a constant factor using a number of group queries that is poly-logarithmic in n (and such an approximation is sufficient for our purposes).

In what follows, when we refer to the *pair query model* (respectively, *neighbor query model* and *group query model*), we mean that only pair queries (respectively, neighbor queries and group queries) are allowed. When both pair queries and neighbor queries (as well as degree queries) are allowed, then we refer to the resulting model as the *combined model*.

3   Related Work on Testing k-Colorability

Testing k-colorability has previously been studied in the pair query model for the case that the graph is dense, that is, $d = \Theta(n)$. For this case k-colorability is testable using a number of queries that is *independent* of the graph size (and polynomial in k and $1/\epsilon$ [7,3]).[2]

Testing k-colorability has previously been studied in the neighbor query model for the case that $k = 3$ and the graph has constant maximum degree. (that is, $d = O(1)$, and furthermore, the maximum degree d_{max} is $O(1)$ as well). In this case Bogdanov et. al. [5] proved that is necessary to perform $\Omega(n)$ queries (that is, there is no algorithm with sublinear query complexity).

Testing k-colorability for $k = 2$ (i.e., testing bipartitness) has previously been studied for general graphs in the combined model [10] where it was shown that $\tilde{\Theta}(\min\{\sqrt{n}, n/d\})$ (pair and neighbor) queries are both sufficient and necessary. The proof of the lower bound in [10] implies that if only neighbor queries are allowed then $\Omega(\sqrt{n})$ queries are necessary for every value of d, and if only pair queries are allowed then $\Omega(n/d)$ queries are necessary.[3]

---

[2] Interestingly, the earlier work of Rödl and Duke [12] implicitly implies that k-colorability is testable using a number of queries that is independent of the graph size, but is a tower function of $1/\epsilon$.

[3] In earlier work [9,8] it was shown that if only neighbor queries are allowed and the distance measure is with respect to $d_{max}n$ rather than dn, then $\tilde{\Theta}(\sqrt{n})$ are both necessary and sufficient.

4 Our Results

In this work we study the power of the different types of queries when testing
k-colorability of general graphs for a fixed $k \geq 3$. In previous work on testing
properties of graphs, the pair query model was studied in the case of dense
graphs, the neighbor query model was studied in the case of bounded-degree
graphs, and for general graphs, the combined model was considered. Here we are
interested in understanding how the query complexity of the problem behaves as
a function of the edge density (and the number of vertices) when the algorithm
is allowed to perform only one type of query, and whether there is a gain when
allowing to combine query types. One motivation for this investigation is that
the type of queries allowed depends on the way the graph is represented. Thus,
allowing both pair queries and neighbor queries (as in the combined model) assumes that the algorithm has access both to an adjacency matrix representation
(that supports pair queries) and to an incidence lists representation (that supports neighbor queries), which is not necessarily the case. Our second motivation
is simply complexity theoretic: understanding the strength of each query type
separately (and possibly combined) for varying edge densities.

In what follows we say that an algorithm has *one-sided error* if it always accepts graphs that are k-colorable, otherwise it has *two-sided error*. Our results
are stated in terms of the dependence on n and d. With a slight abuse of nota-
tion, we write $f = \tilde{O}(g)$ (and similarly, $f = \tilde{\Omega}(g)$) if $f(x) = O(g(x)) \cdot \mathrm{polylog}(n)$
for every x, where n is the number of vertices. In all our upper bounds the depen-
dence on both k and $1/\epsilon$ is polynomial. The bounds are for the query complexity
in the different models. We note that the running time of our algorithms may
be exponential in the number of queries, but the focus of this work is only on
the query complexity.

Theorem 1. *The following holds for testing k-colorability in the pair query model:*

1. *There exists a one-sided error tester that performs $\tilde{O}((\frac{n}{d})^2)$ queries.*
2. *Every one-sided error tester must perform $\Omega((\frac{n}{d})^2)$ queries.*

Theorem 2. *The following holds for testing k-colorability in the neighbor query model:*

1. *There exists a one-sided error tester that performs $O(n)$ queries.*
2. *Every tester must perform $\Omega\left(\max\{\frac{n}{d}, \sqrt{n}\}\right)$ queries.*
3. *Every one-sided error tester must perform $\Omega(n^{1 - \frac{1}{\lceil (k+1)/2 \rceil}})$ queries.*
4. *Every one-sided error tester for $k \geq 6$ must perform $\Omega(n \cdot d^{-\frac{1}{\lceil k/2 \rceil - 1}})$ queries.*

Observe that for one-sided error testers in the neighbor query model, as k increases, our lower bound approaches our upper bound.

Theorem 3. *The following holds for testing k-colorability in the group query model:*

1. *There exists a one-sided error tester that performs $\tilde{O}(\frac{n}{d})$ queries.*
2. *Every tester must perform $\tilde{\Omega}(\frac{n}{d})$ queries.*

By combining Theorems 1, 2, 3 and the fact that neighbor queries can be emulated using a logarithmic number of group queries, we get the next corollary.

Corollary 4. *The following holds for testing k-colorability in the combined query model:*

1. *There exists a one-sided error tester that performs* $\min((\tilde{O}(\frac{n}{d})^2), O(n))$ *queries.*
2. *Every tester must perform* $\tilde{\Omega}(\frac{n}{d})$ *queries.*

The results are summarized in Table 1 and are illustrated in Figure 1.

Table 1. Results for one-sided error testing of k-colorability

	Pair Queries	Neighbor Queries	Pair&Neighbor Queries	Group Queries
Upper Bound	$\tilde{O}((\frac{n}{d})^2)$	$O(n)$	$\min\{\tilde{O}((\frac{n}{d})^2), O(n)\}$	$\tilde{O}(\frac{n}{d})$
Lower Bound	$\Omega((\frac{n}{d})^2)$	$\Omega(n^{1-\frac{1}{\lceil(k+1)/2\rceil}})$ $\Omega(n \cdot d^{-\frac{1}{\lceil k/2\rceil-1}})$ if $k \geq 6$	$\Omega(\frac{n}{d})$ also for 2-sided error	$\Omega(\frac{n}{d})$ also for 2-sided error

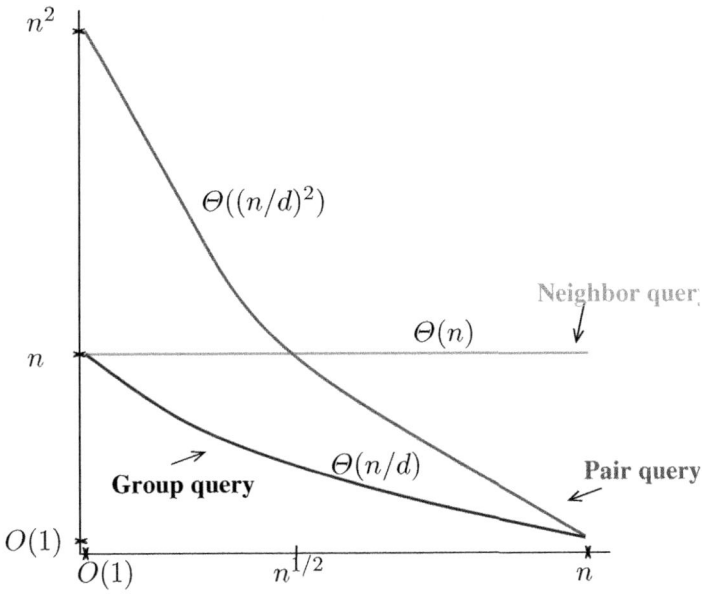

Fig. 1. A schematic illustration of the query complexity for the different query types (and one-sided error). For the sake of simplicity we ignore logarithmic factors in the bounds and furthermore, for the neighbor query model we think of k being large. For the combined model we have that the lower bound on the query complexity coincides with the group query model, and the upper bound coincides with the neighbor query model until $d = n^{1/2}$ and from that point on it coincides with the pair query model.

Discussion of our results and conclusions. We next discuss the main phenomena we observe in our study of testing k-colorablity in the different query models.

- While the query complexity in the pair query model and the query complexity in the group query model are monotone decreasing functions of d, the query complexity in the neighbor query model remains roughly the same for every value of d (and large values of k).
- When comparing the pair query model to the neighbor query model in more detail we see that the query complexity in the pair query model is higher than in the neighbor query model for $d < \sqrt{n}$, while once d passes $\sqrt{n}$ it becomes lower (and continues decreasing). The extreme case is $d = \Theta(n)$, where in the pair query model the query complexity does not depend on n.[4] This coincides with the intution that neighbor queries are useful for sparse graphs and pairs queries are useful for dense graphs.
- When comparing the group query model to the other models we observe that the query complexity in the former model is never higher (upto poly-logarithmic factors) than in the latter models. As noted previously, this holds for other graph properties provided that it is sufficient to use random neighor queries in the neighbor query model (since these can be emulated by group queries). When non-random neighbor queries are required in order to perform degree queries, then one should verify whether approximate degree queries are sufficient (since these can be emulated using group queries).

  Moreover, we note that a rather straightforward modification of the arguments of [10] shows that using group queries one can give strictly better bounds for testing bipartiteness than in the combined model. Therefore, there exist properties for which the group query model is stronger than the combined model. We also show that for the case of testing k-colorability and a certain distribution over graphs, the combined model is strictly stronger than the optimum of the pair query and the neighbor query models.

References

1. Alon, N., Fischer, E., Krivelevich, M., Szegedy, M.: Efficient testing of large graphs. Combinatorica 20, 451–476 (2000)
2. Alon, N., Fischer, E., Newman, I., Shapira, A.: A combinatorial characterization of the testable graph properties: it's all about regularity. In: Proceedings of the 38th ACM STOC, pp. 251–260 (2007)
3. Alon, N., Krivelevich, M.: Testing k-colorability. SIAM Journal on Discrete Math 15(2), 211–227 (2002)
4. Ben-Eliezer, I., Kaufman, T., Krivelevich, M., Ron, D.: Comparing the strength of query types in property testing: the case of k-colorability. In: Proceedings of the 19th SODA, pp. 1213–1222 (2008)
5. Bogdanov, A., Obata, K., Trevisan, L.: A lower bound for testing 3-colorability in bounded degree graphs. In: Proceedings of the 43th IEEE FOCS, pp. 93–102 (2002)

---

[4] In fact, when $d = \Theta(n)$ then many natural properties are testable in the pair query model using a constant number of queries (see e.g. [7,1,2]).

6. Du, D., Hwang, F.: Combinatorial group testing and its applications. World Scientific, Singapore (1993)
7. Goldreich, O., Goldwasser, S., Ron, D.: Property testing and its connection to learning and approximation. JACM 45(4), 653–750 (1998)
8. Goldreich, O., Ron, D.: A sublinear bipartite tester for bounded degree graphs. Combinatorica 19(3), 335–373 (1999)
9. Goldreich, O., Ron, D.: Property testing in bounded degree graphs. Algorithmica 32(2), 302–343 (2002)
10. Kaufman, T., Krivelevich, M., Ron, D.: Tight bounds for testing bipartiteness in general graphs. SIAM Journal on Computing 33(6), 1441–1483 (2004)
11. Parnas, M., Ron, D.: Testing the diameter of graphs. Random Structures and Algorithms 20(2), 165–183 (2002)
12. Rödl, V., Duke, R.: On graphs with small subgraphs of large chromatic number. Graphs Combin. 1, 91–96 (1985)
13. Rubinfeld, R., Sudan, M.: Robust characterization of polynomials with applications to program testing. SIAM Journal on Computing 25(2), 252–271 (1996)

Testing Linear-Invariant Non-linear Properties: A Short Report

Arnab Bhattacharyya[1], Victor Chen[2], Madhu Sudan[3], and Ning Xie[4]

[1] MIT CSAIL, Cambridge, MA
`abhatt@mit.edu`
[2] Institute of Theoretical Computer Science, Tsinghua University, Beijing, China
`victor.vc@gmail.com`
[3] Microsoft Research New England, Cambridge, MA
`madhu@microsoft.com`
[4] MIT CSAIL, Cambridge, MA
`ningxie@csail.mit.edu`

Abstract. The rich collection of successes in property testing raises a natural question: Why are so many different properties turning out to be locally testable? Are there some broad "features" of properties that make them testable? Kaufman and Sudan (STOC 2008) proposed the study of the relationship between the invariances satisfied by a property and its testability. Particularly, they studied properties that were invariant under linear transformations of the domain and gave a characterization of testability in certain settings. However, the properties that they examined were also linear. This led us to investigate linear-invariant properties that are not necessarily linear. Here we describe some of the resulting works which consider natural linear-invariant properties, specifically properties that are described by forbidden patterns of values that a function can take, and show testability under various settings.

Keywords: property testing, regularity lemma, linear invariance.

1 Introduction

The field of property testing, initiated by the works in [BLR93, BFL91] and defined formally in [RS96, GGR98], has seen an enormous and diverse collection of successes lately. The rich collection of properties that turn out to be testable extremely locally (with say, constant number of queries) relative to the size of the object being tested, leads to a natural question: Why are so many properties locally testable? Are there some broad unifying themes in the properties being tested, and the testers being used? In an attempt to explain this richness and diversity, Kaufman and Sudan [KS08] suggested that the "invariances" shown by a property may play a central role in their testability. A property of functions mapping a domain D to a range R is said to be invariant under a map $\pi : D \to D$, if whenever a function f satisfies the property, so does the function $f \circ \pi$. In particular, if π is a permutation, then this says that the property is

O. Goldreich (Ed.): Property Testing, LNCS 6390, pp. 260–268, 2010.
© Springer-Verlag Berlin Heidelberg 2010

invariant if the domain is relabelled according to π. Kaufman and Sudan suggest that many properties that are known to be testable have a rich collection of invariances and often testability is implied by such invariances. They then focus on algebraic properties in particular and notice that the properties in consideration are defined over domains that are vector spaces over some field, and the properties are invariant under linear, and sometimes affine, transformations of the domain. They also show that using very few additional features of the property, one can deduce testability, thus unifying many previous results (including those in [BLR93, RS96, AKK$^+$05, KR06, JPRZ04]).

One of the more restrictive "additional features" of the properties studied by [KS08] is that the property itself is "linear", the range of the functions being considered is a field and the property forms a vector space over this field. While this feature is definitely exhibited by all algebraic properties, it is a very different requirement to the requirement of linear-invariance; and motivated our work [BCSX09] where we attempt to extend the study of "testing based on invariance" beyond this restriction. The principal results of our work is an infinite class of "natural non-linear, linear-invariant properties" which we show to be testable. In the process we describe an even richer class of linear-invariant properties whose testability remains open, which if shown to be locally testable would unify the results of this work, with those studied in the algebraic setting (including those of [KS08]). We describe our problems and results in greater detail below.

2   Definitions: Constraints, Characterizations, Invariance and Orbits

We consider properties of functions mapping some domain D to some range R. We let $\{D \to R\}$ denote the set of all such functions and describe a property by the set of functions $\mathcal{F} \subseteq \{D \to R\}$ that satisfy the property. Throughout this article, we will consider functions mapping the domain $D = \{0,1\}^n$ to the range $\{0,1\}$, where the domain is viewed as the n-dimensional vector space over the binary[1] field $\mathbb{F}_2$. By an abuse of notation, $\mathcal{F}$ will actually refer to an ensemble of properties, one for each value of $n \in \mathbb{Z}^+$.

We measure the distance between functions in the (by now) standard way: the distance between f and g, denoted $\delta(f,g)$, is the quantity $\delta(f,g) = \Pr_{x \in D}[f(x) \neq g(x)]$. The distance of f to a property $\mathcal{F}$ is the quantity $\delta(f,\mathcal{F}) = \min_{g \in \mathcal{F}}\{\delta(f,g)\}$. We say f is δ-close to $\mathcal{F}$ if $\delta(f,\mathcal{F}) \leq \delta$ and δ-far otherwise. A $(k(\delta),\tau(\delta))$-local (one-sided) *tester* for $\mathcal{F}$ is a probabilistic oracle algorithm T that takes as input a parameter δ and queries an oracle for f $k(\delta)$ times and accepts functions in $\mathcal{F}$ with probability one, while rejecting functions that are δ-far with probability at least $\tau(\delta)$. If such a tester exists, $\mathcal{F}$ is said to be *locally testable*. Note that we are interested only in testers where $\tau(\delta) > 0$ for every $\delta > 0$. Also, neither $k(\cdot)$

---

[1] Most notions also extend to the case where the domain is a vector space over an arbitrary finite field, and the range is an arbitrary finite set.

nor $\tau(\cdot)$ is a function of n. If k is furthermore independent of δ, we say that the tester is *proximity-oblivious*, following Goldreich and Ron [GR09].

Next, we turn to the notion of invariance. For a function $\pi : D \to D$, we say that a property $\mathcal{F}$ is *invariant* under π if the function $f \in \mathcal{F}$ implies $f \circ \pi \in \mathcal{F}$. A property $\mathcal{F} \subseteq \{\{0,1\}^n \to \{0,1\}\}$ is *linear-invariant* if for every linear function $L : \{0,1\}^n \to \{0,1\}^n$, $\mathcal{F}$ is invariant under L. Our hope is to describe a large collection of natural linear-invariant properties that are locally testable.

A very broad collection of natural testable properties are what may be described as "locally characterized properties" – we describe these next. A k-local *constraint* $C = (a_1, \ldots, a_k; S)$ is given by a k-tuple $a_1, \ldots, a_k \in D$ and non-empty set $S \subseteq R^k$. We say that a function $f : D \to R$ satisfies the constraint C if $(f(a_1), \ldots, f(a_k)) \notin S$. We say that a property $\mathcal{F}$ satisfies the constraint C if every function $f \in \mathcal{F}$ satisfies C. A collection of constraints $C_1, \ldots, C_m$ k-*locally characterizes* a property $\mathcal{F}$ if each constraint is k-local and $f \in \mathcal{F}$ if and only if f satisfies C_j for every $j \in \{1, \ldots, m\}$. If k does not depend on n, we say the property is *locally characterized*.

It is natural to analyze the testability of locally characterized properties, and indeed the early work of Rubinfeld and Sudan [RS96] does suggest analyzing the "robustness" of characterizations to design and analyze local tests for properties. (Roughly, a characterization is robust if the only functions that satisfy most constraints in the characterization are those that are close to the property.) Characterizations are effectively also a necessary condition for the existence of non-adaptive proximity-oblivious tests [GR09], which are the prevalent ones in the algebraic setting. Finally, for linear-invariant properties, characterizations take on an especially nice form, as we describe next.

Given a k-local constraint $C = (a_1, \ldots, a_k; S)$ on functions mapping D to R and a map $\pi : D \to D$, let the π-rotation of C, denoted $\pi \circ C$, be the k-local constraint $\pi \circ C = (\pi(a_1), \ldots, \pi(a_k); S)$. Note that if $\mathcal{F}$ is invariant under π and $\mathcal{F}$ satisfies C, then it also satisfies $\pi \circ C$. For linear-invariant properties, it follows that the existence of a single constraint C implies an abundance of constraints $L \circ \pi$, one for each linear function L mapping the domain to itself. We refer to the set of constraints $\{L \circ C | L \text{ is linear } \}$ as the *orbit* of the constraint C. Given this abundance of local constraints satisfied by a property $\mathcal{F}$, one could even hope that the family is characterized by the orbit of a single constraint. To this end we say that $\mathcal{F}$ has a k-*local single orbit characterization* if there exists a k-local constraint C such that its orbit characterizes $\mathcal{F}$. At first glance, the existence of a single orbit characterization may seem like a very strong requirement, but not at second glance! In particular the following proposition is easy to show:

Proposition 1. *If $\mathcal{F} \subseteq \{\{0,1\}^n \to \{0,1\}\}$ is k-locally characterized and linear-invariant, then it has a K-local single orbit characterization, for some $K \leq 2^k$.*

(When the domain is a vector space over a field of cardinality q, the bound weakens to $K \leq q^k$.) The existence of such a nice and very "uniform" characterization of $\mathcal{F}$ suggests a very natural test for the propery $\mathcal{F}$ locally characterized by the orbit of a single constraint C: Pick a random linear map L and verify that f satisfies $L \circ C$.

If this test can be shown to be sound, then it would imply that every locally characterized linear-invariant property is locally testable. This question remains open (see more in Section 4.1), and our work [BCSX09] takes some first steps towards understanding the testability of this class of properties (and shows testability of a proper, but infinite, subclass). We describe our specific results next.

3   Our Results in [BCSX09]

To understand the class of properties that are linear-invariant and not linear, it is useful to start with a simple example (that is not already covered by the results of [KS08]). Our work starts with the "triangle-freeness" property introduced by Green [Gre05] and extends it. A function $f : \{0,1\}^n \to \{0,1\}$ is said to be *triangle-free* if the set $f^{-1}(1)$ does not contain a triple of the form $x, y, x + y$. In our language, the property of being triangle-free could be describe by the family $\mathcal{F}$ characterized by the orbit of the constraint $C = (a, b, a+b; S = \{111\})$ where a and b are two (arbitrary) linearly independent vectors over the domain $\{0,1\}^n$. Green [Gre05] shows that the property of being triangle-free is indeed locally testable, though the analysis is quite different from the analyses in algebraic settings. In our work, we extend this test to a broader collection of constraints.

To describe this extension, we need to introduce a few more pieces of notation. We say that a property $\mathcal{F}$ characterized by the orbit of a constraint $C = (a_1, \ldots, a_k; S)$ is *monotone* if S is an upward closed set, i.e., for $x, y \in \{0,1\}^k$, if $x \in S$ and $x_i \leq y_i$ for all $i \in [k]$, then $y \in S$. (In other words, removing elements from the support of a function satisfying a monotone property keeps the function in the property.) We call the constraint $C = (a_1, \ldots, a_k; S)$ a *pattern* if the set S has only one element. Note that if C is a pattern, it is monotone exactly when $S = \{1^k\}$. We refer to the property described by the orbit of a single monotone pattern C as being *C-free*.

Notice that the family described by a constraint $C = (a_1, \ldots, a_k; S)$ is essentially a function of the underlying "matroid". The matroid perspective simply views $a_1, \ldots, a_k$ as an abstract set of k elements and tells us which subsets of these elements are independent and which ones are not. (The exact definition is not important to us, since we retain the linear-algebraic descriptions in our definition below; but the notions is from matroid theory.) We say $a_1, \ldots, a_k$ form a *graphic matroid* if there exists an undirected graph $G = (V, E)$ with k edges $E = \{e_1, \ldots, e_k\}$ such that for every subset $S \subseteq \{1, \ldots, k\}$ the set $\{a_i | i \in S\}$ is linearly independent if and only if the graph $G_S = (V, \{e_i | i \in S\})$ has no cycles. We say that a constraint C is *based on a graphic matroid* if the constraint points $a_1, \ldots, a_k$ form a graphic matroid.

Our main theorem in [BCSX09] can now be stated.

Theorem 1 ([BCSX09]). *For a k-local monotone pattern C based on a graphic matroid, the property of being C-free is locally testable. Specifically, there exists a function $\tau = \tau_k : \Re^+ \to \Re^+$ and a k-query test T that accepts C-free functions with probability one, while rejecting functions that are δ-from being C-free with probability at least $\tau(\epsilon)$.*

As a consequence, any monotone linear-invariant property locally character-ized by the orbit of a constraint C based on a graphic matroid is locally testable with a proximity-oblivious tester.

The bound on τ is quite weak. Let $W(t)$ denote a tower of twos with height $\lceil t \rceil$. Our proof only guarantees that $\tau(\epsilon) \geq W(\text{poly}(1/\epsilon))^{-1}$, a rather fast vanishing function. In fact, all known proofs, even for the property of being triangle-free, have this tower-behavior inherently because they rely on some form of a "regu-larity lemma", which we now describe.

To analyze the triangle-freeness property, Green developed a regularity lemma for groups, which is analogous to Szemerédi's regularity lemma for graphs. In the boolean case, Green's regularity lemma shows how, given any function $f : \{0,1\}^n \to \{0,1\}$, one can find a subgroup H of $\{0,1\}^n$ such that the restriction of f to almost all cosets of H is "regular", where "regularity" is defined based on the "Fourier coefficients" of f.

This lemma continues to play a central role in [BCSX09] as well. To extract a large feasible class of matroids, we also use a notion from a work of Green and Tao [GT06] of the complexity of a linear system (or matroids). The "least com-plex" matroids have complexity 1, and it was shown that the regularity lemma can be applied to all matroids of complexity 1 to show that they are testable.

The presence of the many restrictions on the nature of the constraint C leads to a natural question: Are there many (or any) new properties that can be tested based on Theorem 1? Of course, there are infinitely many different constraints C, but the property of being C-free need not be different for different C's. For example, one can permute the points $a_1, \ldots, a_k$ and the coordinates of S to obtain essentially the same constraint. Alternately, one can replace the constraint C by the constraint $L \circ C$ for an invertible linear map L and get the same family. But equivalence goes beyond such syntactic concerns. For example, suppose C is a constraint based on a graphic matroid, where the underlying graph is one whose biconnected components are triangles. Then being C-free is essentially the same as being triangle-free, in that every triangle-free function is also C-free, while every C-free function is $O(2^{-n})$-close to being triangle free. Thus one needs to prove explicitly that the class of properties being tested include (infinitely many) new ones. (We also advocate that this concern ought to be addressed explicitly in any work in property testing that aims to work for a broad class of properties.)

In [BCSX09], we consider the following two infinite classes of (monotone) patterns based on graphic matroids. For $\ell = 3, 4, \ldots$, let O_ℓ be the constraint $O_\ell = (a_1, \ldots, a_\ell; \{1^\ell\})$ where $a_1, \ldots, a_{\ell-1} \in \{0,1\}^n$ are linearly independent and $a_\ell = \sum_{i < \ell} a_i$. O_ℓ is thus based on the graphic matroid corresponding to the cycle of length ℓ. Similarly for $\ell = 2, 3, \ldots$, let K_ℓ be the constraint on $k = \binom{\ell}{2}$ points based on the graphic matroid of the complete graph on ℓ vertices. Note that O_ℓ-freeness and K_ℓ-freeness are testable for every ℓ, by Theorem 1. The following theorem shows that these (infinite class of properties) are all pairwise distinct (i.e., for every pair, at least one property contains elements which are $\Omega(1)$-far from the other).

Theorem 2 ([BCSX09]). *The class of C-free properties for C being a monotone pattern based on graphic matroids include infinitely many distinct ones. In particular:*

1. *For every odd $\ell \geq 3$, if f is $O_{\ell+2}$-free, then it is also O_ℓ-free. However, there exist functions g that are O_ℓ-free but far from being $O_{\ell+2}$-free.*
2. *For every $\ell \geq 2$, if f is K_ℓ-free, then it is also $K_{\ell+1}$-free. However, for $\ell \geq 3$ there exists a function g that is K_ℓ-free but far from being $K_{\binom{\ell}{2}+2}$-free.*

Theorems 1 and 2 combine to give some room for optimism that one may get an exact understanding of the class of linear-invariant properties that have $O(1)$-query proximity oblivious tests.

Our results show that, at least under severe restrictions, the natural test for such a property does work, and that, despite the restrictions, this does lead to an infinite class of new properties. Fortunately, subsequent work revealed that several of the limitations in Theorem 1 above turned out to be limitations of the proof technique alone, and stronger techniques can be brought to bear on this class of problems. We discuss some of the subsequent work next.

4   Subsequent Work and Open Problems

4.1   Boolean Functions over $\mathbb{F}_2^n$

Our work can be viewed as a step towards the proof of the following conjecture.

Conjecture 1. Suppose $\mathcal{F}$ is a linear-invariant property of functions mapping $\{0,1\}^n$ to $\{0,1\}$. Then, $\mathcal{F}$ is locally testable with a proximity-oblivious tester *if and only if $\mathcal{F}$ is locally characterized.*

It's not hard to show one direction, namely that any linear-invariant property that has a proximity-oblivious local tester is locally characterized. The proof of this is analogous to the proof of the corresponding statement for graphs in Theorem 4.7 of [GR09]. Our work in [BCSX09] makes some progress in the opposite direction but is restricted due to two obstacles. The first restriction is that we have to assume that the characterization of the property corresponds to a graphic matroid, and secondly, we have to assume that the property is monotone.

The restriction that the underlying matroid be graphic was tackled independently by Král et al. [KSV08] and Shapira [Sha09]. (It turns out that such a step also relates closely to a conjecture of Green [Gre05] about solutions to linear systems over the integers.) They showed[2] the following in our terminology.

Theorem 3 ([KSV08, Sha09]). *If a linear-invariant property is locally characterized and monotone, then it is locally testable with a proximity-oblivious tester.*

---

[2] More precisely, Theorem 3 follows from the main result of [KSV08] and [Sha09], along with a twist to handle non-uniformity of the property with respect to n, similar to what is done in the proof of Theorem 4.7 in [GR09].

The techniques used to prove this theorem were somewhat different from those in [BCSX09]. Both [KSV08] and [Sha09] gave ingenious reductions to testing whether a hypergraph is free from a fixed collection of sub-hypergraphs. Powerful tools for tackling the latter problem were already known [FR02, RS04, NRS06, Gow07, AT08], which could then be applied. Unfortunately, it is not at all clear how to reduce to a sub-hypergraph-freeness property if the linear-invariant property is not monotone.

More recently, Bhattacharyya et al. [BGS10] could remove the monotonicity condition but now again had to insist that the underlying matroid be graphic. The restriction to graphic matroids is essentially because of the same reason as in our paper [BCSX09].

Theorem 4 ([BGS10][3]). *If a linear-invariant property is locally characterized by the orbit of a constraint based on a graphic matroid, then it is locally testable with a proximity-oblivious tester.*

It remains open how to combine Theorems 3 and 4. In fact, even the special case when the property is characterized by the orbit of a single non-monotone *pattern* remains unresolved. We note that a positive resolution to questions such as the above would lead to a single unifying result capturing the theorems of Alon et al. [AKK+05] as well as Green [Gre05] – a unification that we don't have yet.

4.2   Finite-Valued Functions over $\mathbb{F}_q^n$

A general open direction is to extend the results of the previous section to arbitrary finite-valued functions over arbitrary, but constant sized, fields. There has been some partial progress for boolean-valued functions over field $\mathbb{F}_q^n$ for fixed prime power q. Theorem 3 is known to hold in this setting. (In fact, [KSV09] even shows testability for certain monotone properties of boolean functions over nonabelian groups!) The authors of [BGS10] conjecture that their techniques can be extended to prove the analog of Theorem 4 for boolean-valued functions over $\mathbb{F}_q^n$. We are not aware of any nontrivial progress for the analogous questions for non-boolean-valued functions.

More generally one could also consider functions over $\mathbb{F}_q^n$ where the field size is not a constant. We note that in such a case, single orbit characterizations do not necessarily capture all locally characterized properties, but understanding the testability of single-orbit characterized properties would remain a challenging first step.

4.3   Improving the Soundness Analysis

One of the intriguing aspects of testing non-linear linear-invariant properties is that the proof techniques employed thus far have been very different from the techniques used in the linear cases. One implication of the difference in

---

[3] We note that the main result of [BGS10] is actually stronger, since it also shows testability for certain properties which are not locally characterized, and so, do not have proximity-oblivious testers.

techniques is that the "soundness" analysis is much weaker. In particular this leads to $\tau(\epsilon)$ being much smaller than any polynomial in ϵ in Theorem 1 (as well as in the stronger forms). In contrast, in the case of linear, linear-invariant properties, the growth of $\tau(\epsilon)$ is linear in ϵ (Theorem 5.20 of [KS08]). This leads to the question: Is such subpolynomial growth inherent? A positive answer to this question would be insightful in that it would explain (somewhat) the need for new proof techniques in the case of testing non-linear, linear-invariant properties. Partial progress in this direction is reported in the work of Bhattacharyya and Xie [BX10]. They show that distinguishing triangle-free functions from those ϵ-far from triangle-free with constant probability requires $(1/\epsilon)^{1+\alpha}$ queries for some positive constant α, thus separating non-linear linear-invariant properties from linear linear-invariant ones.

In the other direction, Fox has recently shown [Fox10] that the use of the Szemerédi regularity lemma can be avoided for the analysis of testing subgraph-freeness, and the soundness analysis can be (very) mildly improved. This translates to a corresponding improvement for the properties considered in [BCSX09] also. Perhaps it is possible to strengthen such an approach to get much better bounds than we currently have.

References

[AKK+05] Alon, N., Kaufman, T., Krivelevich, M., Litsyn, S., Ron, D.: Testing Reed-Muller codes. IEEE Transactions on Information Theory 51(11), 4032–4039 (2005)

[AT08] Austin, T., Tao, T.: On the testability and repair of hereditary hypergraph properties. Random Structures and Algorithms (2008) (to appear), http://arxiv.org/abs/0801.2179

[BCSX09] Bhattacharyya, A., Chen, V., Sudan, M., Xie, N.: Testing linear-invariant non-linear properties. In: Symposium on Theoretical Aspects of Computer Science, pp. 135–146 (2009)

[BFL91] Babai, L., Fortnow, L., Lund, C.: Non-deterministic exponential time has two-prover interactive protocols. Computational Complexity 1(1), 3–40 (1991)

[BGS10] Bhattacharyya, A., Grigorescu, E., Shapira, A.: A unified framework for testing linear-invariant properties. To appear in FOCS (2010)

[BLR93] Blum, M., Luby, M., Rubinfeld, R.: Self-testing/correcting with applications to numerical problems. J. Comp. Sys. Sci. 47, 549–595 (1993); Earlier version in STOC 1990

[BX10] Bhattacharyya, A., Xie, N.: Lower bounds for testing triangle-freeness in boolean functions. In: Proc. 21st ACM-SIAM Symposium on Discrete Algorithms, pp. 87–98 (2010)

[Fox10] Fox, J.: A new proof of the graph removal lemma. Technical report (June 2010), http://arxiv.org/abs/1006.1300

[FR02] Frankl, P., Rödl, V.: Extremal problems on set systems. Random Structures and Algorithms 20(2), 131–164 (2002)

[GGR98] Goldreich, O., Goldwasser, S., Ron, D.: Property testing and its connection to learning and approximation. Journal of the ACM 45, 653–750 (1998)

[Gow07]   Gowers, W.T.: Hypergraph regularity and the multidimensional Szemerédi theorem. Annals of Mathematics 166(3), 897–946 (2007)

[GR09]   Goldreich, O., Ron, D.: On proximity oblivious testing. In: Proc. 41st Annual ACM Symposium on the Theory of Computing, pp. 141–150 (2009)

[Gre05]   Green, B.: A Szemerédi-type regularity lemma in abelian groups, with applications. Geom. Funct. Anal. 15(2), 340–376 (2005)

[GT06]   Green, B., Tao, T.: Linear equations in primes. Annals of Mathematics (2006) (to appear)

[JPRZ04]   Jutla, C.S., Patthak, A.C., Rudra, A., Zuckerman, D.: Testing low-degree polynomials over prime fields. In: Proc. 45th Annual IEEE Symposium on Foundations of Computer Science, pp. 423–432 (2004)

[KR06]   Kaufman, T., Ron, D.: Testing polynomials over general fields. SIAM J. on Comput. 36(3), 779–802 (2006)

[KS08]   Kaufman, T., Sudan, M.: Algebraic property testing: the role of invariance. In: Proc. 40th Annual ACM Symposium on the Theory of Computing, pp. 403–412. ACM, New York (2008)

[KSV08]   Král', D., Serra, O., Vena, L.: A removal lemma for systems of linear equations over finite fields (2008)

[KSV09]   Král, D., Serra, O., Vena, L.: A combinatorial proof of the removal lemma for groups. Journal of Combinatorial Theory 116(4), 971–978 (2009)

[NRS06]   Nagle, B., Rödl, V., Schacht, M.: The counting lemma for regular k-uniform hypergraphs. Random Structures and Algorithms 28(2), 113–179 (2006)

[RS96]   Rubinfeld, R., Sudan, M.: Robust characterizations of polynomials with applications to program testing. SIAM J. on Comput. 25, 252–271 (1996)

[RS04]   Rödl, V., Skokan, J.: Regularity lemma for k-uniform hypergraphs. Random Structures and Algorithms 25(1), 1–42 (2004)

[Sha09]   Shapira, A.: Green's conjecture and testing linear-invariant properties. In: Proc. 41st Annual ACM Symposium on the Theory of Computing, pp. 159–166 (2009)

Optimal Testing of Reed-Muller Codes

Arnab Bhattacharyya[1], Swastik Kopparty[1], Grant Schoenebeck[2],
Madhu Sudan[3], and David Zuckerman[4]

[1] MIT
[2] UC Berkeley
[3] Microsoft Research
[4] UT Austin

Abstract. We consider the problem of testing if a given function $f : \mathbb{F}_2^n \to \mathbb{F}_2$ is close to any degree d polynomial in n variables, also known as the problem of testing Reed-Muller codes. We are interested in determining the query-complexity of distinguishing with constant probablity between the case where f is a degree d polynomial and the case where f is $\Omega(1)$-far from all degree d polynomials. Alon et al. [AKK+05] proposed and analyzed a natural 2^{d+1}-query test T_0, and showed that it accepts every degree d polynomial with probability 1, while rejecting functions that are $\Omega(1)$-far with probability $\Omega(1/(d2^d))$. This leads to a $O(d4^d)$-query test for degree d Reed-Muller codes.

We give an asymptotically optimal analysis of T_0, showing that it rejects functions that are $\Omega(1)$-far with $\Omega(1)$-probability (so the rejection probability is a universal constant independent of d and n). In particular, this implies that the query complexity of testing degree d Reed-Muller codes is $O(2^d)$.

Our proof works by induction on n, and yields a new analysis of even the classical Blum-Luby-Rubinfeld [BLR93] linearity test, for the setting of functions mapping $\mathbb{F}_2^n$ to $\mathbb{F}_2$. Our results also imply a "query hierarchy" result for property testing of affine-invariant properties: For every function $q(n)$, it gives an affine-invariant property that is testable with $O(q(n))$-queries, but not with $o(q(n))$-queries, complementing an analogous result of [GKNR08] for graph properties.

This is a brief overview of the results in the paper [BKS+09].

Keywords: low-degree polynomials, Gowers norm, affine-invariant codes.

1 Introduction

We consider the task of testing if a Boolean function f on n bits, given by an oracle, is close to a degree d multivariate polynomial (over $\mathbb{F}_2$, the field of two elements). This specific problem, also known as the testing problem for the Reed-Muller code, was considered previously by Alon, Kaufman, Krivelevich, Litsyn, and Ron [AKK+05] who proposed and analyzed a natural 2^{d+1}-query test for this task. In this work we give an improved, asymptotically optimal, analysis of their test. Below we describe the problem, its context, our results and some implications.

O. Goldreich (Ed.): Property Testing, LNCS 6390, pp. 269–275, 2010.
© Springer-Verlag Berlin Heidelberg 2010

2   Reed-Muller Codes and Testing

The Reed-Muller codes are parameterized by two parameters: n the number of variables and d the degree parameter. The Reed-Muller codes consist of all functions from $\mathbb{F}_2^n \to \mathbb{F}_2$ that are evaluations of polynomials of degree at most d. We use $\mathrm{RM}(d, n)$ to denote this class, i.e., $\mathrm{RM}(d, n) = \{f : \mathbb{F}_2^n \to \mathbb{F}_2| \deg(f) \leq d\}$.

The proximity of functions is measured by the (fractional Hamming) distance. Specifically, for functions $f, g : \mathbb{F}_2^n \to \mathbb{F}_2$, we let the *distance* between them, denoted by $\delta(f, g)$, be the quantity $\Pr_{x \leftarrow_U \mathbb{F}_2^n}[f(x) \neq g(x)]$. For a family of functions $\mathcal{F} \subseteq \{g : \mathbb{F}_2^n \to \mathbb{F}_2\}$ let $\delta(f, \mathcal{F}) = \min\{\delta(f, g)|g \in \mathcal{F}\}$. We say f is δ-close to $\mathcal{F}$ if $\delta(f, \mathcal{F}) \leq \delta$ and δ-far otherwise.

Let $\delta_d(f) = \delta(f, \mathrm{RM}(d, n))$ denote the distance of f to the class of degree d polynomials. The goal of Reed-Muller testing is to "test", with "few queries" of f, whether $f \in \mathrm{RM}(d, n)$ or if f is far from $\mathrm{RM}(d, n)$. Specifically, for a function $q : \mathbb{Z}^+ \times \mathbb{Z}^+ \times (0, 1] \to \mathbb{Z}^+$, a q-*query tester* for the class $\mathrm{RM}(d, n)$ is a randomized oracle algorithm T that, given oracle access to some function $f : \mathbb{F}_2^n \to \mathbb{F}_2$ and a proximity parameter $\delta \in (0, 1]$, queries at most $q = q(d, n, \delta)$ values of f and accepts $f \in \mathrm{RM}(d, n)$ with probability 1, while if $\delta(f, \mathrm{RM}(d, n)) \geq \delta$ it rejects with probability at least, say, $1/2$. The function q is the *query complexity* of the test and the main goal here is to make q as small as possible, as a function possibly of d, n and δ. We denote the test T run using oracle access to the function f by T^f

This task was already considered by Alon et al. [AKK+05] who gave a tester with query complexity $O(\frac{d}{\delta} \cdot 4^d)$. This tester repeated a simple $O(2^d)$-query test, that we denote T_*, several times. Given oracle access to f, T_* selects a $(d + 1)$-dimensional affine subspace A, and accepts if f restricted to A is a degree d polynomial. This requires 2^{d+1} queries of f (since that is the number of points contained in A). [AKK+05] show that if $\delta(f) \geq \delta$ then T_* rejects f with probability $\Omega(\delta/(d \cdot 2^d))$. Their final tester then simply repeated T_* $O(\frac{d}{\delta} \cdot 2^d)$ times and accepted if all invocations of T_* accepted. The important feature of this result is that the number of queries is independent of n, the dimension of the ambient space. Alon et al. also show that any tester for $\mathrm{RM}(d, n)$ must make at least $\Omega(2^d + 1/\delta)$ queries. Thus their result was tight to within almost quadratic factors, but left a gap open. We close this gap in this work.

3   Main Result

Our main result is an improved analysis of the basic 2^{d+1}-query test T_*. We show that if $\delta_d(f) \geq 0.1$, in fact even if it's at least $0.1 \cdot 2^{-d}$, then in fact this basic test rejects with probability lower bounded by some *absolute constant*. We now give a formal statement of our main theorem.

Theorem 1. *There exists a constant $\epsilon_1 > 0$ such that for all d, n, and for all functions $f : \mathbb{F}_2^n \to \mathbb{F}_2$, we have*

$$\Pr[T_*^f \; rejects] \geq \min\{2^d \cdot \delta_d(f), \epsilon_1\}.$$

Therefore, to reject functions δ-far from $\mathrm{RM}(d, n)$ with constant probability, one can repeat the test T_* at most $O(1/\min\{2^d\delta_d(f), \epsilon_1\}) = O(1 + \frac{1}{2^d\delta})$ times, making the total query complexity $O(2^d + 1/\delta)$. This query complexity is asymptotically tight in view of the earlier mentioned lower bound in [AKK$^+$05].

Our error-analysis is also asymptotically tight. Note that our theorem effectively states that functions that are accepted by T_* with constant probability (close to 1) are (very highly) correlated with degree d polynomials. To get a qualitative improvement one could hope that every function that is accepted by T_* with probability strictly greater than half is somewhat correlated with a degree d polynomial. Such stronger statements however are effectively ruled out by the counterexamples to the "inverse conjecture for the Gowers norm" given by [LMS08, GT07]. Since the analysis given in these works does not match our parameters asymptotically, we show how an early analysis due to the authors of [LMS08] can be used to show the asymptotic tightness of the parameters of Theorem 1.

Our main theorem (Theorem 1) is obtained by a novel proof that gives a (yet another!) new analysis even of the classical linearity test of Blum, Luby, Rubinfeld [BLR93]. Below we explain some of the context of our work and some implications.

4   Query Hierarchy for Affine-Invariant Properties

Our result falls naturally in the general framework of property testing [BLR93, RS96, GGR98]. Goldreich et al. [GKNR08] asked an interesting question in this broad framework: Given an ensemble of properties $\mathcal{F} = \{\mathcal{F}_N\}_N$ where $\mathcal{F}_N$ is a property of functions on domains of size N, which functions correspond to the query complexity of some property? That is, for a given complexity function $q(N)$, is there a corresponding property $\mathcal{F}$ such that $\Theta(q(N))$-queries are necessary and sufficient for testing membership in $\mathcal{F}_N$? This question is interesting even when we restrict the class of properties being considered.

For completely general properties this question is easy to solve. For graph properties [GKNR08] et al. show that for every efficiently computable function $q(N) = O(N)$ there is a graph property for which $\Theta(q(N))$ queries are necessary and sufficient (on graphs on $\Omega(\sqrt{N})$ vertices). Thus this gives a "hierarchy theorem" for query complexity.

Our main theorem settles the analogous question in the setting of "affine-invariant" properties. Given a field $\mathbb{F}$, a property $\mathcal{F} \subseteq \{\mathbb{F}^n \to \mathbb{F}\}$ is said to be affine-invariant if for every $f \in \mathcal{F}$ and affine map $A : \mathbb{F}^n \to \mathbb{F}^n$, the composition of f with A, i.e, the function $f \circ A(x) = f(A(x))$, is also in $\mathcal{F}$. Affine-invariant properties seem to be the algebraic analog of graph-theoretic properties and generalize most natural algebraic properties (see Kaufman and Sudan [KS08]).

Since the Reed-Muller codes form an affine-invariant family, and since we have a tight analysis for their query complexity, we can get the affine-invariant version of the result of [GKNR08]. Specifically, given any (reasonable) query complexity function $q(N)$ consider N that is a power of two and consider the class of functions on $n = \log_2 N$ variables of degree at most $d = \lceil \log_2 q(N) \rceil$. We have that membership in this family requires $\Omega(2^d) = \Omega(q(N))$-queries, and on the other hand $O(2^d) = O(q(N))$-queries also suffice, giving an ensemble of properties $\mathcal{P}_N$ (one for every $N = 2^n$) that is testable with $\Theta(q(N))$-queries.

Theorem 2. *For every* $q : \mathbb{N} \to \mathbb{N}$ *that is at most linear, there is an affine-invariant property that is testable with* $O(q(n))$ *queries (with one-sided error) but is not testable in* $o(q(n))$ *queries (even with two-sided error). Namely, this property is membership in* $\mathrm{RM}(\lceil \log_2 q(n) \rceil, n)$.

5   Gowers Norm

A quantity closely related to the rejection probability for T_* also arises in some of the recent results in additive number theory, under the label of the *Gowers norm*, introduced by Gowers [Gow98, Gow01].

To define this norm, we first consider a related test $T_0^f(k)$ which, given parameter k and oracle access to a function f, picks $x_0, a_1, \ldots, a_k \in \mathbb{F}_2^n$ uniformly and independently and accepts if f restricted to the affine subspace $x_0 + \mathrm{span}(a_1, \ldots, a_k)$ is a degree $k - 1$ polynomial. Note that since we don't require $a_1, \ldots, a_k$ to be linearly independent, T_0 sometimes (though rarely) picks a subspace of dimension $k - 1$ or less. When $k = d + 1$, if we condition on the event that $a_1, \ldots, a_k$ are linearly independent, $T_0(d + 1)$ behaves exactly as T_*. On the other hand when $a_1, \ldots, a_k$ do have a linear dependency, $T_0(k)$ accepts with probability one. It turns out that when $n \geq d + 1$, the probability that $a_1, \ldots, a_{d+1}$ are linearly independent is lower bounded by a constant, and so the rejection probability of $T_0(d + 1)$ is lower bounded by a constant multiple of the rejection probability of T_* (for every function f). The test T_0 has a direct relationship with the Gowers norm.

In our notation, the Gowers norm can be defined as follows. For a function $f : \mathbb{F}_2^n \to \mathbb{F}_2$, the kth-Gowers norm of f, denoted $\|f\|_{U^k}$, is given by the expression

$$\|f\|_{U^k} \overset{\text{def}}{=} (\Pr[T_0^f(k) \text{ accepts}] - \Pr[T_0^f(k) \text{ rejects}])^{\frac{1}{2^k}}.$$

Gowers [Gow01] (see also [GT05]) showed that the "correlation" of f to the closest degree d polynomial, i.e., the quantity $1 - 2\delta_d(f)$, is at most $\|f\|_{U^{d+1}}$. The well-known Inverse Conjecture for the Gowers Norm states that some sort of converse holds: if $\|f\|_{U^{d+1}} = \Omega(1)$, then the correlation of f to some degree d polynomial is $\Omega(1)$, or equivalently $\delta_d(f) = 1/2 - \Omega(1)$. (That is, if the acceptance probability of T_0 is slightly larger than $1/2$, then f is at distance slightly smaller than $1/2$ from some degree d polynomial.) Lovett et al. [LMS08] and Green and Tao [GT07] disproved this conjecture, showing that the symmetric polynomial S_4 has $\|S_4\|_{U^4} = \Omega(1)$ but the correlation of S_4 to any degree 3 polynomial is exponentially small. This still leaves open the question of establishing tighter relationships between the Gowers norm $\|f\|_{U^{d+1}}$ and the maximal correlation of f to some degree d polynomial. The best analysis known seems to be in the work of [AKK$^+$05] whose result can be interpreted as showing that there exists $\epsilon > 0$ such that if $\|f\|_{U^{d+1}} \geq 1 - \epsilon/4^d$, then $\delta_d(f) = O(4^d(1 - \|f\|_{U^{d+1}}))$.

Our results show that when the Gowers norm is close to 1, there is actually a tight relationship between the Gowers norm and distance to degree d. More precisely, there exists $\epsilon > 0$ such that if $\|f\|_{U^{d+1}} \geq 1 - \epsilon/2^d$, then $\delta_d(f) = \Theta(1 - \|f\|_{U^{d+1}})$.

6   XOR Lemma for Low-Degree Polynomials

One application of the Gowers norm and the Alon et al. analysis to complexity theory is an elegant "hardness amplification" result for low-degree polynomials, due to Viola and Wigderson [VW07]. Let $f : \mathbb{F}_2^n \to \mathbb{F}_2$ be such that $\delta_d(f)$ is noticeably large, say ≥ 0.1. Viola and Wigderson showed how to use this f to construct a $g : \mathbb{F}_2^m \to \mathbb{F}_2$ such that $\delta_d(g)$ is significantly larger, around $\frac{1}{2} - 2^{-\Omega(m)}$. In their construction, $g = f^{\oplus t}$, the t-wise XOR of f, where $f^{\oplus t} : (\mathbb{F}_2^n)^t \to \mathbb{F}_2$ is given by:

$$f^{\oplus t}(x_1, \ldots, x_t) = \sum_{i=1}^{t} f(x_i).$$

In particular, they showed that if $\delta_d(f) \geq 0.1$, then $\delta_d(f^{\oplus t}) \geq 1/2 - 2^{-\Omega(t/4^d)}$. Their proof proceeded by studying the rejection probabilities of T_* on the functions f and $f^{\oplus t}$. The analysis of the rejection probability of T_* given by [AKK$^+$05] was a central ingredient in their proof. By using our improved analysis of the rejection probability of T_* from Theorem 1 instead, we get the following improvement.

Theorem 3. *Let ϵ_1 be as in Theorem 1. Let $f : \mathbb{F}_2^n \to \mathbb{F}_2$. Then*

$$\delta_d(f^{\oplus t}) \geq \frac{1 - (1 - 2\min\{\epsilon_1/4, 2^{d-2} \cdot \delta_d(f)\})^{t/2^d}}{2}.$$

In particular, if $\delta_d(f) \geq 0.1$, then $\delta_d(f^{\oplus t}) \geq 1/2 - 2^{-\Omega(t/2^d)}$.

7   Technique

The heart of our proof of the main theorem (Theorem 1) is an inductive argument on n, the dimension of the ambient space. While proofs that use induction on n have been used before in the literature on low-degree testing (see, for instance, [BFL91, BFLS91, FGL$^+$96]), they tend to have a performance guarantee that degrades significantly with n. Indeed no inductive proof was known even for the case of testing linearity of functions from $\mathbb{F}_2^n \to \mathbb{F}_2$ that showed that functions at $\Omega(1)$ distance from linear functions are rejected with $\Omega(1)$ probability. (We note that the original analysis of [BLR93] as well as the later analysis of [BCH$^+$96] do give such bounds - but they do not use induction on n.) In the process of giving a tight analysis of the [AKK$^+$05] test for Reed-Muller codes, we thus end up giving a new (even if weaker) analysis of the linearity test over $\mathbb{F}_2^n$. Below we give the main idea behind our proof.

Consider a function f that is δ-far from every degree d polynomial. For a "hyperplane", i.e., an $(n-1)$-dimensional affine subspace A of $\mathbb{F}_2^n$, let $f|_A$ denote the restriction of f to A. We first note that the test can be interpreted as first picking a random hyperplane A in $\mathbb{F}_2^n$ and then picking a random $(d+1)$-dimensional affine subspace A' within A and testing if $f|_{A'}$ is a degree d polynomial. Now, if on every hyperplane A, $f|_A$ is still δ-far from degree d polynomials then we would be done by the inductive hypothesis. In fact our hypothesis gets weaker as $n \to \infty$, so that we can even afford a few hyperplanes where $f|_A$ is not δ-far. The crux of our analysis is when $f|_A$ is close to some degree d polynomial P_A for several (but just $O(2^d)$) hyperplanes. In this case we manage to "sew" the different polynomials P_A (each defined on some $(n-1)$-dimensional subspace within $\mathbb{F}_2^n$) into a degree d polynomial P that agrees with *all* the P_A's. We then show that this polynomial is close to f, completing our argument.

To stress the novelty of our proof, note that this is not a "self-correction" argument as in [AKK+05], where one defines a natural function that is close to P, and then works hard to prove it is a polynomial of appropriate degree. In contrast, our function is a polynomial by construction and the harder part (if any) is to show that the polynomial is close to f. Moreover, unlike other inductive proofs, our main gain is in the fact that the new polynomial P has degree no greater than that of the polynomials given by the induction.

The proofs of the theorems mentioned above may be found in our paper [BKS+09].

References

[AB01]   Alon, N., Beigel, R.: Lower bounds for approximations by low degree polynomials over $\mathbb{Z}_m$. In: IEEE Conference on Computational Complexity, pp. 184–187 (2001)

[AKK+05]   Alon, N., Kaufman, T., Krivelevich, M., Litsyn, S., Ron, D.: Testing Reed-Muller codes. IEEE Transactions on Information Theory 51(11), 4032–4039 (2005)

[BCH+96]   Bellare, M., Coppersmith, D., Håstad, J., Kiwi, M., Sudan, M.: Linearity testing over characteristic two. IEEE Transactions on Information Theory 42(6), 1781–1795 (1996)

[BCJ+06]   Brown, M.V., Calkin, N.J., James, K., King, A.J., Lockard, S., Rhoades, R.C.: Trivial Selmer groups and even partitions of a graph. INTEGERS 6 (December 2006)

[BFL91]   Babai, L., Fortnow, L., Lund, C.: Non-deterministic exponential time has two-prover interactive protocols. Computational Complexity 1(1), 3–40 (1991)

[BFLS91]   Babai, L., Fortnow, L., Levin, L.A., Szegedy, M.: Checking computations in polylogarithmic time. In: Proceedings of the 23rd ACM Symposium on the Theory of Computing, pp. 21–32. ACM Press, New York (1991)

[BKS+09]   Bhattacharyya, A., Kopparty, S., Schoenebeck, G., Sudan, M., Zuckerman, D.: Optimal testing of Reed-Muller codes. ECCC Technical Report, TR09-086 (October 2009)

[BLR93]   Blum, M., Luby, M., Rubinfeld, R.: Self-testing/correcting with applica-
           tions to numerical problems. J. Comp. Sys. Sci. 47, 549–595 (1993); Earlier
           version in STOC 1990 (1990)
[BM88]   Brent, R.P., McKay, B.D.: On determinants of random symmetric matrices
           over $\mathbb{Z}_m$. ARS Combinatoria 26A, 57–64 (1988)
[FGL⁺96]   Feige, U., Goldwasser, S., Lovász, L., Safra, S., Szegedy, M.: Interac-
           tive proofs and the hardness of approximating cliques. Journal of the
           ACM 43(2), 268–292 (1996)
[GGR98]   Goldreich, O., Goldwasser, S., Ron, D.: Property testing and its connection
           to learning and approximation. Journal of the ACM 45, 653–750 (1998)
[GKNR08]   Goldreich, O., Krivelevich, M., Newman, I., Rozenberg, E.: Hierarchy the-
           orems for property testing. Electronic Colloquium on Computational Com-
           plexity (ECCC) 15(097) (2008)
[Gow98]   Gowers, W.T.: A new proof of Szemerédi's theorem for arithmetic progres-
           sions of length four. Geometric Functional Analysis 8(3), 529–551 (1998)
[Gow01]   Gowers, W.T.: A new proof of Szemerédi's theorem. Geometric Functional
           Analysis 11(3), 465–588 (2001)
[GT05]   Green, B., Tao, T.: An inverse theorem for the Gowers U^3 norm.
           arXiv.org:math/0503014 (2005)
[GT07]   Green, B., Tao, T.: The distribution of polynomials over finite fields, i
           with applications to the Gowers norms. Technical report (November 2007),
           http://arxiv.org/abs/0711.3191v1
[KS08]   Kaufman, T., Sudan, M.: Algebraic property testing: the role of invari-
           ance. In: STOC 2008: Proceedings of the 40th annual ACM symposium
           on Theory of computing, pp. 403–412. ACM, New York (2008)
[LMS08]   Lovett, S., Meshulam, R., Samorodnitsky, A.: Inverse conjecture for the
           Gowers norm is false. In: Ladner, R.E., Dwork, C. (eds.) STOC, pp. 547–
           556. ACM, New York (2008)
[RS96]   Rubinfeld, R., Sudan, M.: Robust characterizations of polynomials with
           applications to program testing. SIAM J. on Comput. 25, 252–271 (1996)
[VW07]   Viola, E., Wigderson, A.: Norms, XOR lemmas, and lower bounds for
           GF(2) polynomials and multiparty protocols. In: Twenty-Second Annual
           IEEE Conference on Computational Complexity, CCC 2007, pp. 141–154
           (June 2007)

Query-Efficient Dictatorship Testing with Perfect Completeness

Victor Chen

Institute of Theoretical Computer Science, Tsinghua University, Beijing,
People's Republic of China
victor.vc@gmail.com

Abstract. The problem of dictatorship testing is often used a starting
in constructing a PCP system. Samorodnitsky and Trevisan in STOC
2006 designed a dictatorship test that makes q queries and has soundness
approximately $O(q \cdot 2^{-q})$. However, their test has imperfect completeness.
We describe some of the progress made in designing dictatorship tests
with perfect completeness.

Keywords: PCPs, dictatorship test.

Linearity and dictatorship testing have been studied in the past decade both for
their combinatorial interest and connection to complexity theory. These tests
distinguish functions which are linear/dictator from those which are far from
being a linear/dictator function. The tests do so by making queries to a function
at certain points and receiving the function's values at these points. The parameters of interest are the number of queries a test makes and the completeness
and soundness of a test.

In this note, we consider boolean functions of the form $f : \{0,1\}^n \to \{-1,1\}$.
We say a function f is *linear* if $f = (-1)^{\sum_{i \in S} x_i}$ for some subset $S \subseteq [n]$. A *dic-
tator* function is simply a linear function where $|S| = 1$, i.e., $f(x) = (-1)^{x_i}$ for
some i. A dictator function is often called a *long code*, and it is first used in [3] for
the constructions of probabilistic checkable proofs (PCPs), see e.g., [2,1]. Since
then, it has become standard to design a PCP system as the composition of
two verifiers, an outer verifier and an inner verifier. In such case, a PCP system
expects the proof to be written in such a way so that the outer verifier, typically
based on the verifier obtained from Raz's Parallel Repetition Theorem [18], selects some tables of the proof according to some distribution and then passes
the control to the inner verifier. The inner verifier, with oracle access to these
tables, makes queries into these tables and ensures that the tables are the encoding of some error-correcting codes and satisfy some joint constraint. The long
code encoding is usually employed in these proof constructions, and the inner
verifier simply tests whether a collection of tables (functions) are long codes
satisfying some constraints. Following this paradigm, constructing a PCP with
certain parameters reduces to the problem of designing a long code test with
similar parameters.

O. Goldreich (Ed.): Property Testing, LNCS 6390, pp. 276–279, 2010.

One question of interest is the tradeoff between the soundness and query complexity of a tester. If a tester queries the functions at every single value, then trivially the verifier can determine all the functions. One would like to construct a dictatorship test that has the lowest possible soundness while making as few queries as possible. One way to measure this tradeoff between the soundness s and the number of queries q is *amortized query complexity*, defined as $\frac{q}{\log s^{-1}}$. This investigation, initiated in [25], has since spurred a long sequence of works [22,20,11,6]. All the testers from these works run many iterations of a single dictatorship test by reusing queries from previous iterations. The techniques used are Fourier analytic, and the best amortized query complexity from this sequence of works has the form $1 + O\left(\frac{1}{\sqrt{q}}\right)$.

The next breakthrough occurs when Samorodnitsky [19] introduces the notion of a *relaxed* linearity test along with new ideas from additive combinatorics. In property testing, the goal is to distinguish objects that are very structured from those that are pseudorandom. In the case of linearity/dictatorship testing, the structured objects are the linear/dictator functions, and functions that are far from being linear/dictator are interpreted as pseudorandom. The recent paradigm in additive combinatorics is to find the right framework of structure and pseudorandomness and analyze combinatorial objects by dividing them into structured and pseudorandom components, see e.g. [24] for a survey. One success is the notion of Gowers norm [7], which has been fruitful in attacking many problems in additive combinatorics and computer science. In [19], the notion of pseudorandomness for linearity testing is relaxed; instead of designating the functions that are far from being linear as pseudorandom, the functions having small low degree Gowers norm are considered to be pseudorandom. By doing so, an optimal tradeoff between soundness and query complexity is obtained for the problem of relaxed linearity testing. (Here the tradeoff is stronger than the tradeoff for the traditional problem of linearity testing.)

In a similar fashion, in the PCP literature since [9], the pseudorandom objects in dictatorship tests are not functions that are far from being a dictator. The pseudorandom functions are typically defined to be either functions that are far from all "juntas" or functions whose "low-degree influences" are $o(1)$. Both considerations of a dictatorship test are sufficient to compose the test in a PCP construction. In [21], building on the analysis of the relaxed linearity test in [19], Samorodnitsky and Trevisan construct a dictatorship test (taking the view that functions with arbitrary small "low-degree influences are pseudorandom) with amortized query complexity $1 + O\left(\frac{\log q}{q}\right)$. Furthermore, the test is used as the inner verifier in a conditional PCP construction (based on unique games [12]) with the same parameters. However, their dictatorship test suffers from an inherent loss of perfect completeness. Ideally one would like testers with one-sided errors. One, for aesthetic reasons, testers should always accept valid inputs. Two, for some hardness of approximation applications, in particular coloring problems (see e.g. [10] or [5]), it is important to construct PCP systems with one-sided errors.

In [4], the following theorem is proved:

Theorem 1. *For every $q \geq 3$, there exists an (adaptive) dictatorship test that makes q queries, has completeness 1, and soundness $\frac{O(q^3)}{2^q}$; in particular it has amortized query complexity $1 + O\left(\frac{\log q}{q}\right)$.*

The tester is a variant of the one given in [21] and is adaptive in the sense that it makes its queries in two stages. It first makes roughly $\log q$ nonadaptive queries into the function. Based on the values of these queries, the tester then selects the rest of the query points nonadaptively. The analysis is based on techniques developed in [11,21,10,8].

Recently, Tamaki and Yoshida in their ECCC preprint [23] designed a completely new dicatorship test. Their test, in contrast to the one in [4], is nonadaptive and has slightly better soundness. Formally, they proved the following:

Theorem 2. *For every $q \geq 3$, there exists a non-adaptive dictatorship test that makes q queries, has completeness 1, and soundness $O(q \cdot 2^{-q})$.*

Unfortunately, it is not clear how to extend the tests in [4,23] to a PCP construction. One possibility might lie in the works of O'Donnell and Wu [15,16], where they first constructed an optimal three bit dictatorship test with perfect completeness and and extended their technique to construct a conditional PCP system. Similar to the 3-bit test [15], it may be possible to extend the query-efficient dictator tests [4,23] to PCPs using Khot's d-to-1 outer verifier [12]. In particular, we leave the following conjecture as a challenging open problem:

Conjecture 1. For infinitely many q, there exists a PCP system that makes q queries, has completeness 1, and soundness $\text{poly}(q) \cdot 2^{-q}$.

References

1. Arora, S., Lund, C., Motwani, R., Sudan, M., Szegedy, M.: Proof verification and the hardness of approximation problems. J. ACM 45(3), 501–555 (1998)
2. Arora, S., Safra, S.: Probabilistic checking of proofs: a new characterization of NP. J. ACM 45(1), 70–122 (1998)
3. Bellare, M., Goldreich, O., Sudan, M.: Free bits, PCPs, and nonapproximability–towards tight results. SIAM Journal on Computing 27(3), 804–915 (1998)
4. Chen, V.: A hypergraph dictatorship test with perfect completeness. In: Dinur, I., Jansen, K., Naor, J., Rolim, J. (eds.) APPROX-RANDOM 2009. LNCS, vol. 5687, pp. 448–461. Springer, Heidelberg (2009)
5. Dinur, I., Mossel, E., Regev, O.: Conditional Hardness for Approximate Coloring. SIAM Journal on Computing 39(3), 843–873 (2009)
6. Engebretsen, L., Holmerin, J.: More Efficient Queries in PCPs for NP and Improved Approximation Hardness of Maximum CSP. In: Diekert, V., Durand, B. (eds.) STACS 2005. LNCS, vol. 3404, pp. 194–205. Springer, Heidelberg (2005)
7. Gowers, W.T.: A new proof of Szemerédi's theorem. Geom. Funct. Anal. 11(3), 465–588 (2001)

8. Guruswami, V., Lewin, D., Sudan, M., Trevisan, L.: A tight characterization of NP with 3 query PCPs. In: FOCS, pp. 8–17 (1998)
9. Håstad, J.: Some optimal inapproximability results. J. of ACM 48(4), 798–859 (2001)
10. Håstad, J., Khot, S.: Query Efficient PCPs with Perfect Completeness. Theory of Computing 1(7), 119–148 (2005)
11. Håstad, J., Wigderson, A.: Simple analysis of graph tests for linearity and PCP. Random Struct. Algorithms 22(2), 139–160 (2003)
12. Khot, S.: On the power of unique 2-prover 1-round games. In: STOC, pp. 767–775 (2002)
13. Khot, S., Kindler, G., Mossel, E., O'Donnell, R.: Optimal Inapproximability Results for MAX-CUT and Other 2-Variable CSPs? SIAM Journal on Computing 37(1), 319–357 (2007)
14. Khot, S., Saket, R.: A 3-Query Non-Adaptive PCP with Perfect Completeness. In: CCC, pp. 159–169 (2006)
15. O'Donnell, R., Wu, Y.: 3-bit dictator testing: 1 vs. 5/8. In: SODA, pp. 365–373 (2009)
16. O'Donnell, R., Wu, Y.: Conditional Hardness for Satisfiable-3CSPs. In: STOC, pp. 493–502 (2009)
17. Parnas, M., Ron, D., Samorodnitsky, A.: Testing Basic Boolean Formulae. SIAM Journal on Discrete Mathematics 16(1), 20–46 (2002)
18. Raz, R.: A Parallel Repetition Theorem. SIAM Journal on Computing 27(3), 763–803 (1998)
19. Samorodnitsky, A.: Low-degree tests at large distances. In: STOC, pp. 506–515 (2007)
20. Samorodnitsky, A., Trevisan, L.: A PCP characterization of NP with optimal amortized query complexity. In: STOC, pp. 191–199 (2000)
21. Samorodnitsky, A., Trevisan, L.: Gowers uniformity, influence of variables, and PCPs. SIAM Journal on Computing 39(1), 323–360 (2009)
22. Sudan, M., Trevisan, L.: Probabilistically checkable proofs with low amortized query complexity. In: FOCS, pp. 18–27 (1998)
23. Tamaki, S., Yoshida, Y.: A query efficient non-adaptive long code test with perfect completeness. ECCC, TR09-074 (2009)
24. Tao, T.: Structure and randomness in combinatorics. In: FOCS, pp. 3–15 (2007)
25. Trevisan, L.: Recycling queries in PCPs and in linearity tests (extended abstract). In: STOC, pp. 299–398 (1998)

Composition of Low-Error 2-Query PCPs Using Decodable PCPs[*]

Irit Dinur[1] and Prahladh Harsha[2]

[1] Weizmann Institute of Science, Israel
irit.dinur@weizmann.ac.il
[2] Tata Institute of Fundamental Research, India
prahladh@tifr.res.in

Abstract. The main result of this paper is a generic composition theorem for low error two-query probabilistically checkable proofs (PCPs). Prior to this work, composition of PCPs was well-understood only in the constant error regime. Existing composition methods in the low error regime were non-modular (i.e., very much tailored to the specific PCPs that were being composed), resulting in complicated constructions of PCPs. Furthermore, until recently, composition in the low error regime suffered from incurring an extra 'consistency' query, resulting in PCPs that are not 'two-query' and hence, much less useful for hardness-of-approximation reductions.

In a recent breakthrough, Moshkovitz and Raz [In *Proc. 49th IEEE Symp. on Foundations of Comp. Science (FOCS)*, 2008] constructed almost linear-sized low-error 2-query PCPs for every language in NP. Indeed, the main technical component of their construction is a novel composition of certain specific PCPs. We give a modular and simpler proof of their result by repeatedly applying the new composition theorem to known PCP components.

To facilitate the new modular composition, we introduce a new variant of PCP, which we call a *decodable PCP (dPCP)*. A dPCP is an *encoding* of an NP witness that is both locally checkable and locally decodable. The dPCP verifier in addition to verifying the validity of the given proof like a standard PCP verifier, also locally decodes the original NP witness. Our composition is generic in the sense that it works regardless of the way the component PCPs are constructed.

Keywords: PCP, composition, locally decodable, low soundness error.

1 Probabilistically Checkable Proofs – Introduction

Probabilistically checkable proofs (PCPs) provide a proof format that enables verification with only a constant number of queries into the proof. This is formally captured by the (by now standard) notion of a probabilistic verifier.

---

[*] A full version of this paper appears in the Electronic Colloquium on Computational Complexity [DH09]. The current extended abstract is a modification of the introduction of the full version for the purposes of the ICS mini-workshop on propert testing.

O. Goldreich (Ed.): Property Testing, LNCS 6390, pp. 280–288, 2010.

Definition 1 (PCP Verifier). *A PCP verifier V for a language L is a polynomial time probabilistic algorithm that behaves as follows: On input x, and oracle access to (proof) string π (over an alphabet Σ), the verifier reads the input x, tosses some random coins r, and based on x and r computes a window $I = (i_1, \ldots, i_q)$ of indices to read from π, and a predicate $f : \Sigma^q \to \{0, 1\}$. The verifier then accepts iff $f(\pi_I) = 1$.*

- *The verifier is* complete *if for every $x \in L$ there is a proof π accepted with probability 1. I.e., $\exists \pi$, $\mathrm{Pr}_{I,f}[f(\pi_I) = 1] = 1$.*
- *The verifier is* sound *with soundness error $\delta < 1$ if for any $x \notin L$, every proof π is accepted with probability at most δ. I.e., $\forall \pi$, $\mathrm{Pr}_{I,f}[f(\pi_I) = 1] \leq \delta$.*

The celebrated PCP Theorem [AS98, ALM+98] states that every language in NP has a verifier that is complete and sound with a constant $\delta < 1$ soundness error while using only a logarithmic number of random coins, and reading only $q = O(1)$ proof bits. Naturally, (and motivated by the fruitful connection to inapproximability due to [FGL+96]), much attention has been given to obtaining PCPs with "good" parameters, such as $q = 2$, smallest possible soundness error δ, and smallest possible alphabet size $|\Sigma|$. These are the parameters of focus in this paper.

How does one construct PCPs with such remarkable proof checking properties? In general, it is easier to construct such PCPs if we relax the alphabet size $|\Sigma|$ to be large (typically super-constant, but sub-exponential). This issue is similar to a well-known issue that arises in coding theory; wherein it is relatively easy to construct codes with good error-correcting properties over a large, super constant sized, alphabet (e.g., Reed-Solomon codes). Codes over a constant-sized alphabet (e.g., GF(2)) are then obtained from these codes by (repeatedly) applying the "code-concatenation" technique of Forney [For66]. The equivalent notion in the context of PCP constructions is the paradigm of "proof composition", introduced by Arora and Safra [AS98]. Informally speaking, proof composition is a recursive procedure applied to PCP constructions to reduce the alphabet size. Proof composition is applied (possibly several times over) to PCPs over the large alphabet to obtain PCPs over a small (even binary) alphabet.

Proof composition is an essential ingredient of all known constructions of PCPs. Composition of PCPs with high soundness error (greater than $1/2$) is by now well understood using the notion of *PCPs of proximity* [BGH+06] (called *assignment testers* in [DR06]). These allow for modular composition, in the high soundness error regime which in turn led to alternate proofs of the PCP Theorem and constructions of shorter PCPs [BGH+06, Din08, BS08]. However, these composition theorems are inapplicable when constructing PCPs with low-soundness error (arbitrarily small soundness error or even any constant less than $1/2$). (See survey on constructing low error PCPs by Dinur [Din08] for a detailed explanation of this limitation).

Our first contribution is a definition of an object which we call a *decodable PCP*, which allows for clean and modular composition in the low error regime.

2 Decodable PCPs (dPCPs)

Consider a probabilistically checkable proof for the language CIRCUITSAT (the language of all satisfiable circuits). The natural NP proof for CIRCUITSAT is simply a satisfying assignment. An intuitive way to construct a PCP for CIRCUITSAT is to *encode* the assignment in a way that enables probabilistic checking. This intuition guides all known constructions, although it is not stipulated in the definition.

In this work, we make the intuitive notion of proof encoding explicit by introducing the notion of a *decodable PCP (dPCP)*. A dPCP for CIRCUITSAT is an encoding of the satisfying assignment that can be both verified and decoded locally in a probabilistic manner. In this setting, the verifier is supposed to both verify that the dPCP is encoding a *satisfying* assignment, as well as to decode a symbol in that assignment. More precisely, we define a *PCP decoder* for CIRCUITSAT to be (along the lines of Definition 1) a probabilistic algorithm that is given an input circuit C, oracle access to a dPCP π, and, in addition, an index i. Based on C, i and the randomness r it computes a window I and a *function* f (rather than a predicate). This function is supposed to evaluate to the i-th symbol of a satisfying assignment for C; or to reject.

- The PCP decoder is *complete* if for every y such that $C(y) = 1$ there is a dPCP π such that $\Pr_{i,I,f}[f(\pi_I) = y_i] = 1$.
- The PCP decoder has *soundness error* δ and list size l if for any (purported) dPCP π there is a list of $\leq l$ valid proofs such that the probability (over the index i and (I, f)) that $f(\pi_I)$ is inconsistent with the list but not reject is at most δ.

The list of valid proofs can be viewed as a "list decoding" of the dPCP π. Since we are interested in the low soundness error regime, list-decoding is unavoidable. Of course, we can define dPCPs for any NP language and not just CIRCUITSAT, but we focus on CIRCUITSAT since it suffices for the purpose of composition.

The notion of dPCPs allows for modular composition in the case of low soundness error (described next) in analogy to the way PCPPs and assignment testers [BGH+06, DR06] allow for modular composition in the case of high soundness error. Moreover, using dPCPs we show a two query composition that yields a completely modular proof of the recent result of Moshkovitz and Raz [MR08b].

Finally, we note that decodable PCPs are not hard to come by. Decodable PCPs or variants of them are implicit in many PCP constructions [AS03, RS97, DFK+99, BGH+06, DR06, MR07, MR08b] and existing PCP constructions can often be adapted to yield decodable PCPs.

3 Composition with dPCPs

There is a natural and modular way to compose a PCP verifier V with a PCP decoder $\mathcal{D}$. The composed PCP verifier V' begins by simulating V on a probabilistically checkable proof Π. It determines a set of queries into Π (a local window I), and a local predicate f. Instead of directly querying Π and testing if $f(\Pi_I) = 1$, V' relies on the inner PCP decoder $\mathcal{D}$ to perform this action. For this task, the inner PCP decoder $\mathcal{D}$ is supplied with a dedicated proof that is supposedly an encoding of the relevant local view Π_I. The main issue is consistency: the composed verifier V' must ensure that the dedicated proofs supposedly encoding the various local views are consistent with the same Π (i.e. they should be encodings of local views coming from a single valid PCP for V). This is achieved easily with PCP decoders: the composed verifier V' asks $\mathcal{D}$ to decode a random value from the encoded local view, and compares it to the appropriate symbol in Π.

The above description of composition already appears[1] to lead to a modular presentation of the composition performed in earlier low-error PCP constructions [AS03, RS97, DFK+99, MR07]. But at the same time, like these compositions, it incurs an additional query per composition, namely the "consistency" query to the outer PCP Π. (The queries made by V' are the queries of $\mathcal{D}$ plus the one additional consistency query to Π).

Nevertheless, inspired by [MR08b] and equipped with a better understanding of composition in the low soundness error case, we are, now, in a position to remove this extra consistency query.

4   Composition with Only Two Queries

Our main contribution is a composition theorem that does not incur an extra query. The extra query above comes from the need to check that all the inner PCP decoders decode to the same symbol. This check was performed by comparing the decoded symbol to the symbol in the outer PCP Π. Instead, we verify consistency by invoking *all* the inner PCP decoders that involve this symbol *in parallel*, and then checking that they all decode to the same symbol. This avoids the necessity to query the outer PCP Π for this symbol and saves us the extra query.

We describe our new composed verifier V' more formally below. As before, let V be a PCP verifier, and $\mathcal{D}$ a PCP decoder.

1. The composed PCP verifier simulates V on a hypothetical PCP Π; it chooses a random index i in Π, and then determines *all* the possible random strings $R_1, \ldots, R_D$ that cause V to query this index.
2. For each random string R_j ($j = 1 \ldots D$), V' needs to check that the corresponding local view of Π would have lead V to accept. This is done by running $\mathcal{D}$, for each $j = 1 \ldots D$, on a dedicated proof $\pi(R_j)$ that is supposedly the encoding of the j-th local view (i.e., the one generated by V on random string R_j) into Π. Furthermore, V' expects $\mathcal{D}$ to decode the symbol Π_i.
3. Finally V' accepts if and only if *all* the D parallel runs of $\mathcal{D}$ accept and output the same symbol.

Observe that the composed verifier V' does not access the PCP for V (i.e., Π) at all, rather only the dedicated proofs for the inner PCP decoders. The outer PCP Π is only "mentally" present in order to compute $R_1, \ldots, R_D$. A few important points are in order.

---

[1] We have not verified the details.

- **Two Queries and Robust Soundness.** As described, V' makes many queries rather than just two. This is fixed by the following easy transformation: the first query will supposedly be answered by the complete local view V' expects to read, and the second query will consist of one random symbol in the local view of V'. The soundness error of the resulting two-query PCP is equal to the *robust soundness error* of V': an upper bound on the average agreement between a local view read by V' and an accepting local view.

  Thus, drawing on the above correspondence, the fact that V' has low robust soundness error implies the required two-query composition. Of course, the composition could have been described entirely in the 2-query PCP language.

- **Size of alphabet or window size.** The purpose of composition is to reduce the alphabet size, or, in the language of robust PCPs, to reduce the window size, that is, the number of queries made by V'. Recall that V' runs $\mathcal{D}$ in parallel on all D local views corresponding to $R_1, \ldots, R_D$. Thus, the window size equals the query complexity of $\mathcal{D}$ multiplied by the number D of local views (which we refer to as the *proof degree* of V). Hence composition is meaningful only if the proof degree is small to begin with (otherwise, the local window of V' is not smaller than that of V and we haven't gained anything from composition). In general PCPs, the proof degree is very high. In fact, this has been one of the obstacles to achieving this result prior to [MR08b]. However, a key observation of [MR08b] is that it is easy to reduce the proof degree using standard tools from derandomization (i.e., expander replacement).

  Viewed alternatively, one can handle V of arbitrarily high proof degree by making the following change to V'. Instead of running $\mathcal{D}$ to verify the local tests corresponding to *all* of $R_1, \ldots, R_D$, V' can *pseudo-randomly* sample a small number of these and run $\mathcal{D}$ only on the selected ones.

  The fact that the query complexity is at least D is an inherent bottleneck in our composition method. Combined with the bound of $D \geq 1/\delta$, this poses a limitation of this technique towards achieving exponential dependence of the error probability on alphabet size, a point discussed later in this introduction.

The new composition is generic in the sense that it works regardless of how the original components V and $\mathcal{D}$ are constructed.

5   Background and Motivation

Let us step back to give some motivation for obtaining PCPs with small soundness error and two queries (for a more comprehensive treatment, see [MR08b]). Two is the absolute minimal number of queries possible for a non-trivial PCP. Thus, it is interesting to find what are the strongest 2-query PCPs that still capture NP. However, the main motivation for two query PCPs is for proving hardness of approximation results.

Two query PCPs with soundness error δ are (more or less) equivalent to LABEL-COVER$_\delta$, which is a promise problem defined as follows[2]: The input is a bipartite graph and an alphabet Σ, and for each edge e there is a function $f_e : \Sigma \to \Sigma$, which we think of as a *constraint* on the labels of the vertices. The constraint is satisfied by values a and b iff $f_e(a) = b$. The problem is to distinguish between two cases: (1) there exists a labeling of the vertices satisfying all constraints, or (2) every labeling satisfies at most δ fraction of the constraints.

LABEL-COVER$_\delta$ is probably the most popular starting point for hardness of approximation reductions. In particular, even though there are 3-query PCPs with much smaller soundness error, they currently have far fewer applications to inapproximability.

The fact that LABEL-COVER$_\alpha$ is NP-hard for some constant $\alpha < 1$ (and constant alphabet size) is nothing but a reformulation of the PCP Theorem [AS98, ALM+98]. Strong inapproximability results, however, require[3] NP-hardness of LABEL-COVER$_\delta$ for arbitrarily small, sometimes even sub-constant soundness error δ. There are two known routes to obtaining hardness results for LABEL-COVER$_\delta$ with small soundness error δ. The first, is via an application of the parallel repetition theorem of Raz [Raz98] to the LABEL-COVER$_\alpha$ instance produced by the PCP Theorem. However, this application of the repetition theorem blows up the size of the problem instance from n to $n^{O(\log(1/\delta))}$ and thus remains polynomial only for constant, though arbitrarily small, δ. One might try to get a polynomial sized construction by carefully choosing a subset of the entire parallel repetition construction. This is known as the problem of "derandomizing the parallel repetition theorem". Feige and Kilian [FK95] showed that such derandomization is impossible under certain (rather general) conditions. Nevertheless, in a recent paper, Impagliazzo et. al. [IKW09] obtained a related derandomization. While their derandomization result applies only to direct products and not to the construction of PCPs, this direction seems promising. Another potential direction is to use the gap-amplification technique of Dinur [Din07], however as shown by Bogdanov [Bog05] gap-amplification fails below a soundness error of $1/2$.

The second route to sub-constant δ goes through the classical (algebraic) construction of PCPs. Indeed, hardness for label cover with sub-constant error can be obtained from the low soundness error PCPs of [RS97, AS03, MR08a], more or less by omitting the composition steps, and carefully combining queries. The following "manifold vs. point" PCP construction has been folklore since [RS97, AS03], and formally described in [MR08b].

Theorem 1 (Manifold vs. Point PCP). *There exists a constant $c > 1$ such that the following holds: For every $\frac{1}{n} \leq \delta \leq \frac{1}{(\log n)^c}$, there exists an alphabet Σ of size at most $\exp(\mathrm{poly}(1/\delta))$ such that LABEL-COVER$_\delta$ over Σ is NP-hard.*

---

[2] We focus on the important special case of projection constraints.

[3] In some cases the hardness gap is inversely proportional to δ, and in others, it is the sum of two terms: a problem-dependent term (e.g. 7/8 in Håstad's hardness result [Hås01] for 3-SAT), and a "low order" term that is polynomial in δ.

The above result is unsatisfactory as the size of the alphabet $|\Sigma|$ is super-polynomial. Combined with the fact that hardness-of-approximation reductions are usually exponential in $|\Sigma|$ (and always at least polynomial in $|\Sigma|$) the super polynomial size of Σ renders the above theorem useless. The situation can be redeemed if the theorem could be extended to the entire range of smaller $|\Sigma|$ (with a corresponding increase in δ).

A natural way to perform this extension would be to apply the composition paradigm to the PCPs constructed in Theorem 1 and reduce the alphabet size. Indeed, this is how one constructs PCPs with sub-constant error and *a constant* number of queries for the entire range of $\Omega(1) \leq |\Sigma| \leq \exp((\log n)^{1-\epsilon})$ [RS97, AS03, DFK$^+$99]. However, the composition a la [RS97, AS03, DFK$^+$99] incurs at least one additional query, which means that the final PCP is no longer "two-query", so it does not lead to a hardness result for label cover. Alternatively, the composition technique of [BGH$^+$06, DR06] using PCPs of proximity or assignment testers is inapplicable in this context as it fails to work for soundness error less than $1/2$. Thus, all earlier composition techniques are either inapplicable in the low error regime or if applicable, incur an extra query and thus, are no longer in the framework of the LABEL-COVER problem.

6   The Two-Query PCP of Moshkovitz and Raz [MR08b]

In a recent breakthrough, [MR08b] show that the above theorem can in fact, be extended to the entire range of δ and $|\Sigma|$ (and maintaining $|\Sigma| \approx \exp(\text{poly}(1/\delta))$). This is done by composing certain specific 2-query PCPs with low soundness error without incurring an additional query per composition.

Theorem 2 ([MR08b]). *For every $\delta \in (1/\text{polylog} n, 1)$, there exists an alphabet Σ of size at most $\exp(\text{poly}(1/\delta))$ such that LABEL-COVER$_\delta$ over Σ is NP-hard (in fact, even under nearly length preserving reductions).*

The main technical component of their construction is a novel composition of certain specific PCPs. However, the construction is so organically tied to the specific algebraic components that are being composed, as to make it extremely difficult to differentiate between the details of the PCP, and what it is that makes the composition go through.

We give a modular and simpler proof of this theorem using our composition theorem. Our proof relies on a PCP system based on the manifold vs. point construction (as in Theorem 1). The parameters we need are rather weak: it is enough that on input size n the PCP decoder / verifier makes n^α queries and has soundness error $\delta = 1/n^\beta$, for small constants α, β. After one composition step the number of queries goes (roughly) from n^α to n^{α^2}, and so on. After each composition step we add a combinatorial step, consisting of degree and alphabet reduction, that prepares the verifier for the next round of composition. After i rounds the number of queries is about n^{α^i}, and the soundness error is about $\delta = 1/n^{O(\alpha^i)}$. Choosing $1 \leq i \leq \log \log n$ appropriately gives us the result.

The modular composition theorem allows us to easily keep track of a super-constant number of steps, thus avoiding the need for another tailor-made Hadamard-based PCP which was required in the proof of [MR08b]. (The later approach could also be implemented in our setting).

Randomness and the length of the PCP: The above discussion completely ignores the randomness complexity of the underlying PCPs. However, it is easy to verify that the composition described above is, in fact, randomness efficient; this is because the same inner randomness can be used for all the D parallel runs of the inner PCP decoder. Thus, if we start from a version of the Theorem 1 (the manifold vs. point PCP) based on an almost linear-size low-degree test (c.f., [MR08a]), we obtain a nearly length preserving version of Theorem 2 (i.e., a reduction taking instances of size n to instances of size almost linear in n). Furthermore, the fact that we account for the input index i separately from the inner randomness r of the PCP decoder leads to an even more randomness-efficient composition, however, we do not exploit this fact in the proof of Theorem 2.

Polynomial dependence of soundness error on alphabet size: Theorem 2 suffers from the following bottleneck: the error probability δ is inverse logarithmic (and not inverse-polynomial) with respect to the size of the alphabet Σ. This limitation is inherent in our composition method as discussed above. Thus, the "sliding-scale conjecture" of Bellare et al. [BGLR93] that for every $|\Sigma| \in (1, n)$, LABEL-COVER$_\delta$ over Σ is NP-hard for $\delta = \text{poly}(1/|\Sigma|)$ remains open.

References

[ALM$^+$98]   Arora, S., Lund, C., Motwani, R., Sudan, M., Szegedy, M.: Proof verification and the hardness of approximation problems. J. ACM 45(3), 501–555 (1998); Preliminary Version in 33rd FOCS (1992), eccc: TR98-008, doi:10.1145/278298.278306

[AS98]   Arora, S., Safra, S.: Probabilistic checking of proofs: A new characterization of NP. J. ACM 45(1), 70–122 (1998); Preliminary Version in 33rd FOCS (1992), doi:10.1145/273865.273901

[AS03]   Arora, S., Sudan, M.: Improved low-degree testing and its applications. Combinatorica 23(3), 365–426 (2003); Preliminary Version in 29th STOC (1997), eccc: TR97-003, doi:10.1007/s00493-003-0025-0

[BGH$^+$06]   Ben-Sasson, E., Goldreich, O., Harsha, P., Sudan, M., Vadhan, S.: Robust PCPs of proximity, shorter PCPs and applications to coding. SIAM J. Computing 36(4), 889–974 (2006); Preliminary Version in 36th STOC (2004), eccc: TR04-021, doi:10.1137/S0097539705446810

[BGLR93]   Bellare, M., Goldwasser, S., Lund, C., Russell, A.: Efficient probabilistically checkable proofs and applications to approximation. In: Proc. 25th ACM Symp. on Theory of Computing (STOC), pp. 294–304. ACM, New York (1993), doi:10.1145/167088.167174

[Bog05]   Bogdanov, A.: Gap amplification fails below 1/2 (2005) (Comment on "Dinur, The PCP theorem by gap amplification), eccc:TR05-046

[BS08]   Ben-Sasson, E., Sudan, M.: Short PCPs with polylog query complexity. SIAM J. Computing 38(2), 551–607 (2008); Preliminary Version in 37th STOC (2005), eccc:TR04-060, doi:10.1137/050646445

[DFK+99]   Dinur, I., Fischer, E., Kindler, G., Raz, R., Safra, S.: PCP characterizations of NP: Towards a polynomially-small error-probability. In: Proc. 31st ACM Symp. on Theory of Computing (STOC), pp. 29–40. ACM, New York (1999), eccc:TR98-066, doi:10.1145/301250.301265

[DH09]   Dinur, I., Harsha, P.: Composition of low-error 2-query PCPs using decodable PCPs. Technical Report TR09-042, Electronic Colloquium on Computational Complexity (2009), eccc:TR09-042

[Din07]   Dinur, I.: The PCP theorem by gap amplification. J. ACM 54(3), 12 (2007); Preliminary Version in 38th STOC (2006), eccc: TR05-046, doi:10.1145/1236457.1236459

[Din08]   Dinur, I.: PCPs with small soundness error. SIGACT News 39(3), 41–57 (2008), doi:10.1145/1412700.1412713

[DR06]   Dinur, I., Reingold, O.: Assignment testers: Towards a combinatorial proof of the PCP Theorem. SIAM J. Computing 36, 975–1024 (2006); Preliminary Version in 45th FOCS (2004), doi:10.1137/S0097539705446962

[FGL+96]   Feige, U., Goldwasser, S., Lovász, L., Safra, S., Szegedy, M.: Interactive proofs and the hardness of approximating cliques. J. ACM 43(2), 268–292 (1996); Preliminary version in 32nd FOCS (1991), doi:10.1145/226643.226652

[FK95]   Feige, U., Kilian, J.: Impossibility results for recycling random bits in two-prover proof systems. In: Proc. 27th ACM Symp. on Theory of Computing (STOC), pp. 457–468. ACM, New York (1995), doi:10.1145/225058.225183

[For66]   David Forney, G.: Concatenated Codes. MIT Press, Cambridge (1966)

[Hås01]   Håstad, J.: Some optimal inapproximability results. J. ACM 48(4), 798–859 (2001); Preliminary Version in 29th STOC (1997), doi:10.1145/502090.502098

[IKW09]   Impagliazzo, R., Kabanets, V., Wigderson, A.: Direct product testing: Improved and derandomized. In: Proc. 41st ACM Symp. on Theory of Computing (STOC), pp. 131–140. ACM, New York (2009), eccc:TR09-090, doi:10.1145/1536414.1536435

[MR07]   Moshkovitz, D., Raz, R.: Sub-constant error probabilistically checkable proof of almost linear size (2007), eccc:TR07-026

[MR08a]   Moshkovitz, D., Raz, R.: Sub-constant error low degree test of almost-linear size. SIAM J. Computing 38(1), 140–180 (2008); Preliminary Version in 38th STOC (2006), eccc:TR05-086, doi:10.1137/060656838

[MR08b]   Moshkovitz, D., Raz, R.: Two query PCP with sub-constant error. In: Proc. 49th IEEE Symp. on Foundations of Comp. Science (FOCS), pp. 314–323. IEEE, Los Alamitos (2008), eccc:TR08-071, doi:10.1109/FOCS.2008.60

[Raz98]   Raz, R.: A parallel repetition theorem. SIAM J. Computing 27(3), 763–803 (1998); Preliminary Version in 27th STOC (1995), doi:10.1137/S0097539795280895

[RS97]   Raz, R., Safra, S.: A sub-constant error-probability low-degree test, and a sub-constant error-probability PCP characterization of NP. In: Proc. 29th ACM Symp. on Theory of Computing (STOC), pp. 475–484. ACM, New York (1997), doi:10.1145/258533.258641

Hierarchy Theorems for Property Testing[*]

Oded Goldreich[1], Michael Krivelevich[2], Ilan Newman[3], and Eyal Rozenberg[4]

[1] Department of Computer Science, Weizmann Institute of Science, Rehovot, Israel
`oded.goldreich@weizmann.ac.il`
[2] School of Mathematical Sciences, Tel Aviv University, Tel Aviv 69978, Israel
`krivelev@post.tau.ac.il`
[3] Department of Computer Science, Haifa University, Haifa, Israel
`ilan@cs.haifa.ac.il`
[4] Department of Computer Science, Technion, Haifa, Israel
`eyalroz@technion.ac.il`

Abstract. Referring to the query complexity of property testing, we prove the existence of a rich hierarchy of corresponding complexity classes. That is, for any relevant function q, we prove the existence of properties that have testing complexity $\Theta(q)$. Such results are proven in three standard domains often considered in property testing: generic functions, adjacency predicates describing (dense) graphs, and incidence functions describing bounded-degree graphs. While in two cases the proofs are quite straightforward, the techniques employed in the case of the dense graph model seem significantly more involved. Specifically, problems that arise and are treated in the latter case include (1) the preservation of distances between graphs under a blow-up operation, and (2) the construction of monotone graph properties that have local structure.

Keywords: Graph Properties, Monotone Graph Properties, Graph Blow-up, One-Sided versus Two-Sided Error, Adaptivity versus Non-adaptivity.

1 Background

In the last decade, the area of property testing has attracted much attention (see, e.g., a couple of recent surveys [R1, R2]). Loosely speaking, property testing typically refers to sub-linear time probabilistic algorithms for deciding whether a given object has a predetermined property or is far from any object having this property. Such algorithms, called testers, obtain local views of the object by making adequate queries; that is, the object is seen as a function and the testers get oracle access to this function (and thus may be expected to work in time that is sub-linear in the length of the object).

Following most work in the area, we focus on the query complexity of property testing, measured as a function of the size of the object as well as the desired proximity (parameter). Interestingly, many natural properties can be tested in

---

[*] A preliminary version has appeared as TR08-097 of ECCC, and an extended abstract has appeared in the proceedings of *RANDOM'09*.

O. Goldreich (Ed.): Property Testing, LNCS 6390, pp. 289–294, 2010.

complexity that only depends on the proximity parameter; examples include linearity testing [BLR], and testing various graph properties in two natural models (e.g., [GGR, AFNS] and [GR1, BSS], respectively). On the other hand, properties for which testing requires essentially maximal query complexity were proved to exist too; see [GGR] for artificial examples in two models and [BHR, BOT] for natural examples in other models. In between these two extremes, there exist natural properties for which the query complexity of testing is logarithmic (e.g., monotonicity [EKK+, GGL+]), a square root (e.g., bipartiteness in the bounded-degree model [GR1, GR2]), and possibly other constant powers (see [FM, PRR]).

2 Our Main Results

One natural question that arises is whether there exist properties of arbitrary query complexity. We answer this question affirmatively, proving the existence of a rich hierarchy of query complexity classes. Such hierarchy theorems are easiest to state and prove in the generic case: Loosely speaking, for every sub-linear function q, *there exists a property of functions over $[n]$ that is testable using $q(n)$ queries but is not testable using $o(q(n))$ queries.*

Similar hierarchy theorems are proved also for two standard models of testing graph properties: the adjacency representation model (of [GGR]) and the incidence representation model (of [GR1]). For the incidence representation model (a.k.a the bounded-degree graph model), we show that, for every sub-linear function q, *there exists a property of bounded-degree N-vertex graphs that is testable using $q(N)$ queries but is not testable using $o(q(N))$ queries.* Furthermore, one such property corresponds to the set of N-vertex graphs that are 3-colorable and consist of connected components of size at most $q(N)$.

The bulk of this paper is devoted to hierarchy theorems for the adjacency representation model (a.k.a the dense graph model), where the complexity is stated as a function of the number of vertices (rather than as a function of the number of all vertex pairs, which is the representation size). Our main results for the adjacency matrix model are:

1. For every sub-quadratic function q, *there exists a graph property Π that is testable in q queries, but is not testable in $o(q)$ queries.* Furthermore, for "nice" functions q, it is the case that Π is in $\mathcal{P}$ and the tester can be implemented in poly(q)-time.
2. For every sub-quadratic function q, there exists a *monotone* graph property Π that is testable in $O(q)$ queries, but is not testable in $o(q)$ queries.

Additional results regarding the adjacency representation model are outlined in Section 4.

Conventions. For sake of simplicity, we state all results while referring to query complexity as a function of a size parameter that is polynomially related to the object's size (i.e., in the case of generic Boolean functions the size parameter is the size of the function's domain, but in the case of graphs the size parameter is the number of vertices). In other words, we consider a fixed (constant) value of the proximity parameter, denoted ϵ. In such cases, we sometimes use the term ϵ-testing, which refers to testing when the proximity parameter is fixed to ϵ. All our lower bounds hold for any sufficiently small value of the proximity parameter, whereas the upper bounds hide a (polynomial) dependence on (the reciprocal of) this parameter. In general, bounds that have no dependence on the proximity parameter refer to some (sufficiently small but) fixed value of this parameter.

A remotely related prior work. In contrast to the foregoing conventions, we mention here a result that refers to graph properties that are testable in (query) complexity that only depends on the proximity parameter. This result, due to [AS], establishes a (very sparse) hierarchy of such properties. Specifically, [AS, Thm. 4] asserts that for every function q there exists a function Q and a graph property that is ϵ-testable in $Q(\epsilon)$ queries but is *not* ϵ-testable in $q(\epsilon)$ queries. (We note that while Q depends only on q, the dependence proved in [AS, Thm. 4] is quite weak (i.e., Q is lower bounded by a non-constant number of compositions of q), and thus the hierarchy obtained by setting $q_i = Q_{i-1}$ for $i = 1, 2, \ldots$ is very sparse.)

3   Our Techniques

The proofs of the hierarchy theorems for the generic case and for the incidence representation graph model, are quite straightforward. In contrast, the treatment of the dense graph model is significantly more involved. We discuss the source of trouble next.

Given that properties of maximal query complexity are known in each of the testing models that we consider, a natural idea towards proving hierarchy theorems is to construct properties that correspond to repetitions of the original properties; that is, each object in the new property consists of an adequate number of objects, each belonging to the original property. Straightforward implementations of this idea work in the generic case and in the incidence representation graph model, but not in the dense graph model. The point is that a naive repetition of a graph, in this model, necessarily creates a graph that is not dense.

Nevertheless, the graph blow-up operation does seem to be the adequate construction that we seek. Loosely speaking, the graph blow-up operation replaces each vertex by an independent set (of a predetermined size), and replaces edges by corresponding complete bipartite graphs. One source of trouble is that the blow-up operation does not necessarily preserve distances; indeed the relative distance between the blow-up of G_1 and G_2 is at most the relative distance between the original graphs, but the naive assumption that it may not be smaller is false. We overcome this difficulty by showing that for certain graphs, which we call dispersed, the blow-up does preserve the original distances (up to a constant factor).[1] Thus, we first reduce the testing of the original property to testing a

---

[1] Our result was superseded by Oleg Pikhurko, who showed that *for any two graphs, the distance is actually preserved up to a constant factor* [P, Sec. 4].)

corresponding property that refers to dispersed graphs. (An n-vertex graph is called dispersed if the neighbor sets of any two vertices differ on at least $\Omega(n)$ elements.)

Using dispersed graphs also allows us to overcome another technical difficulty, which relates to the complexity of our tester. In particular, the use of dispersed graphs allows us to recover the canonical labeling of an unlabeled graph, which is helpful whenever a graph property (viewed as a set of labeled graphs) is obtained by a closure under isomorphism of some set of labeled graphs (cf. [GGR]).

When trying to obtain a result for monotone graph properties, we encounter another technical difficulty. The difficulty is that standard constructions of monotone graph properties (cf. [GT]) tend to lack any local structure, since the property should be preserved under arbitrary edge additions. We demonstrate that the latter conclusion is a bit hasty, by showing that a local structure can be essentially maintained as long as the edge density does not exceed some threshold, whereas we can include in the property all graphs that have edge density that exceeds this threshold.

A third type of difficulty arises when we try to obtain one-sided error testers (see Section 4). Towards this end, we use a different type of graph blow-up, which we call *generalized blow-up*. While under the aforementioned (balanced) blow-up operation each vertex is replaced by an independent set of the same size, in a generalized blow-up these independent sets may have different sizes.

4   Additional Results

The bulk of our work provides hierarchy theorems for graph properties in the adjacency matrix model. In particular, we have already mentioned the basic hierarchy theorem regarding this model and our related theorem for monotone graph properties. (These theorems are incomparable, see discussion below.)

We also address a refined issue that has been ignored above. Specifically, we note that all our lower bounds refer to two-sided error testers, whereas the upper bounds in the generic case and in the bounded-degree graph model are demonstrated using one-sided error testers (which only make these separations stronger). In contrast, the aforementioned upper bounds for the adjacency matrix model use two-sided error testers. Seeking a hierarchy of one-sided error testing also in this model, we modify the basic construction in order to obtain one-sided error testers (while the lower bounds still hold for two-sided error testers). However, the latter theorem loses some features of the former theorems; see discussion below.

We mention that our results for graph properties in the adjacency matrix model use the existence of graph properties that are in $\mathcal{P}$ and have maximal query complexity. We prove the existence of such graph properties, by building on a prior construction of [GGR], which only asserted such properties in $\mathcal{NP}$.

Discussion: Three incomparable results regarding graph properties in the adjacency matrix model. As mentioned above, we proved three hierarchy theorems for testing graph properties in the adjacency matrix model.

1. The basic theorem is established by non-monotone properties (in $\mathcal{P}$), while the tester demonstrating the upper bound is relatively efficient in the sense that its running time is polynomial in its query complexity. Both the lower and upper bounds refer to two-sided testers.
2. The second theorem refers to monotone properties (in $\mathcal{NP}$). Again, both the lower and upper bounds refer to two-sided testers.
3. The third theorem refers to properties in $\mathcal{P}$, but the tester demonstrating the upper bound is not relatively efficient (i.e., its decision predicate is in $\mathcal{NP}$). However, in this case the tester has one-sided error (whereas the lower bound holds also for two-sided testers).

Obtaining a single theorem that combines all good features is left as an open problem.

Acknowledgments

We are grateful to Ronitt Rubinfeld for asking about the existence of hierarchy theorems for the adjacency matrix model. Ronitt raised this question during a discussion that took place at the Dagstuhl 2008 workshop on sub-linear algorithms. We are also grateful to Arie Matsliah, Dana Ron, and Yoav Tzur for helpful discussions. In particular, we thank Arie Matsliah for providing us with a proof that the blow-up operation does not preserve distances in a perfect manner.

O.G. was partially supported by the Israel Science Foundation (grant No. 1041/08). M.K. was partially supported by a USA-Israel BSF Grant, by a grant from the Israel Science Foundation, and by Pazy Memorial Award. I.N. was partially supported by an Israel Science Foundation (grant number 1011/06).

References

[ABI]   Alon, N., Babai, L., Itai, A.: A fast and Simple Randomized Algorithm for the Maximal Independent Set Problem. J. of Algorithms 7, 567–583 (1986)
[AFKS]   Alon, N., Fischer, E., Krivelevich, M., Szegedy, M.: Efficient Testing of Large Graphs. Combinatorica 20, 451–476 (2000)
[AFNS]   Alon, N., Fischer, E., Newman, I., Shapira, A.: A Combinatorial Characterization of the Testable Graph Properties: It's All About Regularity. In: 38th STOC, pp. 251–260 (2006)
[AGHP]   Alon, N., Goldreich, O., Hastad, J., Peralta, R.: Simple constructions of almost k-wise independent random variables. Journal of Random structures and Algorithms 3(3), 289–304 (1992)
[AS]   Alon, N., Shapira, A.: Every Monotone Graph Property is Testable. SIAM Journal on Computing 38, 505–522 (2008)
[BSS]   Benjamini, I., Schramm, O., Shapira, A.: Every Minor-Closed Property of Sparse Graphs is Testable. In: 40th STOC, pp. 393–402 (2008)
[BLR]   Blum, M., Luby, M., Rubinfeld, R.: Self-Testing/Correcting with Applications to Numerical Problems. JCSS 47(3), 549–595 (1993)
[BHR]   Ben-Sasson, E., Harsha, P., Raskhodnikova, S.: 3CNF Properties Are Hard to Test. SIAM Journal on Computing 35(1), 1–21 (2005)

[BOT]   Bogdanov, A., Obata, K., Trevisan, L.: A lower bound for testing 3-colorability in bounded-degree graphs. In: 43rd FOCS, pp. 93–102 (2002)

[EKK+]   Ergun, F., Kannan, S., Kumar, S.R., Rubinfeld, R., Viswanathan, M.: Spot-checkers. JCSS 60(3), 717–751 (2000)

[FM]   Fischer, E., Matsliah, A.: Testing Graph Isomorphism. In: 17th SODA, pp. 299–308 (2006)

[GGL+]   Goldreich, O., Goldwasser, S., Lehman, E., Ron, D., Samorodnitsky, A.: Testing Monotonicity. Combinatorica 20(3), 301–337 (2000)

[GGR]   Goldreich, O., Goldwasser, S., Ron, D.: Property testing and its connection to learning and approximation. Journal of the ACM, 653–750 (July 1998)

[GR1]   Goldreich, O., Ron, D.: Property Testing in Bounded Degree Graphs. Algorithmica 32(2), 302–343 (2002)

[GR2]   Goldreich, O., Ron, D.: A Sublinear Bipartitness Tester for Bounded Degree Graphs. Combinatorica 19(3), 335–373 (1999)

[GT]   Goldreich, O., Trevisan, L.: Three theorems regarding testing graph properties. Random Structures and Algorithms 23(1), 23–57 (2003)

[LNS]   Lachish, O., Newman, I., Shapira, A.: Space Complexity vs. Query Complexity. Computational Complexity 17, 70–93 (2008)

[NN]   Naor, J., Naor, M.: Small-bias Probability Spaces: Efficient Constructions and Applications. SIAM J. on Computing 22, 838–856 (1993)

[PRR]   Parnas, M., Ron, D., Rubinfeld, R.: Testing Membership in Parenthesis Laguages. Random Structures and Algorithms 22(1), 98–138 (2003)

[P]   Pikhurko, O.: An Analytic Approach to Stability (2009) (manuscript), http://arxiv.org/abs/0812.0214

[R1]   Ron, D.: Property Testing: A Learning Theory Perspective. Foundations and Trends in Machine Learning 1(3), 307–402 (2008)

[R2]   Ron, D.: Algorithmic and Analysis Techniques in Property Testing. Foundations and Trends in TCS 5(2), 73–205 (2010)

[RS]   Rubinfeld, R., Sudan, M.: Robust characterization of polynomials with applications to program testing. SIAM Journal on Computing 25(2), 252–271 (1996)

[S]   Shaltiel, R.: Recent Developments in Explicit Constructions of Extractors. In: Current Trends in Theoretical Computer Science: The Challenge of the New Century. Algorithms and Complexity, vol. 1, World Scientific, Singapore (2004); Preliminary version in Bulletin of the EATCS 77, 67–95 (2002)

Algorithmic Aspects of Property Testing in the Dense Graphs Model

Oded Goldreich[1] and Dana Ron[2]

[1] Department of Computer Science, Weizmann Institute of Science, Rehovot, Israel
oded.goldreich@weizmann.ac.il
[2] Department of Electrical Engineering-Systems, Tel-Aviv University, Tel-Aviv, Israel
danar@eng.tau.ac.il

Abstract. In this paper we consider two basic questions regarding the query complexity of testing graph properties in the adjacency matrix model. The first question refers to the relation between adaptive and non-adaptive testers, whereas the second question refers to testability within complexity that is inversely proportional to the proximity parameter, denoted ϵ. The study of these questions reveals the importance of algorithmic design in this model. The highlights of our study are:

- A gap between the complexity of adaptive and non-adaptive testers. Specifically, there exists a natural graph property that can be tested using $\widetilde{O}(\epsilon^{-1})$ adaptive queries, but cannot be tested using $o(\epsilon^{-3/2})$ non-adaptive queries.
- In contrast, there exist natural graph properties that can be tested using $\widetilde{O}(\epsilon^{-1})$ non-adaptive queries, whereas $\Omega(\epsilon^{-1})$ queries are required even in the adaptive case.

We mention that the properties used in the foregoing conflicting results have a similar flavor, although they are of course different.

Keywords: Adaptivity vs. Non-adaptivity, Graph Properties.

This article is an extended abstract of our technical report [GR08]. While the main text assume familiarity with the basic model, all relevant definitions appear in Section A.1.

1 Introduction

In the last couple of decades, the area of property testing has attracted much attention (see, e.g., a couple of recent surveys [R1, R2]). Loosely speaking, property testing typically refers to sub-linear time probabilistic algorithms for deciding whether a given object has a predetermined property or is far from any object having this property. Such algorithms, called testers, obtain bits of the object by performing queries, which means that the object is seen as a function and the testers get oracle access to this function. Thus, a tester may be expected to work in time that is sub-linear in the length of the description of this object.

O. Goldreich (Ed.): Property Testing, LNCS 6390, pp. 295–305, 2010.

Much of the aforementioned work (see, e.g., [GGR, AFKS, AFNS]) was devoted to the study of testing graph properties in the adjacency matrix model, which is also the setting of the current work. In this model, introduced in [GGR], graphs are viewed as symmetric Boolean functions over a domain consisting of all possible vertex-pairs. Namely, an N-vertex graph $G = ([N], E)$ is represented by the function $g : [N] \times [N] \to \{0, 1\}$ such that $\{u, v\} \in E$ if and only if $g(u, v) = 1$. Consequently, an N-vertex graph represented by the function $g : [N] \times [N] \to \{0, 1\}$ is said to be ϵ-far from some predetermined graph property if more than $\epsilon \cdot N^2$ entries of g must be modified in order to yield a representation of a graph that has this property. We refer to ϵ as the proximity parameter, and the complexity of testing is stated in terms of ϵ and the number, N, of vertices in the graph.

Interestingly, many natural graph properties can be tested within query complexity that depends only on the proximity parameter; see [GGR], which presents testers with query complexity $\text{poly}(1/\epsilon)$, and [AFNS], which characterizes the class of properties that are testable within query complexity that depends only on the proximity parameter (where this dependence may be an arbitrary function of ϵ). However, a common phenomenon in all the aforementioned works is that they utilize quite naive algorithms and their focus is on the analysis of these algorithms, which is often quite sophisticated. This phenomenon is no coincidence: As shown in [AFKS, GT], when ignoring a quadratic blow-up in the query complexity, property testing in this model reduces to sheer combinatorics. Specifically, without loss of generality, the tester may just inspect a random induced subgraph (of adequate size) of the input graph.

In this paper we demonstrate that a more refined study of property testing in this model reveals the importance of algorithmic design also in this model. This is demonstrated both by studying the advantage of adaptive testers over non-adaptive ones as well as by studying the class of properties that can be tested within complexity that is inversely proportional to the proximity parameter.

2   Two Related Studies

We start by reviewing the two related studies conducted in the current work.

2.1   Adaptivity vs. Non-adaptivity

A tester is called non-adaptive if it determines all its queries independently of the answers obtained for previous queries, and otherwise it is called adaptive. Indeed, by [AFKS, GT], the benefit of adaptivity (or, equivalently, the cost of non-adaptivity) is polynomially bounded: Specifically, any (possibly adaptive) tester, for any graph property, of query complexity $q(N, \epsilon)$ can be transformed into a non-adaptive tester of query complexity $O(q(N, \epsilon)^2)$. But is this quadratic gap an artifact of the known proofs (of [AFKS, GT]) or does it reflect something inherent?

A recent work by [GR07] suggests that the latter case may hold: For every $\epsilon > 0$, they showed that the set of N-vertex bipartite graphs of maximum degree

$O(\epsilon N)$ is ϵ-testable (i.e., testable with respect to proximity parameter ϵ) by $\widetilde{O}(\epsilon^{-3/2})$ queries, while by [BT] a non-adaptive tester for this set must use $\Omega(\epsilon^{-2})$ queries. Thus, there exists a case where non-adaptivity has the cost of increasing the query complexity; specifically, for any $c < 4/3$, the query complexity of the non-adaptive tester is greater than a c-power of the query complexity of the adaptive tester (i.e., $\widetilde{O}(\epsilon^{-3/2})^c = o(\epsilon^{-2})$). We stress that the result of [GR07] does not refer to property testing in the "proper" sense; that is, the complexity is not analyzed with respect to a varying value of the proximity parameter for a fixed property. It is rather the case that, for every value of the proximity parameter, a different property, which depends on this parameter, is considered. The upper bounds and lower bounds refer to this combination of a property tailored for a fixed value of the proximity parameter. Thus, *the work of [GR07] leaves open the question of whether there exists a single graph property such that adaptivity is beneficial for any value of the proximity parameter* (as long as $\epsilon > N^{-\Omega(1)}$). That is, the question is whether adaptivity is beneficial for the standard asymptotic-complexity formulation of property testing.

2.2   Complexity Linearly Related to the Proximity Parameter

As shown in [GGR], many natural graph properties can be tested within query complexity that is polynomial in the reciprocal of the proximity parameter and independent of the size of the graph. We ask whether a linear complexity is possible at all, and if so which properties can be tested with query complexity that is linear (or almost linear) in the reciprocal of the proximity parameter, that is, with query complexity $\tilde{O}(1/\epsilon)$.[1]

The first question is easy to answer even when avoiding *trivial* properties. We say that a graph property Π is trivial for testing if for every $\epsilon > 0$ there exists $N_0 > 0$ such that for every $N \geq N_0$ either all N-vertex graphs belong to Π or all of them are ϵ-far from Π. Note that the property of being a clique (equiv., an independent set) can be tested by $O(1/\epsilon)$ queries, even when these queries are non-adaptive (e.g., make $O(1/\epsilon)$ random queries and accept if and only if all return 1). Still, we ask whether "more interesting" graph theoretical properties can also be tested within similar complexity, either only adaptively or also non-adaptively. In particular, the property of being a clique (or an independent set) is viewed as "non-interesting" since it contains a single N-vertex graph (per each N) and is represented by a monochromatic function.

3   Our Results

We address the foregoing questions by studying a sequence of natural graph properties, which are defined formally in Section A.2. The first property in the sequence, called clique collection and denoted $\mathcal{CC}$, is the set of graphs such that each graph consists of a collection of isolated cliques. Testing this property corresponds

---

[1] Note that $\Omega(1/\epsilon)$ queries are required for testing any of the graph properties considered in the current work.

to the following natural clustering problem: can a set of possibly related elements be partitioned into "perfect clusters" (i.e., two elements are in the same cluster if and only if they are related)? For this property, $\mathcal{CC}$, we prove a gap between adaptive and non-adaptive query complexity, where the adaptive query complexity is almost linear in the reciprocal of the proximity parameter. That is:

Theorem 3.1. (the query complexity of clique collection):

1. *There exists an adaptive tester of query complexity $\widetilde{O}(\epsilon^{-1})$ for $\mathcal{CC}$. Furthermore, this tester has one-sided error and runs in time $\widetilde{O}(\epsilon^{-1})$.*[2]
2. *Any non-adaptive tester for $\mathcal{CC}$ must have query complexity $\Omega(\epsilon^{-4/3})$.*
3. *There exists a non-adaptive tester of query complexity $O(\epsilon^{-4/3})$ for $\mathcal{CC}$. Furthermore, this tester has one-sided error and runs in time $O(\epsilon^{-4/3})$.*

Note that the complexity gap between Parts 1 and 2 of Theorem 3.1 matches the gap established by [GR07] for "non-proper" testing. A larger gap is established for a property of graphs, called bi-clique collection and denoted $\mathcal{BCC}$, where a graph is in $\mathcal{BCC}$ if it consists of a collection of isolated bi-cliques (i.e., complete bipartite graphs). We note that bi-cliques may be viewed as the bipartite analogues of cliques (w.r.t. general graphs), and indeed they arise naturally in clustering applications that are modeled by bipartite graphs over two types of elements.

Theorem 3.2. (the query complexity of bi-clique collection):

1. *There exists an adaptive tester of query complexity $\widetilde{O}(\epsilon^{-1})$ for $\mathcal{BCC}$. Furthermore, this tester has one-sided error and runs in time $\widetilde{O}(\epsilon^{-1})$.*
2. *Any non-adaptive tester for $\mathcal{BCC}$ must have query complexity $\Omega(\epsilon^{-3/2})$. Furthermore, this holds even if the input graph is promised to be bipartite.*

The furthermore clause in Part 2 of Theorem 3.2 holds also for the model studied in [AFN], where the bi-partition of the graph is given.

Theorem 3.2 asserts that the gap between the query complexity of adaptive and non-adaptive testers may be a power of $1.5 - o(1)$. Recall that the results of [AFKS, GT] assert that the gap may not be larger than quadratic. We conjecture that this upper bound can be matched.

Conjecture 3.3 (an almost-quadratic complexity gap): *For every positive integer $t \geq 5$, there exists a graph property Π for which the following holds:*

1. *There exists an adaptive tester of query complexity $\widetilde{O}(\epsilon^{-1})$ for Π.*
2. *Any non-adaptive tester for Π must have query complexity $\Omega(\epsilon^{-2+(2/t)})$.*
3. *There exists an efficient non-adaptive tester of query complexity $\widetilde{O}(\epsilon^{-2+2t^{-1}})$ for Π.*

---

[2] We refer to a model in which elementary operations regarding pairs of vertices are charged at unit cost.

Furthermore, Π consists of graphs that are each a collection of "super-cycles" of length t, where a super-cycle is a set of t independent sets arranged on a cycle such that each pair of adjacent independent sets is connected by a complete bipartite graph.

We were able to prove Part 2 of Conjecture 3.3, but failed to provide a full analysis of an algorithm that we designed for Part 1. However, we were able to prove a *promise problem version of Conjecture 3.3*; specifically, this promise problem (stated in Theorem A.4) refers to inputs promised to reside in a set $Π' ⊃ Π$ and the tester is required to distinguish graphs in $Π$ from graphs that are $ε$-far from $Π$.

In contrast to the foregoing results that aim at identifying properties with a substantial gap between the query complexity of adaptive versus non-adaptive testing, we also study cases in which no such gap exists. Since query complexity that is linear in the reciprocal of the proximity parameter is minimal for many natural properties, and, in fact, for any property that is "non-trivial for testing" (as defined at the end of Section 2), we focus on non-adaptive testers that approximately meet this bound. Among the results obtained in this direction, we highlight the following one.

Theorem 3.4. (the query complexity of collections of $O(1)$ cliques): *For every positive integer c, there exists a non-adaptive tester of query complexity $\widetilde{O}(ε^{-1})$ for the set of graphs such that each graph consists of a collection of up to c cliques. Furthermore, this tester has one-sided error and runs in time $\widetilde{O}(ε^{-1})$.*

Theorem 3.4 should be viewed as a first step in the study of graph properties that are the simplest to test; that is, the class of graph properties that have a non-adaptive of query complexity $\widetilde{O}(ε^{-1})$. We mention that a second step, which significantly generlaizes Theorem 3.4, has been subsequently taken in [A09, AG].

Discussion. The foregoing results demonstrate that a finer look at property testing of graphs in the adjacency matrix model reveals the role of algorithm design in this model. In particular, in some cases (see, e.g., Theorems 3.1 and 3.2), carefully designed adaptive algorithms outperform any non-adaptive algorithm. Indeed, this conclusion stands in contrast to [GT, Thm. 2], which suggests that a less fine view, which ignores polynomial blow-ups,[3] deems algorithm design irrelevant to this model. We also note that, in some cases (see, e.g., Theorem 3.4 and Part 3 of Theorem 3.1), carefully designed non-adaptive algorithms outperform canonical ones.

As discussed previously, one of the goals of this work was to study the relation between adaptive and non-adaptive testers in the adjacency matrix model. Our results demonstrate that, in this model, the relation between the adaptive and

---

[3] Recall that [GT, Thm. 2] asserts that canonical testers, which merely select a random subset of vertices and rule according to the induced subgraph, have query-complexity that is at most quadratic in the query-complexity of the best tester. We note that [GT, Thm. 2] also ignores the time-complexity of the testers.

non-adaptive query-complexities is not fixed, but rather varies with the computational problem at hand. In some cases (e.g., Theorem 3.4) the complexities are essentially equal, indeed, as in the case of sampling [CEG]. In other cases (e.g., Theorem 3.1), these complexities are related by a fixed power (e.g., 4/3) that is strictly between 1 and 2. And, yet, in other cases (e.g., Theorem A.4) the non-adaptive complexity is quadratic in the adaptive complexity, which is the maximum gap possible (by [AFKS, GT]). Furthermore, by Theorem A.4, for any $t \geq 4$, there exists a promise problem for which the aforementioned complexities are related by a power of $2 - (2/t)$.

Needless to say, the fundamental relation between adaptive and non-adaptive algorithms was studied in a variety of models, and the current work studies it in a specific natural model (i.e., of property testing in the adjacency matrix representation). In particular, this relation has been studied in the context of property testing in other domains. Specifically, in the setting of testing the satisfiability of linear constraints, it was shown that adaptivity offers absolutely no gain [BHR]. A similar result holds for testing monotonicity of sequences of positive integers [F04]. In contrast, an exponential gap between the adaptive and non-adaptive complexities may exist in the context of testing other properties of functions [F04]. Lastly, we mention that an even more dramatic gap exists in the setting of testing graph properties in the bounded-degree model (of [GR02]); see [RS06].

4   A Complexity Theoretic Perspective

Let us start by rephrasing Conjecture 3.3, while recalling that it refers to properties for which testing requires (adaptive) query complexity that is at least linear in the reciprocal of the proximity parameter (see Proposition A.2).

Conjecture 3.3 (rephrased). *For every integer $t \geq 2$, there exists a* (natural) *graph property Π_t such that* non-adaptively *testing Π_t has query complexity* $\widetilde{\Theta}(q^{2-(2/t)})$, *where $q = q(N, \epsilon)$ denotes the the query complexity of* (adaptively) *testing Π_t.*

Recall that it is known that the non-adaptive query complexity of testing any graph property is at most quadratic in the adaptive query complexity. We stress that Conjecture 3.3 not only asserts that this upper bound is essentially tight, but rather asserts an infinite hierarchy of possible functional relations between the non-adaptive and adaptive query complexity.

The results in this work refer to "two and a half" elements in the conjectured hierarchy as well as to a corresponding hierarchy of promise problems. Specifically, denoting the (adaptive) query complexity by $q = q(N, \epsilon)$, we have:

- Theorem 3.4 establishes the conjecture for $t = 2$. Specifically, Theorem 3.4 presents natural graph properties that have non-adaptive query complexity $\widetilde{\Theta}(q)$.
- Theorem 3.1 establishes the conjecture for $t = 3$. Specifically, Theorem 3.1 presents a natural graph property that has non-adaptive query complexity $\widetilde{\Theta}(q^{4/3})$.

- Theorem 3.2 establishes half of the conjecture for $t = 4$. Specifically, Theorem 3.2 presents a natural graph property that has non-adaptive query complexity $\widetilde{\Omega}(q^{3/2})$.
- Theorem A.4 fully establishes the conjecture in the setting of promise problems. We stress that these promise problems are fixed (independently of the proximity parameter).

Indeed, in all our results $q = q(N, e) = \widetilde{\Omega}(1/\epsilon)$. We also mention that in all our results the upper bounds are established by one-sided error testers, whereas the lower bounds hold also for general (i.e., two-sided error) testers.

Open problems. In addition to the resolution of Conjecture 3.3, our study raises many other open problems; the most evident ones are listed next.

1. What is the non-adaptive query complexity of $\mathcal{BCC}$? Note that Theorem 3.2 only establishes a lower bound of $\Omega(\epsilon^{-3/2})$. We conjecture that an efficient non-adaptive algorithm of query complexity $\widetilde{O}(\epsilon^{-3/2})$ can be devised.
2. For which constants $c \in [1, 2]$ does there exist a property that has adaptive query complexity of $q(\epsilon)$ and non-adaptive query complexity of $\widetilde{\Theta}(q(\epsilon)^c)$? Note that Theorem 3.1 shows that $4/3$ is such a constant, and the same holds for the constant 1 (see, e.g., Theorem 3.4). We conjecture (see Conjecture 3.3) that, for any $t \geq 2$, it holds that the constant $2 - (2/t)$ also satisfies the foregoing requirement. It may be the case that these constants are the only ones that satisfy this requirement.
3. Characterize the class of graph properties for which the query complexity of non-adaptive testers is almost linear in the query complexity of adaptive testers.
4. Characterize the class of graph properties for which the query complexity of non-adaptive testers is almost quadratic in the query complexity of adaptive testers.
5. Characterize the class of graph properties for which the query complexity of adaptive (resp., non-adaptive) testers is almost linear in the reciprocal of the proximity parameter.

The last characterization project may be the most feasible among the three foregoing characterization projects. We mention that this is partially addressed in [A09, AG], which significatly extends and build upon Theorem 3.4. Finally, we recall the well-known open problem, partially addressed in [AS], of providing a characterization of the class of graph properties that are testable within query complexity that is polynomial in the reciprocal of the proximity parameter.

Acknowledgments

O.G. was partially supported by the Israel Science Foundation (grants No. 460/05 and 1041/08). D.R. was partially supported by the Israel Science Foundation (grants No. 89/05 and 246/08).

References

[A81]   Alon, N.: On the number of subgraphs of prescribed type of graphs with a given number of edges. Israel J. Math. 38, 116–130 (1981)

[AFKS]   Alon, N., Fischer, E., Krivelevich, M., Szegedy, M.: Efficient Testing of Large Graphs. Combinatorica 20, 451–476 (2000)

[AFN]   Alon, N., Fischer, E., Newman, I.: Testing of bipartite graph properties. SIAM Journal on Computing 37, 959–976 (2007)

[AFNS]   Alon, N., Fischer, E., Newman, I., Shapira, A.: A Combinatorial Characterization of the Testable Graph Properties: It's All About Regularity. In: 38th STOC, pp. 251–260 (2006)

[AS]   Alon, N., Shapira, A.: A Characterization of Easily Testable Induced Subgraphs. Combinatorics Probability and Computing 15, 791–805 (2006)

[A09]   Avigad, L.: On the Lowest Level of Query Complexity in Testing Graph Properties. Master Thesis, Weizmann Institute of Scienc (December 2009)

[AG]   Avigad, L., Goldreich, O.: Testing Graph Blow-Up, http://www.wisdom.weizmann.ac.il/~oded/p_lidor.html

[BHR]   Ben-Sasson, E., Harsha, P., Raskhodnikova, S.: 3CNF properties are hard to test. SIAM Journal on Computing 35(1), 1–21 (2005)

[BT]   Bogdanov, A., Trevisan, L.: Lower Bounds for Testing Bipartiteness in Dense Graphs. In: IEEE Conference on Computational Complexity, pp. 75–81 (2004)

[CEG]   Canetti, R., Even, G., Goldreich, O.: Lower Bounds for Sampling Algorithms for Estimating the Average. IPL 53, 17–25 (1995)

[F04]   Fischer, E.: On the strength of comparisons in property testing. Inform. and Comput. 189(1), 107–116 (2004)

[GGR]   Goldreich, O., Goldwasser, S., Ron, D.: Property testing and its connection to learning and approximation. Journal of the ACM, 653–750 (July 1998)

[GR02]   Goldreich, O., Ron, D.: Property Testing in Bounded Degree Graphs. Algorithmica 32(2), 302–343 (2002)

[GR08]   Goldreich, O., Ron, D.: Algorithmic Aspects of Property Testing in the Dense Graphs Model. ECCC, TR08-039 (2008)

[GR09]   Goldreich, O., Ron, D.: On Proximity Oblivious Testing. In: Extended Abstract in the Proceedings of the 41st STOC (2009)

[GT]   Goldreich, O., Trevisan, L.: Three theorems regarding testing graph properties. Random Structures and Algorithms 23(1), 23–57 (2003)

[GR07]   Gonen, M., Ron, D.: On the Benefit of Adaptivity in Property Testing of Dense Graphs. In: Charikar, M., Jansen, K., Reingold, O., Rolim, J.D.P. (eds.) RANDOM 2007 and APPROX 2007. LNCS, vol. 4627, pp. 525–539. Springer, Heidelberg (2007)

[R1]   Ron, D.: Property Testing: A Learning Theory Perspective. Foundations and Trends in Machine Learning 1(3), 307–402 (2008)

[R2]   Ron, D.: Algorithmic and Analysis Techniques in Property Testing. Foundations and Trends in TCS 5(2), 73–205 (2010)

[RS06]   Raskhodnikova, S., Smith, A.: A note on adaptivity in testing properties of bounded-degree graphs. ECCC,TR06-089 (2006)

[RS96]   Rubinfeld, R., Sudan, M.: Robust characterization of polynomials with applications to program testing. SIAM Journal on Computing 25(2), 252–271 (1996)

A   Appendix

In this section we review the definition of property testing, when specialized to graph properties in the adjacency matrix model. We also define several natural graph properties, which serve as the pivot of our study, and state some additional results.

A.1   Basic Notions

For an integer n, we let $[n] = \{1, \ldots, n\}$. A generic N-vertex graph is denoted by $G = ([N], E)$, where $E \subseteq \{\{u, v\} : u, v \in [N]\}$ is a set of unordered pairs of vertices. Any set of such graphs that is closed under isomorphism is called a graph property. By oracle access to such a graph $G = ([N], E)$ we mean oracle access to the Boolean function that answers the query $\{u, v\}$ (or rather $(u, v) \in [N] \times [N]$) with the bit 1 if and only if $\{u, v\} \in E$.

Definition A.1 (property testing for graphs in the adjacency matrix model): *A* tester *for a graph property Π is a probabilistic oracle machine that, on input parameters N and ϵ and access to an N-vertex graph $G = ([N], E)$, outputs a binary verdict that satisfies the following two conditions.*

1. *If $G \in \Pi$ then the tester accepts with probability at least $2/3$.*
2. *If G is ϵ-far from Π then the tester accepts with probability at most $1/3$, where G is ϵ-far from Π if for every N-vertex graph $G' = ([N], E') \in \Pi$ it holds that the symmetric difference between E and E' has cardinality that is greater than ϵN^2.[4]*

If the tester accepts every graph in Π with probability 1, then we say that it has one-sided error. *A tester is called* non-adaptive *if it determines all its queries based solely on its internal coin tosses* (and the parameters N and ϵ); *otherwise it is called* adaptive.

The query complexity of a tester is the number of queries it makes to any N-vertex graph oracle, as a function of the parameters N and ϵ. We say that a tester is efficient if it runs in time that is polynomial in its query complexity, where basic operations on elements of $[N]$ (and in particular, uniformly selecting an element in $[N]$) are counted at unit cost. We note that all testers presented in this paper are efficient, whereas the lower bounds hold also for non-efficient testers.

We shall focus on properties that can be tested with query complexity that only depends on the proximity parameter, ϵ. Thus, the query complexity upper bounds that we state hold for any values of ϵ and N, but will be meaningful only for $\epsilon > 1/N^2$ or so. In contrast, the lower bounds (e.g., of $\Omega(1/\epsilon)$) cannot possibly hold for $\epsilon < 1/N^2$, but they will indeed hold for any $\epsilon > N^{-\Omega(1)}$. Alternatively, one may consider the query-complexity as a function of ϵ, where for each fixed value of $\epsilon > 0$ the value of N tends to infinity.

---

[4] Indeed, it is more natural to require that this symmetric difference should have cardinality that is greater than $\epsilon \cdot \binom{N}{2}$. The current convention is adopted for the sake of convenience.

Notation and a convention. For a fixed graph $G = ([N], E)$, we denote by $\Gamma(v) = \{u : \{u, v\} \in E\}$ the set of neighbors of vertex v. At times, we look at E as a subset of $V \times V$; that is, we often identify E with $\{(u, v) : \{u, v\} \in E\}$. If a graph $G = ([N], E)$ is not ϵ-far from a property Π then we say that G is ϵ-close to Π; this means that at most ϵN^2 edges should be added and/or removed from G such to yield a graph in Π.

A.2   The Graph Properties to Be Studied

The set of graphs that consists of a collection of isolated cliques is called clique collection and is denoted $\mathcal{CC}$; that is, a graph $G = ([N], E)$ is in $\mathcal{CC}$ if and only if the vertex set $[N]$ can be partitioned into $(C_1, \ldots, C_t)$ such that the subgraph induced by each C_i is a clique and there are no edges with endpoints in different C_i's (i.e., for every $u < v \in [N]$ it holds that $\{u, v\} \in E$ if and only if there exists an i such that $u, v \in C_i$). If $t \leq c$ then we say that G is in $\mathcal{CC}^{\leq c}$; that is, $\mathcal{CC}^{\leq c}$ is the subset of $\mathcal{CC}$ that contains graphs that are each a collection of up-to c isolated cliques.

A bi-clique is a complete bipartite graph (i.e., a graph $G = (V, E)$ such that V is partitioned into $(S, V \setminus S)$ such that $\{u, v\} \in E$ if and only if $u \in S$ and $v \in V \setminus S$). Note that a graph is a bi-clique if and only if its complement is in $\mathcal{CC}^{\leq 2}$. The set of graphs that consists of a collection of isolated bi-cliques is called bi-clique collection and denoted $\mathcal{BCC}$; that is, a graph $G = ([N], E)$ is in $\mathcal{BCC}$ if and only if the vertex set $[N]$ can be partitioned into $(V_1, \ldots, V_t)$ such that the subgraph induced by each V_i is a bi-clique and there are no edges with endpoints in different V_i's (i.e., each V_i is partitioned into $(S_i, V_i \setminus S_i)$ such that for every $u < v \in [N]$ it holds that $\{u, v\} \in E$ if and only if there exists an i such that $(u, v) \in S_i \times (V_i \setminus S_i)$).

Generalizations of $\mathcal{BCC}$ are obtained by considering collections of "super-paths" and "super-cycles" respectively. A super-path (of length t) is a sequence of disjoint sets of vertices, $S_1, \ldots, S_t$, such that vertices $u, v \in \bigcup_{i \in [t]} S_i$ are connected by an edge if and only if for some $i \in [t-1]$ it holds that $u \in S_i$ and $v \in S_{i+1}$. Note that a bi-clique can be viewed as a super-path of length two. We denote the set of graphs that consists of a collection of isolated super-paths of length t by $\mathcal{SP}_t\mathcal{C}$ (e.g., $\mathcal{SP}_2\mathcal{C} = \mathcal{BCC}$). Similarly, a super-cycle (of length t) is a sequence of disjoint sets of vertices, $S_1, \ldots, S_t$, such that vertices $u, v \in \bigcup_{i \in [t]} S_i$ are connected by an edge if and only if for some $i \in [t]$ it holds that $u \in S_i$ and $v \in S_{(i \bmod t)+1}$. Note that a bi-clique that has at least two vertices on each side can be viewed as a super-cycle of length four (by partitioning each of its sides into two parts). We denote the set of graphs that consists of a collection of isolated super-cycles of length t by $\mathcal{SC}_t\mathcal{C}$ (e.g., $\mathcal{SC}_4\mathcal{C} \subset \mathcal{BCC}$, where the strict containment is due to the pathological case of bi-cliques having at most one node on one side).

A.3   Additional Results

In this section we state two simple lower bounds as well as the promise problem version of Conjecture 3.3.

Lower bounds. We first note that $\Omega(1/\epsilon)$ (adaptive) queries are required for testing any graph property that is non-trivial for testing, where a graph property Π is non-trivial for testing if there exists $\epsilon_0 > 0$ such that for infinitely many $N \in \mathbb{N}$ there exist N-vertex graphs G_1 and G_2 such that $G_1 \in \Pi$ and G_2 is ϵ_0-far from Π. We note that all properties considered in this work are non-trivial for testing. On the other hand, the negation of this (non-triviality) condition means that for every $\epsilon > 0$ and all sufficiently large $N \in \mathbb{N}$ either Π contains no N-vertex graph or all N-vertex graphs are ϵ-close to Π. In such a case (for every such ϵ and N), the tester may decide without even looking at the graph.[5] Turning back to properties that are non-trivial for testing, we prove that any tester for such a property must have query complexity $\Omega(1/\epsilon)$.

Proposition A.2. *Let Π be a property that is non-trivial for testing. Then, any tester for Π has query complexity $\Omega(1/\epsilon)$.*

Note that the claim holds also for general properties (i.e., arbitrary sets of functions). To justify the fact that all our testers are inherently non-canonical, we show that (for any property that is non-trivial for testing) canonical testers must use $\Omega(\epsilon^{-2})$ queries.

Proposition A.3. *Let Π be a property that is non-trivial for testing. Then, any canonical tester for Π has query complexity $\Omega(1/\epsilon^2)$.*

The promise problem version of Conjecture 3.3. For every positive integer $t \geq 4$, we consider a promise problem, denoted Π_t, having inputs that are either in $\mathcal{SC}_t\mathcal{C}$ or in some specific subset, denoted $\mathcal{SC}_{2t}\mathcal{C}'$, of $\mathcal{SC}_{2t}\mathcal{C}$. On proximity parameter ϵ, a tester of Π_t is required to accept inputs in $\mathcal{SC}_t\mathcal{C}$ and reject inputs in $\mathcal{SC}_{2t}\mathcal{C}'$ that are ϵ-far from $\mathcal{SC}_t\mathcal{C}$.

Theorem A.4. (an almost-quadratic complexity gap for promise problems): *For every positive integer $t \geq 4$, the promise problem Π_t satisfies the following:*

1. *There exists an adaptive tester of query complexity $\widetilde{O}(\epsilon^{-1})$ for Π_t. Furthermore, this tester runs in time $\widetilde{O}(\epsilon^{-1})$.*
2. *Any non-adaptive tester for Π_t must have query complexity $\Omega(\epsilon^{-2+(2/t)})$.*
3. *There exists a non-adaptive tester of query complexity $O(\epsilon^{-2+(2/t)})$ for Π_t. Furthermore, this tester runs in time $O(\epsilon^{-2+(2/t)})$.*

---

[5] Indeed, there exists natural graph properties that are trivial for testing (e.g., connectivity, non-planarity, having no vertex of odd degree); see [GGR, Sec. 10.2.1].

Testing Euclidean Spanners*

Frank Hellweg, Melanie Schmidt, and Christian Sohler

Department of Computer Science
Technical University of Dortmund
44227 Dortmund, Germany
{frank.hellweg,melanie.schmidt,christian.sohler}@tu-dortmund.de

Abstract. In this paper we develop a property testing algorithm for the problem of testing whether a directed geometric graph is a $(1 + \delta)$-spanner.

Keywords: Geometric properties, sparse graphs.

1 Introduction

Property testing is the computational task of deciding whether a given object (for example, a graph, a function, or a point set) has a predetermined property Π (for example bipartiteness, linearity, or convex position) or is far away from every object with property Π. Thus, property testing can be viewed as a relaxation of the standard decision problem "Does input graph G have property Π or not?". Since in property testing one only needs to solve a relaxed decision problem, this can often be done much faster than solving the exact problem. Therefore, given access to the input object the goal of property testing is to develop very fast randomized algorithms that perform the relaxed decision task by only looking at a small part of the input object. Typically, the running time of a property testing algorithm is sublinear and sometimes even independent of the object's description size.

Property testing has been introduced by Rubinfeld and Sudan [35] and the study of combinatorial properties has been initiated by Goldreich, Goldwasser, and Ron [27]. Since then, property testing algorithms have been developed for properties of functions [26,25,11], properties of distributions [9,8], algebraic properties [12,35,30], graph and hypergraph properties [27,3,4,18,10,17,7,28], and geometric properties [20,22]. In this paper we continue the study of property testing algorithms for *geometric properties*.

The first property testing algorithms for geometric properties were developed independently in [20] and [22]. In [20] the authors studied properties of geometric objects like point sets and geometric graphs. Among other things, the authors proved that it can be tested in $O(\sqrt[d+1]{n^d/\epsilon})$ queries whether a point set in $\mathbb{R}^d$

---

* An extended version of this abstract will appear in the proceedings of the 18th European Symposium on Algorithms (ESA), 2010. This work was supported by DFG project SO 514/3-1.

is in convex position and with $\widetilde{O}(\sqrt{n/\epsilon})$ time whether a given geometric graph whose vertices lie in the $\mathbb{R}^2$ is a Euclidean minimum spanning tree (see also [21]). In [22] the authors showed that one can test in $O(\log n)$ time (for constant ϵ) whether a list of points is a list of vertices of a convex polygon. The property of convexity of subsets of the $\mathbb{R}^d$ can be tested with exponential (in d) query complexity [34]. The property how well-clustered a point set is has also been studied [1]. The authors showed that certain properties corresponding to radius and diameter k-clustering are testable in time independent of the input size.

Most previous testers allow only very simple access to the input data, for example, if the input object is a point set, then only access to random points of the set is given. In [19] the authors introduce different models that allow more complex queries to input point sets, for example, queries for random points inside a specified query range R. Such queries are supported efficiently by basic geometric data structures. Using such queries, they obtain new property testing algorithms, for example, an improved tester for convex position.

Closely related to property testing is the area of sublinear time algorithms. In this field geometric properties have also been studied. It has been shown that it can be tested in $O(\sqrt{n})$ expected time whether two 3D polyhedra intersect and that one can perform a point location in planar convex subdivisions with bounded face size in $O(\sqrt{n})$ time [14]. In both cases, the algorithm is given access to a standard representation of the object, for example, to doubly connected edge lists, and no preprocessing of the data is allowed. It has also been shown that the cost of the Euclidean minimum spanning tree of a point set in $\mathbb{R}^d$ can be approximated within a factor of $(1 + \epsilon)$ in sublinear time, if one allows access to the point set via certain range queries [16].

1.1   Euclidean Spanners

A weighted directed geometric graph (P, E) is a directed graph whose vertex set is a set of points $P = \{p_1, \ldots, p_n\}$ in the Euclidean space $\mathbb{R}^d$ and whose edge weights (lengths) are given by the Euclidean distance of the vertices, i.e. edge $[p, q\rangle$ has length $\|p - q\|_2$. A graph is called $(1 + \delta)$-spanner, if for every pair of vertices p, q the shortest path distance $d_G(p, q)$ in G is at most $(1 + \delta) \cdot \|p - q\|_2$, i.e. the shortest path distance in G is a good approximation of the true distance of the points p and q.

Definition 1. *Let $\delta > 0$ be a parameter. A geometric graph G is called a $(1+\delta)$-spanner, if $d_G(p, q) \leq (1 + \delta)\|q - p\|_2$ for all pairs of vertices $(p, q) \in P^2, p \neq q$.*

Euclidean spanners are a fundamental geometric graph structure as they can be used to approximately solve many geometric proximity problems, and they find applications, for example, in the area of mobile ad-hoc networks. Many different constructions of Euclidean spanners are known. Euclidian $(1 + \delta)$-spanners with a linear number of edges can, for example, be constructed for every constant $\delta > 0$ by using so-called Θ-graphs [15,29] or structures based on the well-separated pair decomposition [13,32]. Also techniques to construct spanners with bounded-degree are known [6]. For more details we refer to the book [32]. We will investigate the question whether a given graph is a Euclidean spanner. The related question of computing the stretch factor $(1 + \delta)$ of a given graph has recently been studied in [5,23,31]. Additionally, Ahn et al. [2] discuss the problem to find an edge whose removal leads to the smallest possible increase in the stretch factor, and Farshi et al. [24] consider the question which edge should be added to receive the best decrease in the stretch factor (both articles consider very special cases only).

1.2   Our Contribution

In this work, we develop property testing algorithms for Euclidean spanners for $0 < \delta < 1$. We are given access to a geometric graph $G = (P, E)$, whose vertex set $P = \{p_1, \ldots, p_n\}$ is a point set in a constant dimensional space $\mathbb{R}^d$, i.e. $p_i \in \mathbb{R}^d$ for all $1 \leq i \leq n$. The algorithm is given n, the number of vertices of G, but not the vertex positions. It can query in $O(1)$ time the coordinates of a point p_i for every index i, $1 \leq i \leq n$. The graph structure of G is given in the non-functional adjacency list model [33], i.e., we assume that for every vertex p_i we can query in $O(1)$

- the degree $\deg(p_i)$ of a vertex p_i for every index i, $1 \leq i \leq n$, and
- the index of the j-th neighbor of p_i for indices i, $1 \leq i \leq n$, and j, $1 \leq j \leq \deg(p_i)$.

We next define the notion of ϵ-far.

Definition 2. *Let G be a directed geometric graph and let $0 < \epsilon < 1$. G is ϵ-far from being a $(1+\delta)$-spanner, if one has to modify (insert, delete or replace) more than ϵn edges to make G a $(1+\delta)$-spanner. G is ϵ-close to being a $(1+\delta)$-spanner, if it is not ϵ-far from it.*

An algorithm $\mathcal{A}$ is called a *property tester with one-sided error* for the property of being a $(1 + \delta)$-spanner, if for any directed geometric graph G it outputs

- *true* with a probability of 1, if G is a $(1 + \delta)$-spanner
- *false* with probability at least $2/3$, if G is ϵ-far from being a $(1 + \delta)$-spanner

when it is given n, δ and ϵ as input and access to the geometric graph $G = (P, E)$ as described above. The *query complexity* of $\mathcal{A}$ is the worst-case number of queries performed by the algorithm (counting queries for vertex positions, degrees and neighbors).

In this paper, we show that the property of being a $(1 + \delta)$-spanner can be tested with $\tilde{O}(D\frac{\sqrt{n} \cdot \log^6 \Delta}{\epsilon^4})$ queries for constant d and δ under the assumption that the points come from $\{1, \ldots, \Delta\}^d$ and that D is the maximum degree of the input graph. We also provide a lower bound of $\Omega(n^{1/3})$ for property testing algorithms with 1-sided error.

2   The Testing Algorithm

Our testing algorithm works as follows. Ignoring the dependence on ϵ, δ and Δ, the algorithm samples $s = \widetilde{\Theta}(\sqrt{n})$ vertices $r_1, \ldots, r_s$ uniformly at random and performs a Dijkstra's algorithm from each of these vertices with respect to the Euclidean lengths of the edges until $\Theta(\log n)$ vertices have been visited. Now let W_i denotes the distance of the furthest vertex visited from starting vertex r_i. The algorithm checks for every sample vertex r_j with $\|r_i - r_j\|_2 < W_i/(1 + \delta)$, whether there is a spanner path from r_i to r_j. If such a path does not exist, our algorithm rejects. Note that, if G is a $(1 + \delta)$-spanner then such a path must exist, because any vertex not seen by the Dijkstra traversal has a graph distance of at least W_i. In order to prove that any input graph that is ϵ-far from a $(1 + \delta)$-spanner is rejected with high (constant) probability, we show that any graph that does not have too many missing edges between closeby vertices can be turned into a spanner by adding these missing edges plus relatively few long edges. This implies that any such graph cannot be ϵ-far from a Euclidean spanner and so any graph that is ϵ-far from a Euclidean spanner has many missing local edges. By combining this observation with a typical birthday paradox argument for the end points of the missing edges we obtain our main result.

Theorem 1. *Let $G = (P, E)$ be a geometric graph with $P \subseteq \{1, \ldots, \Delta\}^d$ and maximum degree D. There is a property testing algorithm with query complexity and running time $\widetilde{O}(\delta^{-5d}\epsilon^{-5}D\log^6 \Delta\sqrt{n})$ that accepts G, if G is a $(1+\delta)$-spanner and rejects G with probability at least $2/3$, if G is ϵ-far from a $(1 + \delta)$-spanner.*

References

1. Alon, N., Dar, S., Parnas, M., Ron, D.: Testing of Clustering. SIAM Journal on Discrete Mathematics 16(3), 393–417 (2003)
2. Ahn, H.-K., Farshi, M., Knauer, C., Smid, M., Wang, Y.: Dilation-Optimal Edge Deletion in Polygonal Cycles. In: Algorithms and Computation, pp. 88–99. Springer, Heidelberg (2007)
3. Alon, N., Fischer, E., Krivelevich, M., Szegedy, M.: Efficient Testing of Large Graphs. Combinatorica 20(4), 451–476 (2000)
4. Alon, N., Fischer, E., Newman, I., Shapira, A.: A combinatorial characterization of the testable graph properties: it's all about regularity. SIAM Journal on Computing 39(1), 143–167 (2009)
5. Agarwal, P.K., Klein, R., Knauer, C., Langerman, S., Morin, P., Sharir, M., Soss, M.: Computing the Detour and Spanning Ratio of Paths, Trees, and Cycles in 2D and 3D. Discrete and Computational Geometry 39(1-3), 17–37 (2007)
6. Arya, S., Das, G., Mount, M., Salowe, J.S., Smid, M.: Euclidean spanners: short, thin, and lanky. In: Proceedings of the 27th Annual ACM Symposium on Theory of Computing (STOC), pp. 489–498 (1995)
7. Avart, C., Rödl, V., Schacht, M.: Every Monotone 3-Graph Property is Testable. SIAM Journal on Discrete Mathematics 21(1), 73–92 (2007)
8. Batu, T., Fortnow, L., Fischer, E., Kumar, R., Rubinfeld, R., White, P.: Testing Random Variables for Independence and Identity. In: Proceedings of the 42nd IEEE Symposium on Foundations of Computer Science (FOCS), pp. 442–451 (2001)

9. Batu, T., Fortnow, L., Rubinfeld, R., Smith, W., White, P.: Testing that distributions are close. In: Proceedings of the 41st IEEE Symposium on Foundations of Computer Science (FOCS), pp. 259–269 (2000)

10. Benjamini, I., Schramm, O., Shapira, A.: Every minor-closed property of sparse graphs is testable. In: Proceedings of the 40th Annual ACM Symposium on Theory of Computing (STOC), pp. 393–402 (2008)

11. Blais, E.: Testing juntas nearly optimally. In: Proceedings of the 41st Annual ACM Symposium on Theory of Computing (STOC), pp. 151–158 (2009)

12. Blum, M., Luby, M., Rubinfeld, R.: Self-Testing/Correcting with Applications to Numerical Problems. In: Proceedings of the 22nd Annual ACM Symposium on Theory of Computing (STOC), pp. 73–83 (1990)

13. Callahan, P.B., Kosaraju, S.R.: Faster algorithms for some geometric graph problems in higher dimensions. In: Proceedings of the 4th ACM-SIAM Symposium on Discrete Algorithms, pp. 291–300 (1993)

14. Chazelle, B., Liu, D., Magen, A.: Sublinear Geometric Algorithms. SIAM Journal on Computing 35(3), 627–646 (2006)

15. Clarkson, K.L.: Approximating algorithms for shortest path motion planning. In: Proceedings of the 19th ACM Symposium on the Theory of Computation, pp. 56–65 (1987)

16. Czumaj, A., Ergun, F., Fortnow, L., Magen, A., Newman, I., Rubinfeld, R., Sohler, C.: Approximating the Weight of the Euclidean Minimum Spanning Tree in Sublinear Time. SIAM Journal on Computing 35(1), 91–109 (2005)

17. Czumaj, A., Sohler, C.: Testing hypergraph colorability. Theoretical Computer Science 331(1), 37–52 (2005)

18. Czumaj, A., Shapira, A., Sohler, C.: Testing hereditary properties of non-expanding bounded-degree graphs. SIAM Journal on Computing 38(6), 2499–2510 (2009)

19. Czumaj, A., Sohler, C.: Property Testing with Geometric Queries. In: Meyer auf der Heide, F. (ed.) ESA 2001. LNCS, vol. 2161, pp. 266–277. Springer, Heidelberg (2001)

20. Czumaj, A., Sohler, C., Ziegler, M.: Property Testing in Computational Geometry. In: Paterson, M. (ed.) ESA 2000. LNCS, vol. 1879, pp. 155–166. Springer, Heidelberg (2000)

21. Czumaj, A., Sohler, C.: Testing Euclidean minimum spanning trees in the plane. ACM Transactions on Algorithms 4(3) (2008)

22. Ergun, F., Kannan, S., Kumar, R., Rubinfeld, R., Viswanathan, M.: Spot-Checkers. Journal of Computer and System Sciences 60(3), 717–751 (2000)

23. Eppstein, D., Wortman, K.A.: Minimum dilation stars. Computational Geometry: Theory and Applications 37(1), 27–37 (2007)

24. Farshi, M., Giannopoulos, P., Gudmundsson, J.: Finding the best shortcut in a geometric network. In: Proceedings of the 21th Annual Symposium on Computational Geometry, pp. 327–335 (2005)

25. Fischer, E., Lehman, E., Newman, I., Raskhodnikova, S., Rubinfeld, R., Samorodnitsky, A.: Monotonicity testing over general poset domains. In: Proceedings of the 34th Annual ACM Symposium on Theory of Computing (STOC), pp. 474–483 (2002)

26. Goldreich, O., Goldwasser, S., Lehman, E., Ron, D., Samorodnitsky, A.: Testing Monotonicity. Combinatorica 20(3), 301–337 (2000)

27. Goldreich, O., Goldwasser, S., Ron, D.: Property Testing and its Connection to Learning and Approximation. Journal of the ACM 45(4), 653–750 (1998)

28. Goldreich, O., Ron, D.: Property Testing in Bounded Degree Graphs. Algorithmica 32(2), 302–343 (2002)

29. Keil, M.: Approximating the complete Euclidean graph. In: Karlsson, R., Lingas, A. (eds.) SWAT 1988. LNCS, vol. 318, pp. 208–213. Springer, Heidelberg (1988)
30. Kaufman, T., Sudan, M.: Algebraic property testing: the role of invariance. In: Proceedings of the 40th Annual ACM Symposium on Theory of Computing (STOC), pp. 403–412 (2008)
31. Narasimhan, G., Smid, M.: Approximating the Stretch Factor of Euclidean Graphs. SIAM Journal on Computing, 978–989 (2000)
32. Narasimhan, G., Smid, M.: Geometric Spanner Networks. Cambridge University Press, Cambridge (2007)
33. Parnas, M., Ron, D.: Testing the diameter of graphs. Random Structures & Algorithms 20(2), 165–183 (2002)
34. Rademacher, L., Vempala, S.: Testing Geometric Convexity. In: Lodaya, K., Mahajan, M. (eds.) FSTTCS 2004. LNCS, vol. 3328, pp. 469–480. Springer, Heidelberg (2004)
35. Rubinfeld, R., Sudan, M.: Robust Characterizations of Polynomials with Applications to Program Testing. SIAM Journal on Computing 25(2), 252–271 (1996)

Symmetric LDPC Codes and Local Testing

Tali Kaufman[1] and Avi Wigderson[2]

[1] Department of Computer Science, Weizmann Institute of Science, Rehovot, Israel
kaufmant@mit.edu
[2] Institute for Advanced Study, Princeton, USA
avi@ias.edu

Abstract. Coding theoretic and complexity theoretic considerations naturally lead to the question of generating symmetric, sparse, redundant linear systems. This paper provides new way of constructions with better parameters and new lower bounds.

Low Density Parity Check (*LDPC*) codes are linear codes defined by short constraints (a property essential for *local testing* of a code). Some of the best (theoretically and practically) used codes are LDPC. *Symmetric* codes are those in which all coordinates "look the same", namely there is some transitive group acting on the coordinates which preserves the code. Some of the most commonly used locally testable codes (especially in PCPs and other proof systems), including all "low-degree" codes, are symmetric. Requiring that a symmetric binary code of length n has large (linear or near-linear) distance seems to suggest a "conflict" between 1/rate and density (constraint length). In known constructions, if one is constant then the other is almost worst possible - $n/poly(\log n)$.

Our main positive result simultaneously achieves *symmetric* low density, constant rate codes generated by a *single* constraint. We present an *explicit* construction of a symmetric and transitive binary code of length n, near-linear distance $n/(\log \log n)^2$, of constant rate and with constraints of length $(\log n)^4$. The construction is in the spirit of Tanner codes, namely the codewords are indexed by the edges of a sparse regular expander graph. The main novelty is in our construction of a transitive (non Abelian!) group acting on these edges which preserves the code. Our construction is one instantiation of a framework we call *Cayley Codes* developed here, that may be viewed as extending zig-zag product to symmetric codes.

Our main negative result is that the parameters obtained above cannot be significantly improved, as long as the acting group is solvable (like the one we use). More specifically, we show that in constant rate and linear distance codes (aka "good" codes) invariant under solvable groups, the density (length of generating constraints) cannot go down to a constant, and is bounded below by $\log^{(\Omega(\ell))} n$ if the group has a derived series of length ℓ. This negative result precludes natural local tests with constantly many queries for such solvable "good" codes.

1 Introduction

The work in this paper is partially motivated from several (related) research directions. Following is a very high level description of these.

O. Goldreich (Ed.): Property Testing, LNCS 6390, pp. 312–319, 2010.

Locally testable codes. Codes in which the proximity to a codeword can be determined by a few coordinate queries have proven a central ingredient in some major results in complexity theory. They appear as low-degree tests in the $IP = PSPACE$, $MIP = NEXP$ and $PCP = NP$ theorems, and indeed the work of [16] (which was later partly derandomized by [8]) elucidates their role as the "combinatorial heart" of PCPs. The quest to simultaneously optimize their coding theoretic parameters and the number of queries used has recently culminated in the combination of [7] and [13] (see also [26]) in a length n binary linear code of linear distance and rate $1/(\log n)^{O(1)}$, testable with a constant number of queries (which are testing linear constraints of constant length). Further improving the rate to a constant is a major open problem. Essential to locally testable codes is having short constraints.

LDPC codes. Low Density Parity Check codes are precisely linear codes with short constraints. Density is the constraints length. These codes were defined in the seminal work of Gallager [14] in the 60's. Only in the 90's, due to works of [22],[30],[32] and others did LDPC codes start to compete with the algebraic constructions in the coding-theory scene. Today these provide some of the best practical and theoretical codes for many noise models, and are extremely efficient to encode and decode. In particular, they can achieve linear distance, constant rate and constant constraint size simultaneously. But their natural potential for local testing was (possibly) devastated by such results as [6], who showed that a general class of LDPC codes, based on expanders, requires a linear number of queries to test, despite having constant-size constraints. We note that possessing short defining constraints is not always an obvious property of a code – e.g. it was only recently discovered in [19] that the sparse dual-BCH codes have such constraints (but unfortunately this code has a very bad rate).

Symmetric codes. Many of the classical codes, e.g. Hamming, Reed-Solomon, Hadamard, Reed-Muller, BCH, and some Goppa codes are symmetric, namely there is a transitive group acting on the coordinates which leaves the code invariant. Symmetry is not only elegant mathematically - it often also implies concise representation of the code as well as tools to analyze its quality parameters, like rate and distance. Huge literature is devoted to such codes within coding theory, but even for cyclic codes (those invariant under cyclic shifts) it is still a major open problem if they can have simultaneously constant rate and linear distance. The conjecture is that this is impossible. A major result of Berman from the seventies [9] shows that there are no good cyclic codes of length n where all the prime divisors of n arc bounded. Interesting progress on this conjecture was made by Babai Shpilka and Stefankovic [5] that extend Berman's result and relax the conditions on the sizes of the prime divisors of the code length. Moreover [5] show that the conjecture is true if one requires the cyclic code to be defined by constraints of constant length (i.e to be LDPC). McElice [25] proved (non constructively) that there are asymptotically good *non-linear* codes invariant under the action of very large groups, however these codes are clearly not LDPC.

Symmetric low-density and locally testable codes. Starting with linearity testing of [10] and the first low-degree tests of [4,28], nearly all locally testable codes appearing in proof systems *are* symmetric. A theory studying the extent to which symmetry can help (or handicap) local testing was initiated by Kaufman and Sudan [18]. They generalized known examples showing that when the acting group is the affine group (and the coordinates are naturally identified with the elements of the vectors space acted upon), then having short constraints that define the code is not only necessary, but also *sufficient* for local testability. Moreover, in these cases the orbit (under the group action) of a *single* constraint suffices to define the code, and a canonical local test is picking a random constraint from that orbit[1]. Again, the rate of all these codes is poor, and [18] challenge reconciling the apparent conflict between rate and density, possibly for other groups.

Expanding Cayley graphs. Gallager's construction [14] of LDPC codes was based on sparse random graphs, and Tanner's construction [35] was based on high girth graphs. Sipser and Spielman [32] identified *expansion* as the crucial parameter of graphs which yield codes with good parameters. This was followed up in almost all subsequent works, using expanders to construct codes. This work motivated further explicit constructions of good expanders. As example, we note that the [32] "belief propagation" decoding algorithm for LDPC was simplest if the underlying graph is a *lossless* expander, and subsequently [11] were able to explicitly construct such expanders. *Unfortunately, all codes constructed this way seem far from symmetric.* But expander *graphs* can certainly be symmetric! Indeed, almost all constructions of expander graphs are Cayley graphs, namely the vertices correspond to the elements of a finite group, and edges are prescribed by a fixed generating set of the group. It is evident that such graphs are symmetric, namely the group itself acts transitively on the vertices and preserves the edges. We note importantly that even the zig-zag product construction of expanders [31], which started as a combinatorial alternative to algebraic constructions, was extended to allow iterative probabilistic constructions of Cayley graphs [2,27] via the semi-direct product of groups. Our codes are partially motivated by making explicit the probabilistic construction of [2,27] Attempts to construct codes iteratively exist, with the best example being Meir's, partially explicit construction [26]. However, again, this code is far from symmetric.

Several natural research directions point to the following question: **To what extent can symmetric LDPC codes attain (or even come close to) the coding theory gold standard of linear distance and constant rate?** To fix ideas, let us consider symmetric codes with linear (or even near-linear) distance, and examine the trade-off between density and 1/rate. In all known

---

[1] We note that the existence of a *single* constraint that generates a code gives rise to a canonical algorithm for local testing the code. An algorithm that picks a random constraint from the orbit. For codes invariant under the affine group, Kaufman and Sudan have shown that such a canonical algorithm is indeed a valid local tester for the code. This motivates the search for other symmetric codes generated by the orbit(s) of one (or few) generators, with the hope that local testing would be implied.

codes if 1/rate or density is constant then the other is *worst* possible, about $n/poly(\log n)$, the code length! Best density/rate trade-offs for known binary high distance symmetric codes are the following. Reed-Muller codes over binary field (say degree-d polynomials), which are invariant under the affine group, have short constraints (2^d-long) but pathetic rate $(\log n)^d/n$. BCH codes, invariant under the cyclic group, have constant rate, but constraints of (worst possible) length $\Omega(n)$. Reed-Muller codes over large fields concatenated with Hadamard achieve density $(\log n)^{1/\epsilon}$ with (1/rate) being $2^{(\log n)^\epsilon}$ [3,34].[2]

Indeed, some believed that the conflict between density and rate in symmetric codes cannot be reconciled. On the other hand, no result precludes the ratio of density/rate from being *best* possible, namely a constant! Our paper addresses both upper and lower bounds on this trade-off.

2 Our Results

Our main positive result allows simultaneous constant rate and polylogarithmic density, and in particular reduces the upper bound on the ratio density/rate to $poly \log n$! More precisely, we provide an explicit construction of length-n symmetric codes of constant rate and distance $n/(\log \log n)^2$ which is defined by constraints of a length $poly(\log n)$. Moreover, these constraints constitute the orbit of a *single* constraint, under the transitive action of a (non Abelian) group.

Our main negative result shows that there is no good code invariant under a solvable group with few low-weight generators. In fact we rule out the possibility of such codes even if the support of their generators is $o(\log^{\Omega(\ell)} n)$ if the group has a derived series of length ℓ and n is the code length. This result exclude the possibility of good solvable locally testable codes with few low weight generators.

3 Our Techniques

In order to prove our upper bound, we develop a framework of "Cayley Codes", which we describe next. They extend Tanner codes in that the coordinates of the code are identified with the edges of a regular expander graph, and constraints are imposed on neighborhoods (namely edges incident on each single vertex) according to a fixed "inner code" B. In Cayley codes we naturally insist that the underlying graph is a Cayley graph, namely the vertices are the elements of a group G, and a set of generators S of the group determine edges in a natural way. While this a graph is symmetric (G acts transitively on its vertices), there is no such guarantee in general for the code. The problem is to find a group that acts on the *edges* of the graph, and preserves all copies of the internal code. We show that if some group H simultaneously acts transitively on the code B and

---

[2] Note that when this code is mostly used to get constant query complexity, it is modified to make coordinates correspond not to the value of the encoded polynomial on a point, but rather its value on an entire line or larger subspace. This has lousy rate, and when derandomized to improve the rate, transitivity of the action is lost.

acts on the group G, then the semi-direct product group $G \rtimes H$ acts transitively on the edges. We note that this action is not standard.

We then turn to find an appropriate instantiation of this idea with good parameters. This paragraph is a bit technical and may be skipped at first reading. The group G is chosen to be the hypercube $\mathbb{F}_2^t$, and S a very specific ϵ-biased set in G (so as to make the associated Cayley graph expanding), which can be identified with the elements of a cyclic group H isomorphic to the multiplicative group of of F_{t4}^*. The inner code B is chosen to be a BCH code on S on which the group H acts transitively. The inferior distance and density of the code B are mitigated since its length is only polylogarithmic in the length of the whole code. Now the action of H on G (whose nature we describe in the technical section) allows the construction of the semi-direct product $G \rtimes H$. We now define the action of this group on directed edges of the graph, and prove that all parts fit: this group acts transitively on the Tanner code of the Cayley graph on $G; S$.

Our lower bound methods extend work of Lubotzky and Weiss [23], who showed a similar lower bound on the number of generators Cayley graphs on these groups to be expanders. The extension is in two directions - we show the same for Schreier graphs, and then extend their argument from finding standard separators to finding ϵ-partitions of the graph to many parts - from which we can deduce information on the distance and rate of the associated Tanner codes.

The proof showing that there are no good solvable codes with few low weight generators has two main parts. First, for a parameter ϵ (later taken to be $o(1)$) we define a new notion that we call an ϵ-partition of a graph, which extends the notion of a small separator, in that we demand that the separating set splits the graph into *many* pieces. More precisely, a graph has an ϵ-partition if one can remove ϵ fraction of its vertices to make all connected components of relative size at most ϵ. We show that a Schreier graph of a solvable group with $d = o(\log^{\Omega(\ell)} n)$ generators has an ϵ-partition where ϵ is sub-constant. In the second part of the proof we associate codes invariant under groups with Schreier graphs over these groups, and show that if the associated Schreier graph has an ϵ-partition then either the rate or the relative distance of the code is bounded by ϵ.

4   Related Work

Alon, Lubotzky and Wigderson [2] provided a randomized construction of high rate high distance codes generated by two orbits. They asked about *explicit* constructions of high rate, high distance codes generated by few orbits (for the group they studied). Our code construction provides such explicit codes generated by *one* orbit!.

A work by Babai,Shpilka and Stefankovic [5] showed that there are no good cyclic codes with low weight constraints (with no restriction on the number of generating constraints). Since low weight constraints are a necessary (but not sufficient) condition for testability, they showed that there are no good cyclic locally testable codes. Our work here shows that there are no good solvable locally testable codes with few low weight generating constraints. i.e. we exclude good locally testable codes over larger groups of symmetry but under the assumption of few low weight generating constraints. As far as we know, it could well be the case that a cyclic code whose dual has a low weight basis must have a basis that is generated by a constant many low-weight constraints.

5   Conclusions and Open Questions

This paper was motivated from by the construction of locally-testable codes of good coding-theoretic parameters. As is well known, Goldreich and Sudan [16] showed how to obtain such codes can be constructed from PCPs with related parameters, and good parameters are achieved by combining the PCPs of Dinur [13] with Ben-Sasson and Sudan [7]. Specifically, they achieve linear binary codes of length n with linear distance, rate $1/(\log n)^c$ and constant-size queries. These codes are completely explicit.

Removing the PCP machinery and obtaining such codes (and even better ones) directly is a basic question, motivated at length in the paper of Meir [26]. He succeeds only partially, in that his construction that is partly probabilistic. Moreover, the construction cleverly retains "proofs of membership" in the code, as part of the code, which make it resemble Dinur's PCP construction.

We take a completely different approach. As all locally-testable codes must be LDPC codes (since low query complexity means low density in the parity check matrix), and moreover many locally-testable codes are symmetric (have a transitive group acting on them), we ask first if the above coding theoretic parameters can be attained by codes that are simultaneously symmetric and low-density. We give the first such construction. Our codes are linear binary codes of length n with near linear distance $n/(\log \log n)^2$, constant rate and both density bounded by $1/(logn)^4$. The group acting transitively is non-Abelian. All previously known symmetric codes with such (or even weaker) distance had either density or $(1/\text{rate})$ close to n, and groups in all cases are Abelian.

There are several open questions that arise from this work.

- Cayley codes and local testing. Are the Cayley codes we construct actually locally testable? We tend to think that they are not, in which case would be the *first* example of a symmetric LDPC code which is not locally testable. As we offer a general framework of Cayley codes, possibly other choices of components in this framework can lead to to locally-testable codes.
- Improving the parameters. Can one get the ultimate – symmetric, constant density *good* codes (namely with linear distance and constant rate)? Our lower bounds imply that for such a result the acting group must be "more noncommutative" than the one we use, namely it cannot be solvable with a constant-length derived series.
- Key to our lower bound is our that Cayley codes of such groups have ϵ-partition, a property which implies in particular that such codes must have two *disjoint* codewords. Interestingly, the question of proving the latter property for similar codes comes up naturally in the work of Lackenby [20,21] on

3-dimensional manifolds. Specifically, he asks if linear codes symmetric under the action of p-groups (which are solvable, but can have constant degree Cayley graphs), which have constant rate, density and normalized distance, must have two codewords with disjoint support. Our lower-bound techniques fails for such groups.

References

1. Alon, N., Kaufman, T., Krivelevich, M., Litsyn, S., Ron, D.: Testing Low Degree Polynomials Over GF(2). In: Arora, S., Jansen, K., Rolim, J.D.P., Sahai, A. (eds.) RANDOM 2003 and APPROX 2003. LNCS, vol. 2764, pp. 188–199. Springer, Heidelberg (2003); Also IEEE Transactions on Information Theory 51(11), 4032–4039 (2005)
2. Alon, N., Lubotzky, A., Wigderson, A.: Semi Direct product in groups and zig-zag product in graphs: connections and applictions. In: Proceedings of the 42nd Annual Symposium on the Foundations of Computer Science (FOCS), pp. 630–637 (2001)
3. Arora, S., Sudan, M.: Improved low degree testing and its applications. Combinatorica 23(3), 365–426 (2003)
4. Babai, L., Fortnow, L., Lund, C.: Non-Deterministic Exponential Time has Two-Prover Interactive Protocols. Computational Complexity 1(1), 3–40 (1991)
5. Babai, L., Shpilka, A., Stefankovic, D.: Locally testable cyclic codes. IEEE Transactions on Information Theory 51(8), 2849–2858 (2005)
6. Ben-Sasson, E., Harsha, P., Raskhodnikova, S.: Some 3CNF Properties are Hard to Test. SIAM Journal on Computing 35(1), 1–21 (2005)
7. Ben-Sasson, E., Sudan, M.: Simple PCPs with poly-log rate and query complexity. In: STOC 2005, 266–275 (2005)
8. Ben-Sasson, E., Sudan, M., Vadhan, S., Wigderson, A.: Randomness-efficient Low Degree Tests and Short PCPs via Epsilon-Biased Sets. In: 35th Annual ACM Symposium, STOC 2003, pp. 612–621 (2003)
9. Berman, S.D.: Semisimple Cyclic and Abelian Codes. Cybernetics 3, 21–30 (1967)
10. Blum, M., Luby, M., Rubinfeld, R.: Self-Testing/Correcting with Applications to Numerical Problems. J. Comp. Sys. Sci. 47(3) (December 1993)
11. Capalbo, M., Reingold, O., Vadhan, S., Wigderson, A.: Randomness Conductors and Constant-Degree Expansion Beyond the Degree/2 Barrier. In: Proceedings of the 34th STOC, pp. 659–668 (2002)
12. Carlitz, L., Uchiyama, S.: Bounds for exponential sums. Duke Math. J. 24, 37–41 (1957)
13. Dinur, I.: The PCP theorem by gap amplification. J. ACM 54(3), 12 (2007)
14. Gallager, R.G.: Low density parity check codes. MIT Press, Cambridge (1963)
15. Grigorescu, E., Kaufman, T., Sudan, M.: Succinct Representation of Codes with Applications to Testing (manuscript)
16. Goldreich, O., Sudan, M.: Locally testable codes and PCPs of almost-linear length. J. ACM 53(4), 558–655 (2006)
17. Holton, D.A., Sheehan, J.: The Petersen Graph. Cambridge University Press, Cambridge (1993)
18. Kaufman, T., Sudan, M.: Algebraic Property Testing: The Role of Invariance. In: Proceedings of the 40th ACM Symposium on Theory of Computing, STOC (2008)
19. Kaufman, T., Litsyn, S.: Almost Orthogonal Linear Codes are Locally Testable. In: FOCS 2005, pp. 317–326 (2005)

20. Lackenby, M.: Large groups, property (τ) and the homology growth of subgroups. Math. Proc. Cambridge Philos. Soc. 146(3), 625–648 (2009)
21. Lackenby, M.: Covering spaces of 3-orbifolds. Duke Math. J. 136(1), 181–203 (2007)
22. Luby, M.G., Mitzenmacher, M., Amin Shokrollahi, M., Spielman, D.A.: Improved Low-Density Parity-Check Codes Using Irregular Graphs. IEEE Transactions on Information Theory 47(2), 585–598 (2001)
23. Lubotzky, A., Weiss, B.: Groups and expanders. In: Friedman, e.J. (ed.) Expanding Graphs. DIMACS Ser. Discrete Math. Theoret. Compt. Sci., vol. 10, pp. 95–109. Amer. Math. Soc., Providence (1993)
24. MacWilliams, F.J., Sloan, N.J.A.: The Theory of Error Correcting Codes. North Holland, Amsterdam (1977)
25. McElice, R.J.: On the Symmetry of Good Nonlinear Codes. IEEE Trans. Inform. Theory IT-16, 609–611 (1970)
26. Meir, O.: Combinatorial Construction of Locally Testable Codes. In: Proceedings of STOC 2008, pp. 285–294 (2008)
27. Meshulam, R., Wigderson, A.: Expanders in Group Algebras. Combinatorica 24(4), 659–680 (2004)
28. Rubinfeld, R., Sudan, M.: Robust characterizations of polynomials with applications to program testing. SIAM Journal on Computing 25(2), 252–271 (1996)
29. Rozenman, E., Shalev, A., Wigderson, A.: A new family of Cayley expanders (?) In: 36th Annual ACM Symposium, STOC 2004, pp. 445–454 (2004)
30. Richardson, T., Urbanke, R.: The Capacity of Low-Density Parity Check Codes under Message-Passing Decoding. IEEE Transactions on Information Theory 47(2), 599–618 (2001)
31. Reingold, O., Vadhan, S., Wigderson, A.: Entropy Waves, the Zig-Zag Graph Product, and New Constant-Degree Expanders. Annals of Mathematics 155(1), 157–187 (2002)
32. Sipser, M., Spielman, D.A.: Expander codes. IEEE Transactions on Information Theory 42(6), 1710–1722 (1996)
33. Sudan, M.: Lecture notes, http://people.csail.mit.edu/madhu/FT01/scribe/bch.ps
34. Sudan, M., Trevisan, L., Vadhan, S.: Pseudorandom generators without the XOR Lemma. Journal of Computer and System Sciences 62(2), 236–266 (2001)
35. Tanner, R.M.: A recursive approach to low complexity codes. IEEE Transactions on Information Theory 27(5), 533–547 (1981)
36. Weil, A.: Sur les courbes algebriques et les varietes qui s'en deduisent. Actualities Sci. et Ind. no. 1041. Hermann, Paris (1948)

Some Recent Results on Local Testing of Sparse Linear Codes[*]

Swastik Kopparty and Shubhangi Saraf

Computer Science and Artificial Intelligence Laboratory, Massachusetts Institute of Technology, Cambridge, USA
swastik@mit.edu, shibs@mit.edu

Abstract. We study the local testability of linear codes. Our approach is based on a reformulation of this question in the language of tolerant linearity testing under a non-uniform distribution. We then study the question of linearity testing under non-uniform distributions directly, and give a sufficient criterion for linearity to be tolerantly testable under a given distribution. We show that several natural classes of distributions satisfy this criterion (such as product distributions and low Fourier-bias distributions), thus showing that linearity is tolerantly testable under these distributions. This in turn implies that the corresponding codes are locally testable.

For the case of random sparse linear codes, we show the testability and decodability of such codes in the presence of very high noise rates. More precisely, we show that any linear code in $\mathbb{F}_2^n$ which is:
- sparse (i.e., has only $\mathrm{poly}(n)$ codewords)
- unbiased (i.e., each nonzero codeword has Hamming weight in $(1/2 - n^{-\gamma}, 1/2 + n^{-\gamma})$ for some constant $\gamma > 0$)

can be locally tested and locally list decoded from $(1/2-\epsilon)$-fraction errors using only $\mathrm{poly}(\frac{1}{\epsilon})$ queries to the received word. This simultaneously simplifies and strengthens a result of Kaufman and Sudan, who gave a local tester and local (unique) decoder for such codes from some constant fraction of errors. For the case of Dual BCH codes, our algorithms can also be made to run in sublinear time.

Building on the methods used for the local algorithms, we also give sub-exponential *time* algorithms for list-decoding arbitrary unbiased (but not necessarily sparse) linear codes in the high-error regime.

Keywords: error correcting codes, list-decoding, random codes.

1 Introduction

We begin by setting up some notation. A linear code $\mathcal{C}$ in $\mathbb{F}_2^N$ is simply a linear subspace of $\mathbb{F}_2^N$. The elements of $\mathcal{C}$ are often referred to as "codewords". We say

---

[*] The current extended abstract is a summary of some of the results that appeared in "Tolerant linearity testing and locally testable codes" that appeared in RANDOM 2009 and "Local list-decoding and testing of random linear codes from high-error" that will appear in STOC 2010.

O. Goldreich (Ed.): Property Testing, LNCS 6390, pp. 320–333, 2010.

that C is sparse if $|C| \leq N^c$ for some constant c. For a string $x \in \mathbb{F}_2^N$, we define its (normalized Hamming) weight $\mathrm{wt}(x)$ to equal $\frac{1}{n} \times$ (the number of nonzero coordinates of x). We define the *bias* of the code C as $\max_{y \in C \setminus \{0\}} \left| \frac{1 - \mathrm{wt}(y)}{2} \right|$. Thus each nonzero codeword y of a code of bias β has $\mathrm{wt}(y) \in [(1-\beta)/2, (1+\beta)/2]$. For two strings $x, y \in \mathbb{F}_2^N$, we define the (normalized Hamming) distance between x and y, $\Delta(x, y)$, to be $\frac{1}{n} \times$ (the number of coordinates $i \in [N]$ where x_i and y_i differ). We then define the distance of a string x from the code C, $\Delta(x, C)$ to be the minimum distance of x to some codeword of C:

$$\Delta(x, C) = \min_{y \in C} \Delta(x, y).$$

The basic algorithmic tasks associated with codes are error-detection and error-correction. Here we are given an arbitrary received word $w \in \mathbb{F}_2^N$, and we want to (1) determine if $\Delta(w, C) > \delta$, and (2) find all codewords $y \in C$ such that $\Delta(w, C) \leq \delta$. In recent years, there has been much interest in developing sublinear time algorithms (and in particular, highly query-efficient algorithms) for these tasks. In what follows, we will describe our results on various aspects of these questions.

2   Locally Testable Codes and Tolerant Linearity Testing

Informally, a local tester for C is a randomized algorithm A, which when given oracle access to a received word $w \in \mathbb{F}_2^N$, makes very few queries to the received word w, and distinguishes between the case where $w \in C$ and the case where w is "far" (in Hamming distance) from all codewords of C. A code C is said to be *locally testable* if there exists a constant query local tester for C.

The first local tester (for any code) came from the seminal work of Blum, Luby and Rubinfeld [BLR93], which gave an efficient, 3-query local tester for the Hadamard code.

In order to study local testability for linear codes, we first reformulate the question in the language of testing under *distributions*. Let $C \subseteq \mathbb{F}_2^N$ be a linear code of dimension n, and let G be an $n \times N$ generator matrix for C. Let $S = \{v_1, v_2, \ldots, v_N\} \subset \mathbb{F}_2^n$ denote the set of columns of G. We associate to C the distribution μ over $\mathbb{F}_2^n$ which is uniform over S. Every word w in $\mathbb{F}_2^N$ can be viewed as a function $f_w : S \to \mathbb{F}_2$, where $f_w(v_i) = w_i$. Under this mapping, every *codeword* of C gets associated with a *linear* function.

Note that via this translation, the problem of testing if w is close to some codeword exactly translates into the problem of testing if f_w is close to some linear function *under the distribution* μ (where the distance of two functions g, h under μ is measured by the probability that g and h differ on a random sample from μ).

In this language, the BLR local tester for the Hadamard code is precisely the problem of testing linearity under μ, where now μ is the uniform distribution over $\mathbb{F}_2^n$.

For a general linear code $\mathcal{C}$, the related distribution μ is uniform over a subset S of $\mathbb{F}_2^n$. For the relationship between testability of the code $\mathcal{C}$ and the testability of linearity under μ to be tight, we must essentially force the tester for linearity under μ to make queries only from the set S. A notion of testing that naturally enforces this requirement is that of *tolerant* linearity testing. We now formally describe this notion and give the connection to locally testable codes.

Let $\mathcal{C}$ be a class of functions from a finite set $\mathcal{D}$ to a finite set $\mathcal{R}$. In the task of *tolerant testing* for $\mathcal{C}$, we are given oracle access to a function $f : \mathcal{D} \to \mathcal{R}$, and we wish to determine using few queries to f, whether f is well approximable by functions in $\mathcal{C}$; equivalently, to distinguish between the case when f is *close* to some element of $\mathcal{C}$, and the case when f is *far* from all elements of $\mathcal{C}$. Tolerant property testing was introduced by Parnas, Ron and Rubinfeld in [PRR06] as a refinement of the problem of property testing [RS96], [GGR98] (where one wants to distinguish the case of f *in* $\mathcal{C}$ from the case when f is *far* from $\mathcal{C}$), and is now widely studied. The usual notion of closeness considered in the literature is via the distance measure $\Delta(f, g) = \Pr_{x \in \mathcal{D}}[f(x) \neq g(x)]$, where $x \in \mathcal{D}$ is picked according to the *uniform distribution* over $\mathcal{D}$.

We propose to study tolerant property testing under general distributions. Given a probability measure μ on $\mathcal{D}$, the μ-distance of f from g, where $f, g : \mathcal{D} \to \mathcal{R}$, is defined by

$$\Delta_\mu(f, g) = \Pr_{x \in \mu}[f(x) \neq g(x)].$$

Then the measure of how well f can be approximated by elements of $\mathcal{C}$ is via the μ-distance

$$\Delta_\mu(f, \mathcal{C}) = \min_{g \in \mathcal{C}} \Delta_\mu(f, g).$$

The new goal in this context then becomes to approximate $\Delta_\mu(f, \mathcal{C})$ using only a few oracle calls to f. In our context, we study a concrete instance of the above framework. We consider the original problem considered in the area of property testing, namely the classical problem of *linearity testing*.

The problem of linearity testing was introduced by Blum, Luby and Rubinfeld in [BLR93]. In this problem, we are given oracle access to a function $f : \mathbb{F}_2^n \to \mathbb{F}_2$, and want to distinguish between the case that f is a *linear* function and the case that f is *far* from the class $\mathcal{L}$ of all linear functions from $\mathbb{F}_2^n$ to $\mathbb{F}_2$. [BLR93] gave a simple 3-query test T that achieves this. In fact, this test also achieves the task of *tolerant linearity testing*; i.e., for any function $f : \mathbb{F}_2^n \to \mathbb{F}_2$, letting $\delta = \Pr[T^f \text{ rejects}]$, we have

$$C_1 \cdot \delta \leq \Delta_{U_n}(f, \mathcal{L}) \leq C_2 \cdot \delta,$$

where C_1 and C_2 are absolute constants, and U_n is the uniform distribution on $\mathbb{F}_2^n$. Hence the test of [BLR93], in addition to *testing* linearity, actually estimates how well f can be *approximated* by functions in $\mathcal{L}$.

Here we study tolerant linearity testing over general probability distributions. Let μ be a probability distribution over $\mathbb{F}_2^n$. In the problem of *tolerant linearity testing under μ*, we wish to estimate the how well f may be approximated under μ by linear functions from $\mathcal{L}$. For a given family $(\mathbb{F}_2^n, \mu_n)_n$, we say *linearity is tolerantly testable* under $\mu = \mu_n$ with q queries, if there exists a q-query tester T_n and constants C_1, C_2 such that for any $f : \mathbb{F}_2^n \to \mathbb{F}_2$, letting $\delta = \Pr[T_n^f \text{ rejects}]$, we have

1. **Perfect completeness:** $\delta = 0$ if and only if $\Delta_\mu(f, \mathcal{L}) = 0$.
2. **Distance approximation:** δ approximates $\Delta_\mu(f, \mathcal{L})$:

$$C_1 \cdot \delta \leq \Delta_\mu(f, \mathcal{L}) \leq C_2 \cdot \delta. \tag{1}$$

The main question is to determine for which μ is linearity tolerantly testable under μ. This seems to be a basic question worthy of further study. Furthermore, the notion of linearity being tolerantly testable under general distributions is intimately connected with the concept of locally testable linear codes, and we now elaborate on this connection.

Now let C be any linear code, and let $s_1, \ldots, s_N \in \mathbb{Z}_2^n$ be the columns of a generator matrix for C. Let μ be the uniform distribution over $\{s_1, \ldots, s_N\}$. Then, if linearity is tolerantly testable under μ, then C is locally testable. Indeed, given any $r : [N] \to \mathbb{Z}_2$, we may define the function $f : \mathbb{Z}_2^n \to \mathbb{Z}_2$ by $f(x) = r(j)$ if $x = s_j$, and $f(x) = 0$ otherwise. By the tolerant testability of linearity under μ, any useful query made by a tolerant linearity tester for μ must be to one of the s_j. The distance of f from linear under μ then translates directly into the Hamming distance of r from C, and the very same tester that tolerantly tests linearity under μ shows that C is locally testable.

2.1 Main Notions and Results

Our main contribution is to highlight a simple criterion, which we call *uniform-correlatability*, that lets us design and analyze tolerant linearity tests under a given distribution. Roughly speaking, a distribution μ over $\mathbb{F}_2^n$ is uniformly-correlatable if one can "design" a distribution of few correlated random variables, with each variable distributed according to μ, while all the sum of the variables if nearly uniformly distributed. In this case, we show that linearity is tolerantly testable under μ with few queries (see Theorem A). We complement this by demonstrating that many natural distributions satisfy this criterion.

Although we state all our results for functions from $\mathbb{F}_2^n$ to $\mathbb{F}_2$, most results carry over to all pairs of abelian groups G, H.

Definition 1 (Uniformly-correlatable distribution). *Let μ be a probability distribution on $\mathbb{F}_2^n$. We say that μ is (ϵ, k)-uniformly-correlatable if there is a random variable $X = (X_i)_{i \in [k]}$ taking values in $(\mathbb{F}_2^n)^k$ such that:*

1. *For each $i \in [k]$, X_i is distributed according to μ.*
2. *The distribution of the random variable $\sum_{i \in [k]} X_i$ is ϵ-close to the uniform distribution over $\mathbb{F}_2^n$.*

Our main result in this setting is that uniformly correlatable distributions are tolerantly testable.

Informal Theorem A. Let μ be a distribution over $\mathbb{F}_2^n$ that is (ϵ, k)-uniformly-correlatable. Then there is a $4k$ query tester T that tolerantly tests linearity over μ.

We supplement the above theorem by showing that several natural classes of distributions are all (ϵ, k)-uniformly-correlatable for suitable ϵ, k, such as product distributions, symmetric distributions and low Fourier-bias distributions, thus showing that linearity is tolerantly testable under these distributions. This in turn implies that the corresponding codes are locally testable. In particular this gives a new and simple proof of a result of Kaufman and Sudan [KS07] showing that sparse unbiased linear codes are locally testable.

3 Overview of Proof for Uniform-Correlatability $\Rightarrow$ Testability

We first give some intuition for the uniform correlatability criterion. For T to be a tester for linearity under μ, it needs to satisfy the following minimum requirements: (1) each query made by the tester needs to be distributed essentially according to μ (so that the probability of rejection is upper bounded by the distance), and (2) the queries need to satisfy some linear relations (so that the tester has something to test). This already indicates that a tester will need to "design" a query distribution very carefully, so that both the above requirements are satisfied. This is where the uniformly-correlatable criterion comes in: given the uniformly-correlated distribution, it allows us to design other correlations quite flexibly, and in particular to produce queries distributed according to μ that satisfy linear relations.

The proof of Theorem A at a very high level follows by using the uniform correlatability criterion to reduce linearity testing over μ to linearity testing under the uniform distribution[1] for which we already know the BLR linearity test. We design two tests, Test 1 and Test 2, and define a "self-corrected" version h of the function f being tested. We show that Test 1 is essentially the BLR test (over the uniform distribution) applied to the function h. Hence if Test 1 passes with high probability, then the BLR analysis implies that h is close to a linear function g under the uniform distribution. Test 2 is designed such that if that also passes with high probability, then it implies that f is actually close to that linear function g under μ.

In designing these two tests, we ensure that each query to f made by the tester is distributed essentially according to μ. Hence it follows that the probability of rejection of the tests is at most fixed multiple (depending on the number of queries made by the tester) of the distance of f from linear, and hence the tester is tolerant.

---

[1] Note that the uniform distribution is $(0, 1)$-uniformly-correlatable, and for this case, the test given by Theorem A essentially reduces to the BLR linearity test.

4   High Error

Our main result in the high-error regime is that random sparse linear codes are locally testable and locally list-decodable in the *high-error* regime with only a *constant* number of queries. More precisely, we show that for all constants $c > 0$ and $\gamma > 0$, and for every linear code $\mathcal{C} \subseteq \{0, 1\}^N$ which is:

- *sparse*: $|\mathcal{C}| \leq N^c$, and
- *unbiased*: each nonzero codeword in $\mathcal{C}$ has weight $\in (\frac{1}{2} - N^{-\gamma}, \frac{1}{2} + N^{-\gamma})$,

$\mathcal{C}$ is locally testable and locally list-decodable from $(\frac{1}{2} - \epsilon)$-fraction worst-case errors using only $\mathrm{poly}(\frac{1}{\epsilon})$ queries to a received word. We also give *sub-exponential time* algorithms for list-decoding arbitrary unbiased (but not necessarily sparse) linear codes in the high-error regime. In particular, this yields the first sub-exponential time algorithm even for the problem of (unique) decoding random linear codes of inverse-polynomial rate from a fixed positive fraction of errors.

Earlier, Kaufman and Sudan had shown that sparse, unbiased codes can be locally (unique) decoded and locally tested from a constant-fraction of errors, where this constant-fraction tends to 0 as the number of codewords grows. Our results significantly strengthen their results, while also having significantly simpler proofs.

At the heart of our algorithms is a natural "self-correcting" operation defined on codes and received words. This self-correcting operation transforms a code $\mathcal{C}$ with a received word w into a simpler code $\mathcal{C}'$ and a related received word w', such that w is close to $\mathcal{C}$ if and only if w' is close to $\mathcal{C}'$. Starting with a sparse, unbiased code $\mathcal{C}$ and an arbitrary received word w, a constant number of applications of the self-correcting operation reduces us to the case of local list-decoding and testing for the *Hadamard code*, for which the well known algorithms of Goldreich-Levin and Blum-Luby-Rubinfeld are available. This yields the constant-query local algorithms for the original code $\mathcal{C}$.

The above mentioned "self correcting" operation was motivated (and is also analysed) using the same shift of viewpoint from testing and decoding of linear codes to testing and decoding with respect to distributions as mentioned in Section 2.

Our algorithm for decoding unbiased linear codes in sub-exponential time proceeds similarly. Applying the self-correcting operation to an unbiased code $\mathcal{C}$ and an arbitrary received word a super-constant number of times, we get reduced to the problem of *learning noisy parities*, for which non-trivial sub-exponential time algorithms were recently given by Blum-Kalai-Wasserman and Feldman-Gopalan-Khot-Ponnuswami. Our result generalizes a result of Lyubashevsky, which gave sub-exponential time algorithms for decoding random linear codes of inverse-polynomial rate, from *random errors*.

4.1   Local Testing

A particular variant of local testability which is of significant interest is local testing in the "high-error" regime. Here, for every constant $\epsilon > 0$, one wants to query-efficiently distinguish between the cases $\Delta(w, \mathcal{C}) < 1/2 - \epsilon$, i.e., w is "close" to $\mathcal{C}$, and $\Delta(w, \mathcal{C}) \approx 1/2$, i.e., w is as far from $\mathcal{C}$ as a random point is (for codes over large alphabets, $1/2$ gets replaced by 1). For the Hadamard code, the existence of such testers follows from the Fourier-analytic proof of the BLR linearity test [BCH+96]. For the code of degree 2 multivariate polynomials over $\mathbb{F}_2$, local testers in the high-error regime were given by Samorodnitsky [Sam07]. For the code of multivariate polynomials over large fields, local-testers in the high-error regime were given by Raz-Safra [RS97], Arora-Sudan [AS03] and Moshkovitz-Raz [MR06]. More recently, Dinur-Goldenberg [DG08] and Impagliazzo-Kabanets-Wigderson [IKW09] gave query-efficient local testing algorithms in the high-error regime for the combinatorial families: the direct-product and XOR codes. These algebraic and combinatorial high-error local-testers led to some remarkable constructions of PCPs with high soundness.

All known locally-testable codes in the high-error regime are for highly structured algebraic or combinatorial codes. Kaufman and Sudan [KS07] showed that a random sparse linear code is locally testable in the low-error regime by studying its weight distribution and the weight distribution of its dual, and in particular their proof was based on the MacWilliams identities and non-trivial information about the roots of Krawtchouk polynomials. In the paper [KS09] (essentially using Theorem A), we gave an alternate (and arguably simpler) proof of this result, as part of a more general attempt to characterize sparse codes that are locally decodable and testable in the low-error regime. Popular belief [Sud09] suggested that local-testability in the *high-error* regime could not be found in random linear codes.

We show that that random codes *can* have query-efficient local testers in the high-error regime. Specifically, sparse and unbiased codes admit high-error local testers with constant query complexity.

Informal Theorem B. For every constant $c, \gamma > 0$, every linear code $\mathcal{C} \subseteq \mathbb{F}_2^N$ with N^c codewords and bias $N^{-\gamma}$ can be locally tested from $(1/2 - \epsilon)$-fraction errors using only $\mathrm{poly}(\frac{1}{\epsilon})$ queries to a received word.

We in fact can show something stronger (called *distance estimation in the high-error regime*): for such codes, using constantly many queries, one can distinguish between $\Delta(w, \mathcal{C}) > 1/2 - \epsilon_1$ and $\Delta(w, \mathcal{C}) < 1/2 - \epsilon_2$ for every constants $0 < \epsilon_1 < \epsilon_2 < 1/2$.

4.2   Local List-Decoding

Informally, a local list-decoder for $\mathcal{C}$ from δ-fraction errors is a randomized algorithm A that, when given oracle access to a received word $w \in \mathbb{F}_2^N$, recovers the list of all codewords c such that $\Delta(c, w) < \delta$, while querying w in very few coordinates. The codewords thus recovered are "implicitly represented" by randomized algorithms $A_1, \dots, A_l$ with oracle access to w. Given a coordinate $j \in [N]$, A_i makes very few queries to w and is supposed to output the jth coordinate of the codeword that it implicitly represents.

A particular case of local list-decoding which is of significant interest is local list-decoding in the "high-error" regime. Specifically, for every constant $\epsilon > 0$, one wants to query-efficiently locally list-decode a code from $(\frac{1}{2} - \epsilon)$-fraction errors (the significance of $\frac{1}{2} - \epsilon$ is that it is just barely enough to to distinguish the received word from a random string in $\mathbb{F}_2^N$; for codes over large alphabets, one considers the problem of decoding from $(1 - \epsilon)$-fraction errors). Local list-decoding in the high-error regime plays a particularly important role in the complexity-theoretic applications of coding theory (see [STV99], for example).

The first known local list-decoder (for any code) came from the seminal work of Goldreich and Levin [GL89], which gave time-efficient, low-query, local list-decoders for the Hadamard code in the high-error regime. In the following years, many highly non-trivial local list-decoders were developed for various codes, including multivariate polynomial based codes (in the works of Goldreich-Rubinfeld-Sudan [GRS00], Arora-Sudan [AS03], Sudan-Trevisan-Vadhan [STV99], and Gopalan-Klivans-Zuckerman [GKZ08]) and combinatorial codes such as direct-product codes and XOR codes (in the works of Impagliazzo-Wigderson and Impagliazzo-Jaiswal-Kabanets-Wigderson [IW97, IJKW08]). Many of these local list-decoders, especially the ones in the high-error regime, play a prominent role in celebrated results in complexity theory on hardness amplification and pseudorandomness [IW97, STV99, IJKW08].

To summarize, all known local list-decoding algorithms were for highly structured algebraic or combinatorial codes. Kaufman and Sudan [KS07] showed that *random* sparse linear codes can be locally (unique-)decoded from a small constant fraction of errors. This was the first result to show that query-efficient decoding algorithms could also be associated with random, unstructured codes. This result was proved by studying the weight distribution of these codes and their duals. Popular belief [Sud09] again suggested that these low-error local decoders for random codes could *not* be extended to the high-error regime.

Here we show that even random codes *can* have query-efficient local list-decoders in the high-error regime. Specifically, we show that linear codes which are *sparse* and *unbiased* (both properties are possessed by sparse random linear codes with high probability) admit high-error local list-decoders with constant query complexity.

Informal Theorem C. For every constant $c, \gamma > 0$, every linear code $\mathcal{C} \subseteq \mathbb{F}_2^N$ with N^c codewords and bias $N^{-\gamma}$ can be locally list-decoded from $(1/2 - \epsilon)$-fraction errors using only $\text{poly}(\frac{1}{\epsilon})$ queries to a received word.

4.3 Subexponential Time List-Decoding

The techniques we develop to address the previous questions turn out to be useful for making progress on another fundamental algorithmic question in coding theory: that of *time-efficient* worst-case decoding of random linear codes. Given a random linear code $\mathcal{C} \subseteq \mathbb{F}_2^N$ and an arbitrary received word $w \in \mathbb{F}_2^N$, we are interested in quickly finding all the codewords $c \in \mathcal{C}$ such that $\Delta(w, c) < \frac{1}{2} - \epsilon$, for constant $\epsilon > 0$. We show that this problem can be solved in *sub-exponential time*, using just the unbiasedness of $\mathcal{C}$. Our algorithm uses some recent breakthroughs on the problem of learning noisy-parities due to Blum-Kalai-Wasserman [BKW03] and Feldman-Gopalan-Khot-Ponnuswami [FGKP06].

Informal Theorem D. For all constants $\alpha, \gamma > 0$, for every linear code $\mathcal{C} \subseteq \mathbb{F}_2^N$ with dimension n, where $N = n^{1+\alpha}$, and bias $N^{-\gamma}$, and for every constant $\epsilon > 0$, $\mathcal{C}$ can be list-decoded from $(\frac{1}{2} - \epsilon)$-fraction errors in time $2^{O(n/\log\log n)}$.

In particular, the above theorem implies that if $\mathcal{C} \subseteq \mathbb{F}_2^N$ is a *random linear code* with dimension $n = N^{\frac{1}{1+\alpha}}$, then for every constant $\epsilon > 0$, $\mathcal{C}$ can be list-decoded from $(\frac{1}{2} - \epsilon)$-fraction errors in time $2^{O(n/\log\log n)}$.

Earlier, it was not even known how to *unique-decode* random linear codes from *0.1-fraction* worst-case errors in time $2^{o(n)}$. For decoding random linear codes of inverse-polynomial rate from *random errors*, Lyubashevsky [Lyu05] gave a sub-exponential time algorithm, also based on algorithms for the Noisy Parity problem. Our result generalizes his in two ways: we decode from worst-case errors, and we give a natural, explicit criterion (namely low-bias) on the code $\mathcal{C}$ which guarantees the success of the algorithm.

A related result (and one that we use in our proof) is the sub-exponential time worst-case decoding of random linear codes in $\mathbb{F}_2^N$, of dimension $n = O(\log N \cdot \log\log N)$, in a weaker model [FGKP06, Theorem 10]. In this model, the adversary first corrupts the received bit associated to $(1/2-\epsilon)$-fraction of the 2^n possible linear encoding functions, after which the code is randomly chosen. Our result has a more natural coding theory interpretation: the random code is chosen first, and then the adversary choses an arbitrary received word at distance $(1/2 - \epsilon)$ from the code. In the language of learning theory, the [FGKP06] result concerns learning parities in the presence of *agnostic noise*, while our result deals with the model of learning parities in the presence of *nasty classification noise* [BEK02].

4.4   Time-Efficient Local Algorithms for Dual-BCH Codes

For the family of *dual-BCH* codes, perhaps the most important family of sparse, unbiased codes, we show that the constant-query local list-decoding and local testing algorithms can be made to run in a *time-efficient* manner too. The dual-BCH codes form a natural family of polynomial-based codes generalizing the Hadamard code. They have a number of extremal properties which give them an important role in coding theory. For example, the dual-BCH code $\mathcal{C} \subseteq \mathbb{F}_2^N$ with N^t codewords has bias as small as $O(t \cdot N^{-1/2})$, which is *optimal* for codes with N^t codewords!

The key to making our earlier query-efficient local list-decoding and local testing algorithms run in a time-efficient manner for dual-BCH codes, is a time-efficient efficient algorithm for a certain sampling problem that arises in the local list-decoder and tester. This sampling problem turns out to be closely related to an algorithmic problem that was considered in the context of low-error testing of dual-BCH codes [KL05], that of sampling constant-weight BCH codewords. A variant of the sampling algorithm of [KL05] turns out to suffice for our problem too, and this leads to the following result.

Informal Theorem E. For every constant c, the dual-BCH code $\mathcal{C} \subseteq \mathbb{F}_2^N$ with N^c codewords, can be locally list-decoded and locally tested from $(1/2 - \epsilon)$-fraction errors in time $\mathrm{poly}(\log N, \frac{1}{\epsilon})$ using only $\mathrm{poly}(\frac{1}{\epsilon})$ queries to a received word.

The original algorithm for sampling constant-weight BCH codewords given in [KL05], and was based [Lit09] on results on the weight distribution of BCH codes [KL95]. We give an alternate (and possibly simpler) analysis of this result.

5   Overview of Proofs in the High Error Regime

In this section, we give an overview of the main ideas underlying our algorithms. Our goal in this section is to stress the simplicity and naturalness of our techniques.

The main component of our algorithms is a certain "self-correcting" operation which transforms a code $\mathcal{C}$ with a received word w into a simpler code $\mathcal{C}'$ and a related received word w', such that w is close to $\mathcal{C}$ if and only if w' is close to $\mathcal{C}'$. Repeated application of this self-correcting operation will allow us to reduce our list-decoding and testing problems for $\mathcal{C}$ to certain kinds of list-decoding and testing problems for a significantly simpler code $\mathcal{C}^*$ (in our case, $\mathcal{C}^*$ will be the Hadamard code). Query-efficient/time-efficient algorithms for the simpler code $\mathcal{C}^*$ then lead to query-efficient/time-efficient algorithms for the original code $\mathcal{C}$.

In order to simplify the description of the self-correcting operation, we first translate our problems into the language of list-decoding and testing under *distributions* just as we did in Section 2. Let $\mathcal{C} \subseteq \mathbb{F}_2^N$ be a linear code of dimension n, and let G be an $n \times N$ generator matrix for $\mathcal{C}$. Let $S = \{v_1, v_2, \ldots, v_N\} \subset \mathbb{F}_2^n$ denote the set of columns of G. We associate to $\mathcal{C}$ the distribution μ over $\mathbb{F}_2^n$ which is uniform over S. Note that if the code $\mathcal{C}$ has low bias, then the resulting distribution μ has small Fourier bias. Every word w in $\mathbb{F}_2^N$ can be viewed as a function $f_w : S \to \mathbb{F}_2$, where $f_w(v_i) = w_i$. Under this mapping, every *codeword* of $\mathcal{C}$ gets associated with a *linear* function.

Note that via this translation, the problem of testing if w is close to some codeword exactly translates into the problem of testing if f_w is close to some linear function *under the distribution* μ (where the distance of two functions g, h under μ is measured by the probability that g and h differ on a random sample from μ). Similarly, the problem of local list-decoding, i.e. the problem of finding all codewords close to w, translates into the problem of finding all linear functions that are close to f_w under the distribution μ.

We now come to the self-correcting operation on f and μ. The operation has the property that it maintains the property "f correlates with a linear function under μ", and at the same time it results in a distribution that is "simpler" in a certain precise sense.

Define $\mu^{(2)}$ to be the convolution of μ with itself; i.e., it is the distribution of the sum of two independent samples from μ. We define $f^{(2)} : \mathbb{F}_2^n \to \mathbb{F}_2$ to be the (probabilistic) function, where for a given x, $f^{(2)}(x)$ is sampled as follows: first sample y_1 and y_2 independently and uniformly from μ conditioned on $y_1 + y_2 = x$, and return $f(y_1) + f(y_2)$ (if there are no such y_1, y_2, then define $f^{(2)}(x)$ arbitrarily).

The following two simple facts are key to what follows:

- $\mu^{(2)}$ is "simpler" than μ: the statistical distance of $\mu^{(2)}$ to the uniform distribution on $\mathbb{F}_2^n$ is significantly smaller than the statistical distance of μ to the uniform distribution on $\mathbb{F}_2^n$ (this follows from the low Fourier bias of μ, which in turn came from the unbiasedness of $\mathcal{C}$).
- If f is $(\frac{1}{2} - \epsilon)$-close to a linear function g under μ, then $f^{(2)}$ is $(\frac{1}{2} - 2\epsilon^2)$-close to g under $\mu^{(2)}$: this is a formal consequence of our definition of $f^{(2)}$. In particular, if f is noticeably-close to g under μ, then so is $f^{(2)}$ under $\mu^{(2)}$.

This leads to a general approach for list-decoding/testing for linear functions under μ. First pick k large, and consider the distribution $\mu^{(k)}$ and the function $f^{(k)}$ (defined analogously to $\mu^{(2)}$ and $f^{(2)}$). If k is chosen large enough, then $\mu^{(k)}$ will in fact be 2^{-10n}-close to the uniform distribution in statistical distance. Furthermore, if k is not too large, then $f^{(k)}$ will be noticeably-close under $\mu^{(k)}$ to the same linear functions that f is close to under μ. Thus, if k is suitable (as a function of the initial bias/sparsity of the code) $f^{(k)}$ is noticeably-close *under the uniform distribution* to the same linear functions that f is close to under μ. Now all we need to do is run a local list-decoding/testing algorithm on $f^{(k)}$ under the uniform distribution.

An important issue that was swept under the rug in this discussion, is the query/time-efficiency of working with $f^{(k)}$ and $\mu^{(k)}$. If we ignore running-time, one can simulate oracle access to $f^{(k)}$ using just a factor k larger number of queries to f. This leads to our query-efficient (but time-inefficient) algorithms for sparse, unbiased linear codes in the high-error regime (in this setting k only needs to be a constant). We stress that our proof of this result is significantly simpler and stronger than earlier analyses of local algorithms (in the low-error regime) of sparse, unbiased codes [KL05, KS07].

The bottleneck for implementing these local, query-efficient algorithms in a time-efficient manner is the following algorithmic "back-sampling" problem: given a point $x \in \mathbb{F}_2^n$, produce a sample from the distribution of $y_1, \ldots, y_k$ picked independently from μ conditioned on $\sum y_i = x$. A time-efficient back-sampling algorithm would allow us to time-efficiently simulate oracle access to $f^{(k)}$ given oracle access to f. For random sparse linear codes, solving this problem in time sublinear in N is impossible; however for specific, interesting sparse unbiased codes, this remains an important problem to address. For the special case of dual-BCH codes, perhaps the most important family of sparse, unbiased codes, we observe that the back-sampling problem can be solved using a small variant of an algorithm of Kaufman-Litsyn [KL05]. Thus for dual-BCH codes, we get poly $\log(N)$-time, constant-query local testing and local list-decoding algorithms in the high-error regime.

For sub-exponential time list-decoding, we follow the same plan. Here too we will self-correct f to obtain a function $f^{(k)}$, such that every linear function that correlates with f under the μ distribution, also correlates with $f^{(k)}$ under the uniform distribution over $\mathbb{F}_2^n$. However, since we are now paying attention to running time (and we do not know how to solve the back-sampling problem for μ efficiently in general), we cannot afford to allow the list-decoder over the uniform distribution over $\mathbb{F}_2^n$ to query the value of $f^{(k)}$ at any point that it desires (since this will force us to back-sample in order to compute $f^{(k)}$ at that point). Instead, we will use some recent remarkable list-decoders ([FGKP06, BKW03]), developed in the context of learning noisy parities, which can find all linear functions close (under the uniform distribution) to an arbitrary function h in sub-exponential time by simply querying the function h at independent uniformly random points of $\mathbb{F}_2^n$! Using the unbiasedness of μ, it turns out to be easy to evaluate $f^{(k)}$ at independent uniformly random points of $\mathbb{F}_2^n$. This leads to our sub-exponential time list-decoding algorithm.

Relationship to the k-wise XOR on codes: Back in the language of codes, what happened here has a curious interpretation. Given a code $\mathcal{C} \subseteq \mathbb{F}_2^N$, the k-wise XOR of $\mathcal{C}$, $\mathcal{C}^{(\oplus k)}$, is the code contained in $\mathbb{F}_2^{N^k}$ defined as follows: for every codeword $c \in \mathcal{C}$, there is a codeword $c^{(\oplus k)} \in F_2^{[N]^k}$ whose value in coordinate $(i_1, \ldots, i_k)$ equals $c_{i_1} \oplus c_{i_2} \oplus \ldots \oplus c_{i_k}$. In terms of this operation, our algorithms simply do the following: given a code $\mathcal{C}$ and received word w, consider the code $\mathcal{C}^{(\oplus k)}$ with received word $w^{(\oplus k)}$. The crucial observation is, that for k chosen suitably as a function of the bias/sparsity of $\mathcal{C}$, the code $\mathcal{C}^{(\oplus k)}$ is essentially, up to repeating each coordinate a roughly-equal number of times, the Hadamard code! Additionally, $w^{(\oplus k)}$ is close to $c^{(\oplus k)}$ for a codeword c if and only if w is close c. Thus decoding/testing $w^{(\oplus k)}$ for the Hadamard code now suffices to complete the algorithm.

The k-wise XOR on codes is an operation that shows up often as a device for hardness amplification, to convert functions that are hard to compute into functions that are even harder to compute. Our algorithms use the XOR operation for "good": here the XOR operation is a vehicle to *transfer* query-efficient/time-efficient algorithms for the Hadamard code to query-efficient/time-efficient algorithms for arbitrary unbiased codes.

References

[AS03]   Arora, S., Sudan, M.: Improved low degree testing and its applications. Combinatorica 23(3), 365–426 (2003); Preliminary version in Proceedings of ACM STOC (1997)

[BCH+96]   Bellare, M., Coppersmith, D., Håstad, J., Kiwi, M., Sudan, M.: Linearity testing over characteristic two. IEEE Transactions on Information Theory 42(6), 1781–1795 (1996)

[BEK02]   Bshouty, N.H., Eiron, N., Kushilevitz, E.: PAC learning with nasty noise. Theor. Comput. Sci. 288(2), 255–275 (2002)

[BKW03]   Blum, A., Kalai, A., Wasserman, H.: Noise-tolerant learning, the parity problem, and the statistical query model. J. ACM 50(4), 506–519 (2003)

[BLR93]   Blum, M., Luby, M., Rubinfeld, R.: Self-testing/correcting with applications to numerical problems. Journal of Computer and System Sciences 47(3), 549–595 (1993)

[DG08]   Dinur, I., Goldenberg, E.: Locally testing direct product in the low error range. In: FOCS, pp. 613–622. IEEE Computer Society, Los Alamitos (2008)

[FGKP06]   Feldman, V., Gopalan, P., Khot, S., Ponnuswami, A.K.: New results for learning noisy parities and halfspaces. In: FOCS, pp. 563–574 (2006)

[GGR98]   Goldreich, O., Goldwasser, S., Ron, D.: Property testing and its connection to learning and approximation. J. ACM 45(4), 653–750 (1998)

[GKZ08]   Gopalan, P., Klivans, A.R., Zuckerman, D.: List-decoding reed-muller codes over small fields. In: Ladner and Dwork [LD08], pp. 265–274

[GL89]   Goldreich, O., Levin, L.: A hard-core predicate for all one-way functions. In: Proceedings of the 21st Annual ACM Symposium on Theory of Computing, pp. 25–32 (May 1989)

[GRS00]   Goldreich, O., Rubinfeld, R., Sudan, M.: Learning polynomials with queries: The highly noisy case. SIAM Journal on Discrete Mathematics 13(4), 535–570 (2000)

[IJKW08]   Impagliazzo, R., Jaiswal, R., Kabanets, V., Wigderson, A.: Uniform direct product theorems: simplified, optimized, and derandomized. In: Ladner, Dwork (eds.) [LD08], pp. 579–588

[IKW09]   Impagliazzo, R., Kabanets, V., Wigderson, A.: New direct-product testers and 2-query pcps. In: STOC, pp. 131–140 (2009)

[IW97]   Impagliazzo, R., Wigderson, A.: P = BPP if E requires exponential circuits: Derandomizing the XOR Lemma. In: Proceedings of the 29th Annual ACM Symposium on Theory of Computing, pp. 220–229 (May 1997)

[KL95]   Krasikov, I., Litsyn, S.: On spectra of BCH codes. IEEE Transactions on Information Theory 41(3), 786–788 (1995)

[KL05]   Kaufman, T., Litsyn, S.: Almost orthogonal linear codes are locally testable. In: Proceedings of the Forty-sixth Annual Symposium on Foundations of Computer Science, pp. 317–326 (2005)

[KS07]   Kaufman, T., Sudan, M.: Sparse random linear codes are locally decodable and testable. In: FOCS, pp. 590–600. IEEE Computer Society, Los Alamitos (2007)

[KS09]   Kopparty, S., Saraf, S.: Tolerant linearity testing and locally testable codes. In: Dinur, I., Jansen, K., Naor, J., Rolim, J. (eds.) APPROX-RANDOM 2009. LNCS, vol. 5687, pp. 601–614. Springer, Heidelberg (2009)

[LD08]   Ladner, R.E., Dwork, C. (eds.): Proceedings of the 40th Annual ACM Symposium on Theory of Computing, Victoria, British Columbia, Canada, 2008, May 17-20. ACM, New York (2008)

[Lit09]   Litsyn, S.: Personal Communication (2009)

[Lyu05]   Lyubashevsky, V.: The parity problem in the presence of noise, decoding random linear codes, and the subset sum problem. In: Chekuri, C., Jansen, K., Rolim, J.D.P., Trevisan, L. (eds.) APPROX 2005 and RANDOM 2005. LNCS, vol. 3624, pp. 378–389. Springer, Heidelberg (2005)

[MR06]   Moshkovitz, D., Raz, R.: Sub-constant error low degree test of almost-linear size. In: Kleinberg, J.M. (ed.) STOC, pp. 21–30. ACM, New York (2006)

[PRR06]   Parnas, M., Ron, D., Rubinfeld, R.: Tolerant property testing and distance approximation. J. Comput. Syst. Sci. 72(6), 1012–1042 (2006)

[RS96]   Rubinfeld, R., Sudan, M.: Robust characterizations of polynomials with applications to program testing. SIAM Journal on Computing 25(2), 252–271 (1996)

[RS97]   Raz, R., Safra, S.: A sub-constant error-probability low-degree test, and a sub-constant error-probability PCP characterization of NP. In: Proceedings of the Twenty-Ninth Annual ACM Symposium on Theory of Computing, pp. 475–484. ACM Press, New York (1997)

[Sam07]   Samorodnitsky, A.: Low-degree tests at large distances. In: STOC, pp. 506–515 (2007)

[STV99]   Sudan, M., Trevisan, L., Vadhan, S.: Pseudorandom generators without the XOR lemma. In: Proceedings of the 31st Annual ACM Symposium on Theory of Computing, pp. 537–546 (1999)

[Sud09]   Sudan, M.: Personal Communication (2009)

Testing (Subclasses of) Halfspaces

Kevin Matulef[1], Ryan O'Donnell[2], Ronitt Rubinfeld[3], and Rocco Servedio[4]

[1] ITCS, Tsinghua University
matulef@csail.mit.edu
[2] Carnegie Mellon University
odonnell@cs.cmu.edu
[3] Massachusetts Institute of Technology
ronitt@csail.mit.edu
[4] Columbia University
rocco@cs.columbia.edu

Abstract. We address the problem of testing whether a Boolean-valued function f is a halfspace, i.e. a function of the form $f(x) = \text{sgn}(w \cdot x - \theta)$. We consider halfspaces over the continuous domain $\mathbf{R}^n$ (endowed with the standard multivariate Gaussian distribution) as well as halfspaces over the Boolean cube $\{-1,1\}^n$ (endowed with the uniform distribution). In both cases we give an algorithm that distinguishes halfspaces from functions that are ϵ-far from any halfspace using only $\text{poly}(\frac{1}{\epsilon})$ queries, independent of the dimension n.

In contrast to the case of general halfspaces, we show that testing natural subclasses of halfspaces can be markedly harder; for the class of $\{-1,1\}$-weight halfspaces, we show that a tester must make at least $\Omega(\log n)$ queries. We complement this lower bound with an upper bound showing that $O(\sqrt{n})$ queries suffice.

Keywords: halfspaces, linear thresholds functions.

This article presents a summary of the results found in [13] and [12] regarding the testability of halfspaces and certain subclasses of halfspaces.

1   Introduction

A *halfspace* is a function of the form $f(x) = \text{sgn}(w_1 x_1 + \cdots + w_n x_n - \theta)$ where $w_1, ..., w_n, \theta \in \mathbf{R}$. The w_i's are called "weights," and θ is called the "threshold." The sgn function is 1 on arguments ≥ 0, and -1 otherwise. The inputs to f can be either Boolean or real. Here we will mainly be concerned with functions over the Boolean cube, i.e. functions of the form $f : \{-1,1\}^n \rightarrow \{-1,1\}$. Halfspaces are also known as *threshold functions* or *linear threshold functions*; for brevity we shall refer to them here as LTFs.

LTFs are a simple yet powerful class of functions, which for decades have played an important role complexity theory, optimization, and perhaps especially machine learning (see e.g. [9,18,2,15,14,17]). A lot of attention has been paid to

O. Goldreich (Ed.): Property Testing, LNCS 6390, pp. 334–340, 2010.
© Springer-Verlag Berlin Heidelberg 2010

the problem of learning LTFs- that is, given examples labeled according to an unknown LTF (either random examples or queries to the function), find an LTF that it is ϵ-close to. However, the question we want to address is that of *testing* LTFs. That is, given query access to a function, we would like to distinguish whether it is an LTF or whether it is ϵ-far from any LTF. Any proper learning algorithm can be used as a testing algorithm (see, e.g., the observations of [8]), but testing potentially requires fewer queries. Indeed, in situations where query access is available, a query-efficient testing algorithm can be used to check whether a function is close to an an LTF, before bothering to run a more intensive algorithm to learn which LTF it is close to.

2   LTFs Are Testable with poly($1/\epsilon$) Queries

The main result in [13] is to show that halfspaces can be tested with a number of queries that is *independent* of n. In fact the dependence is only polynomial in $1/\epsilon$. We note that any learning algorithm — even one with black-box query access to f — must make at least $\Omega(\frac{n}{\epsilon})$ queries to learn an unknown LTF to accuracy ϵ under the uniform distribution on $\{-1,1\}^n$ (this follows easily from, e.g., the results of [11]). So at least in terms of relationship to n, our testing algorithm is a significant improvement over using a learning algorithm. More formally, our main result is the following:

Theorem 1 ([13]). *Let f be a Boolean function $f : \{-1,1\}^n \to \{-1,1\}$, and (as is standard in property testing) we measure the distance between functions with respect to the uniform distribution over $\{-1,1\}^n$. Then there is an algorithm with 2-sided error making* poly($\frac{1}{\epsilon}$) *queries that accepts f with high probability if it is an LTF, and rejects with high probability if it ϵ-far from all LTFs.*

We remark that the class of halfspaces is qualitatively much different than the other classes of Boolean functions that we know how to test. Some previous works have used the method of "implicit learning" to test classes such as s-term DNF formulas and size-s decision trees [4]. However the implicit learning technique only works for classes of functions whose members are close to juntas. This is not the case here, since the class of halfspaces contains, for example, the majority function, which is not at all close to a junta. Other previous works have shown how to test classes with some algebraic structure, like parity functions and low-degree polynomials, but these classes also are quite different from halfspaces.

Characterizations and Techniques. To prove our results, we establish new structural results about LTFs which essentially characterize them in terms of their degree-0 and degree-1 Fourier coefficients. For functions mapping $\{-1,1\}^n$ to $\{-1,1\}$ it has long been known [3] that any linear threshold function f is *completely specified* by the $n+1$ parameters consisting of its degree-0 and degree-1 Fourier coefficients (also referred to as its *Chow parameters*). While this specification has been used to *learn* LTFs in various contexts [1,7,16], it is not clear how it can be used to construct efficient *testers* (for one thing this specification involves $n+1$ parameters, and we want a query complexity independent of n). Intuitively, we get around this difficulty by giving new characterizations of LTFs as those functions that satisfy a particular relationship between just *two* parameters, namely the degree-0 Fourier coefficient and the sum of the squared degree-1 Fourier coefficients. Moreover, our characterizations are robust in that if a function approximately satisfies the relationship, then it must be close to an LTF. This is what makes the characterizations useful for testing.

We first consider functions mapping $\mathbf{R}^n$ to $\{-1, 1\}$ where we view $\mathbf{R}^n$ as endowed with the standard n-dimensional Gaussian distribution. Our characterization is particularly clean in this setting and illustrates the essential approach that also underlies the much more involved Boolean case. On one hand, it is not hard to show that for every LTF f, the sum of the squares of the degree-1 Hermite coefficients[1] of f is equal to a particular function of $\mathbf{E}[f]$ — regardless of *which* LTF f is (we call this function W; it is essentially the square of the "Gaussian isoperimetric" function).

Conversely, we show that if $f : \mathbf{R}^n \to \{-1, 1\}$ is *any* function for which the sum of the squares of the degree-1 Hermite coefficients is within $\pm\epsilon^3$ of $W(\mathbf{E}[f])$, then f must be $O(\epsilon)$-close to an LTF — in fact to an LTF whose n weights are the n degree-1 Hermite coefficients of f. The value $\mathbf{E}[f]$ can clearly be estimated by sampling, and moreover it can be shown that a simple approach of sampling f on pairs of correlated inputs can be used to obtain an accurate estimate of the sum of the squares of the degree-1 Hermite coefficients. We thus obtain a simple and efficient test for LTFs under the Gaussian distribution.

To handle general LTFs over $\{-1, 1\}^n$, we first develop an analogous characterization and testing algorithm for the class of *balanced regular* LTFs over $\{-1, 1\}^n$; these are LTFs with $\mathbf{E}[f] = 0$ for which all degree-1 Fourier coefficients are small. The heart of this characterization is a pair of results which give Boolean-cube analogues of our characterization of Gaussian LTFs. We show that the sum of the squares of the degree-1 Fourier coefficients of any balanced regular LTF is approximately $W(0) = \frac{2}{\pi}$. Conversely, we show that any function f whose degree-1 Fourier coefficients are all small and whose squares sum to roughly $\frac{2}{\pi}$ is in fact close to an LTF — in fact, to one whose weights are the degree-1 Fourier coefficients of f. Similar to the Gaussian setting, we can estimate $\mathbf{E}[f]$ by uniform sampling and can estimate the sum of squares of degree-1 Fourier coefficients by sampling f on pairs of correlated inputs. (An additional algorithmic step is also required here, namely checking that all the degree-1 Fourier coefficients of f are indeed small; it turns out that this can be done by estimating the sum of *fourth* powers of the degree-1 Fourier coefficients, which can again be obtained by sampling f on (4-tuples of) correlated inputs.)

The general case of testing arbitrary LTFs over $\{-1, 1\}^n$ is substantially more complex. Very roughly speaking, the algorithm has three main conceptual steps:

– First the algorithm implicitly identifies a set of $O(1)$ many variables that have "large" degree-1 Fourier coefficients. Even a single such variable cannot be explicitly identified using $o(\log n)$ queries; we perform the implicit

---

[1] These are analogues of the Fourier coefficients for L^2 functions over $\mathbf{R}^n$ with respect to the Gaussian measure.

identification using $O(1)$ queries by adapting an algorithmic technique from [6]. This is similar to the "implicit learning" approach in [4].

- Second, the algorithm analyzes the regular subfunctions that are obtained by restricting these implicitly identified variables; in particular, it checks that there is a single set of weights for the unrestricted variables such that the different restrictions can all be expressed as LTFs with these weights (but different thresholds) over the unrestricted variables. Roughly speaking, this is done using a generalized version of the regular LTF test that tests whether a *pair* of functions are close to LTFs over the same linear form but with different thresholds.
- Finally, the algorithm checks that there exists a single set of weights for the restricted variables that is compatible with the different biases of the different restricted functions. If this is the case then the overall function is close to the LTF obtained by combining these two sets of weights for the unrestricted and restricted variables. (Intuitively, since there are only $O(1)$ restricted variables there are only $O(1)$ possible sets of weights to check here.)

3   Testing a Natural Subclass of Halfspaces Requires More Queries

Complementing the work in [13], in [12] we consider the problem of testing whether a function f belongs to a natural subclass of halfspaces, the class of ± 1-*weight halfspaces*. These are functions of the form $f(x) = \mathrm{sgn}(w_1 x_1 + w_2 x_2 + \cdots + w_n x_n)$ where the weights w_i all take values in $\{-1, 1\}$. Included in this class is the majority function on n variables, and all 2^n "reorientations" of majority, where some variables x_i are replaced by $-x_i$. Alternatively, this can be viewed as the subclass of halfspaces where all variables have the same amount of influence on the outcome of the function, but some variables get a "positive" vote while others get a "negative" vote.

For the problem of testing ± 1-weight halfspaces, we prove two main results:

1. **Lower Bound.** We show that any nonadaptive testing algorithm which distinguishes ± 1-weight halfspaces from functions that are ϵ-far from ± 1-weight halfspaces must make at least $\Omega(\log n)$ many queries. By a standard transformation (see e.g. [5]), this also implies an $\Omega(\log \log n)$ lower bound for adaptive algorithms. Taken together with [13], this shows that testing this natural subclass of halfspaces is more query-intensive then testing the general class of all halfspaces.
2. **Upper Bound.** We give a nonadaptive algorithm making $O(\sqrt{n} \cdot \mathrm{poly}(1/\epsilon))$ many queries to f, which outputs YES with probability at least $2/3$ if f is a ± 1-weight halfspace, and NO with probability at least $2/3$ if f is ϵ-far from any ± 1-weight halfspace.

    We note that it follows from [11] that *learning* the class of ± 1-weight halfspaces requires $\Omega(n/\epsilon)$ queries. Thus, while some dependence on n is necessary for testing, our upper bound shows testing ± 1-weight halfspaces can still be done more efficiently than learning.

Although we prove our results specifically for the case of halfspaces with all weights ± 1, our methods can be used to obtain similar results for other subclasses of halfspaces such as $\{-1, 0, 1\}$-weight halfspaces (± 1-weight halfspaces where some variables are irrelevant).

Techniques. As is standard in property testing, our lower bound is proved using Yao's method. We define two distributions D_{YES} and D_{NO} over functions, where a draw from D_{YES} is a randomly chosen ± 1-weight halfspace and a draw from D_{NO} is a halfspace whose coefficients are drawn uniformly from $\{+1, -1, +\sqrt{3}, -\sqrt{3}\}$. We show that a random draw from D_{NO} is with high probability $\Omega(1)$-far from every ± 1-weight halfspace, but that any set of $o(\log n)$ query strings cannot distinguish between a draw from D_{YES} and a draw from D_{NO}.

Our upper bound is achieved by an algorithm which uniformly selects a small set of variables and checks, for each selected variable x_i, that the magnitude of the corresponding singleton Fourier coefficient $|\hat{f}(i)|$ is close to to the right value. We show that any function that passes this test with high probability must have its degree-1 Fourier coefficients very similar to those of some ± 1-weight halfspace, and that any function whose degree-1 Fourier coefficients have this property must be close to a ± 1-weight halfspace. At a high level this approach is similar to some of what is done in [13], but here we are estimating $\sum_i |\hat{f}(i)|$ rather than $\sum_i \hat{f}(i)^2$. In both instances we are checking that the contribution of the degree-1 Fourier coefficients is "large," but in the second case we are estimating the coefficients more accurately in order to insure to insure that we only pass functions close to ± 1-weight halfspaces.

4   Open Questions

Several questions related to testing halfspaces are still open. Here we point out a just a few:

- First is the question of whether there is a simpler algorithm for testing the general class of halfspaces over the Boolean cube. Although our algorithm makes "only" $\mathrm{poly}(1/\epsilon)$ queries, the exponent of the polynomial is something like 4000. Our algorithm is quite complicated, and hardly seems optimal. Obviously a more efficient algorithm utilizing new ideas would be preferred.
- Our current approach to testing halfspaces makes two-sided error. It is unclear whether this is necessary. In order to get a better handle on testing halfspaces, we might restrict ourselves to the question of one-sided testing. Can we devise a one-sided tester, or show that there is none? We conjecture (albeit without much confidence) that one-sided testing requires a query complexity dependent on n. We make this conjecture based on the fact that for any constant k, there exist boolean functions which are not halfspaces, yet are consistent with a halfspace on any set of less than k examples [10].
- Perhaps the most obvious lingering question is whether we can extend our algorithm for LTFs to test degree-d polynomial threshold functions, or PTFs. This seems to require a significant amount of extra machinery, for example

in relating the size of the degree-d Fourier coefficients to the weights of the corresponding terms inside a PTF, and to the bias of the PTF. Although there are some highly technical obstacles, given all of the recent structural results on PTFs, there is some hope that a testing algorithm can be achieved.

Acknowledgments

K.M. was supported in part by the National Natural Science Foundation of China Grant 60553001, and the National Basic Research Program of China Grant 2007CB807900,2007CB807901.

References

1. Birkendorf, A., Dichterman, E., Jackson, J., Klasner, N., Simon, H.U.: On restricted-focus-of-attention learnability of Boolean functions. Machine Learning 30, 89–123 (1998)
2. Block, H.: The Perceptron: a model for brain functioning. Reviews of Modern Physics 34, 123–135 (1962)
3. Chow, C.K.: On the characterization of threshold functions. In: Proceedings of the Symposium on Switching Circuit Theory and Logical Design (FOCS), pp. 34–38 (1961)
4. Diakonikolas, I., Lee, H., Matulef, K., Onak, K., Rubinfeld, R., Servedio, R., Wan, A.: Testing for concise representations. In: Proc. 48th Ann. Symposium on Computer Science (FOCS), pp. 549–558 (2007)
5. Fischer, E.: The art of uninformed decisions: A primer to property testing. Bulletin of the European Association for Theoretical Computer Science 75, 97–126 (2001)
6. Fischer, E., Kindler, G., Ron, D., Safra, S., Samorodnitsky, A.: Testing juntas. In: Proceedings of the 43rd IEEE Symposium on Foundations of Computer Science, pp. 103–112 (2002)
7. Goldberg, P.: A Bound on the Precision Required to Estimate a Boolean Perceptron from its Average Satisfying Assignment. SIAM Journal on Discrete Mathematics 20, 328–343 (2006)
8. Goldreich, O., Goldwaser, S., Ron, D.: Property testing and its connection to learning and approximation. Journal of the ACM 45, 653–750 (1998)
9. Hajnal, A., Maass, W., Pudlak, P., Szegedy, M., Turan, G.: Threshold circuits of bounded depth. Journal of Computer and System Sciences 46, 129–154 (1993)
10. Hellerstein, L.: On generalized constraints and certificates. Discrete Mathematics 226(211-232) (2001)
11. Kulkarni, S., Mitter, S., Tsitsiklis, J.: Active learning using arbitrary binary valued queries. Machine Learning 11, 23–35 (1993)
12. Matulef, K., Rubinfeld, R., Servedio, R.A., O'Donnell, R.: Testing {-1,1} - Weight Halfspaces. In: Dinur, I., Jansen, K., Naor, J., Rolim, J. (eds.) APPROX-RANDOM 2009. LNCS, vol. 5687, pp. 646–657. Springer, Heidelberg (2009)
13. Matulef, K., O'Donnell, R., Rubinfeld, R., Servedio, R.A.: Testing halfspaces. In: 20th Annual ACM-SIAM Symposium on Discrete Algorithms (SODA), pp. 256–264 (2009)
14. Minsky, M., Papert, S.: Perceptrons: an introduction to computational geometry. MIT Press, Cambridge (1968)

15. Novikoff, A.: On convergence proofs on perceptrons. In: Proceedings of the Symposium on Mathematical Theory of Automata, vol. XII, pp. 615–622 (1962)
16. Servedio, R.: Every linear threshold function has a low-weight approximator. Computational Complexity 16(2), 180–209 (2007)
17. Shawe-Taylor, J., Cristianini, N.: An introduction to support vector machines. Cambridge University Press, Cambridge (2000)
18. Yao, A.: On ACC and threshold circuits. In: Proceedings of the Thirty-First Annual Symposium on Foundations of Computer Science, pp. 619–627 (1990)

Dynamic Approximate Vertex Cover
and Maximum Matching[*]

Krzysztof Onak[1] and Ronitt Rubinfeld[1,2]

[1] Massachusetts Institute of Technology, Cambridge, MA, USA
[2] Tel-Aviv University, Tel Aviv, Israel

Abstract. We consider the problem of maintaining a large matching or
a small vertex cover in a dynamically changing graph. Each update to
the graph is either an edge deletion or an edge insertion. We give the first
randomized data structure that simultaneously achieves a constant approximation factor and handles a sequence of k updates in $k \cdot \mathrm{polylog}(n)$
time. Previous data structures require a polynomial amount of computation per update.

The starting point of our construction is a distributed algorithm of
Parnas and Ron (Theor. Comput. Sci. 2007), which they designed for
their sublinear-time approximation algorithm for the vertex cover size.
This leads us to wonder whether there are other connections between
sublinear algorithms and dynamic data structures.

Keywords: dynamic algorithms, maximum matching, vertex cover.

1 Introduction

Suppose one is given the task of solving a combinatorial problem, such as maximum matching or minimum vertex cover, for a very large and constantly changing graph. In this setting, it is natural to ask, does one need to recompute the
solution from scratch after every update?

Such questions have been asked before for various combinatorial problems—
examples include minimum spanning tree, shortest path length, min-cut, and
many others (some examples include [1,2,3,4,5,6,7]). Classic works for these
problems have shown update times that are sublinear in the input size. For the
problem of maximum matching, Sankowski [8] shows that it can be maintained
with $O(n^{1.495})$ computation per update, which for dense graphs is sublinear in
the number of edges.

For very large graphs, it may be crucial to maintain the maximum matching
with much faster, even polylogarithmic, update time. Note that this might be
hard for maximum matching, since obtaining $o(\sqrt{n})$ update time, even in the case

---

[*] Krzysztof Onak was supported in part by NSF grants 0732334 and 0728645. Ronitt
Rubinfeld was supported in part by NSF grants 0732334 and 0728645, Marie Curie
Reintegration grant PIRG03-GA-2008-231077, and the Israel Science Foundation
grant nos. 1147/09 and 1675/09.

O. Goldreich (Ed.): Property Testing, LNCS 6390, pp. 341–345, 2010.

when only insertions are allowed, would improve on the 30-year-old algorithm of running time $O(m\sqrt{n})$ due to Micali and Vazirani [9], where m is the number of edges in the graph. Therefore, some kind of approximation may be unavoidable. Following similar considerations, Ivković and Lloyd [10] give a factor-2 approximation to both vertex cover and maximum matching, by maintaining a *maximal* matching (which is well known to give the desired approximation for maximum matching and also minimum vertex cover). Their update time is nevertheless still polynomial in n.

In this paper, we concentrate on the setting in which slightly weaker, but still $O(1)$, approximation factors are acceptable, but in which it is crucial that update times be extremely fast, in particular, polylogarithmic in the number of vertices.

Interestingly, our data structure uses a technique that Parnas and Ron [11] designed for their sublinear-time algorithm as a starting point. We think that it is an interesting direction to explore possible connections between sublinear-time algorithms and dynamic data structures.

2   Problem Statement and Our Results

Recall that in the *maximum matching* problem, one wants to find the largest subset of vertex disjoint edges. In the *vertex cover* problem, one wants to find the smallest set of vertices such that each edge of the graph is incident to at least one vertex in the set.

Our goal here is to design a data structure that handles edge removals and edge insertions. The data structure provides access to a list of edges that constitute a large matching and a list of vertices that constitute a small vertex cover. We assume that we start with an empty graph, and n is known in advance.

The main result of the paper is the following:

> There is a randomized data structure for maximum matching and vertex cover that
> (a) achieves a constant approximation factor,
> (b) runs in $O(k \cdot \log^2 n + \min\{k, n^2\} \cdot \log n \cdot \log(1/\delta))$ time for any fixed sequence of k updates with probability $1 - \delta$.

Note that for any sequence of k updates, the expected running time of the data structure is $O(k \cdot \log^2 n)$.

Furthermore, the first step in our presentation is a *deterministic* data structure for vertex cover. The data structure keeps a vertex cover that gives $O(\log n)$ approximation to the minimum vertex cover. The amortized update time of the data structure is $O(\log^2 n)$. Though the approximation factor achieved by this algorithm is relatively weak, the algorithm may be of independent interest because of its relative simplicity and efficient update time.

3   Overview of Our Techniques

We construct our data structure in two stages. We first show a deterministic $O(\log n)$-approximation data structure for vertex cover. Then we modify it, introducing randomization, to achieve a constant approximation factor for both vertex cover and maximum matching.

A Deterministic $O(\log n)$-Approximation Data Structure. We construct a data structure that makes use of a carefully designed partition of vertices into a logarithmic number of subsets. The partition is inspired by a simple distributed algorithm of Parnas and Ron [11]. In [11], the first subset in the partition corresponds to removing vertices of degree approximately greater than n. The second subset corresponds to removing vertices of degree approximately greater than $n/4$ from the modified graph. In general, the i-th subset is a set of vertices that are approximately greater than $n/4^{i-1}$ in the graph with all previous subsets of vertices removed. Finally, after a logarithmic number of steps, the remaining graph has no edges. This implies that the union of all subsets removed so far constitutes a vertex cover. For each of the removed subsets, it is easy to show that the subset size is bounded by $O(\text{VC}(G))$, where $\text{VC}(G)$ is the size of the minimum vertex cover. Hence the total vertex cover is bounded by $O(\text{VC}(G) \cdot \log n)$.

The main idea behind our data structure is to modify the partition of Parnas and Ron in order to allow efficient maintenance of this partition. While this is not possible in the partition of Parnas and Ron, it is possible in our relaxed version of it. As edges are inserted and removed, we want to move vertices between subsets. In order to determine whether to move a vertex, we associate a potential function with every vertex, and we allow a vertex to jump from one set to another only if it has collected enough potential. To do this, we set two thresholds $\tau_1 < \tau_2$ for each subset. A vertex can move into the subset from a subset corresponding to a lower degree if its number of neighbors in a specific graph is at least τ_2. Then the vertex can move back to a subset corresponding to a lower degree only if its number of edges decreases to τ_1 in the same graph. A slight technical difficulty is presented by the fact that moving vertices may increase the potential of other vertices. We overcome this obstacle by carefully selecting constants in the potential function so that the potential of the vertex that moves is spent on increasing the potential of its neighbors whenever needed.

A Randomized $O(1)$-Approximation Data Structure. In this case, we redesign the partition, building upon the previous one. In the process of defining the partition, whenever we remove a large subset W of vertices of degree approximately greater than $n/4^i$, we also show the existence of a matching M which is smaller than W by at most a constant factor. To build the next set of the partition, we not only remove W but also all vertices matched in M. In this way we achieve a matching and a vertex cover of sizes that are within a constant factor of each other. Therefore, both give a constant factor approximation to their respective optimal solutions.

Efficient maintenance of the new partition is more involved, as we are sometimes forced to recompute a new matching. This can happen, for instance, when many edges in the old matchings are deleted from the graph. Unfortunately, the creation of the new matching is expensive, since we have modify the set of the vertices matched in M that are deleted together with W. If the edges in the matching are deleted too quickly, we would have to create a new matching often, in which case we do not know how to maintain small update time. Fortunately, by picking a *random matching*, we can ensure that it is unlikely that many edges from the matching get deleted in a short span of time. Thus, by the time the matching gets deleted, we are likely to have collected enough potential to pay for the creation of a new matching.

4   Other Related Work

A sequence of papers [12,13,14] considers computing a large matching or a large weight matching (in the weighted case) in the semi-streaming model. The stream is a sequence of edges, and the goal of an algorithm is to compute a large matching in a small number of passes over the stream, using $\tilde{O}(n)$ space, and preferably at most polylog(n) update time. Results in this model correspond to results for dynamically changing graphs in which only edge insertions occur, except that the matching is only output once at the end of the processing. To the best of our knowledge, it is not known how to achieve a better approximation factor than 2 in one pass in $\tilde{O}(n)$ space for the maximum matching problem.

Lotker, Patt-Shamir, and Rosén [15] show how to maintain a large matching in a distributed network.

5   Open Problems

The two main questions left open by our paper are:

- Our approximation factors are large constants. How small can they be made with polylogarithmic update time? Can they be made 2? For maximum matching, can the approximation constant be made smaller than 2 for maximum matching?
- Is there a *deterministic* data structure that achieves a constant approximation factor with polylogarithmic update time?

References

1. Eppstein, D., Galil, Z., Italiano, G.F., Nissenzweig, A.: Sparsification—a technique for speeding up dynamic graph algorithms. J. ACM 44(5), 669–696 (1997)
2. Eppstein, D., Galil, Z., Italiano, G.F.: Dynamic graph algorithms. CRC Press, Boca Raton (1997)
3. Henzinger, M.R., King, V.: Randomized fully dynamic graph algorithms with polylogarithmic time per operation. J. ACM 46(4), 502–516 (1999)

4. Holm, J., de Lichtenberg, K., Thorup, M.: Poly-logarithmic deterministic fully-dynamic algorithms for connectivity, minimum spanning tree, 2-edge, and biconnectivity. J. ACM 48(4), 723–760 (2001)
5. Thorup, M.: Worst-case update times for fully-dynamic all-pairs shortest paths. In: STOC, pp. 112–119 (2005)
6. Klein, P.N., Subramanian, S.: A fully dynamic approximation scheme for shortest paths in planar graphs. Algorithmica 22(3), 235–249 (1998)
7. Thorup, M.: Fully-dynamic min-cut. In: STOC, pp. 224–230 (2001)
8. Sankowski, P.: Faster dynamic matchings and vertex connectivity. In: SODA, pp. 118–126 (2007)
9. Micali, S., Vazirani, V.V.: An $O(\sqrt{|V|} \cdot |E|)$ algorithm for finding maximum matching in general graphs. In: FOCS, pp. 17–27 (1980)
10. Ivković, Z., Lloyd, E.L.: Fully dynamic maintenance of vertex cover. In: van Leeuwen, J. (ed.) WG 1993. LNCS, vol. 790, pp. 99–111. Springer, Heidelberg (1994)
11. Parnas, M., Ron, D.: Approximating the minimum vertex cover in sublinear time and a connection to distributed algorithms. Theor. Comput. Sci. 381(1-3), 183–196 (2007)
12. Feigenbaum, J., Kannan, S., McGregor, A., Suri, S., Zhang, J.: On graph problems in a semi-streaming model. Theor. Comput. Sci. 348(2-3), 207–216 (2005)
13. McGregor, A.: Finding graph matchings in data streams. In: Chekuri, C., Jansen, K., Rolim, J.D.P., Trevisan, L. (eds.) APPROX 2005 and RANDOM 2005. LNCS, vol. 3624, pp. 170–181. Springer, Heidelberg (2005)
14. Zelke, M.: Weighted matching in the semi-streaming model. In: STACS, pp. 669–680 (2008)
15. Lotker, Z., Patt-Shamir, B., Rosén, A.: Distributed approximate matching. In: PODC, pp. 167–174 (2007)

Local Property Reconstruction and Monotonicity[*]

Michael Saks[1] and C. Seshadhri[2]

[1] Department of Mathematics, Rutgers University
saks@math.rutgers.edu
[2] IBM Almaden Research Center
csesha@us.ibm.com

Abstract. We propose a general model of local property reconstruction. Suppose we have a function f on domain Γ, which is supposed to have a particular property $\mathcal{P}$, but may not have the property. We would like a procedure that produces a function g that has property $\mathcal{P}$ and is close to f (according to some suitable metric). The reconstruction procedure, called a *filter*, has the following form. The procedure takes as input an element x of Γ and outputs $g(x)$. The procedure has oracle access to the function f and uses a single short random string ρ, but is otherwise deterministic.

This model was inspired by a related model of online property reconstruction that was introduced by by Ailon, Chazelle, Comandur and Liu (2004). It is related to the property testing model, and extends the framework that is used in the model of locally decodable codes. A similar model, in the context of hypergraph properties, was independently proposed and studied by Austin and Tao (2008).

We specifically consider the property of monotonicity and develop an efficient local filter for this property. The input f is a real valued function defined over the domain $\{1, \ldots, n\}^d$ (where n is viewed as large and d as a constant). The function is monotone if the following property holds: for two domain elements x and y, if $x \leq y$ (in the product order) then $f(x) \leq f(y)$. Given x, our filter outputs the value $g(x)$ in $(\log n)^{O(1)}$ time and uses a random seed ρ of the same size. With high probability, the ratio of the Hamming distance between g and f to the minimum possible Hamming distance between a monotone function and f is bounded above by a function of d (independent of n).

1   Online Property Reconstruction

The process of assembling large data sets is prone to varied sources of corruption, such as measurement error, replication error, and communication noise. Error

---

[*] This is an extended abstract of work that will appear as "Local Monotonicity Reconstruction" in SIAM Journal on Computing [30]. A preliminary version of this work appeared as "Parallel Monotonicity Reconstruction" [29]. The work was supported in part by NSF under grants CCF-0515201 and CCF-0832787. It is partly based on material that appeared in the second author's Ph.D. dissertation for the Department of Computer Science, Princeton University.

O. Goldreich (Ed.): Property Testing, LNCS 6390, pp. 346–354, 2010.

correction techniques (i.e. coding) can be used to reduce or eliminate the effects of some sources of error, but often some residual errors may be unavoidable. Despite the presence of such inherent error, the data set may still be very useful.

One problem in using such a data set is that even small amounts of error can significantly change the behavior of algorithms that act on the data. For example, if we do a binary search on an array that is supposed to be sorted, a few erroneous entries may lead to behavior that deviates significantly from the "correct" behavior.

This is an example of a more general situation. We have a data set that ideally should have some specified structural property, i.e., a list of numbers that should be sorted, a set of points that should be in convex position, or a graph that should be a tree. Algorithms that run on the data set may rely on this property. A small amount of error may destroy the property, and result in the algorithm producing wildly unexpected results, or even crashing. In these situations, a small amount of error may be tolerable but only if the structural property is maintained.

These considerations motivated the formulation of the *online property reconstruction* model, which was introduced in [3]. We are given a data set, which we think of as a function f defined on some domain Γ. Ideally, f should have a specified structural property $\mathcal{P}$, but this property may not hold due to unavoidable errors. We wish to construct *online* a new data set g such that:

(1) g has property $\mathcal{P}$ and (2) $d(g, f)$ is small, where $d(g, f)$ is the fraction of values $x \in \Gamma$ for which $g(x) \neq f(x)$.

How small should $d(g, f)$ be in Condition (2)? Define $\varepsilon_f = \varepsilon_f(\mathcal{P})$ to be the minimum of $d(h, f)$ over all h that satisfy $\mathcal{P}$. Of course, ε_f is a lower bound on the deviation of g from f. The *error blow-up* of g is the ratio $d(g, f)/\varepsilon_f$. This error blow-up can be viewed as the price that is paid in order to restore the property $\mathcal{P}$ online, and we want this to be a not too large constant.

An offline reconstruction algorithm explicitly outputs such a g on input f. In the context of large data sets, the explicit construction of g from f requires a considerable amount of computational overhead (at least linear in the size of the data set). For this reason, [3] considered online reconstruction algorithms. Such an algorithm, called a *filter*, gets as input a sequence $x_1, x_2, \ldots$ of elements of Γ presented one at a time and must output the sequence of values $g(x_1), g(x_2), \ldots$ where $g(x_i)$ is produced in response to x_i, before knowing x_{i+1}. The filter can access the function f via an oracle which, given $y \in \Gamma$, answers $f(y)$. The aim is to design a filter that, for any online input sequence of elements in Γ, outputs a function g satisfying (1) and (2) above and furthermore produces each successive $g(x_i)$ quickly, i.e., in time much smaller than $O(|\Gamma|)$.

In [3], a filter for the *monotonicity property* was given. In this setting, the domain Γ is the set $[n]^d = \{(j_1, \ldots, j_d) : j_i \in [n]\}$, where $[n]$ denotes the set $\{1, 2, \ldots, n\}$. The set $[n]^d$ is considered to be partially ordered under the component-wise (product) order: $(i_1, \ldots, i_d) \leq (j_1, \ldots, j_d)$ iff $\forall r, i_r \leq j_r$. A function f defined on Γ is *monotone* if $x \leq y$ implies $f(x) \leq f(y)$. The filter they constructed satisfies Condition (1), has error blow-up that is bounded above by $2^{O(d)}$ (independent of n), and answers each successive query in time $(\log n)^{O(d)}$.

2   Local Property Reconstruction

The filter for monotonicity proposed in [3] has the following general structure. For each successive query x_j, the filter executes a randomized algorithm to compute $g(x_j)$. This algorithm accesses f, and also needs to access the answers $g(x_i)$ for $i < j$ to the queries asked previously. In particular, the function g produced may depend on both the order of the queries and the random bits used by the algorithm.

This general structure for filters has two potential drawbacks: (1) It requires the storage of all previous queries and answers, thus incurring possibly significant space overhead for the algorithm, (2) It does not support a local implementation in which multiple copies of the filter, having read-only access to f, are able to handle queries independently while maintaining mutual consistency.

In this paper, we propose the following strengthened requirements for a filter. A *local filter*[1] for reconstructing property $\mathcal{P}$ is an algorithm A that has oracle access to a function f on domain Γ (the "data set") and to an auxiliary random string ρ (the "random seed"), and takes as input $x \in \Gamma$. For fixed f and ρ, A runs deterministically on input x to produce an output $A_{f,\rho}(x)$. Thus, given f and ρ, $A_{f,\rho}$ specifies a function on domain Γ. We want A to satisfy the following properties:

1. For each f and ρ, $A_{f,\rho}$ satisfies $\mathcal{P}$. [2]
2. For each f, with high probability (with respect to the choice of ρ), the function $A_{f,\rho}$ should be "suitably close" to f.
3. For each x, $A_{f,\rho}$ on x can be computed very quickly.
4. The size of the random seed ρ should be "much smaller" than $|\Gamma|$.

Remark 1: In Condition 2, we say that $A_{f,\rho}$ should be "suitably close" to f. There are various ways to make this precise. Let ε_f denote the minimum distance from f to a function satisfying $\mathcal{P}$ and let $\gamma_f(\rho)$ denote the distance from f to $A_{f,\rho}$. We would like $\gamma_f(\rho)$ to be small compared to ε_f. The error blow-up, which is the ratio of $\gamma_f(\rho)/\varepsilon_f$, works well for the monotoncity property that we study. For other properties, it might be more appropriate to use another criterion: for example, we might consider the difference $\gamma_f(\rho) - \varepsilon_f$. More generally, we could require simply that $\gamma_f(\rho)$ be bounded above by some arbitrary function of ε_f (either independent of $|\Gamma|$ or growing very slowly with $|\Gamma|$).

---

[1] This was originally called a *parallel filter* in the conference version [29]. We made this terminology change since it is more compatible with the existing concepts of locally decodable codes.

[2] In an earlier version of this paper, this condition was replaced by the weaker condition that for each f, $A_{f,\rho}$ should satisfy $\mathcal{P}$ with high probability. Prompted by a question raised by a referee we were able to modify our monotonicity filter to satisfy this stronger property, and so modified the definition accordingly. The weaker condition may be more appropriate for some other properties.

Remark 2: Similarly, for Condition 3, there are various possibilities for interpreting the phrase "very quickly". In this paper, we obtain running times that are polynomial in $\log |\Gamma|$. In Section 3, we will mention some work on other properties where the running time does not depend on the domain size. On the other hand, there may be other properties where it is non-trivial and interesting to obtain running times of the form $|\Gamma|^\delta$.

Remark 3: A local filter can be used, trivially, as an online filter. The space required by the local filter is bounded by the sum of the length of ρ and the running time per query. By keeping these both small (e.g., much smaller than $|\Gamma|$) we obtain an online filter using little auxiliary space.

Remark 4: A local filter can be used to enforce consistent behavior among autonomous processors who each have access to f but do not communicate with each other. We generate one random seed ρ and give the same random seed to each of the processors. Since $A_{f,\rho}$ is deterministic, all processors will reconstruct the same function.

3   Related Work

One case of property reconstruction that has been studied extensively is *error correcting codes*. Suppose $C \subseteq \{0,1\}^n$ is such a code in which all members of C are pairwise at distance at least d. Let $\mathcal{P}$ be the property of being a codeword. The error correction problem for C is to find the *closest codeword* to a given input string x. This can be formulated as a reconstruction problem for the property $\mathcal{P}$.

One variant of the error correction is the problem of local decoding. This problem was explicitly named in [25], but, as noted there, was studied previously in connection with self-correcting computation (e.g., [12, 20]), probabilistically checkable proofs (e.g., [8]), average-case reductions (e.g., [9, 31]), and private information retrieval (e.g., [13]). Here we want a decoding algorithm for a given code that, given oracle access to the bits of an input string x, and given an index $i \in [n]$, finds the ith bit of the closest codeword to x by querying a small (possibly randomly selected) number of bits of x. If we view the local decoding algorithm as a deterministic algorithm that takes input i and a random string r (used to make the decisions) then we require that for each i, most choices of r lead to the correct value for the ith bit of the closest codeword.

This is very similar to (though not quite the same as) the local property reconstruction problem for $\mathcal{P}$; for local property reconstruction we interchange the "for all" and "for most" quantifiers and require that for most choices of r, and for all $i \in [n]$, the algorithm correctly produces the ith bit of the codeword. Also, we pay attention to the length of the random string r, which we want to be suitably small.

In local list decoding, our aim is to find a short *list* of codewords that are all suitably close to the input word. For example, in list decoding of low-degree polynomials [6, 31], the input is a function and the output is a *small list* of low-degree polynomials that are close to the input function.

The monotonicity problem considered in this paper is qualitatively quite different from the local decoding examples. In local decoding there is either *one* correct output, or (in the case of list-decoding) a *sparse list* of possible correct outputs. For monotonicity there may be many (possibly infinitely many) ways to correct a given function to a nearby function with the desired property. One might think that having many possible close corrections (rather than one) makes reconstruction easier but, at least for the monotonicity problem, it does not. The difficulty arises from the requirement that once the random seed is fixed, all query answers provided by the filter must be consistent with a *single* function having the property.

A related notion of reconstruction was discussed in [23], for generalized partition problems in dense graphs. Given an input dense graph G that satisfies some partition property (say k-colorability), we wish to efficiently construct a partition of the vertices that has at most an ε-fraction of violating edges. The algorithms for this problem provided in [23] behaved like local filters. Specifically, there was a constant (function of ε) time algorithm that gave the color class of an input vertex of G, and this could be run independently on all vertices (after fixing a random seed). This coloring was guaranteed to violate at most an ε-fraction of the edges in G.

Independently of our work, a model of *repair* of a property was formulated and studied in [7]. This is closely related to the reconstruction model considered here. The results in [7] primarily considered reconstruction of *hypergraph properties*, and obtained local filters of a very special form that modify an input hypergraph to satisfy a given property. This result can be seen as a generalization of the characterizations of testable properties of dense graphs [4,5]. This does not focus on the exact form of the error blow-up, and only requires that the distance of the reconstructed hypergraph be bounded by some arbitrary function of the minimum distance of the hypergraph to the property.

In general, a local filter for reconstructing a given property can be used to estimate the distance of an input instance to the property. When we fix a random seed and run the filter on f, the filter implicitly outputs a function g that has the desired property and is at distance at most $B\varepsilon_f$ from f (where ε_f is distance of f to $\mathcal{P}$). By choosing a random sample of domain points x and computing the fraction of points where $g(x) \neq f(x)$, we get an estimate of the distance $d(g, f)$. Since $\varepsilon_f \leq d(g, f)$ and with high probability, $\varepsilon_f \geq d(g, f)/B$, we get a multiplicative B-approximation to ε_f in sublinear time.

4   Our Results

In this work, we construct a local filter for monotonicity for functions defined on $[n]^d$ with the following performance:

- The time per query is $(\log n)^{O(d)}$.
- The error blow-up is $2^{O(d^2)}$, independent of n.
- The number of random bits needed to initialize the filter is $(d \log n)^{O(1)}$.

The online filter for monotonicity of [3] has a running time per query of $(\log n)^{O(d)}$ (with a better constant in the exponent) and an error blow-up of 2^d. We see that our filter achieves local behavior while having query time and error blow-up that are similar to (but not quite as good) as those obtained by [3].

Our filter for monotonicity builds on techniques used for property testing of monotonicity. There has been a large amount of work done on property testing, which was defined in [23, 28]. Many testers have been given for a wide variety of combinatorial, algebraic, and geometric problems (see surveys [17, 21, 27]). The related notions of tolerant testing and distance approximation were introduced in [26]. The problem of monotonicity in the context of property testing has been studied in [1, 10, 11, 14, 15, 18, 19, 22, 24]. Sublinear algorithms for *approximating* the distance of a function to monotonicity have been given in [2, 26, 16].

Both the running time and error blow-up of our filter have an exponential dependence on the dimension d. We also prove that this dependence is unavoidable. Specifically we show the following for some constant $0 < \alpha < 1$: given a filter on the boolean hypercube $\{0, 1\}^d$ that answers queries within time $2^{\alpha d}$, there is an input function f such that the filter applied to f has error blow-up $2^{\alpha d}$ with probability close to $1/2$. This shows a complexity gap between testing and reconstruction for the hypercube, since there are monotonicity testers with only a polynomial dependence on d [14, 16, 22].

5   Overview of the Local Filter for Monotonicity

We now discuss some of the ideas used in constructing the looal filter for monotonicity. Details of the construction and analysis can be found in the full paper.

The starting point for the construction of our local filter for monotonicity is the online filter of [3]. We now give the main ideas of their construction, and indicate the difficulties in making their construction local. In the discussion below, when we say an algorithm is "fast", we mean that it runs in time polylogarithmic in $|\Gamma|$.

We start with the case $d = 1$, i.e., the one-dimensional case. The basic idea (implicitly used) in [3] is to classify the domain points as *accepted* and *rejected* in such a way that the following conditions hold:

(1) There is a fast algorithm for testing whether a given point is accepted or rejected.
(2) There are not many rejected points[3].
(3) The function restricted to the set of accepted points is monotone.

The third property ensures that it is possible (though not necessarily efficiently) to change the function only on rejected points and make the function monotone. To do this, define $m(x)$ for $x \in \Gamma$ to be the largest accepted point less than or equal to x, and define $g(x) = f(m(x))$. It is easy to see that this yields a monotone function.

---

[3] The number of rejected points is comparable to the distance of f to monotonicity.

In [3], a point x is rejected if (roughly) there is an interval around x that contains a large fraction of points whose f values are out of order with respect to $f(x)$. With this accepted/rejected classification there seems to be no fast way to compute $m(x)$. Instead, given a query point x, the filter in [3] selects a sample of points less than or equal to x (called the *sample* of x), chooses $m'(x)$ to be the largest accepted point in the sample, and defines $g(x) = g(m'(x))$. The sampling procedure chooses $z < x$ with probability (roughly) inversely proportional to the distance of z from x; in particular the sample includes x itself, so if x is accepted then $m'(x) = x$.

Defining g in this way creates a problem: g need not be monotone. For example, let y be a point and $x = m(y) < y$ be the largest accepted point less than or equal to y. Suppose that a query is made to y and x is not in the sample of y, so $m'(y) < x$. Suppose further that $f(m'(y)) < f(x)$. Suppose that after setting $g(y)$ to $f(m'(y))$, a query is made to index x. Since x is an accepted point we will have $m'(x) = x$ and so $g(x) = f(x)$, but this will violate monotonicity with the already defined $g(y) = f(m'(y))$.

In online reconstruction, this is not a significant problem because the algorithm can save the previously answered queries in a sorted list and impose the condition that future g values be consistent with previously assigned g values. This is what is done in [3].

Local reconstruction does not have this luxury. What we do is to redefine $m'(x)$ so as to guarantee that for any $y > x$, we have $m'(y) \geq m'(x)$. To do this, after sampling the points less than x we identify certain points of the sample which have the potential for creating non-monotonicities and exclude them from the sample. For example, in the scenario above, the point x needs to be excluded from its own sample to avoid the *potential* non-monotonicity with y. Notice that when we exclude x from its own sample we may introduce a new point where $g(x) \neq f(x)$, so we cannot do this too often.

Thus, the main challenge in designing a local filter is to find an efficient way to identify the points that need to be excluded from the sample of x to avoid potential non-monotonicities. We also need to ensure that x is not excluded from its own sample too often.

The difficulties in designing a local filter are greater for the case of higher-dimensional domains ($d \geq 2$). Suppose we had a definition of accepted and rejected satisfying the three conditions stated in the one-dimensional case. In principle, it is still possible to define a monotone g that agrees with f on all accepted points. But explicitly computing such a g is more complicated. Given x, let $M(x)$ be the set of points which are maximal in the set of accepted points less than or equal to x. In the one-dimensional case, $M(x)$ has one element $m(x)$, but in the multi-dimensional case, where the domain is not totally ordered, this is not the case. Still, if we define $g(x)$ to be the maximum of $f(y)$ for $y \in M(x)$, then the resulting g is monotone. To implement this, one would have to find all of the elements of $M(x)$. Even when $M(x)$ has size 1 (as in the one-dimensional case) this is difficult, but here the difficulty is compounded because $M(x)$ might be as large as $\Omega(n^{d-1})$, and we need our computation to run in time polylogarithmic in n. In [3], this is handled by finding a polylogarithmic size sample that is a suitable approximation to $M(x)$, and then defining $g(x)$ to be the maximum of $f(y)$ for y in the sample.

As with the one-dimensional case, using an approximation to $M(x)$ destroys the guarantee that g defined in this way is monotone. Hence, one must save the values of g to all queries, and impose the additional requirement that queries are mutually consistent. Since a local filter cannot save these values, we again need to judiciously exclude points from the sample to avoid non-monotonicities.

The definition of the sample, which we denote REP(x), crucially uses a data structure of nested boxes (products of intervals). Condition (3) is maintained by a careful and efficient scheme for passing crucial information about the distribution of rejected points in a particular box to its sub-boxes.

References

1. Ailon, N., Chazelle, B.: Information theory in property testing and monotonicity testing in higher dimension. Information and Computation 204(11), 1704–1717 (2006)
2. Ailon, N., Chazelle, B., Comandur, S., Liu, D.: Estimating the distance to a monotone function. Random Structures and Algorithms 31(3), 371–383 (2007)
3. Ailon, N., Chazelle, B., Comandur, S., Liu, D.: Property-preserving data reconstruction. Algorithmica 51(2), 160–182 (2008)
4. Alon, N., Fischer, E., Newman, I., Shapira, A.: A combinatorial characterization of the testable graph properties: it's all about regularity. SIAM Journal on Computing 39(1), 143–167 (2009)
5. Alon, N., Shapira, A.: A characterization of the (natural) graph properties testable with one-sided error. SIAM Journal on Computing 37(6), 1703–1727 (2008)
6. Arora, S., Sudan, M.: Improved low-degree testing and its applications. Combinatorica 23(3), 365–426 (2003)
7. Austin, T., Tao, T.: Testability and repair of hereditary hypergraph properties. Random Structures and Algorithms 56(4), 373–463 (2010)
8. Babai, L., Fortnow, L., Levin, L., Szegedy, M.: Checking computations in polylogarithmic time. In: Proceedings of the 23rd Annual Symposium on Theory of Computing (STOC), pp. 21–31 (1991)
9. Babai, L., Fortnow, L., Nisan, N., Wigderson, A.: PP has subexponential time simulations unless EXP-TIME has publishable proofs. Computational Complexity 3, 307–318 (1993)
10. Batu, T., Rubinfeld, R., White, P.: Fast approximate PCPs for multidimensional bin-packing problems. Information and Computation 196(1), 42–56 (2005)
11. Bhattacharyya, A., Grigorescu, E., Jung, K., Raskhodnikova, S., Woodruff, D.: Transitive-closure spanners. In: Proceedings of the 18th Annual Symposium on Discrete Algorithms (SODA), pp. 531–540 (2009)
12. Blum, M., Luby, M., Rubinfeld, R.: Self-testing/correcting with applications to numerical problems. Journal of Computer and System Sciences 47(3), 549–595 (1993)
13. Chor, B., Goldreich, O., Kushilevitz, E., Sudan, M.: Private information retrieval. Journal of the ACM 45, 965–981 (1998)

14. Dodis, Y., Goldreich, O., Lehman, E., Raskhodnikova, S., Ron, D., Samorodnitsky, A.: Improved testing algorithms for monotonicity. In: Hochbaum, D.S., Jansen, K., Rolim, J.D.P., Sinclair, A. (eds.) RANDOM 1999 and APPROX 1999. LNCS, vol. 1671, pp. 97–108. Springer, Heidelberg (1999)
15. Ergun, F., Kannan, S., Kumar, R., Rubinfeld, R., Viswanathan, M.: Spot-checkers. Journal of Computer Systems and Sciences (JCSS) 6(3), 717–751 (2000)
16. Fattal, S., Ron, D.: Approximating the distance to monotonicity in high dimensions. ACN Trans. on Alg. 6(3) (2010)
17. Fischer, E.: The art of uninformed decisions: A primer to property testing. Bulletin of EATCS 75, 97–126 (2001)
18. Fischer, E.: On the strength of comparisons in property testing. Information and Computation 189(1), 107–116 (2004)
19. Fischer, E., Lehman, E., Newman, I., Raskhodnikova, S., Rubinfeld, R., Samorodnitsky, A.: Monotonicity testing over general poset domains. In: Proceedings of the 34th Annual Symposium on Theory of Computing (STOC), pp. 474–483 (2002)
20. Gemmell, P., Lipton, R., Rubinfeld, R., Sudan, M., Wigderson, A.: Self-testing/correcting for polynomials and for approximate functions. In: Proceedings of the 23rd Annual Symposium on Theory of Computing (STOC), pp. 32–42 (1991)
21. Goldreich, O.: Combinatorial property testing - a survey. In: Randomization Methods in Algorithm Design, pp. 45–60 (1998)
22. Goldreich, O., Goldwasser, S., Lehman, E., Ron, D., Samordinsky, A.: Testing monotonicity. Combinatorica 20, 301–337 (2000)
23. Goldreich, O., Goldwasser, S., Ron, D.: Property testing and its connection to learning and approximation. Journal of the ACM 45(4), 653–750 (1998)
24. Halevy, S., Kushilevitz, E.: Testing monotonicity over graph products. Random Structures and Algorithms 33(1), 44–67 (2008)
25. Katz, J., Trevisan, L.: On the efficiency of local decoding procedures for error-correcting codes. In: Proceedings of the 32th Annual Symposium on Theory of Computing (STOC), pp. 80–86 (2000)
26. Parnas, M., Ron, D., Rubinfeld, R.: Tolerant property testing and distance approximation. Journal of Computer and System Sciences 6(72), 1012–1042 (2006)
27. Ron, D.: Property testing. In: Handbook on Randomization, vol. II, pp. 597–649 (2001)
28. Rubinfeld, R., Sudan, M.: Robust characterization of polynomials with applications to program testing. SIAM Journal of Computing 25, 647–668 (1996)
29. Saks, M., Seshadhri, C.: Parallel monotonicity reconstruction. In: Proceedings of 19th Annual Symposium on Discrete Algorithms (SODA), pp. 962–971 (2006)
30. Saks, M.E., Seshadhri, C.: Local monotonicity reconstruction. SIAM Journal on Computing 39(7), 2897–2926 (2010)
31. Sudan, M., Trevisan, L., Vadhan, S.: Pseudorandom generators without the XOR lemma. Journal of Computer and System Sciences 62(2), 236–266 (2001)

Green's Conjecture and Testing Linear Invariant Properties

Asaf Shapira

School of Mathematics and College of Computing,
Georgia Institute of Technology, Atlanta, GA, USA
asafico@math.gatech.edu

Abstract. A system of ℓ linear equations in p unknowns $Mx = b$ is said to have the *removal property* if every set $S \subseteq \{1, \ldots, n\}$ which contains $o(n^{p-\ell})$ solutions of $Mx = b$ can be turned into a set S' containing no solution of $Mx = b$, by the removal of $o(n)$ elements. Green [GAFA 2005] proved that a single homogenous linear equation always has the removal property, and conjectured that every set of homogenous linear equations has the removal property. In this paper we confirm Green's conjecture by showing that every set of linear equations (even non-homogenous) has the removal property. We also discuss some applications of our result in theoretical computer science, and in particular, use it to resolve a conjecture of Bhattacharyya, Chen, Sudan and Xie [4] related to algorithms for testing properties of boolean functions.

Keywords: Property Testing, Linear-Invariance, Hypergraphs, Removal Lemma.

This article is an extended abstract of [19].

1 Background on Removal Lemmas

The (triangle) removal lemma of Ruzsa and Szemerédi [18], which is by now a cornerstone result in combinatorics, states that a graph on n vertices that contains only $o(n^3)$ triangles can be made triangle free by the removal of only $o(n^2)$ edges. Or in other words, if a graph has asymptomatically few triangles then it is asymptotically close to being triangle free. While the lemma was proved in [18] for triangles, an analogous result for any fixed graph can be obtained using the same proof idea. Actually, the main tool for obtaining the removal lemma is Szemerédi's regularity lemma for graphs [21], another landmark result in combinatorics. The removal lemma has many applications in different areas like extremal graph theory, additive number theory and theoretical computer science. Perhaps its most well known application appears already in [18] where it is shown that an ingenious application of it gives a very short and elegant proof of Roth's Theorem [16], which states that every $S \subseteq [n] = \{1, \ldots, n\}$ of positive density contains a 3-term arithmetic progression.

O. Goldreich (Ed.): Property Testing, LNCS 6390, pp. 355–358, 2010.

Recall that an r-uniform hypergraph $H = (V, E)$ has a set of vertices V and a set of edges E, where each edge $e \in E$ contains r distinct vertices from V. So a graph is a 2-uniform hypergraph. Szemeredi's famous theorem [20] extends Roth's theorem by showing that every $S \subseteq [n]$ of positive density actually contains arbitrarily long arithmetic progressions (when n is large enough). Motivated by the fact that a removal lemma for graphs can be used to prove Roth's theorem, Frankl and Rödl [5] showed that a removal lemma for r-uniform hypergraphs could be used to prove Szemeredi's theorem on $(r + 1)$-term arithmetic progressions. They further developed a regularity lemma, as well as a corresponding removal lemma, for 3-uniform hypergraphs thus obtaining a new proof of Szemeredi's theorem for 4-term arithmetic progressions. In recent years there have been many exciting results in this area, in particular the results of Gowers [8] and of Nagle, Rödl Schacht and Skokan [14,15], who independently obtained regularity lemmas and removal lemmas for r-uniform hypergraph, thus providing alternative combinatorial proofs of Szemeredi's Theorem [20] and some of it generalizations, notably those of Furstenberg and Katznelson [6]. Tao [22] and Ishigami [11] later obtained another proof of the hypergraph removal lemma and of its many corollaries mentioned above. For more details see [9].

2   Our Main Result

In this paper we will use the above mentioned hypergraph removal lemma in order to resolve a conjecture of Green [10] regarding the removal properties of sets of linear equations. Let $Mx = b$ be a set of linear equations, and let us say that a set of integers S is (M, b)-free if it contains no solution to $Mx = b$, that is, if there is no vector x, whose entries all belong to S, which satisfies $Mx = b$. Just like the removal lemma for graphs states that a graph that has few copies of H should be close to being H-free, a removal lemma for sets of linear equations $Mx = b$ should say that a subset of the integers $[n]$ that contains few solutions to $Mx = b$, should be close to being (M, b)-free. Let us start be defining this notion precisely.

Definition 2.1 (Removal Property). *Let M be an $\ell \times p$ matrix of integers and let $b \in \mathbb{N}^\ell$. The set of linear equations $Mx = b$ has the removal property if for every $\delta > 0$ there is an $\epsilon = \epsilon(\delta, M, b) > 0$ with the following property: if $S \subseteq [n]$ is such that there are at most $\epsilon n^{p-\ell}$ vectors $x \in S^p$ satisfying $Mx = b$, then one can remove from S at most δn elements to obtain an (M, b)-free set.*

Green [10] has initiated the study of the removal properties of sets of linear equations. His main result was the following:

Theorem 2.1 (Green [10]). *Any single homogenous linear equation has the removal property.*

The main result of Green actually holds over any abelian group. To prove this result, Green developed a regularity lemma for abelian groups, which is somewhat analogous to Szemerédi's regularity lemma for graphs [21]. Although the application of the group regularity lemma for proving Theorem 2.1 was similar to the derivation of the graph removal lemma from the graph regularity lemma, the proof of the group regularity lemma was far from trivial. One of the main conjectures raised in [10] is that a natural generalization of Theorem 2.1 should also hold (Conjecture 9.4 in [10]).

Conjecture 2.2 (Green [10]). *Any system of homogenous linear equations $Mx = 0$ has the removal property.*

Very recently, Král', Serra and Vena [12] gave a surprisingly simple proof of Theorem 2.1, which completely avoided the use of Green's regularity lemma for groups. In fact, their proof is an elegant and simple application the removal lemma for directed graphs [1], which is a simple variant of the graph removal lemma that we have previously discussed. The proof given in [12] actually extends Theorem 2.1 to any single non-homogenous linear equation over arbitrary groups. Král', Serra and Vena [12] also show that Conjecture 2.2 holds when M is a 0/1 matrix, which satisfies certain conditions. But these conditions are not satisfied even by all 0/1 matrices.

In this paper we confirm Green's conjecture for every homogenous set of linear equations. In fact, we prove the following more general result.

Theorem 2.3 (Main Result). *Any set of linear equations $Mx = b$ has the removal property.*

3   Applications to Testing Properties of Boolean Functions

Besides being a natural problem from the perspective of additive number theory, it turns out that Theorem 2.3 has some applications in Theoretical Computer Science, in the area of *Property Testing* [3,17,7]. Property testers are fast randomized algorithms that can distinguish between objects satisfying a certain property $\mathcal{P}$ and objects that are "far" from satisfying it. In an attempt to prove a general sufficient condition that would guarantee that certain properties of boolean functions have efficient testing algorithms, Bhattacharyya, Chen, Sudan and Xie [4] conjectured that certain properties of boolean functions (that are related to the notion of being (M, b)-free) can be efficiently tested. As we show in this paper, our main result gives a positive answer to their open problem.

After our paper appeared on the Arxiv we learned that independently of our work, Král', Serra and Vena managed to improve upon their results in [12,13] and obtain a proof of Conjecture 2.2.

Acknowledgments

A.S. is partially supported by NSF Grant DMS-0901355.

References

1. Alon, N., Shapira, A.: Testing Subgraphs in Directed Graphs. Journal of Computer and System Sciences 69, 354–382 (2004)
2. Austin, T., Tao, T.: On the testability and repair of hereditary hypergraph properties (2008) (manuscript)
3. Blum, M., Luby, M., Rubinfeld, R.: Self-testing/correcting with applications to numerical problems. JCSS 47, 549–595 (1993)
4. Bhattacharyya, A., Chen, V., Sudan, M., Xie, N.: Testing linear-invariant nonlinear properties (2008) (manuscript)
5. Frankl, P., Rödl, V.: Extremal problems on set systems. Random Structures and Algorithms 20, 131–164 (2002)
6. Furstenberg, H., Katznelson, Y.: An ergodic Szemerédi theorem for commuting transformations. J. Analyse Math. 34, 275–291 (1978)
7. Goldreich, O., Goldwasser, S., Ron, D.: Property testing and its connection to learning and approximation. JACM 45(4), 653–750 (1998)
8. Gowers, T.: Hypergraph regularity and the multidimensional Szemerédi theorem. Ann. of Math. 166(3), 897–946 (2007)
9. Gowers, T.: Quasirandomness, counting and regularity for 3-uniform hypergraphs. Combinatorics, Probability and Computing 15, 143–184 (2006)
10. Green, B.: A Szemerédi-type regularity lemma in abelian groups. GAFA 15, 340–376 (2005)
11. Ishigami, Y.: A simple regularization of hypergraphs, http://arxiv.org/abs/math/0612838
12. Král', D., Serra, O., Vena, L.: A combinatorial proof of the removal lemma for groups, arXiv:0804.4847v1
13. Král', D., Serra, O., Vena, L.: A removal lemma for linear systems over finite fields. Jornadas de Matematica Discreta y algortimica (2008)
14. Nagle, B., Rödl, V., Schacht, M.: The counting lemma for regular k-uniform hypergraphs. Random Structures and Algorithms 28, 113–179 (2006)
15. Rödl, V., Skokan, J.: Regularity lemma for k-uniform hypergraphs. Random Structures and Algorithms 25, 1–42 (2004)
16. Roth, K.F.: On certain sets of integers. J. London Math. Soc. 28, 104–109 (1953)
17. Rubinfeld, R., Sudan, M.: Robust characterization of polynomials with applications to program testing. SIAM J. on Computing 25, 252–271 (1996)
18. Ruzsa, I., Szemerédi, E.: Triple systems with no six points carrying three triangles. In: Combinatorics, Keszthely, vol. II. Coll. Math. Soc. J. Bolyai 18, pp. 939–945 (1976)
19. Shapira, A.: Green's conjecture and testing linear invariant properties. In: Proc. of STOC 2009, pp. 159–166 (2009)
20. Szemerédi, E.: Integer sets containing no k elements in arithmetic progression. Acta Arith. 27, 299–345 (1975)
21. Szemerédi, E.: Regular partitions of graphs. In: Bermond, J.C., Fournier, J.C., Las Vergnas, M., Sotteau, D. (eds.) Proc. Colloque Inter. CNRS, pp. 399–401 (1978)
22. Tao, T.: A variant of the hypergraph removal lemma. J. Combin. Theory, Ser. A 113, 1257–1280 (2006)

Author Index